Mycotoxins in Food and Feed

Mycotoxins represent an assorted range of secondary fungal metabolites that extensively occur in numerous food and feed ingredients at any stage during pre- and post-harvest conditions. Mycotoxin contamination in food and feed causes acute and chronic mycotoxicosis, including teratogenic, carcinogenic, estrogenic, neurotoxic, and immunosuppressive effects and several other health issues. *Mycotoxins in Food and Feed* presents an overview of all the major mycotoxins, sources of production, chemistry and biosynthesis, occurrence in food and feed, effect on agriculture, effect on human health, detection techniques, masked mycotoxins, and management and control strategies.

KEY FEATURES

- Provides broad coverage of mycotoxins and their effects on food and feed
- Includes comprehensive information on occurrence, chemistry, detection methods and management strategies for each toxin
- Discusses the recent development in detection technologies for major mycotoxins
- Explores agricultural practices and post-harvest management strategies for managing mycotoxin infestations

This book will be of interest to a variety of professionals and academicians at universities and research institutes, those in the food and feed industries, and policymakers. It will be beneficial for them in managing the growing concerns about mycotoxins in food and feed as well as in designing and implementing uniform rules and regulatory guidelines regarding their limits and detection and management strategies for the safety and security of food and feed from present and emerging mycotoxins.

Mycotoxins in Food and Feed

Detection and Management Strategies

Edited by Pradeep Kumar, Madhu Kamle,
and Dipendra Kumar Mahato

CRC Press
Taylor & Francis Group
Boca Raton London New York

CRC Press is an imprint of the
Taylor & Francis Group, an **informa** business

First edition published 2023
by CRC Press
6000 Broken Sound Parkway NW, Suite 300, Boca Raton, FL 33487-2742

and by CRC Press
4 Park Square, Milton Park, Abingdon, Oxon, OX14 4RN

CRC Press is an imprint of Taylor & Francis Group, LLC

© 2023 Pradeep Kumar, Madhu Kamle, and Dipendra Kumar Mahato

Library of Congress Cataloging-in-Publication Data
Names: Kumar, Pradeep (Professor of biotechnology), editor. | Kamle, Madhu, editor. |
    Mahato, Dipendra Kumar, editor.
Title: Mycotoxins in food and feed : detection and management strategies /
    edited by Pradeep Kumar, Madhu Kamle, and Dipendra Kumar Mahato.
Description: First edition. | Boca Raton : CRC Press, 2023. | Includes bibliographical
    references and index.
Identifiers: LCCN 2022032433 (print) | LCCN 2022032434 (ebook) | ISBN 9781032113920
    (hardback) | ISBN 9781032150352 (paperback) | ISBN 9781003242208 (ebook)
Subjects: LCSH: Mycotoxins. | Food contamination. | Food—Toxicology.
Classification: LCC RA1242.M94 M9644 2023 (print) | LCC RA1242.M94 (ebook) |
    DDC 615.9/5295—dc23/eng/20221114
LC record available at https://lccn.loc.gov/2022032433
LC ebook record available at https://lccn.loc.gov/2022032434

ISBN: 978-1-032-11392-0 (hbk)
ISBN: 978-1-032-15035-2 (pbk)
ISBN: 978-1-003-24220-8 (ebk)

DOI: 10.1201/9781003242208

Contents

Preface   ix
Author Biographies   xi
List of Contributors   xiii

1 Aflatoxins in Food and Feed: Occurrence,
  Detection, and Mitigating Strategies   1
  Namita Ashish Singh, Jyoti, Vidhi Jain

2 Fumonisins in Food and Feed: Their Detection and
  Management Strategies   29
  Shubhangi Srivastava, Ashok Kumar Yadav, Mousumi Ghosh,
  Dipendra Kumar Mahato, Madhu Kamle, Pooja Pandey,
  Sreemoyee Chakraborty, Pradeep Kumar

3 Ochratoxins in Food and Feed: Detection and
  Management Strategies   51
  Arun Kumar Pandey, Rahul Vashishth, Dipendra Kumar Mahato,
  Madhu Kamle, Raman Selvakumar, Monika Mathur, Pradeep Kumar

4 Trichothecene Concerns in Food and Feed:
  Detection and Management Strategies   87
  Madhvi Singh, Chayanika Sarma, Pinky Deka, Ramzan Ahmed,
  Jayabrata Saha, Sourav Chakraborty

5 Detection and Management Strategies for
  Deoxynivalenol in Food and Feed: An Overview   119
  Raman Selvakumar, Dalasanuru Chandregowda Manjunathagowda,
  Arun Kumar Pandey, Dipendra Kumar Mahato, Akansha Gupta,
  Shikha Pandhi, Raveena Kargwal, Madhu Kamle, Pradeep Kumar

6 Occurrence, Detection, and Management of T-2
   Toxin and HT-2 Toxin in Food and Feed   157
   Pooja Yadav, Nabendu Debnath, Shalini Arora, Ashok Kumar Yadav

7 Detection and Management of Ergot Alkaloids and
   Their Therapeutic Applications   189
   Nabendu Debnath, Pooja Yadav, Shalini Arora, Ashok Kumar Yadav

8 Roquefortines in Food and Feed: Detection
   and Management Strategies   213
   Saloni, Dinesh Chandra Rai, Himanshu Kumar Rai,
   Arvind Kumar, Urvashi Vikranta, Shikha Pandhi, Akansha Gupta,
   Dipendra Kumar Mahato

9 PR Toxins: Concerns in Food and Feed with
   Their Detection and Management Strategies   239
   Mousumi Ghosh, Sreemoyee Chakraborty, Sourav Misra,
   Shubhangi Srivastava, Sourabh Bondre, Madhu Kamle,
   Pradeep Kumar, Dipendra Kumar Mahato

10 Occurrence, Production, Determination, Toxicity,
   and Control Strategies of Cyclopiazonic Acid in
   Food Products   265
   Sourav Misra, Sitesh Kumar, Pooja Pandey, Shubham Mandliya,
   Mousumi Ghosh, Shubhangi Srivastava, Dipendra Kumar Mahato

11 Tremorgenic Mycotoxin Concerns in Food and Feed:
   Detection and Management Strategies   287
   Himani, Mitali Madhumita, Mohit Singla, Pramod K Prabhakar

12 Zearalenone in Food and Feed: Occurrence,
   Biosynthesis, Detection and Management Strategies   311
   Shikha Pandhi, Ashok Kumar Yadav, Vidhi Tyagi, Saloni, Akansha
   Gupta, Surabhi Pandey, Dipendra Kumar Mahato, Pradeep Kumar,
   Arun Kumar Pandey, Arvind Kumar

13 Mycotoxins in Stored Foods   333
    Preeti Sharma, Kanishka Chawla, Kamakshi Kalia, Vanshika Saini,
    Aastha Bhardwaj, Vasudha Bansal, Nitya Sharma

14 Climate Change's Impact on Mycotoxin Production
    in Food and Feed   361
    Monika Mathur, Raveena Kargwal, Raman Selvakumar,
    Dipendra Kumar Mahato, Madhu Kamle, Pradeep Kumar

15 Nano-Biosensors for the Monitoring of Toxic
    Contaminants in Food and it's Products   429
    Namita Ashish Singh, Nimisha Tehri, Amit Vashishth,
    Pradeep Kumar

    Index   449

Preface

Twenty-first–century scientific studies emphasize the importance of a good diet for health promotion, disease prevention, and treatment. Natural contaminants in food and feed are a significant source of human and animal health concerns. Mycotoxins are toxic substances produced by fungi that flourish in food or feed, providing a significant health risk to humans and animals. Mycotoxins are well known for creating major health problems in humans and animals when they enter the body.

Traditional strategies for controlling fungus and mycotoxin production are ineffective, as mycotoxins continue to penetrate our food supply. Several research organizations across the world have devised unique techniques to keep mycotoxigenic fungi and their detrimental influence on food and feed products under control.

This book focuses on strategies to control the health risks of associated mycotoxins in food and feed. Each chapter is carefully designed to offer a breadth of information, elucidating various strategies that include physical, chemical, and biological use of phytochemicals and essential oils, wherever applicable. Hence, this book provides a combined approach of advanced detection techniques used against mycotoxigenic species/mycotoxins produced as well as management strategies to help control and guarantee food and feed safety and security.

This book contains 15 chapters that mainly focus on exploring detection and management strategies in various foods and feed from all possible perspectives:

- Aflatoxins in food and feed: Occurrence, detection, and mitigation strategies
- Fumonisins in food and feed: Detection and management strategies
- Ochratoxins in food and feed: Detection and management strategies
- Trichothecenes concerns in food and feed: Detection and management strategies
- Detection and management strategies of deoxynivalenol in food and feed: An overview
- Occurrence, detection, and management of t-2 and ht-2 toxins in food and feed

- Detection and management of ergot alkaloids and their therapeutic applications
- Roquefortines in food and feed: detection and management strategies
- PR toxins: Concerns in food and feed, with detection and management strategies
- Occurrence, production, determination, toxicity, and control strategies for cyclopiazonic acid in food products
- Tremorgenic mycotoxin concerns in food and feed: Detection and management strategies
- Zearalenone in food and food: Occurrence, biosynthesis, and detection and management strategies
- Mycotoxins in stored foods
- Impact of climate change on mycotoxin production in food and feed

This volume is designed to provide an overview of recent and modern detection and management strategies that are in place or in developing stages for applications in food and feed samples. This book will help professionals at universities and research institutes, those at food and feed industries, and policymakers to manage the growing concerns about mycotoxins in food and feed as well as to design and implement uniform rules and regulatory guidelines regarding their limits and detection and management strategies for food and feed safety and security from present and emerging mycotoxins.

Dr. Pradeep Kumar
University of Lucknow, India
Dr. Madhu Kamle
NERIST, India
Dr. Dipendra K Mahato
Deakin University, Australia

Author Biographies

Dr. Pradeep Kumar

Dr. Kumar is currently working as an Associate Professor at Department of Botany, University of Lucknow. He served as Assistant professor in NERIST, international research professor/assistant professor at the Department of Biotechnology, Yeungnam University, South Korea. He was awarded the PBC Outstanding Postdoctoral Fellowship to work for more than three years as a postdoctoral researcher at the Department of Biotechnology Engineering, Ben Gurion University of Negev, Israel. Dr. Kumar's research interests are food microbiology, fungal biology, phytopathology, biocontrol, nanotechnology, phytochemistry, and bioremediation. He has published more than 80 research and review articles in peer-reviewed journals, 21 book chapters, and 7 edited books (Springer-Nature; Taylor & Francis Group), with a total of 4350 citations, h-index 29, and i10-index-56. He was awarded the Early Career Research Award (2017) from the SERB, Government of India, and currently handles four projects as PI and co-PI funded by DBT, DST-SERB, and GBPNIHESD. He is the recipient of several best paper and oral presentation awards and the Narasimhan Award from the Indian Phytopathological Society, India. He was also given the Emerging Scientist Award, Research Excellence Award, and Young Scientist Award by various societies. Dr. Kumar serves as a guest editor for *Evidence-Based Complementary and Alternative Medicine* (Hindawi) and *Microbiology Research* (MDPI), also serving as associate and academic editor and editorial board member of several peer-reviewed journals.

Dr. Madhu Kamle

Dr. Madhu Kamle is currently working as an assistant professor at the Department of Forestry, North Eastern Regional Institute of Science & Technology, Nirjuli, Arunachal Pradesh, India. Her research interests are in plant biotechnology, plant–microbe interactions, microbial genomics, and plant disease diagnosis. She earned her PhD in plant biotechnology from ICAR-CISH, India, and Bundelkhand University, Jhansi, India. She has been awarded the prestigious PBC Outstanding Postdoctoral Fellowship and a postdoctoral fellowship from the Jacob Blaustein Institute of Desert Research, Ben Gurion University, Israel. She also worked as an international research professor at the School of Biotechnology, Yeungnam University, Gyeongsan, Republic of Korea.

Dr. Kamle has ten years of research experience and has published more than 45 research and review articles in peer-reviewed journals, 15 international book chapters, and 3 edited books (Springer-Nature; Taylor & Francis Group), with more than 1950 citations and h-index 19. Currently she is handling three projects as PI and co-PI funded by GBPNIHESD and DBT, Government of India. She is a life member of the Nano-Molecular Society and a member of the American Society of Microbiology.

Dr. Dipendra Kumar Mahato

Dr. Dipendra Kumar Mahato is a PhD graduate from CASS Food Research Centre, Deakin University, Australia. He has a master's degree in food science and technology and a bachelor's degree in biotechnology. He was awarded the Gold Medal for academic excellence during his master's degree. Before beginning his PhD, Dipendra worked as an editorial assistant at Aptara.corp, a publication house. He also had industrial experience with Panacea Biotec. Later, he worked in different research projects at renowned Indian institutes like the Indian Agricultural Research Institute (IARI), New-Delhi, and Indian Institute of Technology (IIT) Kharagpur. He was awarded a prestigious Deakin University Postgraduate Research Scholarship (DUPRS), Australia, to pursue his PhD in food and nutrition. His research areas include food science, microbiology, biotechnology, and innovation in product development in relation to consumer perception, health, and wellbeing. He has published more than 40 research and review articles in peer-reviewed journals and 16 book chapters (Springer-Nature; Taylor & Francis Group; CRC Apple Academic Press), with more than 1700 citations and h-index 16.

Contributors

Ramzan Ahmed
University of Science and Technology
Meghalaya, India

Shalini Arora
Lala Lajpat Rai University of
   Veterinary and Animal Sciences
Hisar, India

Vasudha Bansal
Panjab University
Chandigarh, India

Aastha Bhardwaj
Department of Food Technology
Jamia Hamdard
New Delhi, India

Sourabh Bondre
Banaras Hindu University
Varanasi, India

Sourav Chakraborty
Ghani Khan Choudhury Institute
   of Engineering and Technology
   Narayanpur
Malda, India

Sreemoyee Chakraborty
Department of Food Technology
   & Biochemical Engineering
   Jadavpur University
West Bengal, India

Nabendu Debnath
Central University of Jammu
Bagla, India

Pinky Deka
University of Science and Technology
Meghalaya, India

Mousumi Ghosh
Department of Food Science and
   Technology,
Maulana Abul Kalam Azad University
   of Technology (MAKAUT)
West Bengal, India

Akansha Gupta
Banaras Hindu University
Varanasi, India

Himani
National Institute of Food
   Technology Entrepreneurship and
   Management
Sonepat, India

Vidhi Jain
Department of Microbiology
   Mohanlal Sukhadia University
Udaipur, Rajasthan, India

Jyoti
Department of Microbiology
   Mohanlal Sukhadia University
Udaipur, Rajasthan, India

Madhu Kamle
Applied Microbiology Laboratory,
    Department of Forestry
North-Eastern Regional Institute of
    Science and Technology
Nirjuli, India

Raveena Kargwal
College of Agricultural
    Engineering and Technology
CCSHAU, Hisar, Haryana, India

Arvind Kumar
Banaras Hindu University
Varanasi, India

Pradeep Kumar
Department of Botany, University of
    Lucknow
Lucknow, Uttar Pradesh, India

Sitesh Kumar
Indian Institute of Technology
    Kharagpur
West Bengal, India

Mitali Madhumita
Department of Food Technology, School
    of Health Science and Technology
University of Petroleum and Energy
    Studies
Bidholi, Dehradun, India

Dipendra Kumar Mahato
CASS Food Research Centre, School
    of Exercise and Nutrition Sciences
Deakin University
Burwood, Australia

Shubham Mandliya
Indian Institute of Technology
    Kharagpur
West Bengal, India

**Dalasanuru Chandregowda
Manjunathagowda**
ICAR-Directorate of Onion and
    Garlic Research
Rajgurunagar, India

Monika Mathur
CFST, College of Agricultural
    Engineering and Technology
Chaudhary Charan Singh Haryana
    Agricultural University
Hisar, Haryana, India

Sourav Misra
Indian Institute of Technology Kharagpur
West Bengal, India

Arun Kumar Pandey
MMICT&BM(HM), Maharishi
    Markandeshwar
Haryana, India

Pooja Pandey
Indian Institute of Technology
    Kharagpur
West Bengal, India

Surabhi Pandey
Banaras Hindu University
Varanasi, India

Shikha Pandhi
Banaras Hindu University
Varanasi, India

Pramod K Prabhakar
National Institute of Food
    Technology Entrepreneurship
    and Management
Sonepat, India

Dinesh Chandra Rai
Banaras Hindu University
Varanasi, India

Himanshu Kumar Rai
Banaras Hindu University
Varanasi, India

Jayabrata Saha
University of Science and
  Technology
Meghalaya, India

Saloni Anand
Banaras Hindu University
Varanasi, India

Chayanika Sarma
Indian Institute of Food Processing
  Technology
Thanjavur, Tamil Nadu, India

Raman Selvakumar
ICAR-Indian Agricultural Research
  Institute
New Delhi, India

Nitya Sharma
Indian Institute of Technology
New Delhi, India

Namita Ashish Singh
Department of Microbiology
  Mohanlal Sukhadia University
Udaipur, Rajasthan, India

Madhvi Singh
Institute of Chemical Technology
Bhubaneswar, Odisha, India

Mohit Singla
Tezpur University
Assam, India

Shubhangi Srivastava
University of Hohenheim
Stuttgart, Germany

Nimisha Tehri
Centre for Biotechnology Maharshi
  Dayanand University
Rohtak, Haryana, India

Vidhi Tyagi
Abigail Wexner Research Institute,
  Nationwide Childrens Hospital
Columbus, Ohio, USA

Amit Vashishth
Department of Biotechnology
  Meerut Institute of Engineering
  and Technology Meerut
Uttar Pradesh, India

Rahul Vashishth
Vellore Institute of Technology
Vellore, India

Urvashi Vikranta
Banaras Hindu University
Varanasi, India

Ashok Kumar Yadav
Banaras Hindu University
Varanasi, India

Ashok Kumar Yadav
Central University of Jammu
Bagla, India

Pooja Yadav
Central University of Jammu
Bagla, India

Chapter 1

Aflatoxins in Food and Feed

Occurrence, Detection, and Mitigating Strategies

Namita Ashish Singh, Jyoti, Vidhi Jain

CONTENTS

1.1	Introduction	2
1.2	Aflatoxin Contamination in Food and Feed	4
1.3	Methods Available for Detection of Aflatoxins	4
	1.3.1 Chromatographic Methods	4
	1.3.2 Immunoassays	8
	1.3.2.1 Radioimmunoassay	8
	1.3.2.2 Enzyme Linked Immunosorbent Assay	8
	1.3.2.3 Lateral Flow Immunoassay	9
	1.3.3 Biosensors Based on Immunoassay	9
	1.3.3.1 Immunoassay with Amperometric Technique	9
	1.3.3.2 Electrochemical Immunosensor	9
	1.3.3.3 Impedimetric Immunosensor	10
	1.3.4 Nanobiosensor-Based Method	10
1.4	Management Strategies for Mitigation of Aflatoxin	11
	1.4.1 Pre-harvest Agricultural Management of Aflatoxins	11
	1.4.1.1 Biological Control	11
	1.4.1.2 Improved Plant Resistance against Aflatoxins	11
	1.4.1.2.1 Resistance Breeding against *Aspergillus flavus* and Aflatoxin	11
	1.4.1.2.2 Transgenic Approaches for *Aspergillus flavus*	12
	1.4.2 Post-harvest Agricultural Management of Aflatoxins	12
	1.4.2.1 Degradation of Aflatoxins by Physical Means	12
	1.4.2.1.1 Irradiation	12
	1.4.2.1.2 Cold Plasma	14
	1.4.2.1.3 Treatment with Heat	15

DOI: 10.1201/9781003242208-1

1.4.2.2   Detoxification of Aflatoxins by Chemical Means   15
    1.4.2.2.1   Electrolyzed Oxidizing Water   15
    1.4.2.2.2   Carbon Filtration   15
    1.4.2.2.3   Organic Acids   16
    1.4.2.2.4   Ozone   16
    1.4.2.2.5   Absorption of Nutrients by Dietary
              Clay Minerals   16
1.4.2.3   Biological Decontamination of Aflatoxins   16
    1.4.2.3.1   Microbial Degradation   16
    1.4.2.3.2   Enzymatic Degradation   17
    1.4.2.3.3   Mycotoxin Binders   18
    1.4.2.3.4   Herbal Products   18
    1.4.2.3.5   Neutralization by Specific Antibodies
              Induced by Vaccination   18
    1.4.2.3.6   Molecular Biology Approaches   19
        1.4.2.1.6.1   Crop Engineering for
                       A. flavus Infection Resistance   19
        1.4.2.1.6.2   RNA Interference Technique
                       Based on Host-Induced Gene
                       Silencing (HIGS)   19
1.5   Conclusion   19
Acknowledgments   20

1.1   INTRODUCTION

Aflatoxins (AFs) are a type of fungal toxin produced mostly by *Aspergillus* fungus, predominantly *Aspergillus flavus* and *Aspergillus parasiticus*. If the temperature and humidity conditions are appropriate, these microorganisms infect crops and thrive on foods during storage. Despite having the same geographical ranges, *A. parasiticus* is not broadly dispersed, while *A. flavus* is the most-detected mold in foods. Alatoxins are metabolized by the liver and expelled by ruminants in their bile (Negash, 2018).

There are 20 known fungal metabolites, with at least 14 of these being researched as typical aflatoxins. Only six of these compounds are commonly detected in foods: aflatoxins M1, M2, B1, B2, G1 and G2. The word "aflatoxin" comes from the first letter of *Aspergillus* and the word *flavus*. Aflatoxins are difurocoumarin compounds exhibiting a unique fluorescence when exposed to ultraviolet light (Seid and Mama, 2019). Based on the color of the fluorescence, aflatoxins are categorized as B1 and B2 for blue and G1 and G2 for green florescence. Aflatoxins M1 and M2 are milk aflatoxins, which are hydroxylated versions of aflatoxins B1 and B2, commonly found in both human and animal milk (Li et al., 2019). Aflatoxin B1 is the deadliest fungal toxin among the various types of aflatoxin strains; it raises cattle's apparent protein needs and is a carcinogen (Talebi et al., 2011).

Aflatoxin-producing fungus contaminate feed during development, harvest, and storage, exposing animals to aflatoxins. When lactating cattle eat contaminated feed, aflatoxin B1 (AFB1) is converted into aflatoxin M1 (AFM1). Aflatoxins can enter the body through milk and dairy products, which are major parts of the human diet, providing a risk to consumers, especially infants and young children, leading to numerous health problems (Dohnal et al., 2014; Liu and Wu, 2010). When consumed, aflatoxin attaches to liver proteins, causing human liver cancer. The effects on animals differ according on the dose, exposure length, diet, breed, species, and dietary grade. These toxins are carcinogenic, hepatotoxic, and mutagenic, as well as suppressing immune systems.

Akande et al. (2006) claim that aflatoxin can harm the liver, induce cancer, reduce milk supply, weaken the immune system, and cause anemia in animals. In addition, it has been linked to decreased feed intake and growth as well as development in dairy cattle. Milk production increased by nearly 25% when dairy cattle were provided an aflatoxin-free diet. B1 > M1 > G1 > B2 > M2/G2 and so on are the toxicity profiles of AFs. AFs are the most potent of all known mycotoxins, with toxicological and severe hepatocarcinogenic effects due to their interaction with RNA, DNA, enzymes, and proteins (Enyiukwu et al., 2014). In Malaysia, China, India, and Kenya, chronic aflatoxicosis has been associated with liver cancer, whereas acute aflatoxicosis has been linked to abdominal pain, vomiting, edema, and mortality (Liew and Mohd-Redzwan, 2018). Loss of appetite, vomiting, weakness, and lethargy are some of the acute toxic symptoms of AFB1 poisoning in humans related to hepatotoxicosis and nephrotoxicosis. Per the International Agency for Research on Cancer, AFM1 has about one-tenth the carcinogenic activity of aflatoxin B1 in sensitive species. AFM1 was first classed as a 2B carcinogen with the ability to cause hepatocarcinogenesis. Even after pasteurization, AFM1 remains stable in a variety of dairy products. Because milk is a key component of their diet, newborns are more exposed to cancer risk. As a result, AFM1 is undesirable in milk as well as milk products, posing a danger to humans (Taheur et al., 2019). The quantity of AFM1 excreted in feed is usually between 1 and 3%. Milk can contain the toxin 12 to 24 hours after the first AFB1 ingestion. Seventy-two hours after AFB1 is withdrawn from the food, the concentration of aflatoxin M1 in milk is undetectable (Awad et al., 2011). The highest permitted limit for aflatoxins in all food commodities supplied in India is 30 g/kg (ppb), with aflatoxin M1 in milk with a tolerance value of 0.5 g/kg and aflatoxin B1 with a limit of 20 ppb. According to Food Safety and Standards Regulations 2011, the tolerance value of aflatoxin M1 for milk is 0.5 ppb, while the EU limit is 0.05 ppb (Lawley, 2013).

In EU rules, the maximum aflatoxin M1 level for infant meals is 0.025 g/kg. When it comes to aflatoxin M1, milk is ten times more contaminated in India than the European Union (Sharma and Parisi, 2017).

1.2 AFLATOXIN CONTAMINATION IN FOOD AND FEED

Aflatoxin contamination includes cereals such as maize, rice, and wheat; oilseeds such as sunflower, groundnut, and cotton; spices like chilis, coriander, and turmeric; and many tree nuts, milk, and milk products, as listed in Table 1.1 (Prandini et al., 2009).

1.3 METHODS AVAILABLE FOR DETECTION OF AFLATOXINS

Different methods available for the evaluation of aflatoxins include chromatographic methods and immunoassays, along with nanobiosensor-based methods, which are listed in Figure 1.1.

1.3.1 Chromatographic Methods

Chromatographic methods are mostly utilized for quantifiable monitoring of aflatoxins in food as well as feedstuff. Aflatoxins give off fluorescence when exposed to UV light. Comparison of the intensity of the fluorescence of a sample with a standard is done on the same thin-layer chromatographic (TLC) plate for estimation of the toxin present. The most frequent chromatographic techniques used for aflatoxin analysis are thin-layer chromatography, gas chromatography, high-performance liquid chromatography, and liquid chromatography-tandem mass spectrometry (Hussain et al., 2010; Kim et al., 2010; Hove et al., 2016). Although these techniques are selective, precise, and sensitive, they have some disadvantages, like cost ineffectiveness, the need for experienced personnel, being laboratory bound, and so on.

Manetta et al. (2005) devised a fluorescence detection high-performance liquid chromatography (HPLC) method to monitor aflatoxin M1 in milk and cheese, with a limit of detection of 1 and 5 ng/kg for milk and cheese, respectively. HPLC with fluorescence detection was used in combination with affinity chromatography (IAC) by Lee et al. (2009) with a 2 ppt limit of detection. Using a liquid chromatography-tandem mass spectrometry approach, Kamboj et al. (2020) examined dairy and feed constituents from Kenya. They found aflatoxins in 70% of the samples, ranging from 0.2–318.5 g/kg.

TABLE 1.1   INCIDENCES OF AFLATOXIN IN FOOD AND FEED GLOBALLY

Food Type	Country	Incidence %	Aflatoxin Type	Concentration (ppb)	Method	Reference
Wheat	Malaysia	64.29	AFB1	0.55–5.07	ELISA	Reddy et al. (2011)
Oat		50		0.65–2.85		
Chili		100		0.58–3.5		
Cumin		66.67		1.89–4.64		
Rice		69.23		0.68–3.79		
Sunflower		85.71		1.14–5.33		
Fresh fish	Egypt	33.34	Total aflatoxin	22–70.5	Fluorometer	Hassan *et al.* (2011)
Salted fish		40		18.5–50		
Tulum cheese	Turkey	80	AFM1	<0.378	ELISA	Ertas *et al.* (2011)
Dairy dessert		52		0.0015–0.08		
Yoghurt		56		0.0025–0.078		
Unwashed egg	Saudi Arabia	12	Total aflatoxin	0.61–1.19	ELISA	Bahobail et al. (2012)
Buttermilk	Jordan	100	AFM1	7.97–2027.11	ELISA	Omar (2012)
Sorghum	India	73.04	AFB1	0.01–263.98	ELISA	Ratnavathi et al. (2012)
Broiler feeds	Cameroon	93	AFs	52.00	Fluorometer	Kana et al. (2013)
Peanut meal		100		950.00		
Butter	Pakistan	45	AFM1	0.004–0.41	HPLC	Iqbal and Asi (2013)
Pasteurized milk	Sudan	100	AFM1	0.008–0.765	Fluorometer	Ali et al. (2014)

(Continued)

TABLE 1.1   INCIDENCES OF AFLATOXIN IN FOOD AND FEED GLOBALLY   (CONTINUED)

Food Type	Country	Incidence %	Aflatoxin Type	Concentration (ppb)	Method	Reference
Goat milk	Croatia	100	AFM1	0.003–0.04	ELISA	Bilandzic et al. (2014)
Sorghum malt	Burkina Faso	25	AFB1	46.33–254.73	HPLC	Bationo et al. (2015)
Raw milk	Brazil	72.9	AFM1	0.013–0.708	HPLC-FD	Santili et al. (2015)
Milk products	Serbia	100	AFM1	0.025–1.00	ELISA	Tomasevic et al. (2015)
Breast milk	Mexico	89	AFM1	0.00301–0.0342	ELISA	Cantu-Cornelio et al. (2016)
Corn	Costa Rica	24	Total AFs	0.01–290	ELISA/HPLC	Granados-Chinchilla et al. (2017)
Chilis	United States	64	AFB1	< 2	LFIA	Singh and Cotty (2017)
Corn	Vietnam	33.71	AFB1	1–34.80	ELISA	Lee et al. (2017)
Yoghurt	Burundi	100	AFM1	8.2–63.2	Reveal Q+/ ELISA	Udomkun et al. (2018)
Hazelnut	Italy	16.1	AFB1	56.00	HPLC	Diella et al. (2018)
Cheese	Democratic Republic of Congo	100	AFM1	18.5–261.1	Reveal Q+/ ELISA	Udomkun et al. (2018)

Maize	Ethiopia	100	Total aflatoxin	14.7	LFIA	Worku et al. (2019)
Sorghum malt	Namibia	17	AFG1	0.39–6.95	LC-MS/MS	Nafuka et al. (2019)
Coffee beans	Yemen	100	Total aflatoxin	14.255–23.231	ELISA	Humaid et al. (2019)
Maize bran	Uganda	100	Afs	393.5	Fluorometer	Nakavuma et al. (2020)
Poultry feed ingredients	Kenya	21	AFM1	6.9	LC-MS/MS	Kemboi et al. (2020)
Tea	Morocco	58.9	Total aflatoxin	1.2–116.2	LC-MS/MS	Mannani et al. (2020)
Poultry feeds	Ghana	100	Afs	118.00	LFIA	Aboagye-Nuamah et al. (2021)
Sesame	Nigeria	-	AFB1 AFB2	3.95–11.75 2.35	HPLC	Matthew et al. (2021)

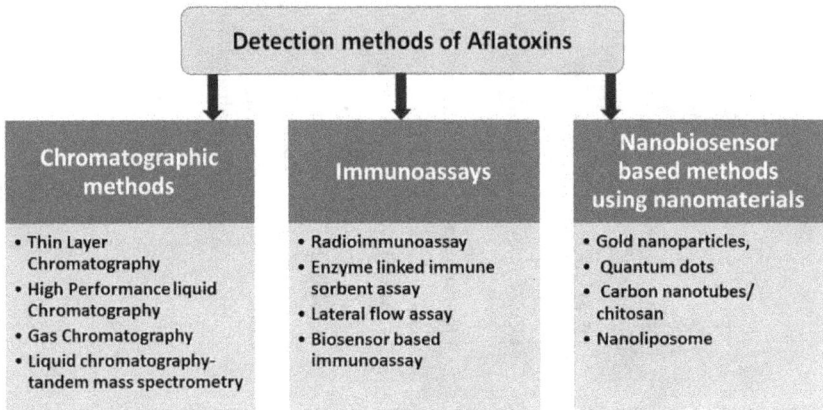

Figure 1.1   Methods available for the monitoring of aflatoxins.

1.3.2 Immunoassays

1.3.2.1 Radioimmunoassay

A radioimmunoassay (RIA) involves incubating a test solution containing an undetermined level of aflatoxin with a fixed amount of tagged aflatoxin and its particular antibody at the same time. Following that, both free and bound radio-labelled aflatoxins are isolated and radioactivity is determined. The test solution's aflatoxin level is evaluated by comparing the results with a standard curve, which is constructed by graphing the ratio of bound and starting amounts of labelled aflatoxin multiplied by 100 vs. the aflatoxin standard level (Qian et al., 1984). The Charm 6602 assay is a radioimmunoassay that is commercially used for monitoring aflatoxin M1 (Offiah and Adesiyun, 2007).

1.3.2.2 Enzyme-Linked Immunosorbent Assay

Enzyme-linked immunosorbent assay (ELISA) is a widely utilized approach for monitoring major aflatoxins B1 and M1 in food and feed materials. Direct, indirect, and competitive inhibition techniques are used in ELISA. In direct inhibition, an enzyme-labeled chief antibody reacts with an antigen, whereas in the competitive method, unlabeled antigens from samples and enzyme conjugate contest to bind with an antibody corresponding to a precise mycotoxin (Pereira et al., 2019). The low detection limit, specificity, negligible cleanup, and ease of use, along with high sample yield, are all advantages of this method; however, the possibility of a false positive as well as negative result, incompatibility with complex mediums, and single use are limitations (Xie et al., 2015).

Intended for the analysis of AFM1 in milk, a sensitive sandwich ELISA has been developed based on the immobilization of rat monoclonal antibody against aflatoxin M1 in 384 microtiter plates to capture aflatoxin M1 antigen. This miniaturized method allowed for the monitoring of aflatoxin M1 in milk at 0.005 pg/mL, which was a significant improvement over previous methods. A magnetic nanoparticle-based ELISA coupled with a micro-plate approach has been published for the evaluation of AFM1 in milk with a limit of detection 0.5 pg/mL. The advantages of this designed test are a compact column, excellent capture efficacy, and low cost (Kanungo et al., 2011). A swift and sensitive super paramagnetic bead- and gold nanoprobe–based method has been reported for detecting AFM1 (Zhang et al., 2013). Aflatoxin concentration is measured in this magnetic bead-based immunosorbent assay (MBISA) through direct competition between aflatoxin M1 and nanoprobes. When compared to traditional ELISA, MBISA has a 15-minute incubation period and eliminates the color development stage completely, with a detection limit of 27.5 ngL^{-1} in milk.

1.3.2.3 Lateral Flow Immunoassay

The swift monitoring of AFB1 in pig feed has been reported using an immunoassay-based lateral flow approach with a limit of detection of 5 microg/kg. A conjugate, absorbent pad, and membrane are the three main components of the test strip. As capture agents, the membrane is mostly coated with AFB1-bovine serum albumin conjugate as well as rabbit anti-mouse antibodies (Delmulle et al., 2005). In maize, a lateral flow–based immunoassay for four main aflatoxins, AFB1, AFB2, AFG1, and AFG2, has been reported (Anfossi et al., 2011).

1.3.3 Biosensors Based on Immunoassay

1.3.3.1 Immunoassay with Amperometric Technique

For aflatoxin M1 monitoring, a flow injection–based immunoassay with amperometric transducer was developed, which involves incubating a sample containing aflatoxin M1 (antigen) with fixed amounts of anti-AFM1 antibody and the tracer antigen (Ag*, aflatoxin M1 covalently coupled to horseradish peroxidase enzyme); sample Ag and tracer antigen (Ag*) compete for antibody binding site. The Ag-Ab combination is then injected into a flow scheme, which is set aside in the column owing to the high affinity of Protein G for the antibody, and the enzyme's action is measured amperometrically with concentrations ranging from 20 to 500 parts per million (Badea et al., 2004).

1.3.3.2 Electrochemical Immunosensors

Micheli et al. (2005) created a disposable electrochemical AFM1 immunosensor by the immobilization of antibodies on the surface of screen-printed electrodes, which competes with free AFM1 as well as aflatoxin coupled with horseradish peroxidase enzyme with a limit of detection of 25 ppt. A low detection limit of 0.01 ppb has been obtained for the evaluation of AFM1 in food products using electrochemical immunosensors (Paniel et al., 2010).

1.3.3.3 Impedimetric Immunosensors

A biosensor based on gold nanoparticles has been reported to detect AFM1 in actual milk samples with concentrations ranging from 1 to 14 ng/mL. Electrochemical impedance spectroscopy along with cyclic voltammetry were employed to observe the assembly processes of cysteamine, gold nanoparticles, and ss-HSDNA (Dinckaya et al., 2011). An impedance-based electrochemical immunosensor has been reported for analysis of AFB1 in pistachio matrices by the immobilization of monoclonal AFB1 antibodies on a gold electrode. In unknown pistachio samples, the developed immunosensor could detect AFB1 concentrations ranging from 4.56 to 50.86 ng/mL (Kaminiaris et al., 2020).

1.3.4 Nanobiosensor-Based Method

Nanobiosensors are emerging as novel detection approaches for their applications in food and agriculture (Singh, 2016). Nanobiosensors are the combination of biology, chemistry, and nanotechnology and utilize different nanoparticles like liposomes, quantum dots, and carbon nanotubes. An aptamer-based colorimetric and chemiluminescence technique has been reported utilizing gold nanoparticles for the monitoring of aflatoxin B1 in rice and peanut samples. The sensitivity of the detection method (0.5 nM) has been increased by using chemiluminescence during the luminol–hydrogen peroxide reaction, which is proportional to the aflatoxin B1 concentration (Hosseini et al., 2015). An aptasensor has been developed employing unmodified gold nanoparticles as colorimetric markers for analysis of alatoxin B1 with a limit of detection of 0.025 ng/mL (Luan et al., 2015).

A nanobiosensor has been designed for the evaluation of AFB1 in peanuts and rice with a limit of detection of 3.4 nM, which was based on the Forster resonance energy transfer method (Sabet et al., 2017). A fluorescent aptasensor using graphene oxide for the quenching of fluorescence of a carboxyfluorescein-labeled aptamer was established for the monitoring of AFM1 in skim milk powder showing a limit of detection of 0.05 μg/kg (Guo et al., 2019). A cadmium sulfide nanoparticle-based biosensor for the assessment of aflatoxin M1 in milk utilizing antibody stripping voltammetry has been published with a detection limit of 30 ppb (Hayat et al., 2019). A colorimetric microfluidic paper-based approach for the evaluation of aflatoxin M1 in milk samples with a detection limit of 10 nM using unaltered gold nanoparticles has been designed (Kasoju et al., 2020). Hamami et al. (2021) described a gold nanoparticle-based biosensor for the monitoring of aflatoxin M1 in milk with a detection limit of 7.14 pg/ml.

An electrochemical immunosensor with a detection limit of 0.07 ng/mL was reported for measuring aflatoxin B1 in food products. This immunosensor is made up of gold nanoparticles immobilized with an aflatoxinB1-bovine serum albumin conjugate (Owino et al., 2008). An electrochemical competitive immu-noassay that uses single-walled carbon nanotubes/chitosan-modified glass carbon electrodes to immobilize AFB1-bovine serum albumin conjugate for AFB1 monitoring in maize powder with lower detection limit, that is, 3.5 pg/mL has been developed (Zhang et al., 2016). A sensitive quartz crystal micro-balance immunosensing approach (resulting in a piezoelectric effect) has been designed for monitoring AFB1 in food utilizing glucose-encapsulated nano liposomes with a detection limit of 0.83 ng/kg (Tang et al., 2018).

1.4 MANAGEMENT STRATEGIES FOR MITIGATION OF AFLATOXINS

Mitigation of aflatoxins can be done at two stages: pre-harvest and post-har-vest. These methods are described in the following.

1.4.1 Pre-Harvest Agricultural Management of Aflatoxins

1.4.1.1 Biological Control

Aspergillus flavus is a fungus that lives in soil and colonizes carbon and nitro-gen-rich environments. *A. flavus* strains S and L have been discovered, with the S strain producing more aflatoxins and sclerotia but fewer conidia. An impor-tant biocontrol method is the introduction of an atoxigenic strain into agri-cultural territory to contest with toxigenic strains. Atoxigenic strains NRRL 21882 that have been employed in the field are non–aflatoxin-generating *A. flavus* strains that outcompete the toxin-generating strain(s) found in the soil (Monda and Alakonya, 2016). Because of competitive exclusion/displacement, co-inoculation of maize along with toxigenic and atoxigenic strains reduces aflatoxin contamination by 80–95%.

1.4.1.2 Improved Plant Resistance against Aflatoxins

Fungal resistance, aflatoxin inhibition, and insect resistance have all been uti-lized to improve maize plant resistance to aflatoxin contamination.

1.4.1.2.1 Resistance Breeding against Aspergillus flavus *and Aflatoxins*

The progress of improved inoculation approaches and in vitro screening proce-dures arose from the necessity for adequate, dependable, and swift screening methodologies for the breeding of aflatoxin accumulation resistance in maize.

The kennel screening assay, for example, has directed the identification of more promising aflatoxin resistance sources (Henry et al., 2012). While aflatoxin-resistant germplasms have been discovered, other breeding techniques like gene pyramiding could be used to improve long-term aflatoxin resistance. The fundamental issue related to breeding for resistance is polygenic characteristics, which means that developing a resistant variety requires several breeding seasons (Medina et al., 2014).

1.4.1.2.2 Transgenic Approaches for Aspergillus flavus

Recombinant insecticidal proteins from *Bacillus thuringiensis*, antifungal peptides, and proteins and host-induced gene-silencing technologies are used to control aflatoxins. The emphasis on insect resistance stems from a link between insect damage and aflatoxin exposure (Barros et al., 2009). Aflatoxin generation and fungal growth are both known to be inhibited by certain proteins. Low *A. flavus* colonization and aflatoxin buildup in maize may be caused by the production of detoxifying enzymes that break down toxins that obstruct the toxin synthesis path. Two amylase enzyme inhibitors from *Aspergillus flavus* have been discovered in maize, and their expression in transgenic maize compact colonization as well as aflatoxin concentration (Monda and Alakonya, 2016).

1.4.2 Post-Harvest Agricultural Management of Aflatoxins

Physical, chemical, and biological strategies for reducing AF content can all be classified as post-harvest methods. Physical as well as chemical methods are commonly much quicker than biological methods and can also be practiced with higher safety and efficiency, which makes them more acceptable for potential consumers (Nagy et al., 2021). Specific microbes cling to and/or convert AFs into less hazardous chemicals in biological detoxification procedures (Bianchini et al., 2009; Tian and Chun, 2017), which are even beneficial with regard to the sensory and nutritive standards of food and offer a secure option, taking into consideration the food safety outlook (Peles et al., 2021). Prevention and various detoxification methods used for removal of aflatoxins are depicted in Figure 1.2.

1.4.2.1 Degradation of Aflatoxins by Physical Means

1.4.2.1.1 Irradiation

Ionizing (electron beam, UV rays, and gamma) and non-ionizing (infrared waves, visible waves, microwaves, etc.) irradiation are frequently used to destroy aflatoxins present in food as well as feed. Electron beam irradiation (EBI) has the ability to reduce aflatoxin levels. Elevated effectiveness, lower equipment cost, dose control, quick processing time, and less heat production

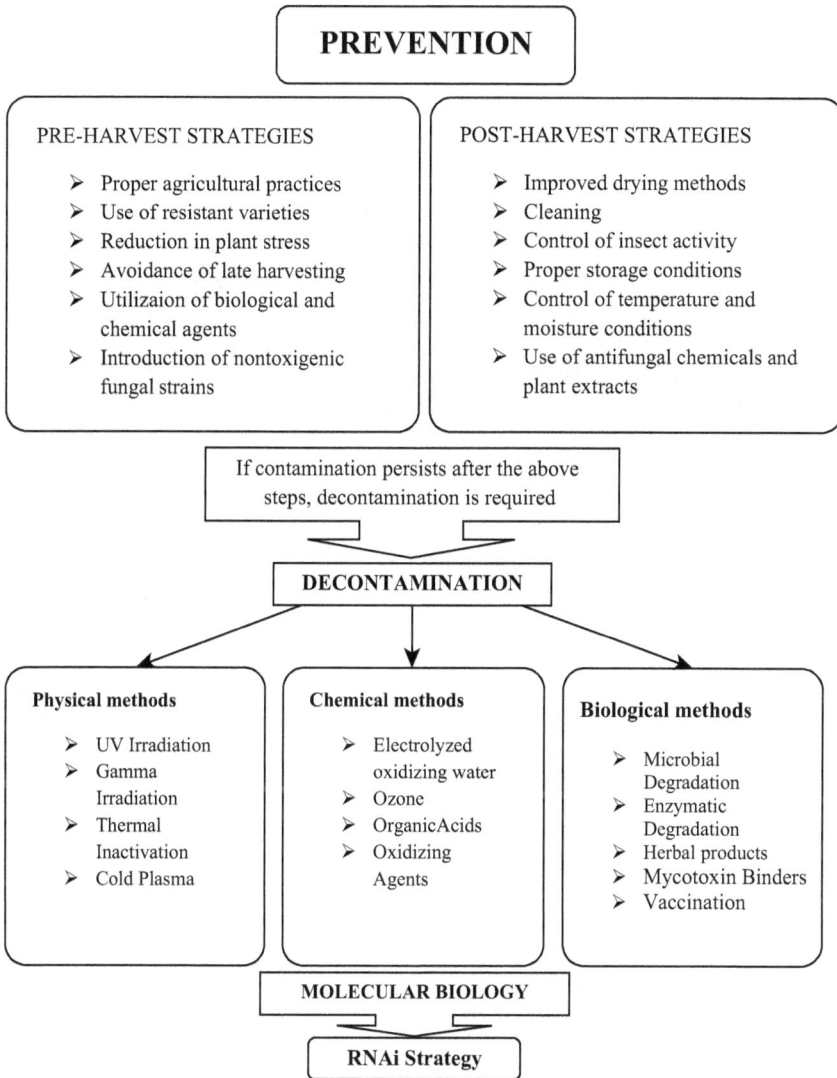

Figure 1.2   Prevention and detoxification methods used for removal of aflatoxins.

are advantages of EBI technology (Assuncao et al., 2015). However, when it comes to aflatoxin destruction, EBI technology is less efficient than radiation. EBI was demonstrated to lower AFB1 levels in Brazil nuts by 65.7% and 53.3% at doses of 10 and 5 kGy, respectively, while irradiation at equal dosages reduced AFB1 content by 84.2% and 70.6%, respectively.

Because of their high penetrability and reactivity, gamma rays have long been the favored radiation source for food. Food treated with gamma radiation up to 10 kGy poses no toxicological or microbiological risks (Patil et al., 2019). Furthermore, irradiation causes high-energy rays to interact with the water in foodstuffs. This generates extremely reactive free radicals, which degrade aflatoxins and target pathogenic microorganisms' DNA (Vita et al., 2014). Photons with high energy from cobalt-60 have been employed to destroy harmful microorganisms by acting on microbial DNA directly (Markov et al., 2015).

UV irradiation is both economical and environmentally friendly. The intensity and period of ultraviolet irradiation have a significant impact on the success of aflatoxin eradication. Short (254 nm) and long (362 nm) wavelengths of UV rays for 30 minutes completely eliminated AFB1 and AFG1 in cereals, whereas 2 hours of contact time with the same wavelengths of UV rays reduced AFB2 in cereals by 50% and 74%, respectively (Da Silva, 2012). For 15 minutes, almonds, groundnuts, and other dry fruits were exposed to UVC at 265 nm. AFG2 was completely removed from all nut samples, and AFG1 was completely degraded in pistachio and almond.

Aflatoxin breakdown in food and feed has been achieved using non-thermal pulsed light technology. The device produces short, high-intensity bursts of light (100–1100 nm) that destroy the nucleic acid as well as the cell wall of bacteria in a matter of seconds. AFB1 in water was degraded by 92.7% after eight pulsed light (PL) flashes of 1 J cm^2 (Di Stefano et al., 2014). A high temperature (more than 130°C) is generated by microwave heating, which is essential for aflatoxin elimination. Corn was microwave heated for 5.5 minutes at 1650 W power, resulting in 36% and 58% reductions in AFB1 and AFB2. Peanut and its derivatives were reported to reduce AFB1 levels by 50–60% by microwave heating according to Mobeen et al. (2011).

A pulsed electric field (PEF) is an advanced approach that alters the permeability of plasmalemmas by holding small pulses of electric field between 80 kV/cm and 100 V/cm. Both aflatoxin B2 and G1 levels in water as well as grape juice samples were reduced by PEF treatments (Hassan and Hussein, 2017). The AFB1 level in maize was reduced by 11% after a 10-minute pulsed ultrasound treatment at 1.65 W/cm^3 power intensity (Liu et al., 2019). When UV radiation was used to treat AFG1, AFB1, and AFB2 in pure water, significant reductions in their concentrations were seen (67.22%, 98.25%, and 29.77%, respectively) (Patras et al., 2017). An increased dose and extensive UV-C treatment, 6.18 kJ/cm^2 for 3 hours, decreased the AFB1 content of rice (Ferreira et al., 2021).

1.4.2.1.2 Cold Plasma

This technique has been utilized to destroy aflatoxins at atmospheric temperature and pressure. When Ouf et al. (2015) subjected *Aspergillus niger* spores and mycotoxin to cold plasma at ambient pressure for 9 mins, they discovered that the concentrations of mycotoxin B2 and Ochratoxin were decreased from 6 and 25g/100 mm^2 correspondingly, and entire spores were killed. Another study indicated that when aflatoxin-infected hazelnuts were detoxified with nitrogen plasma (1150 W), roughly 70% of AFB1 was removed. AFB1 in rice and wheat samples, as well as aflatoxin B1 on glass slides, was treated for 30 minutes with a corona discharge plasma jet (CDPJ), which reduced AFB1 concentration by 56.6%, 45.7%, and 95%, respectively (Puligundla et al., 2019). The food type, plasma system, operational parameters (energy contribution, moisture, functional gas), and exposure length all influence the efficiency of cold plasma technology for aflatoxin destruction (Hertwig et al., 2018).

1.4.2.1.3 Treatment with Heat

With a soybean matrix, temperatures above 160°C have been demonstrated to be efficient in eliminating AFB1. While normal food preparation temperatures, that is, 100°C, have no effect on AFs, the high temperatures used in baking, roasting, frying, and so on are more effective at reducing AF concentration. Extrusion can reduce AF levels up to 80%, depending on the temperature and dampness of the grain, while alkaline treatment can improve the effectiveness of the operation. Furthermore, extrusion alone reduces AF by 23–66% in peanut meal, but when combined with ammonium hydroxide, it can be reduced by up to 87%. Roasting reduces the quantity of AF in peanuts by 50–70% and maize by 40–80% (Karlovsky et al., 2016).

1.4.2.2 Detoxification of Aflatoxins by Chemical Means

1.4.2.2.1 Electrolyzed Oxidizing Water

Electrolysis of 1% sodium chloride water generates electrolyzed oxidizing water, also known as electro-activated water (NaCl). Genotoxicity and cytotoxicity of aflatoxins in HepG2 cell lines were significantly reduced after treating aflatoxin-contaminated corn with neutral electrolyzed oxidizing water at ambient temperature for 15 minutes (Jardon-Xicotencatl et al., 2015). The level of aflatoxin B1 in peanuts was decreased by 85–90% after 15 minutes of soaking in electrolyzed acidic water (Pankaj et al., 2018). Chlorenium and hydroxide ions are responsible for the deactivation process (Escobedo-González et al., 2016). It was discovered that electrolyzed water was effective in eliminating aflatoxin B1 from peanut and olive oil (Yang et al., 2019).

1.4.2.2.2 Carbon Filtration

AF removal from liquid materials can be done by the application of carbon filtration; for example, with coffee samples, a 73–78% reduction was attained. Therefore, it is a fruitful and economical approach for AF control and alleviation (Azam et al., 2021).

1.4.2.2.3 Organic Acids

To decompose aflatoxin, organic acids are typically used. Soaking AFB1-contaminated soyabeans in 1.0 N (tartaric, lactic, and citric acid) for 18 hours at ambient temperature led to reductions in AFB1 levels of 95.1%, 92.7%, and 94.1%, respectively (Rastegar et al., 2017). AFB1 can be reduced by 50.2% in roasted pistachio nuts by providing an acidic environment with lemon juice for 60 mins at 120°C. Aflatoxin breakdown in peanuts was up to 98.3% when acidulation was combined with a pulsed light approach (Abuagela et al., 2019).

1.4.2.2.4 Ozone

Ozone is a potent oxidizer, antibacterial, and disinfectant and can be used directly to decontaminate a variety of foods. Ozone degrades aflatoxins aflatoxinB1 and aflatoxin G1 by electrophilically assaulting the C8-C9 bond of the difuran ring, resulting in the formation of ozonide, but AFB2 and AFG2 are unaffected by ozonization due to the lack of a C8-C9 bond (Jalili, 2016). Ozone treatment enhanced the rate of aflatoxin detoxification in groundnut samples while having no effect on the peroxide and polyphenol concentrations.

1.4.2.2.5 Absorption of Nutrients by Dietary Clay Minerals

Adsorbing agents are chemicals added to foodstuff, fed to animals, or given separately to restrict AF assimilation in the gastrointestinal tract, hence lowering toxin dispersion and metabolism in organs and tissues. Activated charcoal, zeolite, hydrated sodium calcium, and bentonite are among the clay materials that have variable ability to bind AFs. The bentonites' detoxifying abilities were efficient, decreasing contamination to below the European AFM1 standard limits (50 ng/kg) with relatively modest alterations in the milk's nutritional characteristics (Phillips et al., 2008).

1.4.2.3 Biological Decontamination of Aflatoxins

1.4.2.3.1 Microbial Degradation

Aflatoxins in food and feed can be degraded by a variety of microbes isolated from various sources. More than 90% of AFB1 can be degraded by *Rhodococcus* strains. *Bacillus* spp. like *B. subtilis*, *B. licheniformis* BL010, and *Bacillus sp.* TUBF1 have all been reported to break down aflatoxins (Prettl et al., 2017). *Staphylococcus sp.* VGF2 lysate had a 100% AFB1 degradation capacity, but *Pseudomonas fluorescens* and *Pseudomonas anguilliseptica* decreased aflatoxin B1 by 63% and 66.5% individually. The ability of *Streptomyces* strains (no.-124) to degrade aflatoxin B1 was studied, and the results showed that *Streptomyces* strains can degrade AFB1 in 55% of cases. Non-toxigenic *A. niger* strains with aflatoxin decontamination capabilities have been utilized in the food sector. The *A. niger* FS-UV1 strain, which was created by UV irradiation from the wild strain *A. niger* FS-Z1, exhibited improved AFB1 detoxifying ability up to 95.3%

(Wang et al., 2018). The AFB1 toxin was degraded by more than 95% by the microbial consortia TADC7. *Tepidimicrobium* and *Geobacillus* were discovered to have a substantial role in AFB1 breakdown using 16S rRNA sequencing (Shu et al., 2018).

Lactic acid bacteria like *Lactobacillus selangorensis, Lactobacillus acidophilus, Lactococcus lactis, Streptococcus thermophilus, Enterococcus avium,* and *Bifidobacterium animalis* block or eliminate aflatoxins from feed as well as food. The innate properties of lactic acid bacterial (LAB) strains, temperature, incubation period, pH, and matrix all influence their propensity to bind AFs (Peles et al., 2021). Assaf et al. (2019) proposed that biofilm-forming probiotic LAB strains can reduce AFM1 in milk. *Lacticasei bacillus rhamnosus* (previously *Lactobacillus rhamnosus*) GG biofilm was found to reduce AFM1 by 61%.

Antifungal chemicals produced by LAB have been shown to help reduce mycotoxin formation. Organic acids and phenolic compounds, such as phenolic acids, fatty acids, volatile organic compounds, ethanol, hydrogen peroxide, and proteinaceous chemicals, are the most common. Non-lactic acid bacteria like *Enterobacter* spp., *Brachybacterium, Escherichia* spp., *Klebsiella* spp., *Myxococcus, Nocardia* spp., *Pseudomonas, Bacillus* spp., *Rhodococcus* can also impede growth as well as aflatoxin production (Peles et al., 2021).

Yeasts such as *Candida, Pichia, Saccharomycopsis, Schizosaccharomyces, Saccharomycodes, Saccharomyces, Trichosporon,* and *Zygosaccharomyces* species have been shown to considerably limit AF formation in aflatoxigenic molds in several investigations. Yeast supplementation (e.g., *Kluyveromyces* and *Pichia* spp.) has been proven to increase dairy cow performance and boost aflatoxin B1 detoxification in the rumen (Intanoo et al., 2020).

1.4.2.3.2 Enzymatic Degradation

Aflatoxin can be degraded by an array of enzymes, including oxidases, peroxidases, reductases, and laccases. Laccases are multicopper oxidase enzymes that can be found in fungi, plants, bacteria, and insects. Bacterial laccases are more thermolabile and alkalitolerant and have a broader pH and substrate range than fungal laccases, making them better candidates for xenobiotic breakdown. In human liver L-02 cells, the *B. licheniformis* enzyme CotA laccase catalyzed the C3-hydroxylation of aflatoxin B1 and converted dangerous aflatoxin B1 into non-toxic forms (Loi et al., 2016).

Extracellular extract from *Rhodococcus erythropolis* culture, in combination with laccase enzyme from multiple fungal species, demonstrated effective breakdown of AFB1 (Alberts et al., 2009). Peroxidase treatment at 30°C after incubation of 8 hours resulted in significant decline in AFM1 and AFB1 present in milk (Sibaja et al., 2018). Manganese peroxidase (MnP) reduces AFB1 levels by 86.0% while reducing mutagenicity by 69.2%. MADE, an extracellular enzyme isolated from *Myxococcus fulvus* ANSM068, assists in the

breakdown of AFB1, AFM1, and AFG1 at 35°C and pH 6.0. By combining gua-
nosine monophosphate with copper ions at ambient temperature, a laccase-
mimicking nanozyme (nanomaterials with inherent enzyme-like capabilities
that catalyze the substrates of usual enzymes) was created. It has a catalytic
effectiveness similar to the usual laccase and is 2400 times more cost effective,
as well as being resistant to severe temperatures, pH salts, and storage conditions (Loi et al., 2020).

1.4.2.3.3 Mycotoxin Binders

Mycotoxin binders have been used to remove toxins from animal feed by binding with mycotoxins and confining their absorption in the gastrointestinal
tract; further, the bounded toxins can be released through the animal's excreta
or urine (Abdallah et al., 2015; Mgbeahuruike et al., 2018). Inorganic binders
like zeolites, diatomaceous earth, activated charcoal, bentonites, sepiolitic
clay, and organic binders, namely oat fibers and extracted cell wall fraction of
S. cerevisiae, and yeast cells have been used for toxin control in diet (Abdallah
et al., 2015; Yalcin et al., 2018).

1.4.2.3.4 Herbal Products

According to studies, ethanolic extract of *Cassia senna* and methanol extract
of *Cassia tora* reduce aflatoxin B1's mutagenic effect in vitro, while methano-
lic extracts of *Thonninga sanguinea* and *Piper argyrophyllum* reduce AFB1's
genotoxicity and hepatotoxicity (Abd El-Hack et al., 2018). According to a
study, *Curcuma longa* may protect broiler chicks from the destructive effects
of aflatoxin (Rangsaz and Ahangaran, 2011). A study found that aqueous
extracts from 31 medicinal herbs were effective in detoxifying AFB1, while
leaf extract of *Adhatoda vasica Nees* contained partially purified alkaloids
that showed the highest AFB1 breakdown capacity (98%) after 24 hours of
incubation (Vijayanandraj et al., 2014). The efficacy of aqueous extracts of
Corymbia citriodora for decontamination of aflatoxins B1 and B2 in spiked
maize was tested, and it was shown to be reduced by 89.6% and 86.5%, respectively (Iram et al., 2016). Natural plant extracts are employed in the food
industry all over the world as food additives. After 48 hours of incubation
by *Rosmarinus officinalis*, aqueous leaf extract AFB1 was reduced by 60.3%
(Ponzilacqua et al., 2019).

1.4.2.3.5 Neutralization by Specific Antibodies
Induced by Vaccination

Vaccinating dairy cows against aflatoxin B1 could give a nutritional and animal health answer to the problem of AF contamination, enhancing milk safety
and decreasing AFs. Vaccination creates antibodies (Abs) that, by immuno-
interception, prevent AFs from being absorbed, bioactivated, or excreted in

milk or other products (neutralization). AFB1 and AFM1 levels in milk have recently been shown to be reduced by systemic immunization of dairy cows and heifers (Giovati et al., 2014). AFM1 levels in vaccinated cows' milk were substantially lower than in unvaccinated cows' milk.

1.4.2.3.6 Molecular Biology Approaches

1.4.2.1.6.1 Crop Engineering for A. flavus Infection Resistance
The development of anti-fungal transgenic cultivars that provide resistance to aflatoxin-producing fungus would be tremendously advantageous to breeding operations. Transgenic crops carrying genome fragments from plant pathogenic fungus have been widely used to boost or decrease the expression of genes required for antifungal or antitoxin activities (Wani, 2010).

1.4.2.1.6.2 RNA Interference Technique Based on Host-Induced Gene Silencing (HIGS)
The pathogen is regulated by the host plant to suppress the expression of its personal genes without the use of a particular outside protein called host-induced gene silencing (HIGS). Filamentous fungi, like other eukaryotes, have genomes that encode RNA-dependent RNA polymerases implicated in RNA interference (RNAi). It has been discovered that several *Aspergillus* mycoviruses can inhibit RNA silencing in *Aspergillus*. Gene silencing may be the most promising approach for integrating resistance to aflatoxins in a very short period of time, based on recent RNAi research as well as the interface of A*spergillus* mycoviruses with their hosts (Hammond et al., 2008). At harvest and during storage, the altered aflatoxigenic species will be unable to produce aflatoxins (Sharanaiah et al., 2017).

1.5 CONCLUSION

Aflatoxins are major health concern worldwide due to their extensive occurrence in food and feed. As consumer expectations are high, we need sensitive, cost-effective and matrix-friendly detection methods. Reduced risk requires a coordinated approach that manages aflatoxins at all steps of production from farm to table. The European Food Safety Authority and the Food Safety and Standards Authority of India have taken a number of steps to tackle this rising problem, including establishing mitigation techniques before and after harvest along with issuing aflatoxin advisories. Effective agricultural practices in field during harvesting and storage can reduce the risk of aflatoxins. In the future, biological control like mitigation of aflatoxins by microorganisms, mycotoxin binders, herbal products, crop engineering, and RNAi strategies can play a vital role to ensure food safety.

ACKNOWLEDGMENTS

The authors would like to acknowledge the administration of Mohanlal Sukhadia University Udaipur for their encouragement.

REFERENCES

Abdallah, M.F.; Girgin, G.; Baydar, T. Occurrence, prevention and limitation of mycotoxins in feeds. *Animal Nutrition and Feed Technology*. 2015, 15, 471–490.

Abd El-Hack, M.E.; Samak, D.H.; Noreldin, A.E.; El-Naggar, K.; Abdo, M. Probiotics and plant-derived compounds as eco-friendly agents to inhibit microbial toxins in poultry feed: a comprehensive review. *Environmental Science and Pollution Research*. 2018, 25(32). http://doi.org/10.1007/s11356-018-3197-2.

Aboagye-Nuamah, F.; Kwoseh, C.K.; Maier, D.E. Toxigenic mycoflora, aflatoxin and fumonisin contamination of poultry feeds in Ghana. *Toxicon*. 2021, 198, 164–170.

Abuagela, M.O.; Iqdiam, B.M.; Mostafa, H.; Marshall S.M.; Yagiz, Y.; Marshall, M.R.; Sarnoski, P. Combined effects of citric acid and pulsed light treatments to degrade B-aflatoxins in peanut. *Food and Bioproducts Processing*. 2019, 117, 396–403.

Akande, K.E.; Abubakar, M.M.; Adegbola, T.A. Nutritional and health implications of mycotoxins in animal feeds: a review. *Pakistan Journal of Nutrition*. 2006, 5(5), 398–408.

Alberts, J.F.; Gelderblom, W.C.A.; Botha, A.; Van Zyl, W.H. Degradation of aflatoxin B1 by fungal laccase enzymes. *International Journal of Food Microbiology*. 2009, 135(1), 47–52.

Ali, M.A.I.; El Zubeir, I.E.M.; Fadel Elseed, A.M.A. Aflatoxin M1 in raw and imported powdered milk sold in Khartoum state, Sudan. *Food Additives & Contaminants: Part B*. 2014, 7(3), 208–212.

Anfossi, L.; D'Arco, G.; Calderara, M.; Baggiani, C.; Giovannoli, C.; Giraudi, G. Development of a quantitative lateral flow immunoassay for the detection of aflatoxins in maize. *Food Addit Contam Part A Chem Anal Control Expo Risk Assess*. 2011, 28(2), 226–234.

Assaf, J.C.; Khoury, A.; Chokr, A.; Louka, N.; Atoui, A. A novel method for elimination of aflatoxin M1 in milk using *Lactobacillus rhamnosus* G.G. biofilm. *International Journal of Dairy Technology*. 2019, 72, 248–256.

Assuncao, E.; Reis, T.A.; Baquiao, A.C.; Correa, B. Effects of gamma and electron beam radiation on Brazil nuts artificially inoculated with *Aspergillus flavus*. *Journal of Food Protection*. 2015, 78, 1397–1401.

Awad, W.A.; Hess, M.; Twarużek, M.; Grajewski, J.; Kosicki, R.; Böhm, J.; Zentek, J. The impact of the fusarium mycotoxin deoxynivalenol on the health and performance of broiler chickens. *International Journal of Molecular Sciences*. 2011, 12(11), 7996–8012.

Azam, K.; Akhtar, S.; Gong, Y.Y.; Routledge, M.N.; Ismail, A.; Oliveira, C.A.F.; Iqbal, S.Z.; Ali, H. Evaluation of the impact of activated carbon-based filtration system on the concentration of aflatoxins and selected heavy metals in roasted coffee. *Food Control*. 2021, 121, 107583.

Badea, M.; Micheli, L.; Messia, M.C.; Candigliota, T.; Marconi, E.; Velasco-Garcia, M.; Moscone, D.; Palleschi, G. Aflatoxin M1 determination in raw milk using a flow-injection immunoassay system. *Analytica Chimica Acta.* 2004, 520(1–2), 141–148.

Bahobail, A.A.S.; Hassan, S.A.; El-Deeb, B.A. Microbial quality and content aflatoxins of commercially available eggs in Taif, Saudi Arabia. *African Journal of Microbiology Research.* 2012, 6(13), 3337–3342.

Barros, G.; Magnoli, C.; Reynoso, M.; Ramirez, M.; Farnochi, M.; Torres, A.; Dalcero, M.; Sequeira, J.; Rubinstein, C.; Chulze, S. Fungal and mycotoxin contamination in Bt maize and non-Bt maize grown in Argentina. *World Mycotoxin Journal.* 2009, 2(1), 53–60.

Bationo, J.F.; Nikiema, P.A.; Koudougou, K.; Ouedraogo, M.; Bazie, S.R.; Sanou, E.; Barro, N. Assessment of aflatoxin B1 and ochratoxin A levels in sorghum malts and beer in Ouagadougou. *African Journal of Food Science.* 2015, 9(7), 417–420.

Bianchini, A.; Bullerman, L.B. Biological control of molds and mycotoxins in foods. In M. Appell, D.F. Kendra, M.W. Trucksess (Eds.), *Mycotoxin Prevention and Control in Agriculture*, ACS Symposium Series. Washington, DC: American Chemical Society. 2009, pp. 1–16.

Bilandzic, N.; Bozic, D.; Dokic, M.; Sedak, M.; Kolanovic, B.S.; Varenina, I.; Cvetnic, Z. Assessment of aflatoxin M1 contamination in the milk of four dairy species in Croatia. *Food Control.* 2014, 43, 18–21.

Cantu-Cornelio, F.; Aguilar-Toala, J.; de Leon-Rodriguez, C.; Esparza-Romero, J.; Vallejo-Cordoba, B.; Gonzalez-Cordova, A.; Garcia, H.S.; Hernandez-Mendoza, A. Occurrence and factors associated with the presence of aflatoxin M1 in breast milk samples of nursing mothers in central Mexico. *Food Control.* 2016, 62, 16–22.

Da Silva, A.K.A. Sterilization by gamma irradiation. In F. Adrovic (Ed.), *Gamma Radiation.* Vienna, Austria: InTech. 2012, pp. 171e206.

Delmulle, B.S.; De Saeger, S.M.; Sibanda, L.; Barna-Vetro, I.; Van Peteghem, C.H. Development of an immunoassay-based lateral flow dipstick for the rapid detection of aflatoxin B1 in pig feed. *Journal of Agricultural and Food Chemistry.* 2005, 53(9), 3364–3368.

Diella, G.; Caggiano, G.; Ferrieri, F.; Ventrella, A.; Palma, M.; Napoli, C.; Rutigliano, S.; Lopuzzo, M.; Lovero, G.; Montagna, M.T. Aflatoxin contamination in nuts marketed in Italy: preliminary results. *Ann Ig.* 2018, 30(5), 401–409.

Dinckaya, E.; Kınık, O.; Sezginturk, M.K.; Altug, C.; Akkoca, A. Development of an impedimetric aflatoxin M_1 biosensor based on a DNA probe and gold nanoparticles. *Biosensors and Bioelectronics.* 2011, 26(9), 3806–3811.

Di Stefano, V.; Pitonzo, R.; Cicero, N.; D'Oca, M.C. Mycotoxin contamination of animal feeding stuff: detoxification by gamma-irradiation and reduction of aflatoxins and ochratoxin a concentrations. *Food Additives and Contaminants Part A: Chemical Analysis Control Exposure and Risk Assessment.* 2014, 31, 2034–2039.

Dohnal, V.; Wu, Q.; Kuca, K. Metabolism of aflatoxins: key enzymes and interindividual as well as interspecies differences. *Archives of Toxicology.* 2014, 88, 1635–1644.

Enyiukwu, D.N.; Awurum, A.N.; Nwaneri, J.A. Mycotoxins in stored agricultural products: implications to food safety and health and prospects of plant-derived pesticides as novel approach to their management. *Greener Journal of Microbiology and Antimicrobials.* 2014, 2(3), 032–048.

Ertas, N.; Gonulalan, Z.; Yildirim, Y.; Karadal, F. A survey of concentration of aflatoxin M1 in dairy products marketed in Turkey. *Food Control.* 2011, 22, 1956–1959.

Escobedo-González, R.; Méndez-Albores, A.; Villarreal-Barajas, T.; Aceves-Hernández, J.; Miranda-Ruvalcaba, R.; Nicolás-Vázquez, I. A theoretical study of 8-chloro-9-hydroxy-aflatoxin B1, the conversion product of aflatoxin B1 by neutral electrolyzed water. *Toxins.* 2016, 8, 225.

Ferreira, C.D.; Lang, G.H.; da Silva, L.I.; da Silva, T.N.; Hoffmann, J.F.; Ziegler, V.; de Oliveira, M. Postharvest UV-C irradiation for fungal control and reduction of mycotoxins in brown, black, and red rice during long-term storage. *Food Chem.* 2021, 339, 127810.

Giovati, L.; Gallo, A.; Masoero, F.; Cerioli, C.; Ciociola, T.; Conti, S.; Magliani, W.; Polonelli, L. Vaccination of heifers with an aflatoxin improves the reduction of aflatoxin B1 carry over in milk of lactating dairy cows. *PLoS ONE.* 2014, 9(4), e94440. http://doi.org/10.1371/journal.pone.0094440.

Granados-Chinchilla, F.; Molina, A.; Chavarría, G.; Alfaro-Cascante, M.; Bogantes-Ledezma, D.; Murillo-Williams, A. Aflatoxins occurrence through the food chain in Costa Rica: applying the One Health approach to mycotoxin surveillance. *Food Control.* 2017, 82, 217–226.

Guo, X.; Wen, F.; Qiao, Q.; Zheng, N.; Saive, M.; Fauconnier, M.; Wang, J. A novel graphene oxide based aptasensor for amplified fluorescent detection of aflatoxin M1 in milk powder. *Sensors.* 2019, 19, 3840.

Hamami, M.; Mars, A.; Raouafi, N. Biosensor based on antifouling PEG/gold nanoparticles composite for sensitive detection of aflatoxin M1 in milk. *Microchemical Journal.* 2021, 165, Article 106102.

Hammond, T.M.; Andrewski, M.D.; Roossinck, M.J.; Keller, N.P. *Aspergillus* mycoviruses are targets and suppressors of RNA silencing. *Eukaryot Cell.* 2008, 7, 350–357.

Hassan, A.A.; Hassan, M.A.; El Shafei, H.M.; El Ahl, R.M.H.S.; El-Dayem, R.H.A. Detection of aflatoxigenic moulds isolated from fish and their products and its public health significance. *Nature and Science.* 2011, 9(9), 106–114.

Hassan, F.F.; Hussein, H.Z. Detection of aflatoxin M1 in pasteurized canned milk and using of UV radiation for detoxification. *International Journal of Advances in Chemical Engineering and Biological Sciences.* 2017, 4, 130–133.

Hayat, M.; Saepudin, E.; Einaga, Y.; Ivandini, T.A. CdS nanoparticle-based biosensor development for aflatoxin determination. *International Journal of Technology.* 2019, 10, 787–797.

Henry, W.B.; Windham, G.L.; Blanco, M.H. Evaluation of maize germplasm for resistance to aflatoxin accumulation. *Agronomy.* 2012, 2(4), 28–39.

Hertwig, C.; Meneses, N.; Mathys, A. Cold atmospheric pressure plasma and low energy electron beam as alternative nonthermal decontamination technologies for dry food surfaces: a review. *Trends in Food Science and Technology.* 2018, 77, 131–142.

Hosseini, M.; Khabbaz, H.; Dadmehr, M.; Ganjali, M.R.; Mohamadnejad, J. Aptamer-based colorimetric and chemiluminescence detection of aflatoxin B1 in foods samples. *Acta Chimica Slovenica.* 2015, 62, 721–728.

Hove, M.; De Boevre, M.; Lachat, C.; Jacxsens, L.; Nyanga, L.; De Saeger, S. Occurrence and risk assessment of mycotoxins in subsistence farmed maize from Zimbabwe. *Food Control.* 2016, 69, 36–44.

Humaid, A.A.; Alghalibi, S.M.; Al-Khalqi, E.A.A. Aflatoxins and ochratoxin a content of stored Yemeni coffee beans and effect of roasting onmycotoxin contamination. *International Journal of Molecular Microbiology.* 2019, 2(1), 11–21.

Hussain, I.; Anwar, J.; Asi, M.R.; Munawar, M.A.; Kashif, M. Aflatoxin M1 contamination in milk from five dairy species in Pakistan. *Food Control.* 2010, 21, 122–124.

Intanoo, M.; Kongkeitkajorn, M.B.; Suriyasathaporn, W.; Phasuk, Y.; Bernard, J.K.; Pattarajinda, V. Effect of supplemental *Kluyveromyces marxianus* and *Pichia kudria-vzevii* on aflatoxin M1 excretion in milk of lactating dairy cows. *Animals.* 2020, 10, 709.

Iqbal, S.Z.; Asi, M.R. Assessment of aflatoxin M1 in milk and milk products from Punjab, Pakistan. *Food Control.* 2013, 30(1), 235–239.

Iram, W.; Anjum, T.; Iqbal, M.; Ghaffar, A.; Abbas, M. Structural elucidation and toxicity assessment of degraded products of aflatoxin B1 and B2 by aqueous extracts of *Trachyspermum ammi. Frontiers in Microbiology.* 2016, 7, 346–362.

Jalili, M. A review on aflatoxins reduction in food. *Iranian Journal of Health, Safety and Environment.* 2016, 3, 445–459.

Jardon-Xicotencatl, S.; Diaz-Torres, R.; Marroquin-Cardona, A.; Villarreal-Barajas, T.; Mendez-Albores, A. Detoxification of aflatoxin-contaminated maize by neutral electrolyzed oxidizing water. *Toxins (Basel).* 2015, 23(7), 4294–4314.

Kamboj, D.C.; Ochieng, P.E.; Antonissen, G.; Croubels, S.; Scippo, M.-L.; Okoth, S.; Kang'the, E.K.; Faas, J.; Doupovec, B.; Lindahl, J.F. Multi-mycotoxin occurrence in dairy cattle and poultry feeds and feed ingredients from Machakos Town, Kenya. *Toxins.* 2020, 12, 762.

Kaminiaris, M.D.; Mavrikou, S.; Georgiadou, M.; Paivana, G.; Tsitsigiannis, D.; Kintzios, S. An impedance based electrochemical immunosensor for aflatoxin B1 monitoring in pistachio matrices. *Chemosensors.* 2020, 8, 121.

Kana, J.R.; Gnonlonfin, B.G.J.; Harvey, J.; Wainaina, J.; Wanjuki, I.; Skilton, R.A.; Teguia, A. Assessment of aflatoxin contamination of maize, peanut meal and poultry feed mixtures from different agroecological zones in Cameroon. *Toxins.* 2013, 5, 884–894.

Kanungo, L.; Pal, S.; Bhand, S. Miniaturised hybrid immunoassay for high sensitivity analysis of aflatoxin M1 in milk. *Biosensors and Bioelectronics.* 2011, 26(5), 2601–2606.

Karlovsky, P.; Suman, M.; Berthiller, F.; De Meester, J.; Eisenbrand, G.; Perrin, I.; Oswald, I.P.; Speijers, G.; Chiodini, A.; Recker, T. Impact of food processing and detoxification treatments on mycotoxin contamination. *Mycotoxin Research.* 2016, 32, 179–205.

Kasoju, A.; Shahdeo, D.; Khan, A.A.; Shrikrishna, N.S.; Mahari, S.; Alanazi, A.M.; Bhat, M.A.; Jyotsnendu, G.J.; Gnadhi, S. Fabrication of microfluidic device for aflatoxin M1 detection in milk samples with specific aptamers. *Scientific Reports.* 2020, 10, 4627.

Kemboi, D.C.; Ochieng, P.E.; Antonissen, G.; Croubels, S.; Scippo, M.-L.; Okoth, S.; Kang'the, E.K.; Faas, J.; Doupovec, B.; Lindahl, J.F. Multi-mycotoxin occurrence in dairy cattle and poultry feeds and feed ingredients from Machakos Town, Kenya. *Toxins.* 2020, 12, 762.

Kim, H.J.; Lee, J.E.; Kwak, B.; Ahn, J.; Jeong, S. Occurrence of aflatoxin M1 in raw milk from South Korea winter seasons using an immunoaffnity column and high perfor-mance liquid chromatography. *Journal of Food Safety.* 2010, 30, 804–813.

Lawley, R. *Aflatoxins. Food safety watch: the science of safe food.* 2013. www.foodsafety-watch.org/factsheets/aflatoxins/.

Lee, H.S.; Nguyen-Viet, H.; Lindahl, J.; Thanh, H.M.; Khanh, T.N.; Hien, L.T.T. A survey of aflatoxin B1 in maize and awareness of aflatoxins in Vietnam. *World Mycotoxin Journal.* 2017, 10, 195–202.

Lee, J.E.; Kwak, B.; Ahn, J.; Jeon, T. Occurrence of aflatoxin M1 in raw milk in South Korea using an immunoaffinity column and liquid chromatography. *Food Control.* 2009, 20(2), 136–138.

Li, H.; Li, S.; Yang, H.; Wang, Y.; Wang, J.; Zheng, N. l-Proline alleviates kidney injury caused by AFB1 and AFM1 through regulating excessive apoptosis of kidney cells. *Toxins.* 2019, 11(4), 226.

Liew, W.P.P.; Mohd-Redzwan, S. Mycotoxin: its impact on gut health and microbiota. *Frontiers in Cellular and Infection Microbiology.* 2018, 8. http://doi.org/10.3389/fcimb.2018.00060.

Liu, Y.; Li, M.; Bai, F.; Bian, K. Effects of pulsed ultrasound at 20 kHz on the sonochemical degradation of mycotoxins. *World Mycotoxin Journal.* 2019, 12, 357–366.

Liu, Y.; Wu, F. Global burden of aflatoxin-induced hepatocellular carcinoma: a risk assessment. *Environmental Health Perspectives.* 2010, 118(6), 818–824.

Loi, M.; Fanelli, F.; Zucca, P.; Mule, G. Aflatoxin B_1 and M_1 degradation by Lac2 from *Pleurotus pulmonarius* and redox mediators. *Toxins (Basel).* 2016, 8, 245.

Loi, M.; Renaud, J.B.; Rosini, E.; Pollegioni, L.; Vignali, E.; Haidukowski, M.; Sumarah, M.W.; Logrieco, A.F.; Mulè, G. Enzymatic transformation of aflatoxin B1 by Rh_DypB peroxidase and characterization of the reaction products. *Chemosphere.* 2020, 250, 126296.

Luan, Y.; Chen, J.; Xie, G.; Li, C.; Ping, H.; Ma, Z.; Lu, A. Visual and microplate detection of aflatoxin B2 based on NaCl-induced aggregation of aptamer-modified gold nanoparticles. *Microchimica Acta.* 2015, 182, 995–1001.

Manetta, A.C.; Di Giuseppe, L.; Giammarco, M.; Fusaro, I.; Simonella, A.; Gramenzi, A.; Formigoni, A. High-performance liquid chromatography with post-column derivatisation and fluorescence detection for sensitive determination of aflatoxin M1 in milk and cheese. *Journal of Chromatography A.* 2005, 1083(1–2), 219–222.

Mannani, N.; Tabarani, A.; Zinedine, A. Assessment of aflatoxin levels in herbal green tea available on the Moroccan market. *Food Control.* 2020, 108, 106882.

Markov, K.; Mihaljevic, B.; Domijan, A.M.; Pleadin, J.; Delas, F.; Frece, J. Inactivation of aflatoxigenic fungi and the reduction of aflatoxin B1 in vitro and in situ using gamma irradiation. *Food Control.* 2015, 54, 79–85.

Matthew, E.; Oluma, H.O.A.; Ochokwunu, D.I.; Eche, C.O.; Olasan, J.O. Aflatoxin contamination levels in sesame seeds sold in Benue State, North Central Nigeria. *American Journal of Food Science and Health.* 2021, 7(1), 14–23.

Medina, A.; Rodriguez, A.; Magan, N. Effect of climate change on *Aspergillus flavus* and aflatoxin B1 production. *Frontiers in Microbiology.* 2014, 5, 348.

Mgbeahuruike, A.C.; Ejioffor, T.E.; Christian, O.C.; Shoyinka, V.C.; Karlsson, M.; Nordkvist, E. Detoxification of aflatoxin-contaminated poultry feeds by 3 adsorbents, bentonite, activated charcoal, and fuller's earth. *Journal of Applied Poultry Research.* 2018, 27, 461–471.

Micheli, L.; Grecco, R.; Badea, M.; Moscone, D.; Palleschi, G. An electrochemical immunosensor for aflatoxin M_1 determination in milk using screen-printed electrodes. *Biosensors and Bioelectronics.* 2005, 21(4), 588–596.

Mobeen, A.K.; Aftab, A.; Asif, A.; Zuzzer, A.S. Aflatoxins B1 and B2 contamination of peanut and peanut products and subsequent microwave detoxification. *Journal of Pharmacy and Nutrition Sciences.* 2011, 1, 1–3.

Monda, E.O.; Alakonya A.E. A review of agricultural aflatoxin management strategies and emerging innovations in sub-Saharan Africa. *African Journal of Food, Agriculture, Nutrition and Development.* 2016, 16(3), 1684–5374.

Nafuka, S.N.; Misihairabgwi, J.M.; Bock, R.; Ishola, A.; Sulyok, M.; Krska, R. Variation of fungal metabolites in sorghum malts used to prepare Namibian traditional fermented beverages Omalodu and Otombo. *Toxins.* 2019, 11(3), 165.

Nagy, R.; Máthé, E.; Csapó, J.; Sipos, P. Modifying effects of physical processes on starch and dietary fiber content of foodstuffs. *Processes.* 2021, 9, 17.

Nakavuma, J.L.; Kirabo, A.; Bogere, P.; Nabulime, M.M.; Kaaya, A.N.; Gnonlonfin, B. Awareness of mycotoxins and occurrence of aflatoxins in poultry feeds and feed ingredients in selected regions of Uganda. *International Journal of Food Contamination.* 2020, 7, 1–10.

Negash, D. A review of aflatoxin: occurrence, prevention, and gaps in both food and feed safety. *Journal of Nutritional Health & Food Engineering.* 2018, 8(2), 190–197.

Offiah, N.; Adesiyun, A. Occurrence of aflatoxins in peanuts, milk, and animal feed in Trinidad. *Journal of Food Protection.* 2007, 70(3), 771–775.

Omar, S.S. Incidence of aflatoxin M1 in human and animal milk in Jordan. *Journal of Toxicology and Environmental Health, Part A.* 2012, 75(22–23), 1404–1409.

Ouf, S.A.; Basher, A.H.; Mohamed, A.A. Inhibitory effect of double atmospheric pressure argon cold plasma on spores and mycotoxin production of *Aspergillus niger* contaminating date palm fruits. *Journal of Science and Food Agriculture.* 2015, 95, 3204–3210.

Owino, J.H.O.; Arotiba, O.A.; Hendricks, N.; Songa, E.A.; Jahed, N.; Waryo, T.T.; Ngece, R.F.; Baker, P.G.L.; Iwuoha, E.I. Electrochemical immunosensor based on polythionine/gold nanoparticles for the determination of aflatoxin B1. *Sensors.* 2008, 8, 8262–8274.

Paniel, N.; Radoi, A.; Marty, J.L. Development of an electrochemical biosensor for the detection of aflatoxin M1 in milk. *Sensors (Basel).* 2010, 10(10), 9439–9448.

Pankaj, S.K.; Shi, H.; Keener, K.M. A review of novel physical and chemical decontamination technologies for aflatoxin in food. *Trends in Food Science and Technology.* 2018, 71, 73–83.

Patil, H.; Shah, N.G.; Hajare, S.N.; Gautam, S.; Kumar, G. Combination of microwave and gamma irradiation for reduction of aflatoxin B1 and microbiological contamination in peanuts (*Arachis hypogaea* L.). *World Mycotoxin Journal.* 2019, 12, 269–280.

Patras, A.; Julakanti, S.; Yannam, S.; Bansode, R.R.; Burns, M.; Vergne, M.J. Effect of UV irradiation on aflatoxin reduction: a cytotoxicity evaluation study using human hepatoma cell line. *Mycotoxin Research.* 2017, 33, 343–350.

Peles, F.; Sipos, P.; Kovács, S.; Győri, Z.; Pócsi, I.; Pusztahelyi, T. Biological control and mitigation of aflatoxin contamination in commodities. *Toxins.* 2021, 13, 104.

Pereira, S.C.; Cunha, C.S.; Fernandes, J.O. Prevalent mycotoxins in animal feed: occurrence and analytical methods. *Toxins.* 2019, 11(5), 290.

Phillips, T.D.; Afriyie-Gyawu, E.; Williams, J.; Huebner, H.; Ankrah, N.A.; Ofori-Adjei, D.; Jolly, P.; Johnson, N.; Taylor, J.; Marroquin-Cardona, A. Reducing human exposure to aflatoxin through the use of clay: a review. *Food Additives & Contaminants: Part A.* 2008, 25, 134–145.

Ponzilacqua, B.; Rottinghaus, G.E.; Landers, B.R.; Oliveira, C.A.F. Effects of medicinal herb and Brazilian traditional plant extracts on in vitro mycotoxin decontamination. *Food Control.* 2019, 100, 24–27.

Prandini, A.; Tansini, G.; Sigolo, S.; Filippi, L.; Laporta, M.; Piva, G. On occurrence of aflatoxin M1 in milk and dairy products. *Food Chem Toxicol.* 2009, 47, 984–991.

Prettl, Z.; Desi, E.; Lepossa, A.; Kriszt, B.; Kukolya, J.; Nagy, E. Biological degradation of aflatoxin B1 by a *Rhodococcus pyridinivorans* strain in by-product of bioethanol. *Animal Feed Science and Technology.* 2017, 224, 104–114.

Puligundla, P.; Lee, T.; Mok, C. Effect of corona discharge plasma jet treatment on the degradation of aflatoxin B1 on glass slides and in spiked food commodities. *Lwt-Food Science and Technology.* 2019, 124, 108333.

Qian, G.S.; Yasei, P.; Yang, G.C. Rapid extraction and detection of aflatoxin M1 in cow's milk by high-performance liquid chromatography and radioimmunoassay. *Analytical Chemistry*. 1984, 56(12), 2079–2080.

Rangsaz, N.; Ahangaran, M.G. Evaluation of turmeric extract on performance indices impressed by induced aflatoxicosis in broiler chickens. *Toxicology and Industrial Health*. 2011, 27(10), 956–960.

Rastegar, H.; Shoeibi, S.; Yazdanpanah, H.; Amirahmadi, M.; Khaneghah, A.M.; Campagnollo, F.B.; Sant Ana, S.A. Removal of aflatoxin B1 by roasting with lemon juice and/or citric acid in contaminated pistachio nuts. *Food Control*. 2017, 71, 279–284.

Ratnavathi, C.V.; Komala, V.V.; Kumar, B.S.V.; Das, I.K.; Patil, J.V. Natural occurrence of aflatoxin B1 in sorghum grown in different geographical regions of India. *Journal of the Science of Food and Agricultural*. 2012, 92(12), 2416–2420. http://doi.org/10.1002/jsfa.5646.

Reddy, K.R.N.; Raghavender, C.R.; Salleh, B.; Reddy, C.S.; Reddy, B.N. Potential of aflatoxin B1 production by *Aspergillus flavus* strains on commercially important food grains. *International Journal of Food Science and Technology*. 2011, 161–165.

Sabet, F.S.; Hosseini, M.; Khabbaz, H.; Dadmehr, M.; Ganjali, M.R. FRET-based aptamer biosensor for selective and sensitive detection of aflatoxin B1 in peanut and rice. *Food Chem*. 2017, 220, 527–532.

Santili, A.B.N.; de Camargo, A.C.; Nunes, R.D.S.R.; Gloria, E.M.D.; Machado, P.F.; Cassoli, L.D.; Dias, C.T.D.S.; Calori-Domingues, M.A. Aflatoxin M1 in raw milk from different regions of São Paulo state–Brazil. *Food Additives & Contaminants: Part B*. 2015, 8(3), 207–214.

Seid, A.; Mama, A. Aflatoxicosis and occurrence of aflatoxin M1 (AFM1) in milk and dairy products: a review. *Austin Journal of Veterinary Science & Animal Husbandry*. 2019, 6(1), 1054.

Sharanaiah, U.; Honnayakanahalli, M.G.M.; Bhadvelu, C.; Prahlad, S.; Jayanna, S.K.; Sri, R.; Prakasha, A.; Marahel, S.; Tumkur, R.B.; Sollepura, B.R.; Murali, N.; Govinda-Gowda, V.R.; Mohankumar, S.; Harishchandra, S.P. Aflatoxins and food pathogens: impact of biologically active aflatoxins and their control strategies. *Journal of the Science of Food and Agriculture*. 2017, 97, 1698–1707.

Sharma, R.K.; Parisi, S. *Toxins and contaminants in Indian food products*. Springer Nature Switzerland AG. 2017. http://doi.org/10.1007/978-3-319-48049-7_2.

Shu, X.; Wang, Y.; Zhou, Q.; Wu, L. Biological degradation of aflatoxin B$_1$ by cell-free extracts of *Bacillus velezensis* DY3108 with broad PH stability and excellent thermostability. *Toxins (Basel)*. 2018, 10, 330–345.

Sibaja, M.; de Oliveira, K.V.G.; Feltrin, S.A.C.P.; Diaz Remedi, R.; Cerqueira, M.B.R.; Badiale-Furlong, E.; Garda-Buffon, J. Aflatoxin biotransformation by commercial peroxidase and its application in contaminated food. *Journal of Chemical Technology and Biotechnology*. 2018, 94, 1187–1194.

Singh, N.A. Nanotechnology definitions, research, industry and property rights. In *Nanoscience in food and agriculture 1*. Sustainable Agriculture Reviews 20. Switzerland: Springer International Publishing. 2016, pp. 43–64. ISBN: 978-3-319-39303-2.

Singh, P.; Cotty, P.J. Aflatoxin contamination of dried red chilies: contrasts between the United States and Nigeria, two markets differing in regulation enforcement. *Food Control*. 2017, 201880, 374–379.

Taheur, B.F.; Kouidhi, B.; Al Qurashi, Y.M.A.; Ben Salah-Abbès, J.; Chaieb, K. Review: biotechnology of mycotoxins detoxification using microorganisms and enzymes. *Toxicon*. 2019. http://doi.org/10.1016/j.toxicon.2019.02.001.

Talebi, E.; Khademi, M.; Rastad, A. An over review on effect of aflatoxin in animal husbandry. *Asian Journal of Experimental Biological Sciences.* 2011, 2, 754–757.

Tang, Y.; Tang, D.; Zhang, J.; Tang, D. Novel quartz crystal microbalance immunodetection of aflatoxin B1 coupling cargo-encapsulated liposome with indicator-triggered displacement assay. *Analytica Chimica Acta.* 2018, 1031, 161–168.

Tian, F.; Chun, H.S. Natural products for preventing and controlling aflatoxin contamination of food. In *Aflatoxin-control, analysis, detection and health risks.* London: IntechOpen. 2017, pp. 13–44.

Tomasevic, I.; Petrovic, J.; Jovetic, M.; Raicevic, S.; Milojevic, M.; Miocinović, J. Two year survey on the occurrence and seasonal variation of aflatoxin M1 in milk and milk products in Serbia. *Food Control.* 2015, 56, 64–70.

Udomkun, P.; Mutegi, C.; Wossen, T.; Atehnkeng, J.; Nabahungu, N.L.; Njukwe, E.; Vanlauwe, B.; Bandyopadhyay, R. Occurrence of aflatoxin in agricultural produce from local markets in Burundi and Eastern Democratic Republic of Congo. *Food Science & Nutrition.* 2018, 6(8), 2227–2238.

Vijayanandraj, S.; Brinda, R.; Kannan, K.; Adhithya, R.; Vinothini, S.; Senthil, K.; Chinta, R.R.; Paranidharan, V.; Velazhahan, R. Detoxification of aflatoxin B1 by an aqueous extract from leaves of *Adhatoda vasica Nees. Microbiological Research.* 2014, 169, 294–300.

Vita, D.S.; Rosa, P.; Giuseppe, A. Effect of gamma irradiation on aflatoxins and ochratoxin a reduction in almond samples. *Journal of Food Research.* 2014, 3, 113–118.

Wang, Y.; Zhang, H.; Yan, H.; Zhang, Z. Effective biodegradation of aflatoxin B1 using the *Bacillus licheniformis* (BL010) strain. *Toxins (Basel).* 2018, 10, 497–513.

Wani, S.H. Inducing fungus-resistance into plants through biotechnology. *Notulae Scientia Biologicae.* 2010, 2, 14.

Worku, A.F.; Abera, M.; Kalsa, K.K.; Bhadriraju, S.; Habtu, N.G. Occurrence of mycotoxins in stored maize in Ethiopia. *Ethiopian Journal of Agricultural Sciences.* 2019, 29(2), 31–43.

Xie, L.; Chen, M.; Ying, Y. Development of methods for determination of aflatoxins. *Critical Reviews in Food Science and Nutrition.* 2015, 56(16), 2642–2664.

Yalcin, N.F.; Avci, T.; Isik, M.K.; Oguz, H. In vitro activity of toxin binders on aflatoxin B1 in poultry gastrointestinal medium. *Pakistan Veterinary Journal.* 2018, 38(1), 61–65.

Yang, Q. Decontamination of aflatoxin B1. In X.D. Long (Ed.), *Aflatoxin B1 occurrence, detection and toxicological effects.* London: InTechOpen. 2019.

Zhang, X.; Li, C.R.; Wang, W.C.; Xue, J.; Huang, Y.L.; Yang, X.X.; Tan, B.; Zhou, X.P.; Shao, C.; Ding, S.J. A novel electrochemical immunosensor for highly sensitive detection of aflatoxin B1 in corn using single-walled carbon nanotubes/chitosan. *Food Chem.* 2016, 192, 197–202.

Zhang, Z.; Lin, M.; Zhang, S.; Vardhanabhuti, B. Detection of aflatoxin M1 in milk by dynamic light scattering coupled with super paramagnetic beads and gold nanoprobes. *Journal of Agricultural and Food Chemistry.* 2013, 61(19), 4520–4525. http://doi.org/10.1021/jf400043z.

Chapter 2

Fumonisins in Food and Feed
Their Detection and Management Strategies

Shubhangi Srivastava, Ashok Kumar Yadav, Mousumi
Ghosh, Dipendra Kumar Mahato, Madhu Kamle, Pooja
Pandey, Sreemoyee Chakraborty, Pradeep Kumar

CONTENTS

2.1	Introduction	29
2.2	Occurrence of Fumonisins in Food Samples	31
2.3	Sources of Fumonisins	34
2.4	Gene Responsible for Production	34
2.5	Detection Methods	35
2.6	Management and Control Strategies	37
	2.6.1 Chemical and Physical Methods	38
	2.6.2 Biological Methods	39
2.7	Conclusion	41

2.1 INTRODUCTION

Pathogenic fungi, such as *Fusarium proliferatum*, *Fusarium verticillioides*, and similar species, generate *fumonisins* in cereals (Rheeder et al., 2002; Kamle et al., 2019). Furthermore, *Aspergillus nigri* generates fumonisins in grape, peanut, and maize plants (Astoreca et al., 2007a, b; Frisvad et al., 2007; Mogensen et al., 2010; Kumar et al., 2017). Apart from their prevalence in numerous other cereals (rice, barley, wheat, millet, oat, rye) and grain products (tortillas, corn flakes, and chips), products of maize and raw maize itself are more prone to infection with *fumonisins* (Dall'Asta and Battilani, 2016; Cendoya et al., 2018a), which can have a significant impact on one's health. There are more than 15 *fumonisin* homologues identified, of which *fumonisin* A, B, C, and P are the four major groups (Braun et al., 2018; Haschek et al., 2013). The major forms

DOI: 10.1201/9781003242208-2

of *fumonisins* found in food are FB1, FB2, FB3, and B, of which FB2 is the most toxic. FB1 is a propane-1,2, 3-tricarboxylic acid (TCA) diester, where hydroxyl (-OH) groups are at the C-14 and C-15 positions and are involved with the carboxyl groups (-COOH) of TCA to form an ester, while, next, FB2 and FB3 are C-5 and C-10 dehydroxy analogues of FB1 (Kamle et al., 2019). Figure 2.1 depicts the

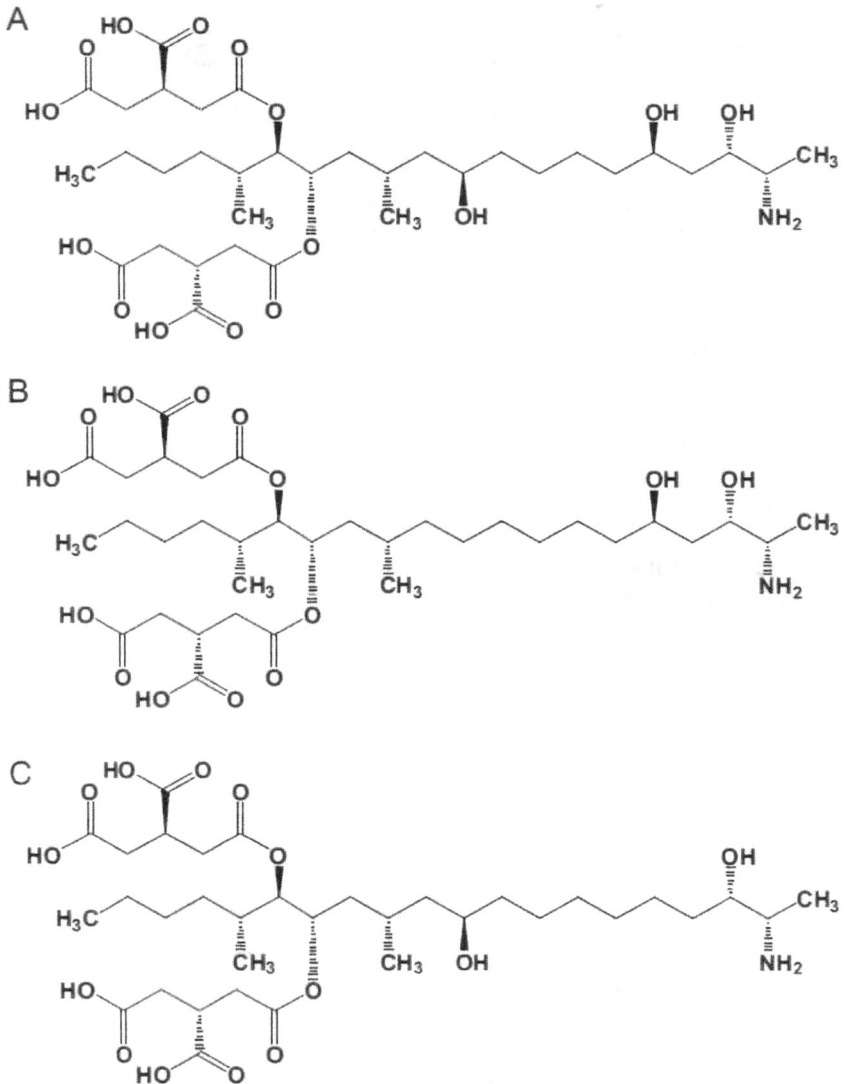

Figure 2.1 Chemical structures of fumonisins FB1 (A), FB2 (B), and FB3 (C).

TABLE 2.1   FORMULA AND MOLECULAR WEIGHT OF DIFFERENT FUMONISINS

Fumonisin Type	Formula	Molecular Weight (g/mol)
FB1	$C_{34}H_{59}NO_{15}$	721.8
FB2	$C_{34}H_{59}NO_{14}$	705.8
FB3	$C_{34}H_{59}NO_{14}$	705.8
FB4	$C_{34}H_{59}NO_{13}$	689.8

chemical structures of *fuminisons* FB1, FB2, and FB3, and Table 2.1 shows the molecular weight (g/mol) of the *fumonisins*.

FB1 is in the category of 2B carcinogenic compounds for human beings per the International Agency for Research on Cancer (IARC). It has also been proved to be harmful for rats and rabbits. Furthermore, the Food and Agriculture Organization (FAO) and World Health Organization (WHO) have established a maximum tolerated daily intake for *fumonisins* of 2 g/kg body weight per day based on the study on nephrotoxicity in rats (Joint FAO-WHO).

As far as toxicity points of concern, *fumonisins* B2 and B3 have toxicological profiles and are quite comparable to *fumonisin* B1. Various *fumonisin* chemical compounds have been studied in various biological tests to reveal the structure-activity interaction. Free amino groups appear to play a vital role in the biological functionality of *fumonisin* B1. A hot and humid atmosphere has been found in studies to enhance FB1 content (Cendoya et al., 2018b; Rheeder et al., 2016; Liverpool-Tasie et al., 2019). Fumonisins were shown to be very stable in maize at temperatures ranging from 28.97 to 32.14°C, humidity levels ranging from 27.29% to 32.14%, and pH levels ranging from 5.5 to 6.0 (Bryla et al., 2017; van Rensburg et al., 2017). As a result, the threat of FB1 to agricultural crops is being reported increasingly frequently in temperate tropical nations.

Fumonisin mycotoxins are generally produced by *F. verticillioides, F. napiforme, F. proliferatum, F. oxysporum, F. anthophilum, F. nygamai,* and *F. dlamini,* which have the potential to endanger both people's and animals' health.

2.2 OCCURRENCE OF FUMONISINS IN FOOD SAMPLES

Contamination by *fumonisins* in agricultural crops and food-related products is affected by agro-climatic conditions. Cereals are the most often contaminated food categories (wheat, rice, sorghum, maize, barley, rye, millet, and oat). FB1 has been found in a variety of dietary items, including garlic, coffee, onion, pea, soybean, asparagus (Seefelder et al., 2002; Irzykowska et al., 2012; Waśkiewicz et al., 2013), dried figs (Karbancıoglu-Güler and Heperkan, 2009;

Heperkan et al., 2012), barley snacks (Park et al., 2002), milk (Gazzotti et al., 2009), and beer (Kawashima et al., 2007). One of the most often contaminated foods by FB1 and FB2 is maize and its products (Stępień et al., 2011). However, due to the rise in temperature during extrusion cooking, contamination by FB2 and FB1 are reduced by 59% during the manufacturing of tortilla chips, 60% during milling for flour production, and 50% during grits and snack manufacturing (Castells et al., 2008; Scudamore et al., 2009).

The *fumonisin* levels in popcorn or sweet corn, milk, meat, and eggs are generally low, while those in corn meal, bran, flour, grits, distiller grains, gluten, milling fractions, and baking mixes are quite high. Studies have reported that *fumonisin* degradation either by distillation or milling is quite insignificant, and some of studies have reported that these *fumonisins* are quite resistant to heat at temperatures frequently used in the food manufacturing industry (Bullerman, 1996; Mac Jr and Valente Soares, 2000; Munkvold et al., 2019).

A study by Yoshizawa et al. (1996) found *fumonisins* in corn grit samples (Thailand), and colonies of *F. proliferatum* and *F. moniliforme* were isolated. A US-based study revealed that FB1 and *moniliformin* infected 34% of samples of maize and 53% of samples of corn items (Gutema et al., 2000). Research done in Brazil to identify *fumonisins* in corn food items found that FB1 was present in 82% and FB2 in 51% of the tested samples (Martins et al., 2012). Table 2.2 shows the *fumonisin* levels (FB1 and FB2) in cornmeal (Bullerman, 1996). Figure 2.2 shows *fumonisin* levels in ppm of different corn-based foods as data collected from Pittet et al. (1992), Doko and Visconti (1994), Stack and Eppley

TABLE 2.2   FUMONISIN LEVELS (FB1 AND FB2) IN CORNMEAL (BULLERMAN, 1996)

Country	Fumonisin Levels (µg/g)	
	FB1	**FB2**
United States	0.0–2.8	0.0–0.9
Canada	0.0–0.05	0.0
Egypt	1.8–3.0	0.5–0.8
Peru	0.0–0.7	0.0–0.1
South Africa	0.0–0.5	0.0–0.1
Switzerland	0.1–0.9	0.005–0.9
Italy	0.1–6.8	0.1–0.8
Portugal	0.1–4.4	0.1–3.2
Zambia	0.1–1.7	0.1–1.7
Europe	0.01–0.12	0.1–0.2

Fumonisin content of corn-based foods

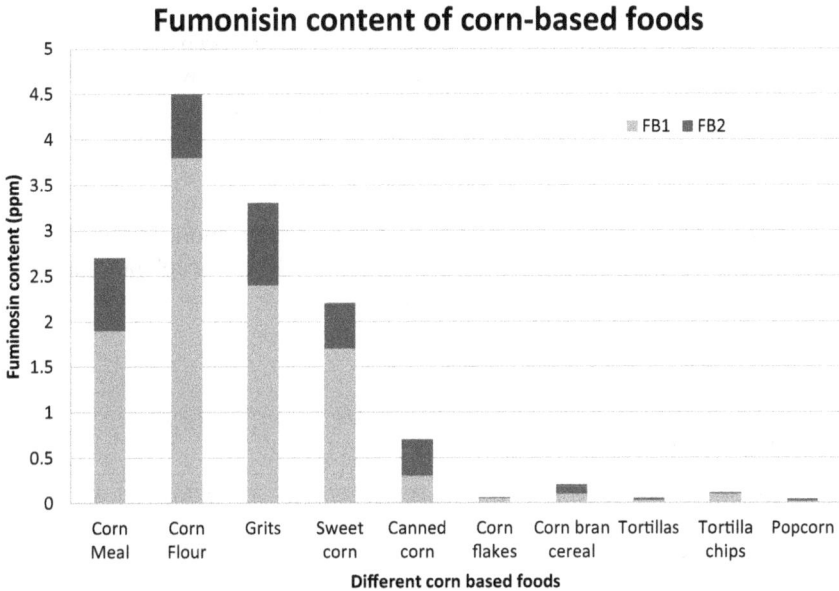

Figure 2.2   The fumonisin content in ppm of different corn-based foods.

(1992), Bullerman (1996), and Sydenham et al. (1993). Per a study on *fumonisin* in corn and its byproducts, the highest quantities of *fumonisins* were found in whole grains, and less was found in corn products that have undergone minimal processing, like corn flour, corn grits, corn meals, and corn starch, which are passed through milling or grinding methods (Figure 2.2). Corn products that are highly processed, like cornflakes and popcorn cereals, are often found to have a negligible amount of or no *fumonisins*. Snacks like corn chips and tortilla chips are likewise high in *fumonisins*, although tortillas and popcorn are low in *fumonisins*. There appear to be relatively low contamination levels. Canned sweet corn appears to have minimal contamination levels as well. *Fumonisins* have been detected from all countries, according to surveys across the world. As corn is grown across the globe, *fumonisins* are also found across the globe. The occurrence of *fumonisins* varies by region. The concentration of *fumonisins* in corn crops grown and produced in Italy is higher than that in the rest of Europe. At this stage, these findings should be regarded with caution because they represent a small number of experiments. Furthermore, these studies have only included surveys of commercially accessible items.

Contamination by FB1- and FB2-type *fumonisins* was also found in chicken broiler and feeding calves of South Korea (Seo et al., 2013). In Thai dried coffee, traces of *fumonisins* were reported (Noonim et al., 2009), while in Spanish coffee, *fumonisin* B1was reported when the method of brewing was altered (García-Moraleja et al., 2015a). The effect of temperature and water activity was evaluated in a study (Cendoya et al., 2018b) on *fumonisin* biosynthesis using *F. proliferatum* in wheat, and it was observed that the maximum growth was at a water activity of 0.99 with 15°C temperature.

Fumonisin dietary intake can result in a variety of negative consequences in agricultural and laboratory animals. These toxins are responsible for pulmonary edema syndrome in pigs (Harrison et al., 1990), leukoencephalomalacia in horses (Ross et al., 1992), hepatotoxicity and nephrotoxicity in rats (Voss et al., 1999), and apoptosis in animal cells (Jones et al., 2001).

2.3 SOURCES OF FUMONISINS

Fumonisins are mostly generated by *F. proliferatum, F. verticillioides*, and other *Fusarium* species. *Fusarium*, a member of the Nectriaceae family, is a saprophyte in plants and soils all over the world (Burgess et al., 1988). *Fusarium* can easily invade and penetrate plant rhizospheres. Moreover, *F. proliferatum* and *F. verticillioides* are identified as the most common maize pathogens (Marasas, 2001). Not just crops but even many popular decorative plants such as carnation, gladiolus, aster begonia, and chrysanthemum are commonly damaged (Gullino et al., 2002) by different *Fusarium* (species: *redolens, hostae, oxysporum*, and *foetens*). In contrast, *Fusarium* can also infect orchids via either pathogenic or non-pathogenic mode. In non-pathogenic mode, *Fusarium* species are either decomposers (Booth, 1971) or aid in seed germination and seedling color development (Vujanovic et al., 2000). Generally, *Fusarium* contaminates maize and generates *fumonisins* mostly during the pre-harvest period, but it can also be reported during the post-harvesting stage when storage conditions are inappropriate.

2.4 GENE RESPONSIBLE FOR PRODUCTION

In *fumonisins*, the biosynthetic gene FUM is in charge of producing *fumonisins*, a transcription factor, and two transport proteins (Proctor et al., 2003). The gene expression of FUM is influenced by environmental circumstances (Desjardins and Proctor, 2007). The first step in *fumonisin* production is catalyzed by the FUM1 gene, which expresses a polyketide synthase enzyme complex that initiates the reaction (Bojja et al., 2004). Moreover, there is a significant relation between the fraction of FUM1 transcripts assessed by RT-PCR and the concentration of *fumonisins* synthesized by *F. proliferatum* and *verticillioides* species

(López-Errasquín et al., 2007). Also, there is another gene called FUM19 that helps in ATP binding, which helps in the export of extracellular *fumonisins* (Seo et al., 2001; Proctor et al., 2003). Furthermore, another gene called FUM8 produces an aminotransferase that serves to keep the FB1 molecule physiologically active (Seo et al., 2001; Baker, 2006). *Fusarium* FUM cluster homologues in the *A. niger* genome mainly include FUM genes 1, 3, 6, 7, 8, 10, 13, 14, 15, 19, and 21, of which FUM 1 is polyketide synthase, FUM 7 is dehydrogenase, FUM 8 is aminotransferase, FUM 10 is acyl-coenzyme A synthetase, FUM 13 is carbonyl reductase, FUM 14 is condensation protein, FUM 15 is hydroxylase, FUM19 is ABC transporter, and FUM 21 is responsible for the transcription factor (Baker et al., 2006; Pel et al., 2007; Kamle et al., 2019). Several other genes are responsible for FB biosynthesis either increasing or decreasing. These genes are encoded with transcription factors with large domains containing several regulators for transcription (Picot et al., 2010). Moreover, some studies also reported that *A. niger* can also produce *fumonisins* such as FB2, FB4, and FB6 when, at the tenth carbon position, a hydroxyl group is lost (Noonim et al., 2009; Mansson et al., 2010; Susca et al., 2014). Furthermore, several environmental variables influence *fumonisin* production and can cause variations in *fumonisin* levels. Source of carbon, nitrogen amount, solute potential, water activity, temperature, and light are among these parameters (Fanelli et al., 2012). Because of differences in culture circumstances, such as static/shaking growth, solid/liquid media, and temperature, the results published by various research groups are not necessarily consistent. The optimal conditions for *fumonisin* growth could not be distinctly defined due to the phenotypic variability. Otherwise, within this species, each strain has a varied adaptation ability to the environment, which is governed by genetic or epigenetic factors, and as a result, secondary metabolism may be variably regulated.

2.5 DETECTION METHODS

From 1980–95, *fumonisins* were detected by thin layer chromatography (TLC) on silica plates. After spraying the plate with p-anisaldehyde (Gelderblom et al., 1988), followed by heating, the *fumonisins* develop as light to dark purple dots. Despite the fact that TLC resulted in the isolation and subsequent identification of *fumonisins*, it lacks the specificity required for a quantitative procedure. Alternatively, *fumonisins* can be seen as fluorescent spots after being exposed to shortwave ultraviolet (UV) radiation by spraying a fluorescamine solution on the produced TLC plate (Rottinghaus et al., 1992).

So, *fumonisins* cannot be consistently identified in culture filtrate extracts at concentrations less than 100–200 µg/g. Derivatization was accomplished by the use of a pre-column containing ortho-pthaldialdehyde (OPA), and detection was done through the HPLC method combined with a fluorescence detector (HPLC-FLD). As *fumonisins* are water soluble, this is well suited for repeated analysis, but the disadvantage of this approach is that it requires large sample size of at least 50 g, an extraction solvent like methanol, and cartridges for solid-phase extraction (Kavanagh, 1981). One study in Iowa tested for the presence of *fumonisins* in corn meal by the use of HPLC and reported significant levels (0.05–1.2 μ/g) of FB1 in corn meal (Rice and Ross, 1994).

Detection of *fumonisins* through gas chromatography/mass spectrometry (GC/MS) mostly gives a positive result and confirmation of *fumonisin* existence. Other MS approaches used in the examination of standards and culture material use either thermos or electro-spray MS. Commercialization of enzyme-linked test kits and affinity columns has resulted from the manufacture of polyclonal as well as monoclonal antibodies. The utility of these items in screening huge quantities of samples appears to be promising.

A quick enzyme-linked immunosorbent test (ELISA) for the quantitative detection of fumonisin in maize B1, B2, and B3 was developed. This test has a measurement range of 0.75–60 ppm, making it ideal for the analysis of samples for human nutrition as well as animal feeding. The detection limit was determined to be 0.75 ppm, while the quantification limit was determined to be 1.00 ppm for the test (Anfossi et al., 2010).

There are quick, easy, cheap, effective, rugged and safe ways (QuEChERS) for detecting FB1 (Arroyo-Manzanares et al., 2015; Bolechová et al., 2015; Nielsen et al., 2015). Matrix solid-phase dispersion (MSPD) (Rubert et al., 2012; Blesa et al., 2014), liquid-liquid extraction (LLE) (Beltrán et al., 2013; García-Moraleja et al., 2015b), solid-liquid extraction (SLE) (Ediage et al., 2015; Jung et al., 2015; Bryła et al., 2016), and dispersive liquid-liquid micro extraction (DLLME) (Arroyo-Manzanares et al., 2013) are some of the widely used *fumonisin* extraction procedures. Recently, it was found that *fumonisin* extraction was higher in finer flours, suggesting the relevance of particle size of any sample for *fumonisin* recovery.

Method validation studies for *fumonisins* previously conducted involved either immunoaffinity clean-up or solid-phase extraction to purify the extracts of the sample, followed by chromatographic separation and transformation into fluorescent derivatives prior to detection. The most often used approach is pre-column derivatization, which does not necessitate the use of any extra equipment or difficult-to-handle substances, and the chromatographic separation of fumonisin derivatives is expected to be straightforward. However, the produced derivatives are not stable, which is one downside of pre-column derivatization, necessitating stringent time-programmed protocols, manual injections, or sophisticated derivatization techniques, such as on-column derivatization, to be used for the analysis of longer sequences of samples. Post-column derivatization, which produces identical derivatives but can handle bigger sample sequences, is an alternative to pre-column derivatization.

Further, many advanced and sophisticated techniques like enzyme-linked immunosorbent assay, lateral flow immunoassay (LFI), surface plasmon resonance (SPR), PCR (Nagaraj et al., 2016), PCR-ELISA (Omori et al., 2018), electronic nose, hyperspectral imaging (Del Fiore et al., 2010; Kimuli et al., 2018), and immune sensors (Lu et al., 2016) for detecting mycotoxins are found to be more efficient (Lee et al., 2004; Ran et al., 2013). Apart from that, the color-encoded lateral flow immunoassay (LFIA) is one of the most popular methods for detecting aflatoxin B1 and type-B *fumonisins* in a single line (Di Nardo et al., 2019). In addition, an immediate and sensitive molecularly imprinted photo electro-chemical (MIP-PEC) sensor-based approach for measuring FB1 has recently been developed (Mao et al., 2019).

2.6 MANAGEMENT AND CONTROL STRATEGIES

Following good agricultural practices (GAPs), good manufacturing practices (GMPs), and proper operation of hazard analysis and critical control points (HACCPs) could improve the health of crops, but assurance of fungus- and mycotoxin-free crops is not possible, although disease-resistant breeding in crops helps maintain the balance between producing disease-resistant crops and keeping agricultural yields high (Alberts et al., 2016). However, due to high rate of production costs, environmental factors, mycotoxin production mechanisms, and several other factors like optimizing agricultural management, techniques are not always viable. Because *fumonisins* infect crop plants like maize throughout their growth in fields (Driehuis et al., 2010), good agricultural practices, good manufacturing practices, and proper storage can reduce the chances of *fumonisin* contamination (Okabe et al., 2015). Crop harvesting at an earlier time could be one strategy for controlling *fumonisin* contamination (Bush et al., 2004), but this strategy cannot be applicable for all agricultural crops, as most of them are harvested only after full maturity. Only forage maize could be harvested early to improve silage digestion. Farmers favor a delayed harvest due to technological advancements; therefore, these practices require careful examination. For example, using kernel processors to collect fodder results in the formation of silage from maize.

The Codex Alimentarius Commission (CAC) has defined maximum *fumonisin* limits for raw maize and maize flour as 4000 and 2000 mg/kg, respectively, and these have been applied in South Africa. Reducing *fumonisin* exposure among subsistence farmers requires a multifaceted strategy that cannot be done merely by regulatory means (Shephard et al., 2019). Nanotechnology and genetic engineering could be used to develop *fusarium*-resistant crops that must be contamination and infection free. As fungi, insects, and inappropriate moisture also promote the growth of *fumonisin*, drought- and insect-resistant crops could also be used in management of *fumonisins*. Furthermore, farmers could be aware regarding drying and sorting techniques of seeds and crops to reduce the risk of infection, and this could be controlled to some extent (Van der Westhuizen et al., 2011). There are various abiotic and biotic factors that influence the growth and proliferation of *fumonisins*. One of the abiotic factors, water stress, is a prevalent environmental factor that influences *fumonisin* development and production. Other stressors that have been linked to the formation of mycotoxins include osmotic stress, pH, and fungicides. Marín et al. (2010) hypothesized that environmental factors that promote water stress (drought) could enhance the possibility of fumonisin contamination in maize by *F. verticillioides*. *Fusarium* infection is favored by drought conditions and excessive watering. Drought stress should be minimized during the growing season and maturation of wheat seeds (Bernhoft et al., 2012). High moisture in plants and crops during their flowering seasons and grain formation at a premature stage of the crops promote *fusarium* infection (Lemmens et al., 2004). There are various approaches like appropriate implementation of hazard analysis critical control points, selecting the right critical points, good agricultural practices, good manufacturing practices, proper storage conditions, and some biological and chemical treatment methods that can be administered in the system for effective reduction and degradation of *fumonisins*. Another control strategy that could be adopted is the application of mycotoxin binders. These are adsorbents that prevent *fumonisins* and other mycotoxins from being absorbed in the gut and entering in the circulatory system. When other mold and mycotoxin prevention techniques fail, mycotoxin binders can be useful. Mycotoxin binders' main goal is to prevent mycotoxins from being absorbed into animals' intestinal tracts by absorbing the toxin to their surface. Clay and yeast-derived compounds are examples of organic and inorganic binders, respectively (Jacela et al., 2010; Kolosova and Stroka, 2011). Mycotoxin modifiers (microbiological in origin), on the other hand, change the chemical structure of mycotoxins and reduce the toxicity effect. Modifiers generally contain complete bacterial and yeast cultures as well as particular extracts such as enzymes (Kabak and Dobson, 2009). To reduce *fumonisins*, several management and control methods are applied either at the pre-harvest, harvest, or processing stage, that is, regulated agricultural methods; a variety of physical, chemical, and biological treatment procedures, and genetic engineering approaches (Wild and Gong, 2010).

2.6.1 Chemical and Physical Methods

Some chemical and physical methods for *fumonisin* reduction have been marketed entailing solvent extraction, sorting and flotation, detoxification by chemical alkalization (such as sulfur dioxide, ammonia, and sodium hydroxide treatments), irradiation, oxidation via ozone, and pyrolysis (He and Zhou, 2010; Alberts et al., 2016). However, there are various limits, problems, and issues with these chemical and physical methods (Schatzmayr et al., 2006; Alberts et al., 2016). Physical procedures have poor effectiveness and specificity in general, but chemical approaches are not always successful, are costly, and may reduce the nutritious content of foods (He and Zhou, 2010; Alberts et al., 2016). Moreover, fungicide-based treatments represent a possible safe approach, but many antifungal agents are harmful chemicals and are not biodegradable in nature and thus can pollute water and soil. The harmful effect of these fungicides on the environment and human health is of concern (da Cruz Cabral et al., 2013). Chemical treatment of grains for longer period of time could result in accumulation of fungal strains in food crops, which then might be transferred to the human body and also cause cross-contamination of other crops during storage (Alberts et al., 2016). As consumers are much more aware now, they are looking for foods free of harmful chemicals, so several regulations have been imposed in terms of the usage of chemical control technologies. There are also an environmental considerations and a cultural drive for natural and safe food that is free of chemical treatments (Edlayne et al., 2009).

2.6.2 Biological Methods

Over the last 29 years, research has revealed widespread support for agricultural management approaches, as well as a growing interest in biological control measures as the most preferable approaches. Several approaches for suppressing fungal development that use herbal extracts, containing a wide range of microbes (biocontrol microbes); essential oils; and natural clay minerals have been commercialized (He and Zhou, 2010). There are many efficient, practical, and innovative techniques that have been developed in rural-area farming groups (Van der Westhuizen et al., 2010). Some important factors in successfully controlling microorganisms are capacity to colonize plant components when infected by any pathogenic microorganism, environmental circumstances, and interaction with other control approaches (Liu et al., 2013). Microorganisms that are naturally connected with and suited to the vegetative components of a certain plant, as well as disease microorganisms, may have benefits as biocontrol agents. *B. subtilis*, for example, has a similar ecological niche as *F. verticillioides*, which restricts fungal development through competitive inhibition (Bacon et al., 2001). In vitro, *Pediococcus pentosaceus*, a lactic acid bacterium, inhibits the development of *F. proliferatum* and *F. verticillioides* (Dalie et al., 2010). Food-grade yeasts are also regarded as suitable biocontrol measures, as they are biologically and genetically stable, are easy to produce, work at low concentrations, are compatible with commercial processing, are capable of surviving in unfavorable environments, and are pesticide resistant. The *Trichoderma* spp. have an inhibitory impact on *F. verticillioides* growth via extracellular enzymes, volatile chemical formation, and antibiotics (Alberts et al., 2016). Several plant-derived phenolic compounds contain antioxidants that have antimicrobial properties that suppress the activity of fungal enzymes. Synthetic food-grade antioxidants like butylated hydroxyanisole (BHA) and propylparaben (PP) have been demonstrated in vitro to have the capacity to limit *F. proliferatum* and *F. verticillioides* fumonisin production at different incubation temperatures and water activities (Etcheverry et al., 2002). A phenolic antioxidant, tetrahydrocurcuminoids (THCs), derived from the plant *Curcuma longa* (commonly known as turmeric), suppresses FB1 synthesis and *F. proliferatum* growth (Coma et al., 2011). Some other plant phenolic compounds such as caffeic acid, chlorophorin, ferulic acid, iroko, maakianin, and vanillic acid inhibit *F. verticillioides* growth (Beekrum et al., 2003). Some oleoresins extracted from *Zingiber officinale*, carbon tetrachloride oleoresins, and essential oil (ginger oil) rhizomes possess antimicrobial activity against *F. moniliforme* and *F. verticillioides*. Furthermore, essential oils extracted from oregano, palmarosa, clove, lemongrass, and cinnamon can successfully inhibit growth/production of *F. proliferatum* and *F. verticillioides* (Velluti et al., 2003). Genetic engineering and crop breeding studies are primarily targeted at reducing insect invasion, detoxifying mycotoxins by molecular techniques, and reducing contamination by mycotoxigenic fungus (Cleveland et al., 2003). More data on genomic resources are required for research into the pathogenicity of fungal–plant interactions, biochemical/regulatory mechanisms of mycotoxins, and development of creative techniques for breeding/designing resistant crops (Desjardins, 2006). In comparison to non-*Bt* hybrids, genetically engineered *Bt* maize and *cry* proteins produced from *Bacillus thuringiensis* have the ability to lower microbial infection and *fumonisin* levels by cloning and expression of genes producing secondary metabolites that have antifungal characteristics and are dominant over the pathway of the limiting enzymes (Duvick, 2001). It should be noted, however, that diverting metabolic pathways may jeopardize other critical biosynthetic routes.

Mycotoxin adsorption methods have been adopted by the use of natural clay as adsorbent media in the food processing sector, resulting in the detoxification of food (Robinson et al., 2012). By this method, the bioavailability of mycotoxins for animals through their feed could be lowered, and the entry of hazardous chemicals in the gastro-intestinal tracts of animals could be minimized. Phyllosilicate and montmorillonites are the clay minerals that adsorb organic molecules via the cation exchange method and hence can easily reduce FB1 in vitro (Aly et al., 2004). Some clay minerals, such as aluminum oxides work on the basis of structure-selective parameters for some mycotoxins. The degree of adsorption is dependent on the polarity of the molecules, clay particle size, and binding affinity (He and Zhou, 2010). A mixture of clay minerals

(1–10%) and yeast (90–99%) has been found useful for the adsorption of several mycotoxins, including *fumonisins* (Howes and Newman, 2000). Despite the rising interest in biological control strategies, considerable attention should be paid to finding natural chemicals capable of suppressing fungal growth and mycotoxins. However, the rising body of information on this issue should be expanded for use in the field prior to harvesting and storage.

2.7 CONCLUSION

Fumonisin contamination in food and feed poses a severe hazard and leads to epidemic disease across the world. To reduce *fumonisin* contamination in foods, different strategies like biochemical, natural, physical, and genetic engineering can be used effectively. A major concern, however, is the invention and development of fungus- and insect-resistant crops that can fight fungal infection and *fumonisin* contamination. Naturally occurring microorganisms have been shown to be capable of degrading and lowering *fumonisin* production and contamination in a variety of agricultural crops. The considerable danger of *fumonisin* contamination of cereal-based diets and feeds draws the greatest attention since it is a possible carcinogen of worldwide concern. To track the prevalence of these mycotoxins, well-developed technologies for the early and quick detection of toxigenic *fumonisins* have been used. Only through continuing study into the consequences and modalities of mycotoxin activity in numerous species have restrictions been developed. Furthermore, the use of genetic engineering and nanotechnology should be promoted in order to generate fungus- and mycotoxin-resistant agricultural crops and ensure food security for future generations.

REFERENCES

Alberts, J. F., Van Zyl, W. H., & Gelderblom, W. C. (2016). Biologically based methods for control of fumonisin-producing *Fusarium* species and reduction of the fumonisins. *Frontiers in Microbiology, 7,* 548.

Aly, S. E., Abdel-Galil, M. M., & Abdel-Wahhab, M. A. (2004). Application of adsorbent agents technology in the removal of aflatoxin B1 and fumonisin B1 from malt extract. *Food and Chemical Toxicology, 42*(11), 1825–1831.

Anfossi, L., Calderara, M., Baggiani, C., Giovannoli, C., Arletti, E., & Giraudi, G. (2010). Development and application of a quantitative lateral flow immunoassay for fumonisins in maize. *Analytica Chimica Acta, 682*(1–2), 104–109.

Arroyo-Manzanares, N., Huertas-Pérez, J. F., Gámiz-Gracia, L., & García-Campaña, A. M. (2013). A new approach in sample treatment combined with UHPLC-MS/MS for the determination of multiclass mycotoxins in edible nuts and seeds. *Talanta, 115,* 61–67.

Arroyo-Manzanares, N., Huertas-Pérez, J. F., Gámiz-Gracia, L., & García-Campaña, A. M. (2015). Simple and efficient methodology to determine mycotoxins in cereal syrups. *Food Chemistry, 177*, 274–279.

Astoreca, A., Magnoli, C., Barberis, C., Chiacchiera, S. M., Combina, M., & Dalcero, A. (2007a). Ochratoxin a production in relation to ecophysiological factors by Aspergillus section Nigri strains isolated from different substrates in Argentina. *Science of the Total Environment, 388*(1–3), 16–23.

Astoreca, A., Magnoli, C., Ramirez, M. L., Combina, M., & Dalcero, A. (2007b). Water activity and temperature effects on growth of *Aspergillus niger, A. awamori* and *A. carbonarius* isolated from different substrates in Argentina. *International Journal of Food Microbiology, 119*(3), 314–318.

Bacon, C. W., Yates, I. E., Hinton, D. M., & Meredith, F. (2001). Biological control of *Fusarium moniliforme* in maize. *Environmental Health Perspectives, 109*(Suppl 2), 325–332.

Baker, S. E. (2006). *Aspergillus niger* genomics: Past, present and into the future. *Medical Mycology, 44*(Supplement_1), S17–S21.

Beekrum, S., Govinden, R., Padayachee, T., & Odhav, B. (2003). Naturally occurring phenols: A detoxification strategy for fumonisin B1. *Food Additives & Contaminants, 20*(5), 490–493.

Beltrán, E., Ibáñez, M., Portolés, T., Ripollés, C., Sancho, J. V., Yusà, V., . . . & Hernández, F. (2013). Development of sensitive and rapid analytical methodology for food analysis of 18 mycotoxins included in a total diet study. *Analytica Chimica Acta, 783*, 39–48.

Bernhoft, A., Torp, M., Clasen, P. E., Løes, A. K., & Kristoffersen, A. B. (2012). Influence of agronomic and climatic factors on *Fusarium* infestation and mycotoxin contamination of cereals in Norway. *Food Additives & Contaminants: Part A, 29*(7), 1129–1140.

Blesa, J., Moltó, J. C., El Akhdari, S., Mañes, J., & Zinedine, A. (2014). Simultaneous determination of *Fusarium* mycotoxins in wheat grain from Morocco by liquid chromatography coupled to triple quadrupole mass spectrometry. *Food Control, 46*, 1–5.

Bojja, R. S., Cerny, R. L., Proctor, R. H., & Du, L. (2004). Determining the biosynthetic sequence in the early steps of the fumonisin pathway by use of three gene-disruption mutants of *Fusarium verticillioides*. *Journal of Agricultural and Food Chemistry, 52*(10), 2855–2860.

Bolechová, M., Benešová, K., Běláková, S., Čáslavský, J., Pospíchalová, M., & Mikulíková, R. (2015). Determination of seventeen mycotoxins in barley and malt in the Czech Republic. *Food Control, 47*, 108–113.

Booth, C. (1971). The genus *Fusarium*. *The Genus Fusarium*. Available online: www.mycobank.org/BioloMICS.aspx?TableKey=14682616000000061&Rec=744&Fields=All (accessed on 7 September 2021).

Braun, M. S., & Wink, M. (2018). Exposure, occurrence, and chemistry of fumonisins and their cryptic derivatives. *Comprehensive Reviews in Food Science and Food Safety, 17*(3), 769–791.

Bryła, M., Roszko, M., Szymczyk, K., Jędrzejczak, R., & Obiedziński, M. W. (2016). Fumonisins and their masked forms in maize products. *Food Control, 59*, 619–627.

Bryła, M., Waśkiewicz, A., Szymczyk, K., & Jędrzejczak, R. (2017). Effects of pH and temperature on the stability of fumonisins in maize products. *Toxins, 9*(3), 88.

Bullerman, L. B. (1996). Occurrence of fusarium and fumonisins on food grains and in foods. *Fumonisins in Food*, 27–38.

Burgess, L. W., Nelson, P. E., Toussoun, T. A., & Forbes, G. A. (1988). Distribution of *Fusarium* species in sections Roseum, Arthrosporiella, Gibbosum and Discolor recovered from grassland, pasture and pine nursery soils of eastern Australia. *Mycologia, 80*(6), 815–824.

Bush, B., Carson, M., Cubeta, M., Hagler, W., & Payne, G. (2004). Infection and fumonisin production by *Fusarium verticillioides* in developing maize kernels. *Phytopathology, 94,* 88–93.

Castells, M., Marín, S., Sanchis, V., & Ramos, A. J. (2008). Distribution of fumonisins and aflatoxins in corn fractions during industrial cornflake processing. *International Journal of Food Microbiology, 123*(1–2), 81–87.

Cendoya, E., Chiotta, M. L., Zachetti, V., Chulze, S. N., & Ramirez, M. L. (2018a). Fumonisins and fumonisin-producing *Fusarium* occurrence in wheat and wheat by products: A review. *Journal of Cereal Science, 80*, 158–166.

Cendoya, E., del Pilar Monge, M., Chiacchiera, S. M., Farnochi, M. C., & Ramirez, M. L. (2018b). Influence of water activity and temperature on growth and fumonisin production by *Fusarium proliferatum* strains on irradiated wheat grains. *International Journal of Food Microbiology, 266*, 158–166.

Cleveland, T. E., Dowd, P. F., Desjardins, A. E., Bhatnagar, D., & Cotty, P. J. (2003). United States Department of Agriculture–Agricultural Research Service research on pre-harvest prevention of mycotoxins and mycotoxigenic fungi in US crops. *Pest Management Science: Formerly Pesticide Science, 59*(6–7), 629–642.

Coma, V., Portes, E., Gardrat, C., Richard-Forget, F., & Castellan, A. (2011). In vitro inhibitory effect of tetrahydrocurcuminoids on *Fusarium proliferatum* growth and fumonisin B1 biosynthesis. *Food Additives and Contaminants, 28*(2), 218–225.

da Cruz Cabral, L., Pinto, V. F., & Patriarca, A. (2013). Application of plant derived compounds to control fungal spoilage and mycotoxin production in foods. *International Journal of Food Microbiology, 166*(1), 1–14.

Dalie, D. K. D., Deschamps, A. M., Atanasova-Penichon, V., & Richard-Forget, F. (2010). Potential of *Pediococcus pentosaceus* (L006) isolated from maize leaf to suppress fumonisin-producing fungal growth. *Journal of Food Protection, 73*(6), 1129–1137.

Dall'Asta, C., & Battilani, P. (2016). Fumonisins and their modified forms, a matter of concern in future scenario? *World Mycotoxin Journal, 9*(5), 727–739.

Del Fiore, A., Reverberi, M., Ricelli, A., Pinzari, F., Serranti, S., Fabbri, A. A., . . . & Fanelli, C. (2010). Early detection of toxigenic fungi on maize by hyperspectral imaging analysis. *International Journal of Food Microbiology, 144*(1), 64–71.

Desjardins, A. E. (2006). *Fusarium mycotoxins: Chemistry, genetics, and biology* (Vol. 531). St. Paul, MN: APS Press.

Desjardins, A. E., & Proctor, R. H. (2007). Molecular biology of *Fusarium* mycotoxins. *International Journal of Food Microbiology, 119*(1–2), 47–50.

Di Nardo, F., Alladio, E., Baggiani, C., Cavalera, S., Giovannoli, C., Spano, G., & Anfossi, L. (2019). Colour-encoded lateral flow immunoassay for the simultaneous detection of aflatoxin B1 and type-B fumonisins in a single test line. *Talanta, 192*, 288–294.

Doko, M. B., & Visconti, A. (1994). Occurrence of fumonisins B1 and B2 in corn and corn-based human foodstuffs in Italy. *Food Additives & Contaminants, 11*(4), 433–439.

Driehuis, F., Te Giffel, M. C., Van Egmond, H. P., Fremy, J. M., & Blüthgen, A. (2010). Feed-associated mycotoxins in the dairy chain: Occurrence and control. *FIL-IDF Bulletin: Federation Internationale de Laiterie International Dairy Federation, 444*, 2.

Duvick, J. (2001). Prospects for reducing fumonisin contamination of maize through genetic modification. *Environmental Health Perspectives, 109*(suppl 2), 337–342.

Ediage, E. N., Van Poucke, C., & De Saeger, S. (2015). A multi-analyte LC–MS/MS method for the analysis of 23 mycotoxins in different sorghum varieties: The forgotten sample matrix. *Food Chemistry, 177*, 397–404.

Edlayne, G., Simone, A., & Felicio, J. D. (2009). Chemical and biological approaches for mycotoxin control: A review. *Recent Patents on Food, Nutrition & Agriculture, 1*(2), 155–161.

Etcheverry, M., Torres, A., Ramirez, M. L., Chulze, S., & Magan, N. (2002). In vitro control of growth and fumonisin production by *Fusarium verticillioides* and *F. proliferatum* using antioxidants under different water availability and temperature regimes. *Journal of Applied Microbiology, 92*(4), 624–632.

Fanelli, F., Schmidt-Heydt, M., Haidukowski, M., Susca, A., Geisen, R., Logrieco, A., & Mulè, G. (2012). Influence of light on growth, conidiation and fumonisin production by *Fusarium verticillioides*. *Fungal Biology, 116*(2), 241–248.

Frisvad, J. C., Smedsgaard, J., Samson, R. A., Larsen, T. O., & Thrane, U. (2007). Fumonisin B2 production by *Aspergillus niger*. *Journal of Agricultural and Food Chemistry, 55*(23), 9727–9732.

García-Moraleja, A., Font, G., Mañes, J., & Ferrer, E. (2015a). Analysis of mycotoxins in coffee and risk assessment in Spanish adolescents and adults. *Food and Chemical Toxicology, 86*, 225–233.

García-Moraleja, A., Font, G., Mañes, J., & Ferrer, E. (2015b). Development of a new method for the simultaneous determination of 21 mycotoxins in coffee beverages by liquid chromatography tandem mass spectrometry. *Food Research International, 72*, 247–255.

Gazzotti, T., Lugoboni, B., Zironi, E., Barbarossa, A., Serraino, A., & Pagliuca, G. (2009). Determination of fumonisin B1 in bovine milk by LC–MS/MS. *Food Control, 20*(12), 1171–1174.

Gelderblom, W. C., Jaskiewicz, K., Marasas, W. F., Thiel, P. G., Horak, R. M., Vleggaar, R., & Kriek, N. (1988). Fumonisins—novel mycotoxins with cancer-promoting activity produced by *Fusarium moniliforme*. *Applied and Environmental Microbiology, 54*(7), 1806–1811.

Gullino, M. L., Minuto, A., Gilardi, G., & Garibaldi, A. (2002). E_cacy of azoxystrobin and other strobilurins against *Fusarium* wilts of carnation, cyclamen and Paris daisy. *Crop Protection, 21*, 57–61.

Gutema, T., Munimbazi, C., & Bullerman, L. B. (2000). Occurrence of fumonisins and moniliformin in corn and corn-based food products of US origin. *Journal of Food Protection, 63*(12), 1732–1737.

Harrison, L. R., Colvin, B. M., Greene, J. T., Newman, L. E., & Cole Jr, J. R. (1990). Pulmonary edema and hydrothorax in swine produced by fumonisin B1, a toxic metabolite of *Fusarium moniliforme*. *Journal of Veterinary Diagnostic Investigation, 2*(3), 217–221.

Haschek, W. M., Rousseaux, C. G., Wallig, M. A., Bolon, B., & Ochoa, R. (Eds.). (2013). *Haschek and Rousseaux's handbook of toxicologic pathology*. Amsterdam: Academic Press.

He, J., & Zhou, T. (2010). Patented techniques for detoxification of mycotoxins in feeds and food matrices. *Recent Patents on Food, Nutrition & Agriculture, 2*(2), 96–104.

Heperkan, D., Güler, F. K., & Oktay, H. I. (2012). Mycoflora and natural occurrence of afla-toxin, cyclopiazonic acid, fumonisin and ochratoxin A in dried figs. *Food Additives & Contaminants: Part A, 29*(2), 277–286.

Howes, A. D., & Newman, K. E. (2000). *U.S. Patent No. 6,045,834*. Washington, DC: U.S. Patent and Trademark Office.

Irzykowska, L., Bocianowski, J., Waśkiewicz, A., Weber, Z., Karolewski, Z., Goliński, P., . . . & Irzykowski, W. (2012). Genetic variation of *Fusarium oxysporum* isolates forming fumonisin B 1 and moniliformin. *Journal of Applied Genetics, 53*(2), 237–247.

Jacela, J. Y., DeRouchey, J. M., Tokach, M. D., Goodband, R. D., Nelssen, J. L., Renter, D. G., & Dritz, S. S. (2010). Feed additives for swine: Fact sheets–flavors and mold inhibi-tors, mycotoxin binders, and antioxidants. *Journal of Swine Health and Production, 18*(1), 27–32.

Joint FAO-WHO Expert Committee on Food Additives, World Health Organization. *Safety evaluation of certain food additives and contaminants: Prepared by the Seventy fourth meeting of the Joint FAO/WHO Expert Committee on Food Additives (JECFA)*. Available online: https://apps.who.int/iris/bitstream/handle/10665/171781/9789240693982_eng.pdf;jsessionid=79EFFBD803B75293027EDA351F998A18?sequence=3 (accessed on 25 September 2021).

Jones, C., Ciacci-Zanella, J. R., Zhang, Y., Henderson, G., & Dickman, M. (2001). Analysis of fumoni-sin B1-induced apoptosis. *Environmental Health Perspectives, 109*(suppl 2), 315–320.

Jung, S. Y., Choe, B. C., Choi, E. J., Jeong, H. J., Hwang, Y. S., Shin, G. Y., & Kim, J. H. (2015). Survey of mycotoxins in commonly consumed Korean grain products using an LC-MS/MS multimycotoxin method in combination with immunoaffinity clean-up. *Food Science and Biotechnology, 24*(4), 1193–1199.

Kabak, B., & Dobson, A. D. (2009). Biological strategies to counteract the effects of myco-toxins. *Journal of Food Protection, 72*(9), 2006–2016.

Kamle, M., Mahato, D. K., Devi, S., Lee, K. E., Kang, S. G., & Kumar, P. (2019). Fumonisins: Impact on agriculture, food, and human health and their management strategies. *Toxins, 11*(6), 328.

Karbancıoglu-Güler, F., & Heperkan, D. (2009). Natural occurrence of fumonisin B1 in dried figs as an unexpected hazard. *Food and Chemical Toxicology, 47*(2), 289–292.

Kavanagh, F. (1981). *Official methods of analysis of the AOAC*. Edited by William Horwitz. The Association of Official Analytical Chemists, 1111 N. 19th St., Arlington, VA 22209. 1980. 1038 pp. 22× 28 cm. 2.4 kg.

Kawashima, L. M., Vieira, A. P., & Soares, L. M. V. (2007). Fumonisin B1 and ochratoxin A in beers made in Brazil. *Food Science and Technology, 27*, 317–323.

Kimuli, D., Wang, W., Lawrence, K. C., Yoon, S. C., Ni, X., & Heitschmidt, G. W. (2018). Utilisation of visible/near-infrared hyperspectral images to classify aflatoxin B1 contaminated maize kernels. *Biosystems Engineering, 166*, 150–160.

Kolosova, A., & Stroka, J. (2011). Substances for reduction of the contamination of feed by mycotoxins: A review. *World Mycotoxin Journal, 4*(3), 225–256.

Kumar, P., Mahato, D. K., Kamle, M., Mohanta, T. K., & Kang, S. G. (2017). Aflatoxins: A global concern for food safety, human health and their management. *Frontiers in Microbiology, 7*, 2170.

Lee, N. A., Wang, S., Allan, R. D., & Kennedy, I. R. (2004). A rapid aflatoxin B1 ELISA: Development and validation with reduced matrix effects for peanuts, corn, pista-chio, and soybeans. *Journal of Agricultural and Food Chemistry, 52*(10), 2746–2755.

Lemmens, M., Buerstmayr, H., Krska, R., Schuhmacher, R., Grausgruber, H., & Ruckenbauer, P. (2004). The effect of inoculation treatment and long-term application of moisture on *Fusarium* head blight symptoms and deoxynivalenol contamination in wheat grains. *European Journal of Plant Pathology, 110*(3), 299–308.

Liu, J., Sui, Y., Wisniewski, M., Droby, S., & Liu, Y. (2013). Utilization of antagonistic yeasts to manage postharvest fungal diseases of fruit. *International Journal of Food Microbiology, 167*(2), 153–160.

Liverpool-Tasie, L. S. O., Turna, N. S., Ademola, O., Obadina, A., & Wu, F. (2019). The occurrence and co-occurrence of aflatoxin and fumonisin along the maize value chain in southwest Nigeria. *Food and Chemical Toxicology, 129*, 458–465.

López-Errasquín, E., Vázquez, C., Jiménez, M., & González-Jaén, M. T. (2007). Real-time RT-PCR assay to quantify the expression of FUM1 and FUM19 genes from the fumonisin-producing *Fusarium verticillioides. Journal of Microbiological Methods, 68*, 312–317.

Lu, L., Seenivasan, R., Wang, Y. C., Yu, J. H., & Gunasekaran, S. (2016). An electrochemical immunosensor for rapid and sensitive detection of mycotoxins fumonisin B1 and deoxynivalenol. *Electrochimica Acta, 213*, 89–97.

Mac Jr, M., & Valente Soares, L. M. (2000). Fumonisins B1 and B2 in Brazilian corn-based food products. *Food Additives & Contaminants, 17*(10), 875–879.

Mansson, M., Klejnstrup, M. L., Phipps, R. K., Nielsen, K. F., Frisvad, J. C., Gotfredsen, C. H., & Larsen, T. O. (2010). Isolation and NMR characterization of fumonisin B2 and a new fumonisin B6 from *Aspergillus niger. Journal of agricultural and food chemistry, 58*(2), 949–953.

Mao, L., Ji, K., Yao, L., Xue, X., Wen, W., Zhang, X., & Wang, S. (2019). Molecularly imprinted photoelectrochemical sensor for fumonisin B1 based on GO-CdS heterojunction. *Biosensors and Bioelectronics, 127*, 57–63.

Marasas, W. F. (2001). Discovery and occurrence of the fumonisins: a historical perspective. *Environmental Health Perspectives, 109*(suppl 2), 239–243.

Marín, P., Magan, N., Vázquez, C., & González-Jaén, M. T. (2010). Differential effect of environmental conditions on the growth and regulation of the fumonisin biosynthetic gene FUM1 in the maize pathogens and fumonisin producers *Fusarium verticillioides* and *Fusarium* proliferatum. *FEMS Microbiology Ecology, 73*(2), 303–311.

Martins, F. A., Ferreira, F. M. D., Ferreira, F. D., Bando, É., Nerilo, S. B., Hirooka, E. Y., & Machinski Jr, M. (2012). Daily intake estimates of fumonisins in corn-based food products in the population of Parana, Brazil. *Food Control, 26*, 614–618.

Mogensen, J. M., Frisvad, J. C., Thrane, U., & Nielsen, K. F. (2010). Production of fumonisin B2 and B4 by *Aspergillus niger* on grapes and raisins. *Journal of Agricultural and Food Chemistry, 58*(2), 954–958.

Munkvold, G. P., Arias, S., Taschl, I., & Gruber-Dorninger, C. (2019). Mycotoxins in corn: Occurrence, impacts, and management. In *Corn* (pp. 235–287). Amsterdam: AACC International Press.

Nagaraj, D., Adkar-Purushothama, C. R., & Yanjarappa, S. M. (2016). Multiplex PCR for the early detection of fumonisin producing *Fusarium verticillioides. Food Bioscience, 13*, 84–88.

Nielsen, K. F., Ngemela, A. F., Jensen, L. B., De Medeiros, L. S., & Rasmussen, P. H. (2015). UHPLC-MS/MS determination of ochratoxin A and fumonisins in coffee using QuEChERS extraction combined with mixed-mode SPE purification. *Journal of Agricultural and Food Chemistry, 63*(3), 1029–1034.

Noonim, P., Mahakarnchanakul, W., Nielsen, K. F., Frisvad, J. C., & Samson, R. A. (2009). Fumonisin B2 production by *Aspergillus niger* in Thai coffee beans. *Food Additives and Contaminants, 26*(1), 94–100.

Okabe, I., Hiraoka, H., & Miki, K. (2015). Influence of harvest time on fumonisin contamination of forage maize for whole-crop silage. *Mycoscience, 56*(5), 470–475.

Omori, A. M., Ono, E. Y. S., Bordini, J. G., Hirozawa, M. T., Fungaro, M. H. P., & Ono, M. A. (2018). Detection of *Fusarium* verticillioides by PCR-ELISA based on FUM21 gene. *Food Microbiology, 73*, 160–167.

Park, J. W., Kim, E. K., Shon, D. H., & Kim, Y. B. (2002). Natural co-occurrence of aflatoxin B1, fumonisin B1 and ochratoxin A in barley and corn foods from Korea. *Food Additives & Contaminants, 19*(11), 1073–1080.

Pel, H. J., de Winde, J. H., Archer, D. B., Dyer, P. S., Hofmann, G., Schaap, P. J., . . . & Stam, H. (2007). Genome sequencing and analysis of the versatile cell factory *Aspergillus niger* CBS 513.88. *Nature Biotechnology, 25*(2), 221–231.

Picot, A., Barreau, C., Pinson-Gadais, L., Caron, D., Lannou, C., & Richard-Forget, F. (2010). Factors of the *Fusarium verticillioides*-maize environment modulating fumonisin production. *Critical Reviews in Microbiology, 36*(3), 221–231.

Pittet, A., Parisod, V., & Schellenberg, M. (1992). Occurrence of fumonisins B1 and B2 in corn-based products from the Swiss market. *Journal of Agricultural and Food Chemistry, 40*(8), 1352–1354.

Proctor, R. H., Brown, D. W., Plattner, R. D., & Desjardins, A. E. (2003). Co-expression of 15 contiguous genes delineates a fumonisin biosynthetic gene cluster in *Gibberella moniliformis*. *Fungal Genetics and Biology, 38*(2), 237–249.

Ran, R., Wang, C., Han, Z., Wu, A., Zhang, D., & Shi, J. (2013). Determination of deoxynivalenol (DON) and its derivatives: current status of analytical methods. *Food Control, 34*(1), 138–148.

Rheeder, J. P., Marasas, W. F., & Vismer, H. F. (2002). Production of fumonisin analogs by *Fusarium* species. *Applied and Environmental Microbiology, 68*(5), 2101–2105.

Rheeder, J. P., Van der Westhuizen, L., Imrie, G., & Shephard, G. S. (2016). *Fusarium* species and fumonisins in subsistence maize in the former Transkei region, South Africa: A multi-year study in rural villages. *Food Additives & Contaminants: Part B, 9*(3), 176–184.

Rice, L. G., & Ross, P. F. (1994). Methods for detection and quantitation of fumonisins in corn, cereal products and animal excreta. *Journal of Food Protection, 57*(6), 536–540.

Robinson, A., Johnson, N. M., Strey, A., Taylor, J. F., Marroquin-Cardona, A., Mitchell, N. J., . . . & Phillips, T. D. (2012). Calcium montmorillonite clay reduces urinary biomarkers of fumonisin B1 exposure in rats and humans. *Food Additives & Contaminants: Part A, 29*(5), 809–818.

Ross, P. F., Rice, L. G., Osweiler, G. D., Nelson, P. E., Richard, J. L., & Wilson, T. M. (1992). A review and update of animal toxicoses associated with fumonisin-contaminated feeds and production of fumonisins by *Fusarium* isolates. *Mycopathologia, 117*(1–2), 109–114.

Rottinghaus, G. E., Coatney, C. E., & Minor, H. C. (1992). A rapid, sensitive thin layer chromatography procedure for the detection of fumonisin B1 and B2. *Journal of Veterinary Diagnostic Investigation, 4*(3), 326–329.

Rubert, J., Dzuman, Z., Vaclavikova, M., Zachariasova, M., Soler, C., & Hajslova, J. (2012). Analysis of mycotoxins in barley using ultra high liquid chromatography high resolution mass spectrometry: Comparison of efficiency and efficacy of different extraction procedures. *Talanta, 99*, 712–719.

Schatzmayr, G., Zehner, F., Täubel, M., Schatzmayr, D., Klimitsch, A., Loibner, A. P., & Binder, E. M. (2006). Microbiologicals for deactivating mycotoxins. *Molecular Nutrition & Food Research*, *50*(6), 543–551.

Scudamore, K., Scriven, F., & Patel, S. (2009). *Fusarium* mycotoxins in the food chain: Maize-based snack foods. *World Mycotoxin Journal*, *2*(4), 441–450.

Seefelder, W., Gossmann, M., & Humpf, H. U. (2002). Analysis of fumonisin B1 in *Fusarium* proliferatum-infected asparagus spears and garlic bulbs from Germany by liquid chromatography–electrospray ionization mass spectrometry. *Journal of Agricultural and Food Chemistry*, *50*(10), 2778–2781.

Seo, D. G., Phat, C., Kim, D. H., & Lee, C. (2013). Occurrence of *Fusarium* mycotoxin fumonisin B 1 and B 2 in animal feeds in Korea. *Mycotoxin Research*, *29*(3), 159–167.

Seo, J. A., Proctor, R. H., & Plattner, R. D. (2001). Characterization of four clustered and coregulated genes associated with fumonisin biosynthesis in *Fusarium verticillioides*. *Fungal Genetics and Biology*, *34*(3), 155–165.

Shephard, G. S., Burger, H. M., Rheeder, J. P., Alberts, J. F., & Gelderblom, W. C. (2019). The effectiveness of regulatory maximum levels for fumonisin mycotoxins in commercial and subsistence maize crops in South Africa. *Food Control*, *97*, 77–80.

Stack, M. E., & Eppley, R. M. (1992). Liquid chromatographic determination of fumonisins B1 and B2 in corn and corn products. *Journal of AOAC International*, *75*(5), 834–837.

Stępień, Ł., Koczyk, G., & Waśkiewicz, A. (2011). Genetic and phenotypic variation of *Fusarium proliferatum* isolates from different host species. *Journal of Applied Genetics*, *52*(4), 487–496.

Susca, A., Proctor, R. H., Butchko, R. A., Haidukowski, M., Stea, G., Logrieco, A., & Moretti, A. (2014). Variation in the fumonisin biosynthetic gene cluster in fumonisin-producing and nonproducing black aspergilli. *Fungal Genetics and Biology*, *73*, 39–52.

Sydenham, E. W., Shephard, G. S., Thiel, P. G., Marasas, W. F., Rheeder, J. P., Peralta Sanhueza, C. E., ... & Resnik, S. L. (1993). Fumonisins in Argentinian field-trial corn. *Journal of Agricultural and Food Chemistry*, *41*(6), 891–895.

Van der Westhuizen, L., Shephard, G. S., Burger, H. M., Rheeder, J. P., Gelderblom, W. C., Wild, C. P., & Gong, Y. Y. (2011). Fumonisin B1 as a urinary biomarker of exposure in a maize intervention study among South African subsistence farmers. *Cancer Epidemiology and Prevention Biomarkers*, *20*(3), 483–489.

Van der Westhuizen, L., Shephard, G. S., Rheeder, J. P., Burger, H. M., Gelderblom, W. C. A., Wild, C. P., & Gong, Y. Y. (2010). Simple intervention method to reduce fumonisin exposure in a subsistence maize-farming community in South Africa. *Food Additives and Contaminants*, *27*(11), 1582–1588.

van Rensburg, B. J., McLaren, N. W., & Flett, B. C. (2017). Grain colonization by fumonisin-producing *Fusarium* spp. and fumonisin synthesis in South African commercial maize in relation to prevailing weather conditions. *Crop Protection*, *102*, 129–136.

Velluti, A., Sanchis, V., Ramos, A. J., Egido, J., & Marın, S. (2003). Inhibitory effect of cinnamon, clove, lemongrass, oregano and palmarose essential oils on growth and fumonisin B1 production by *Fusarium proliferatum* in maize grain. *International Journal of Food Microbiology*, *89*(2–3), 145–154.

Voss, K. A., Porter, J. K., Bacon, C. W., Meredith, F. I., & Norred, W. P. (1999). Fusaric acid and modification of the subchronic toxicity to rats of fumonisins in *F. moniliforme* culture material. *Food and Chemical Toxicology*, *37*(8), 853–861.

Vujanovic, V., St-Arnaud, M., Barabé, D., & Thibeault, G. (2000). Viability testing of orchid seed and the promotion of colouration and germination. *Annals of Botany, 86*(1), 79–86.

Waśkiewicz, A., Stępień, Ł., Wilman, K., & Kachlicki, P. (2013). Diversity of pea-associated *F. proliferatum* and *F. verticillioides* populations revealed by FUM1 sequence analysis and fumonisin biosynthesis. *Toxins, 5*(3), 488–503.

Wild, C. P., & Gong, Y. Y. (2010). Mycotoxins and human disease: A largely ignored global health issue. *Carcinogenesis, 31*(1), 71–82.

Yoshizawa, T., Yamashita, A., & Chokethaworn, N. (1996). Occurrence of fumonisins and aflatoxins in corn from Thailand. *Food Additives & Contaminants, 13*(2), 163–168.

Chapter 3

Ochratoxins in Food and Feed

Detection and Management Strategies

Arun Kumar Pandey, Rahul Vashishth, Dipendra
Kumar Mahato, Madhu Kamle, Raman Selvakumar,
Monika Mathur, Pradeep Kumar

CONTENTS

3.1	Introduction	52
3.2	Sources of Ochratoxins	55
3.3	Gene Responsible for Production	55
3.4	Mechanism of Toxicity and Health Effects	56
	3.4.1 Oxidative Stress	57
	3.4.2 Cell Apoptosis	57
	3.4.3 Cell Autophagy	58
	3.4.4 Calcium Homeostasis	58
	3.4.5 DNA Adducts	58
	3.4.6 Protein Synthesis Inhibition	59
3.5	Occurrence in Food and Feed	59
3.6	Detection Methods	62
	3.6.1 Sample Preparation	63
	3.6.2 Thin Layer Chromatography	63
	3.6.3 Liquid Chromatography	64
	3.6.4 Gas Chromatography	64
	3.6.5 Enzyme-Linked Immunosorbent Assay	65
	3.6.6 Membrane Immunoassay	65
	3.6.7 Fluorescent Immunoassay	65
	3.6.8 Molecular Imprinted Polymers	66
	3.6.9 Immunosensors	66
	3.6.10 Electrochemical Immunosensors	66
	3.6.11 Optical Immunosensors	67

DOI: 10.1201/9781003242208-3

3.7   Management and Control Strategies   68
      3.7.1   Pre-Harvest Conditions   69
      3.7.2   Post-Harvest Conditions   69
      3.7.3   Physical Methods   70
      3.7.4   Physico-Chemical Methods   71
      3.7.5   Chemical Methods   72
              3.7.5.1   Microbiological Methods   72
3.8   Conclusion   72

3.1 INTRODUCTION

Different mycotoxins are produced naturally in agricultural food and feed by several species of fungi, such as aflatoxin, fumonisin, ochratoxin, patulin, zearalenone, trichothecene, deoxynivalenol, and ergot. Ochratoxin is one of the major groups of mycotoxins, which include ochratoxin A, ochratoxin B, and ochratoxin C (OTA, OTB, and OTC) (Figure 3.1). Chemically, ochratoxin B is a non-chlorinated form of ochratoxin A ((R)-N-((5-Chlor-3,4-dihydro-8-hydroxy-3-methyl-1-oxo-1H-benzo[c]-pyranyl)-carbonyl)-3-phenylalanin), whereas ochratoxin C is an ethyl ester of ochratoxin A (Table 3.1). Generally, ochratoxin B and ochratoxin C are assumed to be of less interest compared to ochratoxin A

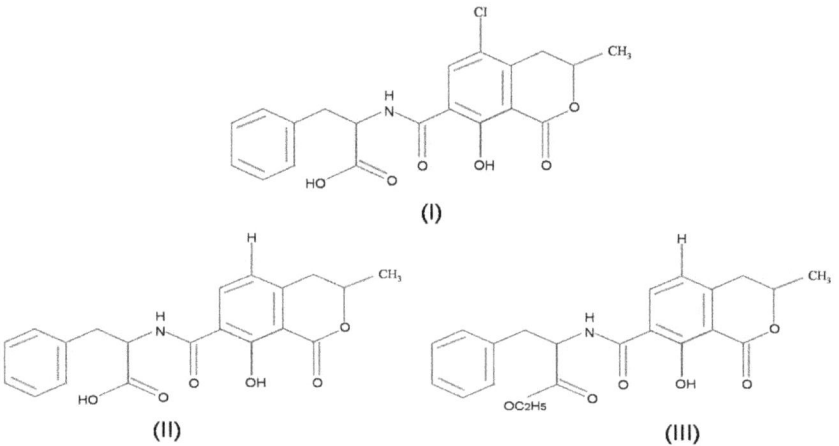

Figure 3.1   Ochratoxin structure: (I) ochratoxin A, (II) ochratoxin B, and (III) ochratoxin C.

Source: Kumar et al. 2020

TABLE 3.1   AN OVERVIEW OF DIFFERENT FORMS OF OCHRATOXINS

Toxins	CAS Number	Chemical Formula	Molecular Mass (g/mol)
Ochratoxin A	303-47-9	$C_{20}H_{18}ClNO_6$	403.8
Ochratoxin B	4825-86-9	$C_{20}H_{19}NO_6$	369.4
Ochratoxin C	4865-85-4	$C_{22}H_{22}ClNO_6$	431.9

Source: Heussner and Bingle 2015

(Heussner and Bingle 2015). van der Merwe and his co-workers first described ochratoxin A when they were working on a new isolated toxic fungal metabolite of *Aspergillus ochraceus* (Van der Merwe et al. 1965). However, there are different fungal species of *Aspergillus* and *Penicillium*, including *A. carbonarius, A. ochraceus, A. niger*, and *P. verrucosum* that produce ochratoxin in food and feed as a toxic secondary metabolite under favorable conditions. Ochratoxin A is very prevalent in agricultural produce, and its contamination has been observed in various food and feeds such as cereals, wine, tea, herbs, cocoa, coffee, milk, fish, eggs, poultry, pork, fruits, vegetables, beans, dehydrated products, infant foods, and animal feed (Heussner and Bingle 2015; Kumar et al. 2020). Exposure to OTA can cause multiple toxicities such as hepatotoxicity, immunotoxicity, nephrotoxicity, genotoxicity, teratogenicity, mutagenicity, and carcinogenicity, which can lead to serious health risks in mammals (Kumar et al. 2020). Studies on in vivo models show that ochratoxin primarily affects kidney function (Bui-Klimke and Wu 2015). It is also a suspected cause of Balkan endemic nephropathy, also known as a malignant kidney disease, in rural states of southeast Europe. Unlike aflatoxin, which is categorized as a group 1 carcinogenic substance for humans, ochratoxin A is listed as a group 2B carcinogenic substance for humans by the International Agency for Research on Cancer (IARC) (IARC 1993; Min et al. 2011; Kumar et al. 2017). However, it should not be assumed that ochratoxin A is less toxic compared to aflatoxin; this shows that the evidence for supporting ochratoxin A–induced carcinogenicity in humans is insufficient compared to that for aflatoxin. A report of the European Commission on the contribution of different agricultural commodities in ochratoxin A exposure to human adults is given in Figure 3.2. To address the probable health concerns, the Nordic Expert Group on Food Safety defined the maximum tolerable daily intake (TDI) up to 5 ng/kg.bw/day in 1991. However, the European Food Safety Authority (EFSA) proposed a provisional tolerable weekly intake (PTWI) for humans of 120 ng/kg.bw/week in 2006. The European Union Scientific Committee of Food (SCF) defined a provisional tolerable daily intake (PTDI) of up to 5 ng/kg.wb/day. Furthermore, the European Union also imposed a maximum permissible limit of ochratoxin A

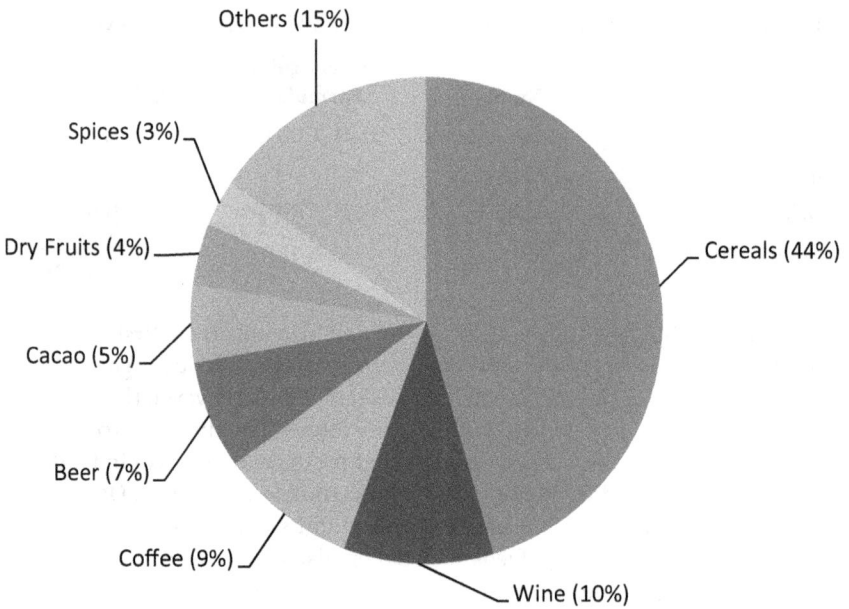

Figure 3.2 Contribution of different agricultural commodities to OTA exposure in human adults.

Source: EC 2002

in various foods and feeds: 5 µg/kg of ochratoxin A in raw cereals, 2 µg/kg in wine, and 5 µg/kg in coffee products (Alhamoud et al. 2019). In 2007, the FAO/WHO Joint Expert Committee on Food Additives (JECFA) revised the PTWI of ochratoxin A to a level of 100 ng/kg.bw/week. However, Health Canada in 2010 fixed the PTWI and negligible cancer risk intake (NCRI) for ochratoxin A at up to 3 and 4 ng/kg.bw/day (Bui-Klimke and Wu 2015; Zhu et al. 2017).

Even though several measures have been adopted to prevent or minimize ochratoxin A levels in food, complete eradication of contamination seems unavoidable. The high affinity of ochratoxin A to bind cell proteins, especially the albumin present in serum, promotes its accumulation in biological tissues such as animal organs and thus leads to carry-over of the contaminant. Bio-accumulation of OTA increases the chance of contamination in animal-derived products, including meat and meat by-products, milk, and eggs (Duarte et al. 2012). Several epizootiological and feeding studies show that exposure to OTA in poultry and pigs results in severe kidney failure (Heussner and Bingle 2015).

3.2 SOURCES OF OCHRATOXINS

Different species of *Aspergillus* and *Penicillium* fungi are mainly responsible for producing ochratoxins in food and feed. The crops get infected with *Aspergillus* sp. such as *A. niger, A. carbonarius, A. ochraceus, A. terreus, A. steynii, A. alliaceus*, and *A. westerdijkiae* during both the pre-and post-harvest periods. Studies show that the most common species of *Aspergillus* is *A. niger*, but most of its isolates are non-ochratoxin producers in general and hence considered less significant (Frusvad et al. 2006; Serra et al. 2006). However, a major ochratoxin A producer is *A. ochraceus*, which also produces several other toxic metabolites, such as penicillic acid, vioxanthin, viomellein, and xanthomegning. It is a slow-growing mesophilic xerophile fungus that usually requires a temperature range of 8–37°C for growth and water activity (aw) of 0.83 to 0.98 for OTA production (Kozakiewicz et al. 1996). Furthermore, in tropical regions, *A. carbonarius* is considered another major ochratoxin A–producing fungus. The growth of *A. carbonarius* is optimal at temperature ranging from 32–35°C, whereas for the germination of spores, optimal aw is 0.83 at 30°C. Apart from this, *A. sclerotiorum, A. sulphurew,* and *A. melleus* are a few other species of *Aspergillus* found to be capable of producing ochratoxin A (Kozakiewicz et al. 1996). In the case of *Penicillium* fungi, *P. verrucosum* and *P. nordicum* are recognized as the most authoritative species responsible for the production of ochratoxin A (Huertas-Perez et al. 2017). However, the other *Penicillum* species, including *P. crustosum, P. brevicompactum, P. chrysogenum, P. oxalicum* and *P olsonii*, are also studied as ochratoxin A producers (Paterson et al. 2004; Vega et al. 2006).

The previous fungal species are capable of producing ochratoxin A even under extreme environmental conditions. *Aspergillus* sp. is capable of producing ochratoxin A in both tropical and subtropical regions; however, some *Penicillium* sp. are capable of producing ochratoxin A at temperatures of about 5°C (Bhat et al. 2010; Gupta et al. 2018). Geisen (2004) found that variation in climacteric factors such as pH, NaCl concentration, and temperature greatly affects the proliferation of the OTA-producing *pksPN* gene in *P. nordicum* and therefore influences the production of ochratoxin A.

3.3 GENE RESPONSIBLE FOR PRODUCTION

Huff and Hamilton (1979) suggested the most probable pathway for biosynthesis of ochratoxin A. According to their study, the synthesis of ochratoxin A starts from acetate and malonate and forms intermediate compounds, including ochratoxin A, ochratoxin B, and ochratoxin C. Several enzymes are involved in catalyzing these steps, including a non-ribosomal peptide synthetase (NRPS) also known as polyketide synthase (PKS), a cytochrome p450 monooxygenase (P450), and a chloroperoxidase/halogenate that includes a chloride atom

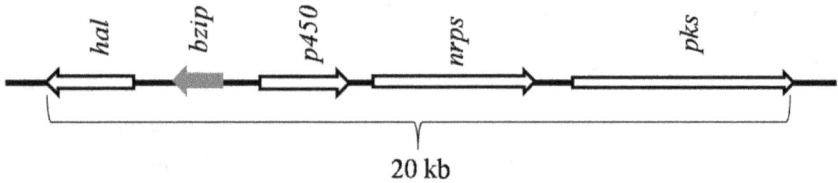

Figure 3.3 A schematic of the clustered genes responsible for biosynthesis of ochratoxin in *A. carbonarius.* The direction of arrows shows the gene transcription. White arrows represent structural genes, while the grey one represents regulatory genes.

Source: Gil-Serna et al. 2020

to ochratoxin B, and at the last step, esterase enzyme catalyzes the pathway. Complete genomic sequencing of ochratoxin A–producing species including *A. carbonarius, P. nordicum, A. steynii, A. niger,* and *A. westerdijkiae* showed a region containing five open reading frames (ORFs) related to five genes, P450, HAL, NRPS, bZIP, and PKS, and their manifestation was correlated with ochratoxin A biosynthesis (Gil-Serna et al. 2018). The genes responsible for enzyme encoding are present in the cluster, except for the gene encoding the esterase enzyme. The four structural genes, PSK, NRPS, P450, and Halogenase, mutually with a bZIP transcription factor, are confirmed to be involved in a cluster of genes responsible for ochratoxin A biosynthesis and have a specific regulatory role in the manifestation of structural genes in both *Aspergillus* and *Penicillium* species (Figure 3.3) (Ferrara et al. 2016). Similar synteny has also been observed in *A. carbonarius, A. niger, A. welwitschiae, A. westerdijkiae,* and *P. nordicum* (Susca et al. 2016; Gil-Serna et al. 2018). In all the clustered genes, ATGACGTGTA or TACACGTCAT palindromic bZIP binding motifs were found upstream, even twice in the case of halogens, which under permissive conditions manifested most during production of ochratoxin A. This shows a strong relationship between the clustered bZIP transcription factor and OTA synthesis regulation (Gil-Serna et al. 2018). However, apart from the core cluster, additional PKS-encoding genes have also been reported in species like *A. carbonarius, P. nordicum, A. westerdijkiae,* and *P. verrucosum.* Gene disruption has been confirmed by PKS-encoding gene involvement in ochratoxin A synthesis, which might complement the manifestation of the PKS-encoding gene cluster under specific circumstances and suppress toxin synthesis (Gallo et al. 2017).

3.4 MECHANISM OF TOXICITY AND HEALTH EFFECTS

OTA exposure in humans can result in multiple health issues due to its carcinogenic, hepatotoxic, immunosuppressive, mutagenic, nephrotoxic, and teratogenic effects (Pfohl-Leszkowicz et al. 2007). The main mechanisms of ochratoxin A–induced toxicity are as follows.

3.4.1 Oxidative Stress

Oxidative stress is a potential factor associated with the induction of several diseases. It induces adverse effects on biological cells through damaging DNA, protein, and lipid content (Kamp et al. 2005; Cavin et al. 2009; Marin-Kuan 2011). It has been recognized by researchers as one of the major modes of ochratoxin A action. In vitro studies show that the concentration of lipid peroxides and malondialdehyde rises in kidney cells by the induction of ochratoxin A due to its accumulation in the proximal tubule of the kidney. Ochratoxin A exposure to primary rat PT and LLC-PK1 cells induced an increase in the level of reactive oxygen species (ROS) concentration and 8-oxoguanine formation while decreasing cellular glutathione (GSH) levels. In renal tubular cells, cellular GSH levels make an important contribution to restricting ochratoxin-induced short-term toxicity (Petrik et al. 2003; Schaaf et al. 2002). Ochratoxin A induces the formation of ROS and reduces superoxide dismustase (SOD) activity in HepG2 and Vero cells (Zheng et al. 2013; Costa et al. 2016). In 2001, ochratoxin A–induced oxidative stress was found during in vivo studies in a rat model. The comparison from the control group showed that the lipid peroxide level in the rat liver was increased, while GSH levels and activity of CAT, SOD, GR, and glutathione peroxidase enzymes in the rat liver were decreased at a significant level due to ochratoxin treatment (Meki and Hussein 2001). The primary mechanism of ochratoxin A toxicity includes the inhibition of *Nrf*2 activation and *Nrf*2 gene transcription, which later induces an increase in ROS. The increase in ROS later results in an increase in peroxidation of lipids, proteotoxic stress, and ultimately DNA oxidative damage (Limonciel and Jennings 2014).

3.4.2 Cell Apoptosis

Several studies have shown that the cell apoptosis could be one of the probable causes of ochratoxin A–induced cytotoxicity. During in vivo studies, the kidneys of rat and mice showed cell apoptosis similar to in vitro studies, including in human HepG2, HeLa, HEK293, CV-1, V79, MDCK-C7, PRK, and PK15 cells (Atroshi et al. 2000; Kamp et al. 2005; Seegers et al. 1994; Bouaziz et al. 2008; Klaric et al. 2008; Gekle et al. 2000). ERK1/2 and the JNK/SAPK have also been reported as major pathways involved in fragments in ochratoxin A–induced cell death (Schramek et al. 1997; Ozcan et al. 2015). Signals from cell receptors to DNA are propagated through the MAPK/ERK pathway. MAP3K, which is also known as MAP kinase kinases, contains an activated MAPK three-tiered cascade. ASK belongs to the family MAP3K, where ASK1 plays an important role in promoting ochratoxin A–induced prohibition of nucleotide metabolism, mRNA splicing, DNA repair, activation of lipid metabolism, the cell cycle, and ultimately renal cytotoxicity (Liang et al. 2015).

3.4.3 Cell Autophagy

Cell autophagy is an accommodative response of a cellular process during stress in diseases. In a cellular process, the role of autophagy is to influence changes in the microenvironment. Principally, autophagy and mitophagy play an adaptive role during diverse pathological conditions to protect organisms. Mitochondrial dysfunction usually occurs during the initial stage of ochratoxin A–induced toxicity, whereas Nix works as a regulated receptor of mitophagy (Novak 2012). Shen et al. (2014) studied ochratoxin A–induced mitophagy and autophagy and found that Nix-deficient HEK293 cells have died after OTA treatment. Their study revealed that ochratoxin up-regulated the pro-apoptotic Bad and AIF proteins, whereas Nix served a central role in mitophagy and autophagy to protect biological cells against induced renal toxicity.

3.4.4 Calcium Homeostasis

A continuous increase in the level of cytosolic calcium is the result of calcium homeostasis breakdown, which is associated with various agents found in different types of cells and causing cytotoxicity (Zhu et al. 2017). Khan et al. (1989) found that the rate of calcium uptake is ATP dependent and can be inhibited by 42 to 45% by ochratoxin A. This ochratoxin A–induced disruption of calcium homeostasis results in damage to the membrane of the endoplasmic reticulum, probably due to an increase in lipid peroxidation. Moreover, an in vivo study showed that the administration of ochratoxin A in an animal model increased the activity of the renal endoplasmic reticulum calcium pump and modulated the intracellular calcium level of fibroblasts present in Syrian hamster embryo (SHE). Such modulation further resulted in mitotic disturbances, leading to cytotoxicity (Zhu et al. 2017).

3.4.5 DNA Adducts

A DNA adduct is a portion/segment of DNA most likely to bond with cancer-causing agents, and they have been studied to understand the mechanism of genotoxicity induced by ochratoxin A (Pfohl-Leszkowicz et al. 1991; Miljkovic et al. 2003). Manderville (2005) reported three probable pathways of DNA adduct formation induced by ochratoxin A. The implicitness of ochratoxin A–induced DNA adducts in the kidney, liver, and spleen was found first through the

32P-post-labeling method. Mantle et al. (2010) reported structural data for the principal adduct through the interaction of ochratoxin A and DNA during an in vitro study. However, the evidence of ochratoxin A–induced DNA adducts as a probable mechanism of ochratoxin A–induced carcinogenicity is controversial.

3.4.6 Protein Synthesis Inhibition

Ochratoxin A–induced protein synthesis inhibition is also reported as a major mechanism of toxicity. Inhibition of protein synthesis influences cell growth and proliferation, as well as interfering with normal cell metabolism. Creppy et al. (1984) demonstrated ochratoxin A–induced inhibition of protein synthesis in spleen, kidney, and liver during in vivo studies and found that it was associated with the prohibition of valyl-tRNA synthetase, aminoacyl-tRNA synthetase, and phenylalanyl-tRNA synthetase expression.

3.5 OCCURRENCE IN FOOD AND FEED

The phenomenon of mycotoxin contamination in food and feed is one of the major challenges today, as it results in significant agricultural losses and severely affects consumer health (Pinotti et al. 2016). Ochratoxin A is the most common toxin among all OTs responsible for contamination of food and feed around the world (Table 3.2). Its contamination has been found in almost all types of raw and processed agricultural products, including cereals like wheat and rice, beans like coffee, cocoa, dry fruits, milk, eggs, fish, meat, spices, medicinal herbs, wine, and so on (Kumar et al. 2020). The contamination of such food items can manifest at any stage of the life cycle during the pre-harvest (in-field), harvest, and/or post-harvest stages. The application of different processing techniques significantly reduces the concentration of ochratoxin A contamination in semi- and/ or fully processed food products compared to unprocessed raw materials. For instance, malting of grains, fermentation of malt, and processing of white and whole breads can effectively reduce ochratoxin A content by 56% (Baxter et al. 2001), 21% (Scott et al. 1995), 80%, and 40% (Scudamore et al. 2003), respectively.

Agricultural products often become contaminated either in the field during harvesting or after harvesting. For instance, grapes are usually affected in the field, whereas commodities like cereals, dry fruits, coffee, and cocoa beans are infected during storage (Bhat et al. 2010). Ochratoxin A contamination of agricultural produce is influenced after harvesting by several environmental conditions, including moisture content, water activity, temperature, and type and integrity of seeds (Madhyastha et al. 1990). Other biotic factors influence the biosynthesis of ochratoxin A, such as amino acids, lactose, and urea. Factors like glucose and sucrose also suppress its occurrence (Abbas et al. 2009).

TABLE 3.2   THE PREVALENCE OF OCHRATOXIN A AND ITS LEVEL OF CONTAMINATION IN CEREALS AND CEREAL PRODUCTS

S. No	Matrix	Nation	Year of Production	No. of Samples	Occurrence (%)	Maximum (μg/kg)
1	Rye	Poland	2017–2019	60	3	2.75
2	Cereal-based baby food	Iran	2017–2018	64	41	1.1
3	Maize	Pakistan	2016–2017	46	71	218.25
4	Flour	Serbia	2012–2016	114	29	23.04
5	Wheat	100 nations	2008–2017	74,821	15	2000
6	Corn, wheat	Pakistan	2015	40	27.5	360
7	Cereal-based food	Portugal	2015	20	50	0.263
8	Infant cereals	United States	2012–2014	155	30	22.1
9	Breakfast cereal (wheat, rice, corn, and oat)	United States	2012–2014	489	41	9.3
10	Wheat and derived samples	Algeria	2012–2013	81	76.65	34.75
11	Wheat	Canada	2011–2014	232	2.2	–
12	Corn, oat, wheat, and rice	United States	2012–2013	144	53	7.43
13	Barley and wheat	United States	2011–2012	262	12.2	185.24
14	Rice, corn, and corn products	Pakistan	2011–2012	275	32	–
15	Rice	China	2009–2011	370	4.9	3.2

Source: Li et al. 2021b

Cereals and cereals-based products are categorized as crucial food crops and have huge religious, cultural, historical, mythological, and economical survival aspects for the human race (Duarte et al. 2010). Due to environmental conditions and storage practices, cereals grains like barley, oats, maize, rye, rice, and wheat most often suffer from the occurrence of fungal growth and become a major source of ochratoxin A exposure to humans (Duarte et al. 2010). The prevalence of ochratoxin A and its level of contamination in cereals and cereal-based products is given in Table 3.2.

Wine is an important beverage traded worldwide. Reports show that ochratoxin A contamination in wine is the most common problem worldwide. Moreover, the prevalence of ochratoxin A in red wine is much higher as compared to white and rosé wine. Wine is reported to be the second major source of ochratoxin A contamination in the diet of Europeans after cereal-based products (Quintela et al. 2013). Its contribution to total ochratoxin A intake by the European population is about 10%. The prevalence of ochratoxin A in wine is mainly correlated with fungal infection of grapes. Studies show that there is a strong association between must maceration with grape skin and the solubilization of ochratoxin A in the must. Although *A. ochraceus* and *P. verrucosum* are the major ochratoxin A–producing fungi, strong evidence is reported for ochratoxin A prevalence in wine by *A. carbonarius* species. The average concentration of ochratoxin A in red wine is usually reported as between 0.026 and 0.9122 µg/l, while in the case of white and rosé wine, it ranges from 0.007 to 0.363 µg/l (Otteneder and Majerus 2000). In 2005, the European Commission fixed the maximum permissible limit of ochratoxin A in wine at 2 µg/l (Quintela et al. 2013).

Green coffee beans are another important commodity used for brewing purposes around the world. Reports show that green coffee beans are more vulnerable to fungal attack, especially by *Aspergillus* species, including *A. carbonarius*, *A. niger*, *A. ochraceus*, and *A. westerdijkiae* when growing in tropical and semi-tropical conditions, while in temperate conditions, attack by *Penicillium* species, especially *P. olsonii*, *P. brevicompactum*, *P. crustosum*, *P. verruculosum*, and *P. oxalicum* is more profound (Alvindia and de Guzman 2016; Vega et al. 2006). Casas-Junco et al. (2018) reported that *Byssochlamys spectabilis*, a strain isolated from roasted coffee beans from the Mexican market, is also able to produce ochratoxin A. However, there are several factors affecting the mycotoxigenic potential to contaminate coffee beans, such as fungal species and strain type; water activity; climatic, storage, and transport conditions; and processing method used (dry or wet process) (Bucheli and Taniwaki 2002). To avoid ochratoxin A–induced toxicity in humans, the European Union has fixed the maximum permissible limit for ochratoxin A in roasted coffee and ground roasted coffee beans at a level of 5 µg/kg. Furthermore, based on toxicological data, the maximum tolerable weekly intake (TWI) was fixed at 120 ng/kg body weight by the European Safety Authority (EFSA 2006).

The prevalence of ochratoxin A in animal tissues, including products obtained from animals such as milk and milk products, poultry and poultry products, meat and meat products, is also indicated by the European Food Safety Authority. This is due to the uptake of ochratoxin A–contaminated feed by animals, including the rapid assimilation of ochratoxin A in blood as well as very slow elimination from the body (EFSA 2006). According to Vettorazzi et al. (2009), an oral dose of ochratoxin A (500 µg/kg) can induce maximum plasma concentration within 2 h. Studies show that the amount of ochratoxin A in animal feed changes from one country to another, where the highest concentration was found in Northern Europe and North America, including Canada, Denmark, and Yugoslavia (above 5000 µg/kg) (Speijers and Van Egmond 1993). The level of toxicity is associated with the concentration and/or period of exposure to the toxic substance. Monogastric animals like poultry and pigs are most susceptible to ochratoxin A–contaminated feeds. In particular, it is believed that pigs are highly susceptible to nephrotoxicity, followed by a decrease in growth and productivity rate. Malagutti et al. (2005) reported that feed efficiency, daily weight gain, and total body weight of pigs were reduced significantly when fed a feed contaminated with 25 µg/kg of ochratoxin A. In the case of poultry, especially chicken and turkeys, ochratoxicosis could result in nephrotoxicity, loss of weight loss, decrease in production of eggs, and lower quality of eggshells (Denli and Perez 2010). Studies reported that when poultry animals of different ages were fed with OTA-contaminated feed (1–5 mg/kg), it resulted in a significant influence on their cellular components, such as an increase in total protein, globulin, albumin, triglyceride, cholesterol, and potassium levels and a decrease in the level of creatine and uric acid, including the activity of serum alkaline phosphatase (ALP) and gamma glutamine transpeptidase (GGT) content (Bailey et al. 1989; Denli et al. 2008; Huff et al. 1988). However, the effect of ochratoxin A toxicity in ruminants was observed to be comparatively low due to its breakdown into less toxic metabolites through rapid detoxification of rumen protozoa and bacterial enzymes (Muller et al. 1998).

Ochratoxin A is resistant to methods used during the processing of cereals and cereal products like milled, baked, and fermented products, and consumers could be exposed to ochratoxin A by consuming contaminated food products (Bullerman and Bianchini 2007; Khaneghah et al. 2018a).

3.6 DETECTION METHODS

Developments in analytical technologies have also been used in the development of more efficient and accurate methods to determine the toxicity of a particular mycotoxin, including isolation and identification. Methods such as thin layer chromatography (TLC), liquid chromatography (LC), thin layer chromatography coupled with mass spectrophotometry (TLC-MS), high-pressure liquid chromatography (HPLC), aptamer-based sensor technology, and biosensors are some of the well-established methodologies used currently for detection, identification, and quantification of ochratoxin in food samples.

3.6.1 Sample Preparation

In general, ochratoxin analysis requires prior sample preparation involving unit operations such as grounding, dissolving, extraction, separation, and concentration. Due to the non-uniform growth of ochratoxins on food/feed, an efficient method should be followed to obtain a true representative analytical food sample. The isolation of OTA can be achieved using any desirable method resulting in efficient extraction of toxin. Water, including other solvents, can be used directly for the extraction of ochratoxin A from contaminated food samples to avoid interference during analysis (identification/quantification). The selection of extraction method generally varies with the type of food and feed matrix and the analytical method to be followed for the recognition, detection, and quantification of toxins. Usually, acidic and alkaline buffers are said to be suitable for the preparation of ochratoxin A detection samples in the case of TLC, HPLC, LC, LC-MS, HPLC-MS, and ultra performance liquid chromatography (UPLC) analysis. On the other hand, buffers with varying compositions are utilized for the preparation of samples involving solid phase extraction (SPE) columns.

Immuno-affinity chromatography (IAC) is a well-established method that is utilized widely for the pre-treatment of samples. It involves the coupling of specific antibodies with the solid phase and then passing through the sample for efficient extraction. During the process, the toxin will bind to the column due to its high affinity, whereas the solid material of the food will pass through the column during washing. The targeted toxin is collected in a later stage with the help of an organic solvent with a higher affinity or solubility of toxin in it (Yu and Lai 2010; Huertas-Perez et al. 2017; Al-Jaal 2019). When developing/following any method, it should be validated per national and international regulatory standards such as Association of Official Agricultural Chemists (AOAC), European Union (EU), Food Safety and Standard Authority of India (FSSAI), and Food and Agriculture Organization (FAO).

3.6.2 Thin-Layer Chromatography

Thin-layer chromatography is one of the oldest and easiest methods for the determination of ochratoxin. However, the limitations of this method involve its sensitivity to the presence of these toxins at lower levels in foods. The efficiency of the detection technique can be improved by following an efficient pretreatment and isolation method to collect highly pure extract from contaminated food and feed. Caputo et al. (2007) developed a rapid and highly sensitive method (0.1 ng) for the determination of ochratoxin that involves the use of a photosensor and silica high-performance thin-layer chromatography (HPTLC) plate. Currently, this method is considered one of the most cost-effective, rapid, and efficient methods for the determination of ochratoxin. Studies show that validated HPTLC methods can be used for robust and efficient extraction of mycotoxins, including OTA in various foods and feed samples such as wines, dairy products, livestock feed, and so on up to a 0.015 μg/kg limit of quantification (LOQ) (Antep and Merdivan 2012; Welke et al. 2010; Kupski and Badiale-Furlong 2015; Kotinagu et al. 2015).

3.6.3 Liquid Chromatography

Liquid chromatography is considered one of the most modern and developed methods with high sensitivity and accuracy to detect, identify, isolate, and quantify the presence of mycotoxins, including ochratoxin. The process can involve single or multiple extraction steps, with and without a cleanup procedure.

Sulyok et al. (2006) successfully used a single-step extraction method without any cleanup step to isolate toxins, whereas Jung et al. (2012) used a double extraction process for the isolation and detection of several mycotoxins, such as aflatoxin B1, B2, M1, M2, and G1 including OTA, and achieved LOPs as low as 0.1–6.1 μg/kg and an LOQ of 0.3–18.4 μg/kg for OTA.

Wei et al. (2018) studied OTA contamination in grape samples before and after processing and achieved a LOQ level as low as 0.1 ug/kg. Their method involved the use of acetonitrile as the extracting solvent (d-SPE) cleanup and UHPLC system along with a Xero-TQS-triple quadrupole mass spectrometer boosted with step wave ion-guide technology.

Pakshir et al. (2021) determined OTA in different types of coffee using HPLC. They utilized an immunoaffinity column for the extraction of coffee samples using HPLC and found that 50% of the 50 analyzed samples had OTA levels higher than the permitted limit (2 μg/kg).

Kyei-Baffour et al. (2021) validated a method based on HPLC-FLD for the determination of OTA in cocoa beans. They analyzed the reduction in ochratoxin levels when the coffee beans were treated with cola nut powder and extract at various levels. They found a reduction of OTA in cocoa beans treated with cola nut powder from 8.32 to 2.85 μg/kg.

3.6.4 Gas Chromatography

Similar to other chromatographic techniques, the separation and identification of compounds can be performed using gas chromatography (GC). The analysis of ochratoxin and other related mycotoxins is limited through GC, as it requires the derivatization of non-volatile OTA to a volatile form, which also increases the cost of estimation (Rahi et al. 2019).

3.6.5 Enzyme-Linked Immunosorbent Assay

Enzyme-linked immunosorbent assay (ELISA) is a rapid analytical method for the determination of ochratoxins involving immunoassay. Other immunoassay-based methods include radioimmunoassay (RIA), chemoluminescence technique (CL-IS), time-resolved fluorescent immunoassay (TR-FIA), and florescence resonance energy transfer immunoassay (FRET). ELISA involves a competitive binding of labeled toxins and toxins to be analyzed. It can analyze several samples in a single run and can be divided into two forms based on direct and indirect binding. The ELISA method is mostly performed in closed-door laboratories; however, the development of immune-chromatographic tests allows analysis of the contaminants of food/feed on site with one-time use disposable kits (Li et al. 2016; Sun et al. 2015; Khataee et al. 2021).

Schneider et al. (1995a, 1995b) developed a quick OTA detection method (lateral flow immunoassay) that involves the use of a strip on which the antibody is coupled with a chromo-logical compound, usually colloidal gold or latex, along with immobilized mycotoxins to be analyzed. These labeled antibodies bind with the immobilized targeted toxin present on the strip and result in the development of an extra-colored band (Meulenberg 2012; Rahi et al. 2019). The recent development in lateral flow assay has seen the utilization of nanoparticles. Quesada-Gonzalez and Merkoçi (2015) developed a nitrocellulose-assisted lateral flow assay kit that can be used to perform the detection of ochratoxin (Santos et al. 2017; Ren et al. 2014).

3.6.6 Membrane Immunoassay

The membrane immunoassay is very similar to the IAC assay, although it has a different flow-through format. This method is used for rapid immune-chemical testing. In this method, the secondary antibody along with an immune-chemical reactant medium consisting of anti-OTA antibody/sample/tracer/substrate is passed through a membrane. The binding of OTA with antibody results in the development of color, and the intensity of the color is estimated using a colorimeter. The use of a portable colorimeter along with a testing kit makes complete onsite analysis possible. The efficiency of the method is quite high and suitable for agricultural produce with a sensitivity level of 4 µg/kg in coffee samples (Malir et al. 2016).

3.6.7 Fluorescent Immunoassay

Fluorescent immunoassay (FIA) involves the use of a fluorescent tag, contrary to the use of various enzymes and a cocktail of enzymes in other immunoassays. FIA is one of the most common and the most-utilized fluorescent immunoassays and functions similarly to ELISA, but its detection method involves the quantification of fluorescence developed during analysis. The common immunoassay detection method for ochratoxin involves a separation step before detection analysis, whereas fluorescence techniques such as TR-FIA, FP-IA, and FRET do not involve such a separation step due to their high specificity, resulting in express detection of toxin (Huang et al. 2009; Maragos 2006; Zezza 2009; Taihua et al. 2011).

3.6.8 Molecular Imprinted Polymers

With the progress of analytical methods, the concentration of toxin analysis shifted from antibody dependent to non-biological material-based elements such as molecularly imprinted polymers (MIPs). In this method, a molecular template is created with the process of co-polymerization using cross-linking monomers of derived functional characteristics in the presence of the target analyte (the impact molecule). MIPs with higher affinity can easily be used in place of antibody-based immunoassays. It possesses higher resistance against different solvents, changing pH, ionic strength, and so on, which further proves its superiority over common immunoassay-based methods. Several studies have shown the successful application of MIP methodology to analyze the presence of ochratoxin in food/feed material (Lerda 2015).

3.6.9 Immunosensors

Immunosensor analysis is based on the detection of signals produced during the binding of antibodies and antigens through a signal transducer. Electrochemical analysis involves the recording of electroactive signals, and an optical immunosensor performs the recording of signals by surface plasmon resonance and piezoelectric quartz crystal microbalances (QCMs). This analysis involves the estimation of mass change when the immobilized antibodies on quartz crystal interact with an antigen (Spinella et al. 2013; Princci et al. 2018; Van der Gaag et al. 2003; Li et al. 2016; Badea et al. 2016).

3.6.10 Electrochemical Immunosensors

Electrochemical immunosensors are based on the principle of recording electroactive signals generated during the interaction between antibody and antigen by means of voltammetry, chronoamperometry, electrochemical impedance spectroscopy, or voltammetry transducer. The process involves the estimation of the potential difference generated using interactions taking place in the reaction medium. Currently, most electrochemical immunosensing techniques involve the use of an enzyme mixture as a medium to generate an electrochemical signal during analysis.

Li et al. (2021c) evolved a novel label-free impedimetric technology-based electrochemical sensor using hydrogel/chitosan for the detection of ochratoxin A. This nano-technology uses chitosan/dipeptide hydrogel as a sensing interface with a detection recovery of 96–102.8% and a detection limit of 0.03 ng/ml^{-1}.

Kirlangiç et al. (2021) worked on the development of an aptamer-based EC detector with analytical electrodes coated with a transitional metal oxide layer (MeOx). They immobilized ochratoxin A clonal aptamer 5^1 aminohexyl on pencil graphite electrode (PGE) through carbetamide chemistry and achieved detection of ochratoxin A up to a detection limit of 0.03 nM.

Hou et al. 2022 reported the development of a label force EC aptasensor for the rapid determination of ochratoxin at ultrasensitive levels. The unique feature of the analytical system involves no coupling medium and high sensitivity with low signal amplification. A linear relationship between R_{ct} and the OTA concentration logarithm was observed in the range of 0.05–10 ng/ml along with a detection limit of 0.05 ng/ml for OTA, making this method much more suitable for agricultural produce.

An organometallic-based EC sensor was developed and tested for its selectivity and sensitivity for the determination of ochratoxin A by Xiang et al. (2021). They used chloroauric acid (HAuCl$_4$) and cetyltrimethylammonium bromide (CTAB) for the development of organometallic catalysts. The catalyst-fabricated electrode showed extended anodic potential in comparison with other Au-based electrodes and also had greater electrolytic activity and selectivity of ochratoxin A. When working under optimum parameters, the electrode exhibits a linearity range of 1.0×10^{-7} to 1.0×10^{-5} mol/l and LOD 2.9×10^{-8} mol/l (SIN = 3), along with a recovery potential of 93.5–98.2%.

3.6.11 Optical Immunosensors

Optical immunosensors involve the recognition of signals produced by surface plasmon resonance, optical waveguides, light spectroscopy, and so on. Viter et al. (2018) designed a photoluminescence immunosensor and conducted analytical, thermodynamic, and kinetic studies to determine the selectivity and sensitivity of the immunosensor for detection of ochratoxin in agriculture produce. The designed PL immunosensor consists of zinc oxide–based nanorods (ZnO-NRs), with a selective layer based on OTA antibody for the detection of ochratoxin A levels in the analyzed sample. Their study shows that the immunosensor can be successfully used for the detection of a wide range of ochratoxin levels ranging from 10^{-4} ng/ml to 20 ng/ml with a sensitivity level of 0.1ng/ml.

Myndrul et al. (2018) built a porous silicon-based photoluminescence immunosensor in which Psi film was developed using a metal-assisted chemical etching process. The selective film on Psi was fabricated using OTA antibodies. The proposed and validated range of ochratoxin A determined was 0.001–100 ng/ml, whereas the selectivity and unit of detection for the level were 0.1 ng/ml and 0.02 ng/m, respectively, with a respective time of 500–700 seconds.

Singh et al. (2021) illustrated the development and investigation of a bioactive free optical sensing system for the detection of ochratoxin A levels. This MnO_2-based nano-system can detect aflatoxin and ochratoxin A produced by *Asperigillusflavus* and other related molds in agriculture products such as cereals, legumes, and beans.

Vitera et al. (2018) worked on a ZnO-nano particle-based immunosensor based on a glass substrate that was later coated by protein A, providing sites for OCTA binding during the analysis of contaminated samples (glass/ZnO-NRs/protein A/anti-OTA). The sensor was developed with a sensitivity range of 0.1–1 ng/ml and a limit of detection of 0.01 ng/ml, along with a response time in the range of 500–800 s against ochratoxin A.

3.7 MANAGEMENT AND CONTROL STRATEGIES

To protect the health of humans as well as farm animals, several countries have implemented regulations to control OTA in food and feed. However, legislation to control OTA varies from country to country, mainly due to environmental conditions (Zhu et al. 2017). Some of the major organizations that are actively participating in preparing worldwide and country-oriented regulations for ochratoxins are CEN, FSSAI, HACCP, AOAC, and the EU.

The establishment of regulation concerning the limits of contamination in food/feed is based on various factors

1. Toxicity-related data availability
2. Mode of occurrence of toxicity
3. Information regarding the distribution of toxin concentration
4. Code of conduct and traceability procedures, which will help in gathering knowledge about the trade contact that existed before the outbreak
5. Food supply sufficiency

As is evident from the previous, the world is moving along with the development of various methods for the determination of ochratoxin levels in food/feed. The major aim for the development of a newer method is to get the best suitable method that will be rapid, cost effective, possibly portable, and most importantly able to measure OTA levels within the range of permitted levels of ochratoxin in regulations. This will help in boosting the self-regulation of countries in terms of safeguarding their respective populations and further helping in trade with other countries. Any of the previously mentioned methods can only be applied to samples when food/feed is already contaminated with a toxin. Therefore, a major shift has taken place in determining and coming up with technologies that can efficiently prevent the contamination or growth of OTA in the agriculture produced. This is attributed to the high stability of OTA against pre- and post-harvest conditions, which forces the application of technologies that can prevent the growth of OTA on food/feed rather than determining its levels in food/feed.

Strategies intended to protect agricultural produce from the contamination of ochratoxin are mainly based on three principles:

1. Prevention of agriculture contamination from ochratoxins
2. Detoxification/decontamination of produce contaminated with ochratoxins
3. Absorption/inhibition of toxins in the gastrointestinal tract during consumption of contaminated food

The contamination of food/feed can occur at two stages

1. Pre-harvest conditions
2. Post-harvest conditions

3.7.1 Pre-Harvest Conditions

The prevention of contamination in food/feed during pre-harvest conditions is considered best, as this can save processing time, cost, and labor. The prevention of contamination before harvesting involves conditions such as the use of Good Agricultural Practices (GAPs), fungicides and other related mycotoxigenic agents, and conditions that are not conducive to the growth of OTA-producing fungi. The use of fungicides to prevent OTA in pre-harvest conditions is prohibited and limited due to their associated side effects. Some researchers have suggested the use of favorable and competitive microflora that will compete with OTA-producing fungi for nutrition and restrict their growth. Change of crops in periodic sessions are also suggested to prevent the chances of OTA contamination. Another efficient method in pre-harvest prevention is the sorting operation. Agricultural produce that is already contaminated or has traces of contamination can be separated from produce that is uncontaminated.

3.7.2 Post-Harvest Conditions

Control of ochratoxin contamination after harvesting of produce can be achieved by several methods, such as physical, chemical, and biological. Further, correct storage after harvesting can immensely help in preventing the contamination of agricultural produce by ochratoxin. It is a well-known fact that the OTA producers flourish well in environments with humidity and heat.

Food/feed can be stored under dry and cool conditions to prevent the growth of OTA producers during storage. The detoxification/decontamination approach to control contamination by OTA involves the utilization of physical, chemical, and biological methods that satisfy the following needs:

1. They must inactivate, kill, or remove OTA-producing fungi.
2. They should not leave any residue with toxic or carcinogenic potential.
3. They should not alter the characteristic morphology and sensorial properties of produce.
4. They should possess the potential to eliminate fungal mycelium or spores to prevent their reoccurrence under suitable conditions for growth
5. They should be efficient and cost effective.

3.7.3 Physical Methods

Physical operations that can be utilized to remove or reduce OTA levels in agricultural products can be classified as sorting, cleaning, washing, roasting, irradiation, microwave processing, baking, milling, and so on. Several researchers have reported the utilization of temperature ranges from 80–200°C and observed a reduction in OTA levels ranging from 0 to 100% (Table 3.3).

TABLE 3.3   REDUCTION IN OTA LEVELS IN COFFEE DURING ROASTING

OTA Origin	Roasting Parameters	% Reduction in OTA	References
Inoculation	200°C, 10–20 min	0–12	Tsubouchi et al. (1987)
Inoculation	180°C, 10 min	31.1	Nehad et al. (2005)
Natural	200 ± 5°C, 20 min	77–87	Levi et al. (1974)
Natural	200 ± 5°C, 20 min	80–90	Gallaz and Stalder (1976)
Natural	200°C, 3 min	65–100	La Pera et al. (2008)
Inoculation	180–240°C, 5–12 min	8–98	Ferraz et al. (2010)
Inoculation	5–6 min, dark roasting	48–87	Micco et al. (1989)
Natural	5–6 min, dark roasting	90–100	Micco et al. (1989)
Natural	250°C, 150 sec	14–62	Studer-Rohr et al. (1994)
Inoculation	250°C, 150 sec	2–28	Studer-Rohr et al. (1994)
Natural	223°C, 4 min	84	Blanc et al. (1998)
Natural	175–204°C, 7–9 min	>90	Romani et al. (2003)
Inoculation	200–220°C, 10–15 min	22.5–93.9	Urbano et al. (2004)

Source: Varga et al. 2010

The application of various temperature ranges in processing such as roasting, baking, irradiation, and microwave processing has shown variation in OTA reduction. The varying reduction during such operations was attributed to the different modes of heating, localized heating, raw sample condition, degree of contamination, moist or dry heating, and so on. The greatest reduction in OTA level involves moist heating of samples at a high temperature (Table 3.3). In certain cases, the freeze–thaw cycle also has been found to impact the levels of OTA. Temperatures as low as –20°C (freezing) and 26°C (thawing) during the periodic cycle were investigated to see the impact of the freeze-thaw cycle on toxin levels (Zhang et al. 2019).

3.7.4 Physico-Chemical Methods

Physico-chemical operations that can be utilized to remove or reduce OTA levels in agricultural produce can be classified based on the use of adsorbents, absorbents, and chelating agents. Absorbents such as activated charcoal, heavy metals, minerals, and synthetic adsorbents are some examples of physico-chemical agents that can be used to restrict or kill OTA producers or reduce the level of produced OTA. The use of functional ingredients such as activated charcoal, potassium caseinate, aluminosilicates, quaternary ammonium grafting, yeast, bacterial cell or vegetal fiber, and synthetic fiber has also shown a huge effect on reducing OTA levels in food/feed (Goryacheva et al. 2007). Furthermore, the use of potassium caseinate was illustrated to achieve reduced levels of up to 80% in a wine sample. Similar results were also obtained

TABLE 3.4   DEGRADATION IN OTA LEVELS USING CHEMICAL AGENTS

	Chemical Agent	Reference
1	Finning agents	Visconti et al. (2008)
2	Polyvinylpyrrolidone, cholestyramine	Castellari et al. (2001)
3	Activated carbon	Dumeau and Trione (2000)
4	Enological decolorizing carbon	Gambuti et al. (2005)
5	Oak wood	Savino et al. (2007)
6	Activated carbon	Olivares-Marín et al. (2009)
7	Chitosan beads	Belajova et al. (2007)
8	Bentonite	Kurtbay et al. (2008)
9	Cellulose acetate esthers	Visconti et al. (2008)
10	Enological decolorizing carbon	Gambuti et al. (2005)
11	Activated carbon and sodium bentonite	Var et al. (2008)

using enological decolorizing carbon (Varga et al. 2010). In the case of physico-chemical methods to achieve a reduction in OTA levels, the use of activated charcoal was found to be the best, with a reduction of up to 90% (Meulenberg 2012).

3.7.5 Chemical Methods

Chemical methods to control OTA levels in agricultural produce involve anti-mitotic agents, oxidizing agents, antioxidants, chlorination, salting, and methods like formaldehyde treatments. In the early times of using these agents, the use of ammonization was accepted as one of the most favorable chemical methods to reduce levels of OTA. The use of the ammonization process to regulate the OTA contamination level in food and feeds shows that it is capable of decomposing OTA in corn, barley, and wheat (Peraica et al. 2002).

Other chemical methods such as bleaching powder, alkaline hydrogen phosphate, treatment with charcoal, ozonation, and so on are also to be found effective in reducing the level of ochratoxin. Studies show that the reduction of aflatoxins using supplementary methods can also aid in the reduction of OTA. Certain organic solvents such as dichloromethane; methyl chloride; and acids like formic acid, acetic acid, and propionic acid are also found to be effective against OTA contamination (Visconti et al. 2008).

3.7.5.1 Microbiological Methods

The use of direct microbes or microbial enzymes involves compounds that act on the toxin without changing the original characteristics of the solution. Microbiological methods involve the immobilization of the desired microorganism on a suitable non-reactant surface. In the majority of the cases, OTA is not removed from the food/feed; instead its activity is modified by the action of certain enzymes, causing the hydrolysis of native OTA molecules, resulting in the inactivation of OTA or its metabolites' toxin compounds. Further, recent times have seen surges in the utilization of saccharification and protease enzymes, which can be attributed to their specificity and selectivity of action against various substrates (Varga et al. 2010). A list of certain microorganisms used in the detoxification of agricultural produce is given in Table 3.5.

3.8 CONCLUSION

The presence of ochratoxins in crops is observed during both pre- and post-harvest conditions, including processing and storage. It causes serious health risks to humans as well as animals due to frequent contamination of food and feeds worldwide. Human exposure to ochratoxins could result in multiple health

TABLE 3.5   DEGRADATION IN OTA LEVELS USING MICROBES AND ENZYMES

	Microbe or Enzyme	Reference
	Bacteria	
1	Rumen microbes	Hult et al. (1976); Varga et al. (2000)
2	*Butyrivibriofibrisolvens*	Westlake et al. (1987)
3	*Lactobacillus Streptococcus Bifidobacterium* sp.	Skrinjar et al. (1996)
4	*Bacillus subtilis B. licheniformis*	Petchkongkaew et al. (2008); Bohm et al. (2000)
5	*Acinetobactercalcoaceticus*	Hwang and Draughon (1994)
6	*Phenylobacterium immobile*	Wegst and Lingens (1983)
7	*Nocardiacorynebacterioides Rhodococcuserythropolis Mycobacterium* sp·	Holzapfel et al. (2002)
8	*Lactobacillus* sp.	Piotrowska and Zakowska (2000); Fuchs et al. (2008)
9	*Eubacteriumcallenderi E. ramulus Streptococcus pleomorphus Lactobacillus vitullinus Sphingomonaspaucimobilis S. saccharolytica Stenotrophomonasnitritreducens Ralstoniaeutropha R. basilensis Ochrobactrum* sp., *Agrobacterium* sp.	Schatzmayr et al. (2002, 2006)
10	*Pseudomonas cepacia P. putida Rhodococcuserythropolis Agrobacterium tumefaciens Comomonasacidovorans*	Schatzmayr et al. (2009)
11	**Protozoa**	Kiessling et al. (1984); Ozpinar et al. (2002)
	Fungi	
12	*Aspergillus niger A. fumigatus*	Varga et al. (2000)
13	*Aspergillus niger A. versicolor A. wentii A. ochraceus*	Abrunhosa et al. (2002)
14	*Aspergillus niger A. japonicus*	Bejaoui et al. (2006)
15	*Pleurotusostreatus*	Engelhardt (2002)
16	*Saccharomyces cerevisiae*	Piotrowska and Zakowska (2000)
17	*Saccharomyces cerevisiae S. bayanus*	Bejaoui et al. (2004)
18	*Rhizopusstolonifer R. microsporus R. homothallicus R. oryzae*	Varga et al. (2005)
19	*Trichosporonmycotoxinivorans*	Molnar et al. (2004)
20	*Phaffiarhodozyma Xanthophyllomycesdendrorhous*	Peteri et al. (2007)
21	*Saccharomyces cerevisiae Kloeckeraapiculata*	Angioni et al. (2007)

(Continued)

TABLE 3.5   DEGRADATION IN OTA LEVELS USING MICROBES AND ENZYMES (CONTINUED)

	Microbe or Enzyme	Reference
22	*Aureobasidium pullulans*	De Felice et al. (2008)
23	*Cryptococcus flavus, C. laurentii, C. curvatus, C. humicolus, Trichosporonovoides, T. dulcitum, T. guehoae, T. mucoides, T. coremiiforme, T. cutaneum, T. laibachii, T. monilifotrme, Rhodotorulamucilaginosa, R. fujisanensis*	Schatzmayr et al. (2009); Schatzmayr et al. (2003)
	Enzymes	
24	Carboxypeptidase A	Stander et al. (2001)
25	Commercial proteases (Pancreatin from porcine pancreas, Protease A, and Prolyve PAC from *A. niger*)	Abrunhosa et al. (2006)
26	Commercial hydrolases (Amano A, crude lipase preparation from *A. niger*)	Stander et al. (2000)

Source: Varga et al. 2010

issues due to their carcinogenic, hepatotoxic immunosuppressive, mutagenic, nephrotoxic, and teratogenic effects. Therefore, management and control strategies have been developed for ochratoxins, including detection, identification, and quantification techniques. Current immunoassays and biosensor methods have the advantage of detecting toxin contamination in food/feed on site. Furthermore, today, decontamination techniques involving physical, chemical, physico-chemical, enzymatic, microbial, and gene modification tools have found wide application in the reduction of levels of ochratoxin contamination in food/feed. The efficient reduction of ochratoxin in food/feed is important to ensure the safety and security of humans and animals.

REFERENCES

Abbas A, Valez H, Dobson ADW (2009) Analysis of the effect of nutritional factors on OTA and OTB biosynthesis and polyketide synthase gene expression in *Aspergillus ochraceus*. *International Journal of Food Microbiology*, 135: 22–27.

Abrunhosa L, Santos L, Venancio A (2006) Degradation of ochratoxin A by proteases and by a crude enzyme of Aspergillus niger. *Food Biotechnology*, 20: 231–242.

Abrunhosa L, Serra R, Venancio A (2002) Biodegradation of ochratoxin A by fungi isolated from grapes. *Journal of Agricultural and Food Chemist*, 50: 7493–7496.

Abrunhosa L, Venancio A (2007) Isolation and purification of an enzyme hydrolyzing ochratoxin A from *Aspergillus niger. Biotechnology Letter*, 29: 1909–1914.

Alhamoud Y, Yang D, Kenston SSF, Liu G, Liu L, Zhou H, Ahmed F, Zhao J (2019) Advances in biosensors for the detection of ochratoxin A: Bio-receptors, nanomaterials, and their applications. *Biosensers and Bioelectronics*, 141: 111418.

Al-Jaal B, Salama S, Al-Qasmi N, Jaganjac M (2019) Mycotoxin contamination of food and feed in the Gulf Cooperation Council countries and its detection. *Toxicon*, 171: 43–50.

Alvindia DG, de Guzman MF (2016) Survey of Philippine coffee beans for the presence of ochratoxigenic fungi. *Mycotoxin Research*, 32: 61–67.

Angioni A, Caboni P, Garau A, Farris A, Orro D, Budroni M, Cabras P (2007) In vitro interaction between ochratoxin A and different strains of Saccharomyces cerevisiae and Kloeckera apiculata. *Journal of Agricultural and Food Chemist*, 55: 2043–2048.

Antep HM, Merdivan M (2012) Determination of ochratoxin A in grape wines after dispersive liquid–liquid microextraction using high performance thin layer and liquid chromatography-fluorescence detection. *Journal of Biological Chemistry*, 40: 155–163.

Atroshi F, Biese I, Saloniemi H, Ali-Vehmas T, Saari S, Rizzo A, Veijalainen P (2000) Significance of apoptosis and its relationship to antioxidants after ochratoxin A administration in mice. *Journal of Pharmacy & Pharmaceutical Sciences*, 3: 281–291.

Badea M, Floroian L, Restani P, Codruta S, Cobzac A, Moga M (2016) Ochratoxin A detection on antibody immobilized on BSA-functionalized gold electrodes. *PLoS One*, 11: e0160021.

Bailey CA, Gibson RM, Kubena LF, Huff WE, Harvey RB (1989) Ochratoxin A and dietary protein. 2. Effects on hematology and various clinical chemistry measurements. *Poultry Science*, 68: 1664–1671.

Baxter ED, Slaiding IR, Kelly B (2001) Behavior of ochratoxin A in brewing. *Journal of the American Society of Brewing Chemists*, 59: 98–100.

Bejaoui H, Mathieu F, Taillandier P, Lebrihi A (2004) Ochratoxin A removal in synthetic and natural grape juices by selected oenological Saccharomyces strains. *Journal of Applied Microbiology*, 97: 1038–1044.

Bejaoui H, Mathieu F, Taillandier P, Lebrihi A (2006) Biodegradation of ochratoxin A by Aspergillus section Nigri species isolated from French grapes: A potential means of ochratoxin A decontamination in grape juices and musts. *FEMS Microbiology Letter*, 255: 203–208.

Belajova E, Rauova D, Dasko L (2007) Retention of ochratoxin A and fumonisin B1 and B2 from beer on solid surfaces: Comparison of efficiency of adsorbents with different origin. *European Food Research and Technology*, 224: 301–308.

Bhat R, Rai RV, Karim AA (2010) Mycotoxins in food and feed: Present status and future concerns. *Comprehensive Review in Food Science and Food Safety*, 9: 57–81.

Blanc M, Pittet A, Munoz-Boksz R, Viani R (1998) Behavior of ochratoxin A during green coffee roasting and soluble coffee manufacture. *Journal of Agricultural and Food Chemistry*, 46: 673–675.

Bohm J, Grajewski J, Asperger H, Cecon B, Rabus B, Razzazi E (2000) Study on biodegradation of some A- and B-trichothecenes and ochratoxin A by use of probiotic microorganisms. *Mycotoxin Research*, 16A: 70–74.

Bouaziz C, Sharaf El Dein O, El Golli E, Abid-Essefi S, Brenner C, Lemaire C, Bacha H (2008) Different apoptotic pathways induced by zearalenone, T-2 toxin and ochratoxin A in human hepatoma cells. *Toxicology*, 254: 19–28.

Bucheli P, Taniwaki MH (2002) Research on the origin, and on the impact of post-harvest handling of manufacturing on the presence of ochratoxin A in coffee. *Food Additives & Contaminants*, 19: 655–665.

Bui-Klimke TR, Wu F (2015) Ochratoxin A and human health risk: A review of the evidence. *Critical Review in Food Science and Nutrition*, 55: 1860–1869.

Bullerman LB, Bianchini A (2007) Stability of mycotoxins during food processing. *International Journal of Food Microbiology*, 119: 140–146.

Caputo D, De Cesare G, Fanelli C, et al. (2007) Innovative detection system of ochratoxin A by thin film photodiodes. *Sensors*, 7: 1317–1322.

Casas-Junco PP, Ragazzo-Sanchez JA, Ascencio-Valle FJ, Calderon-Santoyo M (2018) Determination of potentially mycotoxigenic fungi in coffee (*Coffea arabica* L.) from Nayarit. *Food Science and Biotechnology*, 27: 891–898.

Castellari M, Versari A, Fabiani A, Parpinello GP Galassi S (2001) Removal of ochratoxin A in red wines by means of adsorption treatments with commercial fining agents. *Journal of Agricultural and Food Chemistry*, 49: 3917–3921.

Cavin C, Delatour T, Marin-Kuan M, Fenaille F, Holzhauser D, Guignard G, Bezencon C, Piguet D, Parisod V, Richoz-Payot J (2009) Ochratoxin A mediated DNA and protein damage: Roles of nitrosative and oxidative stresses. *Toxicological Sciences*, 110: 84–94.

Chen W, Li C, Zhang B, Zhou Z, Shen Y, Liao X, Yang J, Wang Y, Li X, Li Y, Shen XL (2018) Advances in biodetoxification of ochratoxin A—A review of the past five decades. *Frontier in Microbiology*, 26. https://doi.org/10.3389/fmicb.2018.01386.

Costa JG, Saraiva N, Guerreiro PS, Louro H, Silva MJ, Miranda JP, Castro M, Batinic-Haberle I, Fernandes AS, Oliveira NG (2016) Ochratoxin A-induced cytotoxicity, genotoxicity and reactive oxygen species in kidney cells: An integrative approach of complementary endpoints. *Food and Chemical Toxicology*, 87: 65–76.

Creppy EE, Roschenthaler R, Dirheimer G (1984) Inhibition of protein synthesis in mice by ochratoxin A and its prevention by phenylalanine. *Food and Chemical Toxicology*, 22: 883–886.

De Felice DV, Solfrizzo M, De Curtis F, Lima G, Visconti A, Castoria R (2008) Strains of Aureobasidium pullulans can lower ochratoxin A contamination in wine grapes. *Phytopathology*, 98: 1261–1270.

Denli M, Blandon JC, Guynot ME, Salado S, Perez JF (2008) Efficacy of a new ochratoxin-binding agent (OcraTox) to counteract the deleterious effect of ochratoxin A in laying hens. *Poultry Science*, 87: 2266–2272.

Denli M, Perez J (2010) Ochratoxins in feed, a risk for animal and human health: Control strategies. *Toxins*, 2: 1065–1077.

Duarte S, Lino C, Pena A (2012) Food safety implications of ochratoxin A in animal-derived food products. *Veterinary Journal*, 192: 286–292.

Duarte S, Pena A, Lino CM (2010) A review on ochratoxin A occurrence and effects of processing of cereal and cereal derived food products. *Food Microbiology*, 27: 187–198.

Dumeau F, Trione D (2000) Influence of different treatments on the concentration of ochratoxin A in red wines. *Revue Oenology of France*, 27: 37–38.

EC (2002) Assessment of dietary intake of ochratoxin A by the population of EU Member States. *Report of the Scientific Cooperation, Task 3.2.7; Directorate-General Health and Consumer Protection*, European Commission: Rome, Italy.

EFSA (2006) Opinion of the scientific panel on contaminants in the food chain on a request from the commission related to ochratoxin A in food. *EFSA Journal*, 365: 1–56.

Engelhardt G (2002) Degradation of ochratoxin A and B by the white rot fungus *Pleurotus ostreatus*. *Mycotoxin Research*, 18: 37–43.

Ferrara M, Perrone G, Gambacorta L, Epifani F, Solfrizzo M, Gallo A (2016) Identification of a halogenase involved in the biosynthesis of ochratoxin A in *Aspergillus carbonarius*. *Applied and Environmental Microbiology*, 82: 56315641.

Ferraz MBM, Farah A, Iamanaka BT, Perrone D, Copetti MV, Marques VX, Vitali AA, Taniwaki MH (2010) Kinetics of ochratoxin A destruction during coffee roasting. *Food Control*, 21: 872–877.

Frusvad JC, Thrane U, Samson RA, Pitt JI (2006) Important mycotoxins and the fungi which produce them. In *Advances in Food Mycology*. Springer, 3–31.

Fuchs S, Sontag G, Stidl R, Ehrlich V, Kundi M, Knasmuller S (2008) Detoxification of patulin and ochratoxin A, two abundant mycotoxins, by lactic acid bacteria. *Food Chemistry and Toxicology*, 46: 1398–1407.

Gallaz L, Stalder R (1976) Ochratoxin A in coffee. *Journal of Microbiology Technology and Chemistry*, 4: 147–149.

Gallo A, Ferrara M, Perrone G (2017) Recent advances on the molecular aspects of ochratoxin A biosynthesis. *Current Opinion in Food Science*, 17: 49–56.

Gambuti A, Strollo D, Genovese A, Ugliano M, Ritieni A, Moio L (2005) Influence of enological practices on ochratoxin A concentration in wine. *The American Journal of Enology and Viticulture*, 56: 155–162.

Geisen R (2004) Molecular monitoring of environmental conditions influencing the induction of ochratoxin A biosynthesis gene in *Penicillium nordicum*. *Molecular Nutrition & Food Research*, 48: 532–540.

Gekle M, Schwerdt G, Freudinger R, Mildenberger S, Wilflingseder D, Pollack V, Dander M, Schramek H (2000) Ochratoxin A induces JNK activation and apoptosis in MDCK-C7 cells at nanomolar concentrations. *Journal of Pharmacology and Experimental Therapeutics*, 283: 1460–1468.

Gil-Serna J, Garcia-Diaz M, Gonzalez-Jaen MT, Vazquez C, Patino B (2018) Description of an orthologus cluster of ochratoxin A biosynthetic gene in *Aspergillus* and *Penicillum* species. A comparative analysis. *International Journal of Food Microbiology*, 268: 3543.

Gil-Serna J, Vazquez C, Ratino B (2020) Genetic regulation of aflatoxin, ochratoxin A, trichothecene and fumonisin biosynthesis: A review. *International Microbiology*, 23: 89–96.

Goryacheva IY, De-Saeger S, NesterenkoIS, et al. (2007) Rapid all-in-one three-step immunoassay for non-instrumental detection of ochratoxin A in high-coloured herbs and spices. *Talanta*, 72: 1230–1234.

Gupta RC, Srivastava A, Lall R (2018) *Ochratoxins and Citrinin, Veterinary Toxicology*. Elsevier, 1019–1027.

Heussner AH, Bingle LEH (2015) Comparative ochratoxin toxicity: A review of the available data. *Toxins*, 7: 4253–4282.

Holzapfel W, Brost I, Farber P, Geisen R, Bresch H, Jany KD, Mengu M, Jakobsen M, Steyn PS, Teniola D, Addo P (2002) *Actinomycetes for Breaking Down Aflatoxin B1, Ochratoxin A, and/or Zearalenon*. Patent number WO02/099142 A3.

Hou Y, Long N, Jia B, Liao X, Yang M, Fu L, Zhou L, Sheng P, Kong W (2022) Development of a label free electrochemical aptasensor for ultrasensitive detection of ochratoxin. *Food Control*, 135: 108833.

Huang B, Xiao H, Zhang J, et al. (2009) Dual-label time-resolved fluoroimmunoassay for simultaneous detection of aflatoxin A and ochratoxin A. *Archives of Toxicology*, 83: 619–624.

Huertas-Perez JF, Arroyo-Manzanares N, Garcia-Campana AM and Gamiz-Gracia L (2017) Solid phase extraction as sample treatment for the determination of ochratoxin A in food: A review. *Critical Reviews in Food Science and Nutrition*, 57: 3405–3420.

Huff WE, Hamilton PB (1979) Mycotoxins—their biosynthesis in fungi: Ochratoxins—metabolites of combined pathways. *Journal of Food Protection*, 42: 815–820.

Huff WE, Kubena LF, Harvey RB (1988) Progression of ochratoxicosis in broiler chickens. *Poultry Science*, 67: 1139–1146.

Hult K, Teiling A, Gatenbeck S (1976) Degradation of ochratoxin A by a ruminant. *Applied Environmental Microbiology*, 32: 443–444.

Hwang CA, Draughon FA (1994) Degradation of ochratoxin A by Acinetobacter calcoaceticus. *Journal of Food Protection*, 57: 410–414.

IARC (1993) *IARC Monographs on the Evaluation of Carcinogenic Risks to Humans: Some Naturally Occurring Substances: Food Items and Constituents, Heterocyclic Aromatic Amines and Mycotoxins*, vol. 56. International Agency for Research on Cancer, UK.

Jung S, Choe B, Shin G, Kim J, Chae Y (2012) Analysis of roasted and ground grains on the Seoul (Korea) market for their contaminants of aflatoxins, ochratoxin A and *Fusarium* toxins by LC-MS/MS. *World Academy of Science Engineering Technology*, 6: 12–23.

Kamp HG, Eisenbrand G, Schlatter J, Eurth K, Janzowski C (2005) Ochratoxin A: Induction of (oxidative) DNA damage, cytotoxicity and apoptosis in mammalian cell lines and primary cells. *Toxicology*, 206: 413–425.

Khan S, Martin M, Bartsch H, Rahimtula AD (1989) Perturbation of liver microsomal calcium homeostasis by ochratoxin A. *Biochemical Pharmacology*, 38: 67–72.

Khaneghah AM, Fakhri Y, Raeisi S, Armoon B, Sant'Ana AS (2018a) Prevalence and concentration of ochratoxin A, zearaleone, deoxynivalenol, and total aflatoxin in cereal-based products: A systematic review and meta-analysis. *Food and Chemical Toxicology*, 118: 830–848.

Khaneghah AM, Fakhri Y, Sant'Ana AS (2018b) Impact of unit operations during processing of cereal-based products on the levels of deoxynivalenol, total aflatoxin, ochratoxin A, and zearalenone: A systematic review and meta-analysis. *Food Chemistry*, 268: 611–624.

Khatee A, Sohrabi H, Arbabzadeh O, Khaaki P, Majidi MR (2021) Frontiers in conventional and nanomaterials based electrochemical sensing and biosensing approaches for ochratoxin A analysis in foodstuffs: A review. *Food and Chemical Toxicology*, 149: 112030.

Kiessling KH, Pettersson H, Sandholm K, Olsen M (1984) Metabolism of aflatoxin, ochratoxin, zearalenon, and three trichothecenes by intact rumen fluid, rumen protozoa, and rumen bacteria. *Applied Environmental Microbiology*, 47: 1070–1073.

Kirlangiç IA, Kara P, Ertas FN (2021) Development of transition metal oxide film coated platforms for aptamer based electrochemical detection of ochratoxin A. *Journal of the Electrochemical Society*, 168: 057516.

Klaric MS, Rumora L, Ljubanovic D, Pepeljnjak S (2008) Cytotoxicity and apoptosis induced by fumonisin B1, beauvericin and ochratoxin A in porcine kidney PK15 cells: Effects of individual and combined treatment. *Archives of Toxicology*, 82: 247–255.

Kotinagu K, Mohanamba T, Rathna Kumari N (2015) *Assessment of Aflatoxin B1 in Livestock Feed and Feed Ingredients by High-Performance Thin Layer Chromatography.* Veterinary World. EISSN: 2231–0916.

Kozakiewicz Z, Highley E, Johnson GI (1996) *Occurrence and Significance of Storage Fungi and Associated Mycotoxins in Rice and Cereal Grains*, vol. 37. Eds. Highley E, Johnson GI. Springer, Technical Report, 18–26.

Kumar P, Mahato DK, Kamle M, Mohanta TK, Kang SG (2017) Aflatoxins: A global concern for food safety, human health and their management. *Frontiers of Microbiology*, 7: 2170.

Kumar P, Mahto DK, Sharma B, Borah R, Haque S, Mahmud MMC, Shah AK, Rawal D, Bora H, Bui S (2020) Ochratoxins in food and feed: Occurrence and its impact on human health and management strategies. *Toxicon*, 187: 151–162.

Kupski L, Badiale-Furlong E (2015) Principal components analysis: An innovative approach to establish interferences in ochratoxin A detection. *Food Chemistry*, 177: 354–360.

Kurtbay HM, Bekci Z, Merdivan M, Yurdakoc K (2008) Reduction of ochratoxin a levels in red wine by bentonite, modified bentonites, and chitosan. *Journal of Agricultural and Food Chemistry*, 56: 2541–2545.

Kyei-Baffour VO, Kongor JE, Anyebuno G, Budu AS, Firibu SK, Afoakwa EO (2021) A validated HPLC-FLD method for the determination of mycotoxin levels in sun dried fermented cocoa beans: Effect of cola nut extract and powder. *LWT*, 148: 111790.

La Pera L, Avellone G, Lo Turco V, Di Bella G, Agozzino P, Dugo G (2008) Influence of roasting and different brewing processes on the ochratoxin A content in coffee determined by high-performance liquid chromatography-fluorescence detection (HPLC-FLD). *Food Additives and Contaminants,* Part A, 25: 1257–1263.

Lerda D (2015) Ochratoxin A (OTA) and public health. In Porter D (Ed.), *Ochratoxins*, 1st ed. Nova Scientific Publisher, Inc.

Levi CP, Trenk HL, Mohr HK (1974) Study of the occurrence of ochratoxin A in green coffee beans. *Journal of the Association of Official Analytical Chemists International*, 57: 866–870.

Li P, Zhou Q, Wang T, Zhou H, Zhang W, Ding X, Zhang Z, Chang PK, Zhang Q (2016). Development of an enzyme-linked immunosorbent assay method specific for the detection of g-group aflatoxins. *Toxins*, 8: 5. https://doi.org/10.3390/toxins8010005.

Li X, Falcone N, Hossain MN, Kraatz HB, Chen X, Huang H (2021c) Development of a novel label free impedimetric electrochemical sensor based on hydrogel/chitosan for the detection of ochratoxin. *Talanta*, 226: 122183.

Li X, Ma W, Ma X, Zhang Q, Li H (2021a) Recent progress in determination of ochratoxin A in foods by chromatography and mass spectrometer method. *Critical Review in Food Science and Nutrition*, 1: 1–7.

Li X, Ma W, Ma Z, Zhang Q, Li H (2021b) The occurrence and contamination level of ochratoxin A in plant and animal derived food commodities. *Molecules*, 26: 6928.

Liang R, Shen XL, Zhang B, Li Y, Xu W, Zhao C, Luo Y, Huang K (2015) Apoptosis signal-regulating kinase 1 promotes ochratoxin A-induced renal cytotoxicity. *Scientific Report*, 5: 8087.

Limonciel A, Jennings P (2014) A review of the evidence that ochratoxin A is an *Nrf*2 inhibitor: Implications for nephrotoxicity and renal carcinogenicity. *Toxins*, 6: 371–379.

Madhyastha SM, Marquardt RR, Frohlich AA, Platford G, Abramson D (1990) Effects of different cereal and oilseed substrates on the growth and production of toxins by *Aspergillus alutaceus* and *Penicillium verrucosum*. *Journal of Agriculture and Food Chemistry*, 38: 1506–1510.

Malagutti L, Zannotti M, Scampini A, Sciaraffia F (2005) Effects of ochratoxin A on heavy pig production. *Animal Research*, 54: 179–184.

Malir F, Ostry V, Pfohl-Leszkowicz A, Malir J, Toman J (2016) Ochratoxin A: 50 years of research. *Toxins*, 4: 8.

Manderville RA (2005) A case for the genotoxicity of ochratoxin A by bioactivation and covalent DNA adduction. *Chemical Research in Toxicology*, 18: 1091–1097.

Mantle PG, Faucet-Marquis V, Manderville RA, Squillaci B, Pfohl-Leszkowicz A (2010) Structures of covalent adducts between DNA and ochratoxin A: A new factor in debate about genotoxicity and human risk assessment. *Chemical Research in Toxicology*, 23: 89–98.

Maragos CM (2006) Fluorescence polarization for mycotoxin determination. *Mycotoxin Research*, 2: 96–99.

Marin-Kuan M, Ehrlich V, Delatour T, Cavin C, Schilter B (2011) Evidence for a role of oxidative stress in the carcinogenicity of ochratoxin A. *Journal of Toxicology*, 2011: 645361.

Meki AR, Hussein AA (2001) Melatonin reduces oxidative stress induced by ochratoxin A in rat liver and kidney. *Comparative Biochemistry and Physiology Part C: Toxicology & Pharmacology*, 130: 305–313.

Meulenberg, E. P. (2012) Immunochemical methods for ochratoxin a detection: A review. *Toxins*, 4: 244–266. https://doi.org/10.3390/toxins4040244.

Micco C, Grossi M, Miraglia M, Brera CA (1989) Study of the contamination by ochratoxin A of green and roasted coffee beans. *Food Additives and Contaminants*, 6: 333–339.

Miljkovic A, Pfohl-Leszkowicz A, Dobrota M, Mantle PG (2003) Comparative responses to mode of oral administration and dose of ochratoxin A or nephrotoxic extract of *Penicillium polonicum* in rats. *Experimental and Toxicologic Pathology*, 54: 305–312.

Min WK, Kweon DH, Park K, Park YC, Seo JH (2011) Characterisation of monoclonal antibody against aflatoxin B1 produced in hybridoma 2C12 and its single-chain variable fragment expressed in recombinant *Escherichia coli*. *Food Chemistry*, 126: 1316–1323.

Molnar O, Schatzmayr G, Fuchs E, Prillinger H (2204) Trichosporon mycotoxinivorans sp. nov., a new yeast species useful in biological detoxification of various mycotoxins. *Systematic Applied Microbiology*, 27: 661–671.

Muller HM, Lerch C, Muller K, Eggert W (1998) Kinetic profiles of ochratoxin A and ochratoxin α during in vitro incubation in buffered forestomach and abomasal contents from cows. *Natural Toxins*, 6: 251–258.

Myndrul V, Viter R, Savchuk N, Erts D, Jevdokimovs D, Silamikelis V, Smyntyna V, Ramanavicius A, Latsunskyi I (2018) Porous silicon based photoluminescence immunosensor for rapid and highly-sensitive detection of ochratoxin A. *Biosensor and Bioelectic*, 102: 661–667.

Nehad EA, Farag MM, Kawther MS, Abdel-Samed AK, Naguib K (2005) Stability of ochratoxin A (OTA) during processing and decaffeination in commercial roasted coffee beans. *Food Additives and Contaminants*, 22: 761–767.

Novak I (2012) Mitophagy: A complex mechanism of mitochondrial removal. *Antioxidants & Redox Signaling*, 17: 794–802.

Olivares- Marín M, Del Prete V, Garcia-Moruno E, Fernandez-Gonzalez C, Macias-Garcia A, Gomez-Serrano A (2009) The development of an activated carbon from cherry stones and its use in the removal of ochratoxin A from red wine. *Food Control*, 20: 298–303.

Otteneder H, Majerus P (2000) Occurrence of ochratoxin A (OTA) in wines: Influence of the type of wine and its geographical origin. *Food Additives and Contaminants*, 17: 793–798.

Ozcan Z, Gul G, Yaman I (2015) Ochratoxin A activates opposing c-MET/PI3K/AKt and MAPK/ERK 1–2 pathways in human proximal tubule HK-2 cells. *Archives of Toxicology*, 89: 1313–1327.

Ozpinar H, Bilal T, Abas I, Kutay C (2002) Degradation of ochratoxin A in rumen fluid in vitro. *Medical Biology*, 9: 66–69.

Pakshir K, Dehghani A, Nouraei H, Zareshahrabadi Z, Zomorodian K (2021) Evaluation of fungal contamination and ochratoxin A detection in different types of coffee by HPLC based method. *Journal of Clinical Laboratory Analysis*, 35: e24001.

Paterson RRM, Venancio A, Lima N (2004) Solutions to *Penicillium* taxonomy crucial to mycotoxin research and health. *Research in Microbiology*, 155: 7–13.

Peraica M, Domijan AM, Jurjevic Z, Cvjetkovic B (2002) Prevention of exposure to myco-toxins from food and feed. *Archives of Industrial Hygiene and Toxicology*, 53(3): 229–237.

Perez de Obanos A, Gonzalez-Penas E, Lopez de Cerain A (2005) Influence of roasting and brew preparation on the ochratoxin A content in coffee infusion. *Food Additives and Contaminants*, 22: 463–471.

Petchkongkaew A, Taillandier P, Gasaluck P, Lebrihi A (2008) Isolation of Bacillus spp. from Thai fermented soybean (Thua-nao): Screening for aflatoxin B1 and ochra-toxin A detoxification. *Journal of Applied Microbiology*, 104: 1495–1502.

Peteri Z, Teren J, Vagvolgyi C, Varga J (2007) Ochratoxin degradation and adsorption caused by astaxanthin-producing yeasts. *Food Microbiology*, 24: 205–210.

Petrik J, Zanic-Grubisic T, Barisic K, Pepeljnjak S, Radic B, Ferencic Z, Cepelak I (2003) Apoptosis and oxidative stress induced by ochratoxin A in rat kidney. *Archives of Toxicology*, 77: 685–693.

Pfohl-Leszkowicz A, Chakor K, Creppy EE, Dirheimer G (1991) DNA adduct formation in mice treated with ochratoxin A. *IARC Scientific Publications*, 115: 245–253.

Pfohl-Leszkowicz A, Manderville RA (2007) Ochratoxin A: An overview on toxicity and carcinogenicity in animals and humans. *Molecular Nutrition & Food Research*, 51: 61–99.

Pinotti L, Ottoboni M, Giromini C, Dello'Orto V, Cheli F (2016) Mycotoxin contamination in the EU feed supply chain: A focus on cereal byproducts. *Toxins (Basel)*, 8: 45.

Piotrowska M, Zakowska Z (2000) The biodegradation of ochratoxin A in food products by lactic acid bacteria and baker's yeast. In Bielecki S, Tramper J, Polak J (Eds.), *Food Biotechnology*. Elsevier, 307–310.

Pitout MJ (1969) The hydrolysis of ochratoxin A by some proteolytic enzymes. *Biochemical Pharmacology*, 18: 485–491.

Princci SS, Ertekin O, Laguna DE, Ozen FS, Ozturk ZZ, Ozturk S (2018) Label-Free QCM immunosensor for the detection of ochratoxin A. *Sensors*, 18: 1161. https://doi.org/10.3390/s18041161.

Quesada-Gonzalez D, Merkoçi A (2015) Nanoparticle-based lateral flow biosensors. *Biosensors Bioelectronics*, 73: 47–63.

Quintela S, Villaran MC, de Armentia IL, Elejalde E (2013) Ochratoxin A removal in wine: A review. *Food Control*, 30: 439–445.

Rahi S, Choudhari P, Ghormade M (2019) Aflatoxin and ochratoxin A detection traditional and current methods. In Satyanarayana T, Deshmukh SK, Deshpande MV (Eds.), *Advancing Frontiers in Mycology and Mycotechnology*. Springer. http://doi.org/10.1007/978-981-13-9349-5_15.

Ren M, Xu H, Huang X, Kuang M, Xiong Y, Xu H, Xu Y, Chen H, Wang A (2014) Immunochromatographic assay for ultrasensitive detection of aflatoxin B1 in maize by highly luminescent quantum dot beads. *ACS Applied Materials and Interfaces*, 6: 14215–14222.

Romani S, Pinnavaia GG, Dalla Rosa M (2003) Influence of roasting levels on ochratoxin A content in coffee. *Journal of Agricultural and Food Chemistry*, 51: 5168–5171.

Santos VO, Pelegrini PB, Mulinari F, Lacerda AF, Moura, RS, Cardoso LPV, Buhrer-Sekula S, Miller RNG, Grossi-de-Sa MF (2017) Development and validation of a novel lateral flow immunoassay device for detection of aflatoxins in soy-based foods. *Analytical Methods*, 9: 2715–2722.

Savino M, Limosani P Garcia-Moruno E (2007) Reduction of ochratoxin A contamination in red wines by oak wood fragments. *American Journal of Enology and Viticulture*, 58: 97–101.

Schaaf GJ, Nijmeijer SM, Maas RF, Roestenberg P, de Groene EM, Fink-Gremmels J (2002) The role of oxidative stress in the ochratoxin A-mediated toxicity in proximal tubular cells. *Biochimica et Biophysica Acta*, 1588: 149–158.

Schatzmayr G, Heidler D, Fuchs E, Binder EM (2009) *Microorganisms for Biological Detoxification of Mycotoxins, Namely Ochratoxins and/or Zearalenons, as Well as Method and Use thereof.* US Patent number 2009/0098244 A1.

Schatzmayr G, Heidler D, Fuchs E, Loibner AP, Braun R, Binder EM (2002) Evidence of ochratoxin A-detoxification ctivity of rumen fluid, intestinal fluid and soil samples as well as isolation of relevant microorganisms from these environments. *Mycotoxin Research*, 18(Suppl. 2): 183–187.

Schatzmayr G, Heidler D, Fuchs E, Nitsch S, Mohnl M, Tiubel M, Loibner AP, Braun R, Binder EM (2003) Investigation of different yeast strains for the detoxification of ochratoxin A. *Mycotoxin Research*, 19: 124–128.

Schatzmayr G, Zehner F, Taubel M, Schatzmayr D, Klimitsch A, Loibner AP, Binder EM (2006) Microbiologicals for deactivating mycotoxins. *Molecular Nutrition and Food Research*, 50: 543–551.

Schneider E, Usleber E, Martlbauer E (1995a) Rapid detection of fumonisin B1 in corn-based food by competitive direct dipstick enzyme immunoassay/enzyme-linked immunofiltration assay with integrated negative control reaction. *Journal of Agriculture and Food Chemistry*, 43: 2548–2552.

Schneider E, Usleber E, Martlbauer E, Terplan G (1995b) Multimycotoxin dipstick enzyme immunoassay applied to wheat. *Food Additives and Contaminants*, 12: 387–393.

Schramek H, Ellflingseder D, Pollack V, Freudinger R, Mildenberger S, Gekle M (1997) Ochratoxin A-induced stimulation of extracellular signal-regulated kinases ½ is associated with madin-darby canine kidney-C7 cell dedifferentiation. *Journal of Pharmacology and Experimental Therapeutics*, 283: 1460–1468.

Scott PM, Kanhere SR, Lawrence GA, Daley EF, Farber JM (1995) Fermentation of wort containing added ochratoxin A and fumonisins B1 and B2. *Food Additives & Contaminants*, 12: 31–40.

Scott PM, Kanhere SR, Lawrence GA, Daley EF, Farber JM (1995) Fermentation of wort containing added ochratoxin A and fumonisins B1 and B2. *Food Additives and Contaminants*, 12: 31–40.

Scudamore KA, Banks J, MacDonald SJ (2003) Fate of ochratoxin A in the processing of whole wheat grains during milling and bread production. *Food Additives & Contaminants*, 20: 1153–1163.

Seegers JC, Bohmer LH, Kruger MC, Lottering ML, de Kock MA (1994) A comparative study of ochratoxin A-induced apoptosis in hamster kidney and HeLa cells. *Toxicology and Applied Pharmacology*, 129: 1–11.

Serra R, Mendonca C, Venancio A (2006) Fungi and ochratoxin A detected in healthy grapes for wine production. *Letters in Applied Microbiology*, 42: 42–47.

Shen XL, Zhang B, Liang R, Cheng WH, Xu W, Luo Y, Zhao C, Huang K (2014) Central role of Nix in the autophagic response to ochratoxin A. *Food and Chemical Toxicology*, 69: 202–209.

Singh AK, Dhiman TK, Kaushik A, Solanki PR (2021) Bio-Active free direct optical sensing of aflatoxin B1 and ochratoxin A using a manganese oxide nano-system. *Frontier*, 1: 1–15.

Skrinjar M, Rasic JL, Stojicic V (1996) Lowering ochratoxin A level in milk by yoghurt bacteria and bifidobacteria. *Folia Microbiology*, 41: 26–28.

Speijers GJA, Van Egmond HP (1993) Worldwide ochratoxin A levels in food and feeds. In Castegnaro M, Creppy E, Dirheimer G (Eds.), *Human Ochratoxicosis and Its Pathologies*. Hohn Libbey Eurotext Ltd. Colloques Institut National De La Sante Et De La Recherche Medicale Colloques Et Seminaires, 85–100.

Spinella K, Mosiello L, Palleschi G, Vitali F (2013) Development of a QCM (quartz crystal microbalance) biosensor to the detection of aflatoxin B1. *Open Journal of Applied Biosensors*, 2: 112–119.

Stander MA, Bornscheurer UT, Henke E, Steyn PS (2000) Screening of commercial hydrolases for the degradation of ochratoxin A. *Journal of Agricultural and Food Chemist*, 48: 5736–5739.

Stander MA, Steyn PS, Van der Westhuizen FH, Payne BE (2001) A kinetic study into the hydrolysis of the ochratoxins and analogs by carboxypeptidase A. *Chemical Research Toxicology*, 14: 302–304.

Studer-Rohr I, Dietrich DR, Schlatter J, Schlatter C (1994) Ochratoxin A in coffee: New evidence and toxicology. *Lebensmittel-Wissenschaft and Technology*, 27: 435–441.

Sulyok M, Berthiller F, Krska R, Schuhmacher R (2006) Development and validation of a liquid chromatography/tandem mass spectrometric method for the determination of 39 mycotoxins in wheat and maize. *Rapid Communication and Mass Spectrometry*, 20: 2649–2659.

Sun D, Gu X, Li JG, Yao T, Dong YC (2015) Quality evaluation of five commercial enzyme linked immunosorbent assay kits for detecting aflatoxin B1 in feedstuffs. *Asian Australas Journal of Animal Sciences*, 28: 691–696.

Susca A, Proctor RH, Morelli M, Haidukowski M, Gallo A, Logrieco AF, Moretti A (2016) Variation in fumonisin and ochratoxin production associated with differences in biosynthetic gene content in *Aspergillus niger* and *A. welwitschiae* isolates from multiple crop and geographic origin. *Frontiers of Microbiology*, 7: 1412.

Taihua L, Jeon KS, Suh YD, KimMG (2011) A label-free, direct and competitive FRET immunoassay for ochratoxin A base don intrinsic fluorescence of an antigen and antibody complex. *Chemical Communications*, 47: 9098–9100.

Tsubouchi H, Yamamoto K, Hisada K, Sakabe Y, Udagawa S (1987) Effect of roasting on ochratoxin A level in green coffee beans inoculated with *Aspergillus ochraceus*. *Mycopathologia*, 97: 111–115.

Urbano GR, de Freitas Leitao MF, Vicentini MC, Taniwaki MH (2004) Preliminary studies on the destruction of ochratoxin A in coffee during roasting. *Food Science and Technology International*, 10: 45–49.

Van der Gaag B, Spath S, Dietrich H, et al. (2003) Biosensors and multiple mycotoxin analysis. *Food Contaminants*, 14: 251–254.

Van der Merwe K, Steyn P, Fourie L, Scott D, Theron J (1965) Ochratoxin A, a toxic metabolite produced by *Aspergillus ochraceus* Wilh. *Nature*, 205: 1112–1113.

Van der Stegen GHD, Essens PJM, van der Lijn J (2001) Effect of roasting conditions on reduction of ochratoxin A in coffee. *Journal of Agricultural and Food Chemistry*, 49: 4713–4715.

Var Isil, Kabak B, Erginkaya Z (2008) Reduction in ochratoxin A levels in white wine, following treatment with activated carbon and sodium bentonite. *Food Control*, 19(6): 592–598.

Varga J, Kocsube S, Peteri Z, Vagvolgyi C, Toth B (2010) Chemical, physical and biological approaches to prevent ochratoxin induced, toxicoses in humans and animals. *Toxins*, 2: 1718–1750.

Varga J, Peteri Z, Tabori K, Teren J, Vagvolgyi C (2005) Degradation of ochratoxin A and other mycotoxins by Rhizopus isolates. *International Journal of Food Microbiology*, 99: 321–328.

Varga J, Rigo K, Teren J (2000) Degradation of ochratoxin A by Aspergillus species. *International Journal of Food Microbiology*, 59: 1–7.

Vega FE, Posada F, Peterson SW, Gianfagan TJ, Chaves F (2006) *Penicillium* species endophytic in coffee plants and ochratoxin A production. *Mycologia*, 98: 31–42.

Vettorazzi A, Gonzalez-Penas E, Troconiz IF, Arbillaga L, Corcuera LA, Gil AG, de Cerain AL (2009) A different kinetic profile of ochratoxin A in mature male rats. *Food and Chemical Toxicology*, 47: 1921–1927.

Visconti A, Perrone G, Cozzi G, Solfrizzo M (2008) Managing ochratoxin A risk in the grape-wine food chain. *Food Additives and Contaminants*, 25: 193–202.

Vitera R, Savchuk M, Iatsunskyic I, Pietralik Z, Starodub N, Shpyrka N, Ramanaviciene A, Ramanavicius A (2018) Analytical, thermodynamical and kinetic characteristics of photoluminescence immunosensor for the determination of ochratoxin A. *Biosensors and Bioelectronics*, 99: 237–243.

Wegst W, Lingens F (1983) Bacterial degradation of ochratoxin A. *FEMS Microbiology Letter*, 17: 341–344.

Wei D, Wu X, Xu J, Dong F, Liu X, Zheng Y, Ji M (2018) Determination of ochratoxin A contamination in grapes, processed grape products and animal-derived products using ultra-performance liquid chromatography-tandem mass spectroscopy system. *Scientific Report*, 8: 2051.

Welke JE, Hoeltz M, Dottori HA, Noll IB (2010) Determination of ochratoxin A in wine by high performance thin-layer chromatography using charged coupled device. *Journal of Brazilian Chemical Society*, 21: 441–446.

Westlake K, Mackie RI, Dutton MF (1987) Effects of several mycotoxins on specific growth rate of Butyrivibrio fibrisolvens and toxin degradation in vitro. *Applied Environmental Microbiology*, 53: 613–614.

Xiang Y, Huang H, Wang D, Du J, Wu D, Xiong W, Hong Y, Chen J, Liao X (2021) Organometallic Au(III) based electrochemical sensor with wide anodic potential window for sensitive detection of ochratoxin A. *Elecctroanalysis*, 33: 2278–2285.

Yu JCC, Lai EPC (2010) Molecularly imprinted polymers for ochratoxin A extraction and analysis. *Toxins*, 2: 1536–1553.

Zezza F, Longobardi F, Pascale M, et al. (2009) Fluorescence polarization immunoassay for rapid screening of ochratoxin A in red wine. *Analytical and Bioanalytical Chemistry*, 395: 1317–1323.

Zhang Z, Fan Z, Nie D, Zhao Z and Han Z (2019) Analysis of the carry-over of ochratoxin A from feed to milk, blood, urine, and different tissues of dairy cows based on the establishment of a reliable LC-MS/MS method. *Molecules*, 24(15): 2823.

Zheng J, Zhang Y, Xu W, Luo Y, Hao J, Shen XL, Yang X, Li X, Huang K (2013) Zinc protects HepG2 cells against the oxidative damage and DNA damage induced by ochratoxin A. *Toxicology and Applied Pharmacology*, 268: 123–131.

Zhu L, Zhang B, Dai Y, Li H, Xu W (2017) A review: Epigenetic mechanism in ochratoxin A toxicity studies. *Toxins*, 9: 113. http://doi.org/10.3390/toxins9040113.

FURTHER READING

Koszegi T, Poor M (2016) Ochratoxin A: Molecular interactions, mechanisms of toxicity and prevention at the molecular level. *Toxins*, 8: 111.

Chapter 4

Trichothecene Concerns in Food and Feed

Detection and Management Strategies

Madhvi Singh, Chayanika Sarma, Pinky Deka, Ramzan Ahmed, Jayabrata Saha, Sourav Chakraborty

CONTENTS

4.1	Introduction	87
4.2	Occurrence in Food and Feed	90
4.3	Source of Toxins	93
4.4	Genes Responsible for Production of Trichothecenes	94
4.5	Impact of Climate Change on Production	95
4.6	Chemistry and Biosynthesis of Trichothecenes	97
4.7	Detection Methods	98
	4.7.1 Chromatographic Methods	98
	4.7.2 Ion Mobility Spectrometry	102
	4.7.3 Immunochemical Methods	102
4.8	Trichothecene Control Strategies	103
	4.8.1 Physical Treatments	103
	4.8.2 Chemical Control	105
4.9	Conclusion	106

4.1 INTRODUCTION

Food safety and mycotoxin contamination are some major issues faced by the rural population in developing countries. On the other hand, health hazards can also occur in public health through chemical contamination in foods and feeds. In this regard, mycotoxins can be considered the most predominant in the food chain (Khaneghah et al., 2020). Mycotoxicosis occurs in humans and

DOI: 10.1201/9781003242208-4

animals due to food contamination caused by secondary metabolites produced in nature (Bhat & Reddy, 2017). The secondary metabolites are generally produced by fungi belonging to *Aspergillus, Penicillium, Fusarium*, and *Claviceps*. It is reported that the most plentiful genus in temperate regions responsible for causing worldwide contamination in cereals and feeds is *Fusarium* (Streit et al., 2012). Thus, mycotoxin contamination is one of the vital concerns regarding food security. Various studies have been carried out on mycotoxin contamination regarding maize and maize-based foods (Andrade et al., 2017). Common mycotoxins are aflatoxin, trichothecenes, fumonisins, zearalenone, and ochratoxin A. The various health problems caused by trichothecenes play a vital role in the agricultural field. Metabolism and elimination take place after ingestion of the mycotoxin, which yields around 20 metabolites, among which hydroxy trichothecenes-2 toxin is the major metabolite (Zaki, 2012). Based on the structure, the compounds are classified as A, B, C, and D. Trichothecenes are a group of chemically similar toxins produced by filaments of fungi like *Fusarium, Myrothecium,* and *Stachybotrys*. Representatives of type A trichothecenes include *T-2 toxin* (T-2), *HT-2 toxin* (HT-2), *neosolaniol* (NEO), *diacetoxyscirpenol* (DAS), *monoacetoxyscirpenol* (MAS), *cerrucarol* (VER), *scirpentriol* (SCP), and their derivatives. The strains *F. sporotrichioids* and *F. poae* mainly produce the toxins. Deoxynivalenol (DON) and its acetyl derivatives are mainly produced by strains of *F. culmorum* and *F. graminearum* (Kachuei et al., 2014), but the C- and D-type trichothecenes not produced by *Fusarium* species differ chemically. Thus, type A and B trichothecenes are the most toxic of all the naturally occurring trichothecenes (Feltrin et al., 2018; Meneely et al., 2011). Extensive research is going on as the existing research reveals various data on the mycotoxicity of foods. Emerging mycotoxins are also produced by *Fusarium* species such as fusaproloferin, which causes contamination of animal feed and human foods in all stages of production, leading to adverse health effects in human and animals (Anfossi et al., 2016; De Santis et al., 2017; Pierron et al., 2016). The variation of climate in Europe supports rapid growth of mycotoxins. Trichothecenes are very stable organisms that are not degraded by factors like light, temperature, milling, or any other processing factors (Agriopoulou et al., 2016). These mycotoxins are non-volatile in nature and can be effectively deactivated by various alkaline or acidic conditions. T-2 toxin is the most explored mycotoxin, as it highly toxic and is easily accessible (Meneely et al., 2011). Trichothecenes are low-molecular-weight mycotoxins that are soluble in acetone, ethyl acetate, chloroform, ethanol, dimethyl sulfoxide, methanol, and propylene glycol but not in water (Cooper et al., 2007). Purified trichothecenes have a lower vaporization capacity, but when heated in organic solvents, they evaporate. This chapter seeks to explore the methods of detection of these mycotoxins and their occurrence in food and feed, level of toxicity, and influence on human health, taking into account the toxicity of the mycotoxins and numerous adverse health concerns.

In many developing countries, especially in rural areas, people are faced with many food quality problems, as they mainly live their lives on products that are locally available. They face some food security and food contamination problems related to mycotoxins. Thus, mycotoxins have become one of the most common contaminants in the food chain. These mycotoxins are secondary metabolites produced by a variety of fungi, mostly belonging to genera like *Fusarium, Aspergillus, Penicillium,* and *Claviceps.* They cause mycotoxicosis in human being as well as animals. It has been seen that temperate regions are often flooded with species of the *Fusarium* genus that are responsible for the cereal and feed contamination, along with the production of toxins. Therefore, steps need to be taken for food security. Contamination of maize and maize-based products is one of the most common. Aflatoxins (AFs), fumonisins (FUMs), zearalenone (ZEA), ochratoxin A (OTA), and trichothecene (TCT) are some common classes of mycotoxin (Coppa et al., 2019; Ji et al., 2014; Zaki, 2012; Haque et al., 2020). Out of these, the trichothecene mycotoxins cause serious hazards in terms of agricultural products. They are produced and removed after ingestion and thus produce nearly 20 metabolites, out of which hydroxyl trichothecenes-2 toxin is one of the major metabolites (EFSA, 2011). These groups are further classified into four classes, A, B, C, and D (EFSA, 2011; EFSA, 2014; Pinton, 2014; McCormick, 2011).

Trichothecenes are classes of toxins that are produced by some of the filamentous species of fungus like *Myrothecium, Stachybotrys*, and *Fusarium.* The classes and their production strains are described in Table 4.1.

Types A and B are found to be most available and are more toxic in nature, but comparatively type A is the most toxic group (T-2 and HT-2). Examples are NIV, FUS-X, and DON (Hussein & Brasel, 2001; Rocha et al., 2005; Agriopoulou et al., 2020). Trichothecenes produced by *Fusarium* species can contaminate feed and food regardless of the stage of production, which causes hazards for human and animals (Berthiller et al., 2013; Berthiller et al., 2009; Tericiolo et al., 2018). Trichothecenes are not affected by environmental factors such as light or temperature during any processes (Ji et al., 2019), but they can be affected by the presence of a strongly acidic or alkaline environment. Since T-2 is highly available and the most toxic in nature, it is the most extensively studied trichothecene mycotoxin (EFSA, 2011). It is identified by low-molecular-weight compounds ranging between 250 and 550 MW (Ji et al., 2019) that are non-volatile; partially soluble in water; and highly soluble in acetone, dimethyl sulfoxide, ethanol, methanol, ethyl acetate, chloroform, and propylene glycol. Moreover, they vaporize when heated in spite of having low vapor pressure.

TABLE 4.1   TYPES AND THEIR PRODUCTION STRAINS OF FUNGUS FOR THE FORMATION OF TRICHOTHECENES

Sl no·	Type	Toxin Produced	Strains Responsible	Reference
1	A	T-2 toxin (T-2), HT-2 neosolaniol, diacetoxyscirpenol, monoacetoxyscirpenol scirpentriol, verrucarol, and their derivatives	*Fusarium sporotrichioides* and *Fusarium poae*	Chen et al., 2020 and Thran et al., 2004
2	B	Deoxynivalenol, 3-AcDON, 15-acetyldeoxynivalenol, nivalenol, and fusarenon	*Fusarium culmorum* and *Fusarium graminearum*	Piec et al., 2016
3	Type C	Crotocin	*Cephalosporium crotocinigenum* and *Trichothecium roseum*	Cole & Cox, 1981
	Type D	Verrucarol	*Myrothecium* spp. *roridum, leucotrichum,* and *verrucaria*	Gupta, 2012

4.2 OCCURRENCE IN FOOD AND FEED

The mycotoxins that contaminate the food and feedstuff are likely to be affected by climatic changes (Paterson & Lima, 2010). Also depending on the geographical area and climate, different fungus attack different types of food and feedstuffs. Some of the major mycotoxin producers that lead to contaminated food are *Aspergillus, Penicillium,* and *Fusarium* species. Out of these, *Aspergillus* and *Penicillium* survive best in areas with high temperatures and low water activity, but *Fusarium* species survive best in areas with high water activity and low temperatures (Bhat et al., 2010).

Trichothecene production is affected by genetic factors as well as the environment in which they develop (He et al., 2010; Ji et al., 2019). In many studies it has been noted that, as compared to temperate zones, in tropical and subtropical areas, mycotoxin contamination is more prevalent in crops, including contamination by trichothecenes. This is mainly because in tropical areas, the temperature and humidity remain high. This leads to high production of toxins because of optimized conditions.

It has been found that with differences in continents, climatic conditions, and weather, the contamination of feed and food by trichothecenes also varies (Toregeani-Mendes et al., 2011; Fink-Gremmels and Van der Merwe, 2019; Ojuri et al., 2019). It also varies if the soil management and soil type differ, as well as with varieties of crop, differences in agricultural practices, and storage and processing (Toregeani-Mendes et al., 2011). TCT is a class of structurally related toxins (Singh & Mehta, 2020), but only Types A and B have high toxicity and are thus of special interest.

Type A trichothecenes produce T-2 and HT-2 toxins, which have an esterified moiety at a position of C-8. Usually T-2 and HT-2 are found in wheat, barley, oats, maize, rice, soybeans, and so on (EFSA, 2011; EFSA, 2014; Krska et al., 2014), but the grains smaller in size like wheat and rice are mostly attacked by ochratoxin A, deoxynivalenol, and zearalenone (Zaki, 2012; Haque et al., 2020).

A study was done in Europe regarding the occurrence of T-2 and HT-2 toxins in food and feedstuff. It found that T-2 and HT-2 are mainly found in cereal grains in Europe, usually from countries like France and the United Kingdom. An investigation regarding the occurrence of T-2 and HT-2 during the time period 2001–2005 was done in barley, wheat, and oats (Edwards, 2009a; Edwards, 2009b; Edwards, 2009c). In the study, post-harvested samples were collected: 1624 (wheat), 446 (barley), and 458 (oats). The level of detection (LOD) was fixed at 10 µg kg^{-1}. It was found that T-2 was found in oats at a very high concentration in almost 84% of the sample collected with mean, median, and maximum concentrations of 84, 140, and 2406 µg kg^{-1}, respectively. But in wheat and barley, T-2 was found in 16% and 12% of the collected samples, respectively. It was also observed that the mean and medium of the concentration in wheat and barley were lower compared to oats: below 10 µg kg^{-1}. Another study was carried out during 2004–2007 to find the concentration and incidence of T-2 and HT-2 toxins in wheat, oats, and maize at UK mills. There were 60 wheat samples, all from the United Kingdom; 27 oat samples, out of which 21 were from the United Kingdom and 6 from Scandinavia; and 86 samples of maize, out of which 56 were from France and 30 from Argentina (Scudamore et al., 2009). It was observed that only 3 wheat samples out of 60 tested positive for T-2 toxin, with the highest concentration at 13 µg kg^{-1}. Out of 27 samples of oats, T-2 was detected in all 21 samples from the United Kingdom in the range of 20–49 µg kg^{-1} in 5 samples, 50–499 µg kg^{-1} in 14 samples; 500–999 µg kg^{-1} in 1 sample, and 1610 µg kg^{-1} in 1 sample. In the Scandinavian samples, the range of T-2 was between 5 and 499 µg kg^{-1}, with the highest value at 221 µg kg^{-1} (Van der Fels-Klerx, 2010).

If we study the contamination rate of trichothecenes in different cereal grains, the level of type atrichothecens varies. In a study carried out by Morcia and coworkers (2016), it was observed that in Italy, T-2 and HT-2 trichothecenes were prevalent in many malting barleys. The amounts of T-2 and HT-2 were detected were 22% to 53% with a value between 26 and 787 µg/kg. Another study was done by Barthel et al. (2012). Trichothecenes were explored in different barleys such as dehulled barley, barley kernels, winter barley, and pearl barley from Germany. Different trichothecenes were found, like DAN, HT-2, and T-2, and the T-2 and HT-2 were found to be 63% and 71%, with 3 and 6.8 µg/kg, respectively. Therefore, the type A trichothecenes, T-2 and HT-2, were predominantly found in barley.

In a study by Bankole et al. (2010), 32 maize sample were collected from Nigeria from 2005–2006, and only one sample has been attacked by MAS and two samples by T-2 tetraol with concentrations of 4, 73, and 280 µg/kg, respectively. There was no positive test for DAS, NEO, T-2 triol, T-2, and HT-2. Another investigation showed that 13% and 1% of the total sample were contaminated with Das and HT-2 with concentrations of 2.5–8 and 7.5–20 µg/kg, respectively (Chilaka et al., 2016). Here there was also no any evidence of contamination by T-2 or NEO. Another study showed that out of 95 maize samples collected from Zimbabwe, only 1% was contaminated by DAS, and there was no evidence of T-2, HT-2, and NEO (Hove et al., 2016). Therefore, it was concluded that there was much less evidence of concentration and occurrence of DAS, T-2, and HT-2 in the maize sample.

Given that rice is a cereal crop cultivated in warmer areas of the world with higher humidity and temperatures, its contamination by fungi is common. A study was carried out by Rodríguez-Carrasco et al. (2012) in which, out of 23 samples of rice collected from Spain, only 1 was infected by NEO. There was no evidence of contamination by HT-2, T-2, and DAS. Similarly another study by Majeed and coworkers (2018) found that out of 180 samples of rice from Pakistan, only 10% were contaminated with HT-2 and 23% with DAS, and there was no evidence found for T-2 and NEO. Whatever the concentration of DAS and HT-2 found in the rice sample, the PMTDI value was 100 ng/kg bw/day. Thus it can be concluded that concentrations of these toxins are always less in rice, and their incidence are also not consistent.

A study showed that when an investigation regarding type A trichothecenes was carried out in Finland, 63.3% of HT-2, 43.3.% of HT-2-3-glucoside, and 46.7% of T-2 were found in a wheat sample with very low concentrations (Nathanail et al., 2015). In another study carried out in wheat grown on organic farms, three wheat samples were contaminated by HT-2, and T-2 had contaminated two samples out of the total. The researchers did not find any evidence of DAS or NEO contamination (Juan et al., 2013). Another study showed durum wheat was contaminated with 115–486 µg/kg of HT-2 and 10–149 µg/kg of T-2. Therefore, it was seen that there was not a significant difference in contamination between wheat grown conventionally and organically, and also T-2, HT-2, and their derivatives are mostly associated with wheat and its products.

A study of an oats sample from Switzerland carried out by Schöneberg and coworkers found that HT-2 and T-2 were the most commonly found type

A trichothecenes in Swiss oats. In the 2015, HT-2 has a contamination rate of 46% and T-2 74%. Also, it was found that the concentration of HT-2 and T-2 in 11 samples exceeded the guideline value of 1000 µg/kg for unprocessed oats by 7%. In another study by Gottschalk and coworkers (2007), 70 samples of oats were collected. Out of these, most of the oats were contaminated with type A trichothecenes. Therefore, oats were found to be highly contaminated with type A trichothecenes, that is, T-2 and HT-2.

A study was carried out by Scudamore and co-workers (2009) in which it was observed that hulling during the milling process of cereals led to the production of many co-products, and these contained T-2. When oats were de-hulled in the study, they found that all the samples but one contained T-2 at a concentrations of 100 µg kg^{-1}, but in two samples, the concentration of T-2 was found to be more than 1000 µg kg^{-1}. Another study by Peterson and others found that by-products of oats from Europe showed mean, median, and maximum T-2 levels of 122, 66, and 595 µg kg^{-1}, respectively.

4.3 SOURCES OF TOXINS

Trichothecenes are found to be produced by various genera of fungi, such as *Fusarium, Stachybotrys, Myrothecium, Trichothecium, Trichoderma, Cephalosporium, Cylindrocarpon, Verticimonosporium,* and *Phomopsis* (Scott, 1989). These mycotoxins are also seen to be extracted from Brazilian plants belonging to *Baccahris* species. Whatever the number, most of the important sources of trichothecenes are fungal, especially *Fusarium.* This is one of the important pathogens of agricultural plants that cause head blight in plants of temperate regions such as triticate, barley, and wheat. *Fusarium graminearum* can grow at a temperatures of 26–28°C with optimum water activity more than 0.88. *Fusarium culmorum* can grow at 21°C with water activity of 0.87. It was found that with an increase in rainfall, the head blight produced by *Fusarium* also increased. The moisture at the anthesis is seen to affect head blight if the temperature remains optimal (Miller, 2002). In corn, *Fusarium graminearum* causes Gibberella or pink ear rot in the presence of moisture. Depending upon different strains, the toxicity of *Fusarium* species differs and also varies with geographical location. Also, fusarium species rely on different factors such as osmotic tension, oxygen, environmental pH, and temperature. It has been observed that in cool weather during early winter, along with heavy rainfall in some areas of the United States, infestation by *Fusarium* is common, along with production of mycotoxins.

Chemically the trichothecenes are grouped into four classes on the basis of structural differences (substitutions of groups at five positions of the backbone of trichothecenes). These groups are type A, which includes T-2 and HT-2. Type B contains nivalenol and DON. Type C includes crotocin, and type D contains macrocyclics. Some of the major toxic trichothecenes are included in type A, such as T-2, along with the deacetylated metabolites, HT-2, and DAS. They have mostly been developed for research purposes by *Fusarium sporotrichioides* and *Fusarium poae* cultures. They have the ability to develop toxins even in light or dark and at low temperatures, but if there is a variation in the temperature, then it leads to an increase in the toxicity of the metabolites. Also, low temperature leads to the development of moldy corn toxicosis, contamination of moldy bean hull, moldy cereal emesis, fusariotoxicosis, and dendrodochiotoxicosis (myrotheciotoxicosis). Type B has a keto group at a position C-8 and bears a hydroxyl group at C-7. It contaminates grain commonly found in the field. It takes into account DON, its acetylated derivatives, nivalenol, and fusarenon-X, which are developed by fungi like *F. culmorum*, *F. graminearum*, and other related fungi. As compared to other groups of trichothecenes with no substitution at the C-2 position, type B is less toxic. Type C is not produced by *Fusarium*. It contains a second epoxide ring at the position of C-7,8. It is also not involved in any adverse effects in livestock, but crotocin is another type C trichothecene that is developed by *Cephalosporium crotocinigenum* and *Trichothecium roseum* and showed less toxicity in mice (Cole & Cox, 1981). Type D trichothecenes are effective cytotoxic compounds with a macrocyclic ring that connects carbon at a position of C-4 to C-15 on the backbone of trichothecenes. It has been seen that the genus *Myrothecium*, which consists of *M. roridum*, *M. leucotrichum*, and *M. verrucaria*, has the capacity to produce verrucarins and roridins. These are the diesters of the trichothecene verrucarol. Moreover, they are severely toxic to the cells of mammals in vitro as well as various animals under in vivo conditions (Mantle, 1991). Many studies have reported that *S. alternans Bonorden*, a fungus that grows in cellulosic vegetation, especially on mildewed wet straw, makes macrocyclic trichothecene mycotoxins. These are highly stable and toxic in nature and also lead to serious cytotoxic effects (Bata et al., 1985).

4.4 GENES RESPONSIBLE FOR PRODUCTION OF TRICHOTHECENES

Many studies have looked at the production of trichothecenes, with many of them using complete genetic profiling. The enzymes and regulatory components involved in the formation of trichothecene are encoded by 15 genes scattered over three loci (Merhej et al., 2011). The core TRI cluster contains ten co-regulated genes, including tri5, the first cloned gene that encodes trichodiene synthase. All of the clustered genes are ordered in the same way in *F. graminearum* and *F. sporotrichioides* (Desjardins & Proctor, 2007; Moretti et al., 2013). These ten genes were found to encode seven structural pathway genes, two regulatory genes, and one transporter (Kimura et al., 2007). In *F. sporotrichioides* and *F. graminearum*, there are two additional genomic loci within genes implicated in trichothecene biosynthesis: the two-gene Tri1-Tri16 locus and the single-gene Tri101 locus (Merhej et al., 2011). These three loci are present in diverse locations on chromosomes, according to data obtained from several genome studies (Moretti et al., 2013). However, a detailed investigation of the TRI cluster revealed the presence of Tri1 and Tri101 in the core cluster of some members of the *Fusarium incarnatum–Fusarium equiseti* group (Proctor et al., 2009), confirming the trichothecene-producing *Fusarium* species' complex evolutionary history.

    F. sporotrichioides has a collection of genes involved in the manufacture of trichothecene (McCormick et al., 1996). Following Keller & Hohn's (1997) research on the gene cluster, nine genes were identified inside the 25-kb area, with eight of them having specific functions. Two of these genes, Tri 3 and Tri 4, were found after complementation of UV-induced mutants that suppressed the generation of trichothecene T-2 toxin. Tri 3 catalyzes the conversion of 15-decalonectrin to calonectrin (McCormick et al., 1996). In the first phase of this method, Tri 4 is a cytochrome P-450 monooxygenase that converts trichodiene to an unidentified oxygenated product. Tri 11, a second cytochrome P-450 monooxygenase, has also been found, and it oxygenates the trichothecene ring at C-15 (Alexander et al., 1998). It is possible that the gene cluster contains two more specialized acetyl-transferases for hydroxylation at the C-3 and C-4 locations (McCormick et al., 1996). Tri 5 encodes the trichodiene synthase enzyme, which converts farnesyl pyrophosphate to trichodiene. The Tri 6 gene's product, a Cys2, His2 zinc-finger protein, is in charge of the biosynthetic process (Proctor et al., 1996).

    The Tri 101 gene, which encodes a protein that catalyzes the acetylation of the trichothecene ring at the C-3 position, was recently identified from *F. graminearum*. A mapping investigation using two of the fewest overlapping cosmid clones encoding Tri 101 found that the Tri 101 gene is sandwiched between a constructive UTP-ammonialigase gene and a phosphate permease gene. Tri 4, Tri 5, or Tri 6 cannot be used to generate this gene (Kimura et al., 1998). In general, it is clear that gene assembly for trichothecene production is conserved in closely related fungi, as opposed to changes in gene assembly that are implicated in significant genetic rearrangements (Trapp et al., 1998).

4.5 IMPACT OF CLIMATE CHANGE ON PRODUCTION

It has been seen that trichothecenes are greatly affected by environmental factors such as temperature and moisture. Environmental factors also affect the growth of fungi as well as their capacity to produce mycotoxins (Tanaka et al., 1988; Homdork et al., 2000). Mycotoxin production can occur either in the field itself or in the storage area of grain. It has been observed that if cereal grains are not properly stored there, may be a chance of moisture migration and increase in temperature, which lead to further growth of mycotoxins, and thus the quality of the grain is reduced (Schwabe & Krämer, 1995; Birzele et al., 2000).

A study was carried out to investigate how temperature, water activity, and fungal strains of *F. graminearum* and *F. culmorum* in crops collected from Spain affect mycotoxin production (Llorens et al., 2004). The growth rate of three isolates of the species *F. graminearum* and *F. culmorum* was inspected at different temperatures, and it was found that similar growth occurred at temperatures of 32°C, 28°C, and 20°C, whereas much less growth occurred at 15°C. Also, temperature played an important role in the production of DON, but the fungal strains and moisture did not show any effect. DON was found to grow at its highest at 28°C with a mean value for the two species of 1.47 Ag/g dry culture. There was no evidence of growth of DON at 32°C but a little growth at 15°C and 20°C: 6.3% and 20.3% of the levels observed at 28°C, respectively. Thus, it was found that the production of DON is favorable at a temperature of around 28°C with an optimal range between 20°C and 28°C irrespective of species. For the production of nivalenol (NIV), fungal isolates and temperature played a significant role, but moisture does not show a significant effect. The production of NIV was higher at a temperature of 20°C, but no evidence was found at 32°C, and similar production was found at 15°C and 28°C, that is, 26.3% and 27.3% of level found at 20°C, which showed that the production of NIV decreased at temperatures other than 20°C. Another study was done by Rybecky and coworkers (2018) where the effect of temperature and moisture content on the production of toxins by *F. meridionale* in soybean was investigated. When the effect of water activity and temperature on lag phase was studied, it was found that all the strains of *F. meridionale* B2300 (A), F5043 (B), and F5048 (C) had a shorter lag phase under high water activity, except at water activity of 0.99 and temperature of 30°C, where strains *of F. meridionale* such as B2300 and F5048 had longer lag phases as compared to those at water activity 0.98 and 0.96. Also, strains of *F. meridionale* B2300 and F5048 had short lag phases at temperatures of 20°C and 25°C, while at 30°C, the longest lag phase was observed for all strains. When growth rate was observed, it was found that the optimal conditions of growth for *F. meridionale* strains were at a temperature of 25°C and water activity with a range of 0.98–0.99. Also, maximum growth was observed at 20°C rather than 30°C for all strains, with an increased growth rate with high water activity, except at 0.99 and 30°C for strain B2300 of *F. meridionale* when toxin production such as DON and NIV was studied under different temperatures and water activity. It was observed that optimal condition for the production of toxins by *F. meridionale*

F5043 and F5048 s were 25°C and 0.96 water activity after 21 days of incubation as compared to at 30°C and 20°C. Maximum production of NIV was seen at 20°C and 0.98 aW when incubated for 21 and 14 days, respectively, for strains like *F. meridionale* F5043 and F5048 when the temperature was kept at 20°C. *F. meridionale* F5043 showed production of DON at low water activity, but *F. meridionale* F5048 showed production of DON at high water activity. There was no evidence of the production of toxins by *F. meridionale* B2300 at this temperature. The highest production of NIV by *F. meridionale* F5043 and F5048 was found to be at 20°C and 0.98 water activity when incubated for 21 and 14 days, respectively. Therefore, it was concluded that temperature and water activity play a very important role in the production of DON and NIV by *F. meridionale*.

4.6 CHEMISTRY AND BIOSYNTHESIS OF TRICHOTHECENES

The sesquiterpenoids, or trichothecenes, are a category of chemical compounds that are closely linked. Nearly 200 trichothecene species have been identified. They have low-molecular-weight (MW 250–550) chemicals and are nonvolatile. Sesquiterpenoids are soluble in chloroform, acetone, dimethyl sulfoxide (DMSO), ethyl acetate, methanol, ethanol, and propylene glycol but not in water. T-2 is the most important mycotoxin, recognized for its ease of use and great toxicity. Trichothecene is extracted from fungi as a golden-brown liquid that evaporates to form a crystalline material. When heated in organic solvents, refined trichothecene has a low vapor pressure and vaporizes.

The cydization of farnesyl pyrophosphate produces trichodiene, which is not produced normally in the body. Trichodiene goes through a series of oxygenations, cydizations, isomerizations, and esterifications to make bioactive trichothecenes such deoxynivalenol and acetylated DON (Desjardins et al., 1993). Gene-encoding enzymes that catalyze the majority of these steps have been discovered and assigned to a gene cluster over the last ten years (Hohn et al., 1993; Keller & Hohn, 1997). In two closely related species, *F. graminearum* and *F. sporotrichioides*, more than eight genes from the trichothecene biosynthetic gene cluster have been discovered (Brown et al., 2001). TR/5 denotes trichodiene synthase (Hohn & Beremand, 1989); TR/3, an acetylase (McCormick et al., 1996); TR/4 and TRill, cytochrome P450 monooxygenases (Alexander et al., 1998; Hohn et al., 1995); TR/6, a transcription factor (Hohn et al., 1999); and TR/7, a transcription factor (Alexander et al., 1999a). A trichothecene biosynthetic gene, TRIJ 01, has been discovered outside of the trichothecene gene cluster.

4.7 DETECTION METHODS

Various rapid, accurate, highly sensitive methods are required to detect, quantify, and identify the adverse effects of mycotoxin contamination. The methods developed for trichothecenes can be applied for the cereal-based foods, feeds, and various organic samples. Quantity assessment of deoxynivalenol, the acetylated form, DON 3 glucoside, and other trichothecenes is done by liquid chromatography coupled with mass spectrometry (LC-MS)(Sarrocco et al., 2019). In vitro analysis for trichothecenes has still not been carried out, and the proficiency has not been verified. Direct and indirect approaches can be used to determine the DON 3 glucoside, and the direct approach is reported to be the preferred method. There are performance criteria available for analytical methods along with reference matrices and reference calibrants. 3-AcDON, 15-AcDON, and DON-3-glucoside follow non-certified calibrants. Immunochemical methods provide a fast and economical step to chromatographic methods (Meneely et al., 2011). Each technique used has various merits and demerits; therefore, determination and investigation of mycotoxins in foods and feeds are of utmost importance in centers all over the world. Various methods of detection along with the occurrence of trichothecenes around the world are represented in Table 4.2.

4.7.1 Chromatographic Methods

Mycotoxins are usually detected using several instrumentation techniques, such as thin layer chromatography (TLC), and qualitative or semi-quantitative techniques have been used for several years. Other instruments; high-performance liquid chromatography coupled with various detectors such as ultraviolet (UV), diode array detectors (DADs), fluorescence detectors (FLDs), or mass spectrometric detectors; and gas chromatography coupled with an electron capture detector, flame ionization detector, or mass spectrometry detector can be used for analysis of mycotoxins (Anfossi et al., 2016; Dellafiora & Dall'Asta, 2016; Kachuei et al., 2014).

Trichothecene mycotoxin screening in feed is usually carried out using gas chromatography by extracting the toxins with acetonitrile water and purifying using Florosil, charcoal-alumina celite, and silica mini-columns. Nivalenol (NIV), diacetoxyscirpenol, deoxynivalenol, T-2, and their fungal metabolites were hydrolyzed by alkali to their parent alcohols, derivatized to pentafluoropropionyl analogs, and quantified with electron capture detection by capillary gas chromatography. The sensitivity of detection is increased by negative chemical ionization mass spectroscopy, and identity is confirmed. DAS, DON, and T-2 were recovered from corn at 80%, 65%, and 85%; from soybeans at 84%, 65%, and 88%; and from mixed feeds at 70%, 57%, and 96% and 0.1 to 2.0 ppm concentration. The

TABLE 4.2   DISTRIBUTION OF TRICHOTHECENES IN FOOD MATRICES WORLDWIDE

Region	Food Matrix	Trichothecene	Range (µg/kg)	Method of Detection	Reference
China	Wheat	DON	0.50–604.00	LC-MS/MS	Han et al., 2014
China	Maize	DON	5.90–9843.30	HPLC	Xing et al., 2017
Romania	Wheat	DON	110.00–1787.00	LC-MS/MS	Habler et al., 2017
Pakistan	Rice	DON	190.00	LC-MS/MS	Habler et al., 2017
Brazil	Wheat	DON	<LOD	HPLC	Majeed et al., 2018
Italy	wheat	DON	1329.00–3937.00	GC-MS	De Lima Rocha et al., 2017
Brazil	Wheat	DON	56.00–27,088.00	GC-MS	Bertuzzi et al., 2014
China	Wheat	DON	243.00–2281.00	HPLC	Savi et al., 2014
Iran	Wheat	DON	240.00–1129.00	LC-MS/MS	Liu et al., 2015
Finland	Wheat	DON	23.00–1270.00	—	Darsanaki et al., 2015
Brazil	Wheat	DON	NA–5510.00	LC-MS/MS	Nathanail et al., 2015
Brazil	Wheat	DON	NA–1310.00	HPLC	De Almeida et al., 2016
Brazil	Wheat	DON	NA–8501.00	HPLC	Calori-Domingues et al., 2016
Hungary	Wheat	DON	NA–2419.00	HPLC	Calori-Domingues et al., 2016
Argentina	Wheat	DON	183.00–2150.00	HPLC	Tralamazza et al., 2016
Switzerland	Wheat	DON	NA–1880.00	ELISA	Tima et al., 2016
Brazil	Wheat	DON	NA–9480.00	LC-MS/MS	Palacios et al., 2017
Poland	Wheat	DON	NA–10,600.00	LC-MS/MS	Vogelgsang et al., 2017

(Continued)

TABLE 4.2   DISTRIBUTION OF TRICHOTHECENES IN FOOD MATRICES WORLDWIDE   (CONTINUED)

Region	Food Matrix	Trichothecene	Range (μg/kg)	Method of Detection	Reference
China	wheat	DON	33.00–3030.00	HPLC	Silva et al., 2018
Hungary	maize	DON; T-2	225.00–2963.00; NA–146.00	ELISA	Tima et al., 2016
Serbia	Corn flour; cornflakes	DON	NA–931.00; NA–8178.00	HPLC	Palacios et al., 2017
Brazil	Barley products	DON	1700.00–7500.00	LC-MS/MS	Vogelgsang et al., 2017
Brazil	Bakery products	DON	60.00–1720.00	HPLC	Silva et al., 2018
India	Infant foods	DON	23,800.00	LC-MS/MS	Gummadidala et al., 2018
Germany	Noodles and pasta	DON	60.00–1609.00	HPLC	Gammadidala et al., 2019
Finland	Oat	DON	23,800.00	LC-MS/MS	De Almeida et al., 2016
United Kingdom	Oat	DON	NA–1866.00	LC-MS/MS	Edwards, 2017
Finland	Oat	DON	NA–21,608.00	GC-MS	Hietaniemi et al., 2016
Poland	Oat	DON	NA–2975	HPLC/HRMS	Bryla et al., 2018
Tunisia	Cereal products	HT-2	LC-MS/MS	LC-MS/MS	Queslati et al.,
Lebanon	Wheat grains, wheat flour, bread	HT-2; T-2	—	LC-MS/MS	Elaridi et al., 2019
Spain	Cereal-based beer	HT-2	NA	GC-ECD	Cano-Sancho et al., 2011
Spain	Wheat semolina	T-2; HT-2	67.00–15.20	GC-QqQ-MS/MS	Rodríguez-Carrasco et al., 2012

NA: Not available.

15-monoacetoxyscirpenol, NIV, HT-2, and T-2 tetraol were recovered at 97%, 97%, 86%, and 56% from corn at 0.25 ppm concentration. The concentration limit of 0.05 ppm in silage and 0.2 ppm in soybeans, corn, and mixed feeds was estimated.

These detection methods were used for both qualitative and quantitative detection with high precision and accuracy. But the main drawback of this method is it requires highly skilled personnel, it is expensive and time consuming, and it requires sophisticated sample preparation (Meneely et al., 2011). Chromatographic techniques with UV and FID detectors are in compliance with regulations and employed for confirmatory assays (De Santis et al., 2017; Pallarés et al., 2017). These instrumentation techniques are sometimes taken as reference methods to validate immunochemical assays. The importance of mass spectrometry (MS) is increasing worldwide due to its high selectivity, sensitivity, accuracy, and ability to analyze multiple compounds (Andrade et al., 2018; Girolamo et al., 2020). This technique allows analysis of wide range of analytes and matrices that allows simultaneous extraction of mycotoxins owing to its quick, easy, cheap, effective, rugged, and safe (QuEChERS) approach. Preconcentration steps are required, as the QuEChERS approach reduces the sensitivity. Instead of preconcentration, sensitivity can be increased by isotope dilution quantification (Anfossi et al., 2016). Unknown compounds can be analyzed by structural confirmation of compounds through high-resolution MS (HRMS) and tandem MS/MS. The new masked mycotoxins were identified using non-selective extraction protocols using HRMS or tandem MS/MS (Feltrin et al., 2018; Gonçalves et al., 2020). Mycotoxins in food and feed have also been analyzed using the LC-MS/MS rapid multiresidue technique. The mycotoxins in the sample characterized were detected by target analysis, but it is a time-consuming technique with limited scope (Anfossi et al., 2016). While this is the most dominant technique for analysis of mycotoxins and resides. Innovations in mass spectrometry have developed analysis in detecting target to non-target compounds and made simultaneous improvements in the technology. The instrumentation became sensitive, and libraries expanded with less abundant mycotoxins. The need for non-target analysis increased, and scientists developed techniques for screening specific compounds (Anfossi et al., 2016). DON can be converted to a deoxynivalenol 3-glucoside called masked mycotoxin by plant detoxification (Singh & Mehta, 2020). Due to the large number of mycotoxins and wide range of physicochemical properties, detection of mycotoxins is challenging. Food and feed matrices are highly complex and contaminated with mycotoxins at low concentrations (Luo et al., 2018). Solid–liquid extraction, liquid–liquid extraction, solid-phase extraction, immunoaffinity chromatography, and QuEChERS are common approaches followed for sample preparation for mycotoxin analysis. Polarity and differences in solubility of mycotoxins in polar trichothecenes make analysis complicated (Freire & Sant'Ana, 2018; Luo et al., 2018; Singh & Mehta, 2020). The matrix components in sample extracts negatively affect the detection system. Extraction and clean-ing techniques need to be developed for high recovery of polar trichothecenes and determination of mycotoxins by minimizing sample matrix effects (Freire & Sant'Ana, 2018; Luo et al., 2018).

4.7.2 Ion Mobility Spectrometry

Ion-mobility spectrometry (IMS), which is based on the velocity of gas-phase ions in an electric field, can be used to label compounds. Ion mobility spec-trometry is comparable to Fourier transform near-infrared (FT-NIR) spectros-copy, but it has the advantages of being simple, quick, and cost effective, and with a low detection limit. Righetti et al. created an ion-immobility application to detect mycotoxins such as HT-2, DON, and T-2 in wheat grains (2018). The sample matrix had no effect on the high repeatability, with a relative standard deviation of less than 2% under varied instrument conditions. The advantages of IMS over traditional LC and GC techniques were reduced background noise and higher sensitivity in detecting mycotoxins. The retention time and addi-tional mass spectrum data called a collision cross-section were provided, and compounds were detected with confidence (Singh & Mehta, 2020).

4.7.3 Immunochemical Methods

Immunochemical assays based on antibody–antigen reactions are useful for routine analysis since they are a simple and quick detection method (Dzantiev et al., 2014). Enzyme-linked immunosorbent assays, chemiluminescence immunoassays, enzyme-linked aptamer assays, fluorescence immunoassays, fluorescence resonance energy transfer immunoassays, time-resolved immu-nochromatographic assays, and metal-enhanced fluorescence assays are examples of immunological techniques (Chauhan et al., 2016). The important parameter of detection techniques is an aptamer that can bind with amino acids, peptides, proteins, and organic and inorganic molecules with high speci-ficity and affinity (Torres-Chavolla & Alocilja, 2009). An electrochemical mag-neto immunosensor was developed by Jodra et al. (2015) to detect FB1 and FB2. The sensor was made with disposable carbon-printed electrodes and magnetic beads. A mesoporous carbon and trimetallicnano rattle with an Au core-based ultrasensitive immunosensor was developed by Liu et al. (2014). The assay exhibited good stability and reproducibility to detect ZEN with a lower limit of 1.7 pg/mL. The identification of new mycotoxins and simultaneous detection of several mycotoxins was difficult due to its strong selectivity and molecular recognition mechanism. To detect target compounds separately in spatially distinct regions, an analytical array technique was developed by Oswald et al. (2013). An immuno-chromatographic strip test device was developed by Song et al. (2014) that can detect ten different toxins such as AFs, DON, ZON, and analogs thereof simultaneously. A unique special address that detects several mycotoxins, including AFB1, DON, ZON, and T-2, in peanuts was reported by Wang et al. (2008). Immunochemical methods have greater selectivity in detecting mycotoxins to ensure food safety in developing countries in comparison to chromatographic methods. The level of contamination of foods by fungi and mycotoxins will increase in the future with changes in the climate and environment. The application of immunoassays and control programs is required for efficient risk management (Wang et al., 2008).

4.8 TRICHOTHECENE CONTROL STRATEGIES

Mycotoxin contamination is a worldwide threat, as it causes damage to agricultural commodities and adverse health effects in humans and animals that require various decontamination and prevention measures (Ayofemi Olalekan Adeyeye, 2020). Pre-harvest measures are usually carried out to avoid growth of toxic fungus or mycotoxins (Luo et al., 2018). But after the growth of mycotoxins in the food matrix, decontamination focuses on post-harvest management. Good agricultural practices (GAPs), good manufacturing practices (GMPs), good environmental factors, good storage practices. Execution of crop changing programs, registered herbicides, fungicides, insecticides, fungal infection, weed removal, treatment of seed bed, analysis of soil, genetic modification (Adebiyi et al., 2019; Alberts et al., 2017). In cereals, grapes, and apples, mycotoxin management is carried out by antagonistic fungi as biological control (Sarrocco et al., 2019). A collective approach is always better in this process, such as GMPs along with GAPs, and a hazard analysis critical control point (HACCP) gives better result (Sarrocco & Vannacci, 2018). Temperature and humidity are vital environmental factors for the production of mycotoxins. Other factors such as storage, moisture, and warehouse humidity are also important for the growth of mold and mycotoxins (Luo et al., 2018). Other measures such as thermal, radiation, plasma, and chemical methods use natural methods like oxidation, reduction, hydrolysis, alcoholysis, absorption, and biological processes (Lyagin & Efremenko, 2019). Loss of quality attributes occurs due to chemical and physical treatment, making these processes limited and ineffective. However, biological treatments are considered more effective and eco-friendly (Sarrocco et al., 2019).

4.8.1 Physical Treatments

Numerous physical treatments such as grading, drying, milling, microwave, peeling, roasting, sorting, cleaning, segregation, and boiling are usually carried out for mycotoxin detoxification. Incorporation of HACCP practices in post-harvest treatments may contribute to the prevention of mycotoxins. Cleaning and sorting are the initial steps of the natural disinfection process (Shi et al., 2018). Removing and sorting low-quality fruits in a lot may reduce the patulin content in fruits up to 99%, and these two steps do not possess any risk for the production of degradable products (Luo et al., 2018). Aflatoxin infection can also be reduced by separating damaged nuclei. Ultraviolet radiation treatment was also reported to be used in order to reduce aflatoxin content in sorting of cereals (Karlovsky et al., 2016).

Processing can lessen mycotoxin content but cannot remove it. Softening is another technique that can reduce mycotoxin contamination, as the fungi accumulate on the surface of the granule (Neme & Mohammed, 2017). It is reported that peeling also reduces aflatoxin levels. Final products are affected by factors like temperature and time. Mycotoxins are thermally stable compounds, but conventional methods like baking and frying above 100°C reduce mycotoxin levels. During extrusion, moisture and temperature affect the reduction of aflatoxin up to 50–80% (Shanakhat et al., 2018).

Storage conditions also play a major role in controlling mycotoxin growth, as they affect the growth of fungus. Humidity and temperature are two important factors that promote the growth of mycotoxin and fungi. Twenty to 50% of crop loss was reported in various countries due to improper storage practices (Neme & Mohammed, 2017). Packaging conditions, temperature control, proper ventilation, and aeration control the growth of mycotoxins (Gonçalves et al., 2019).

Various storage processes for cereals include ionizing or non-ionizing radiation. Removal of pathogenic microbes is carried out by radiation, but it also partially eliminates mycotoxins. It is reported to be practiced in industries that change the reactions of food matrices (Shanakhat et al., 2018). ZEA radiation was reported to be safe up to 10 kGy on water and orange, pineapple, and tomato juice, but higher doses were observed to affect the quality attributes of fruit juices. Due to the various adverse effects on physical, chemical, and biological properties of foods, radiation technology is still questionable (Gonçalves et al., 2019).

Cold plasma is used in processing of food due to its high microbe-inhibiting potential (Karlovsky et al., 2016). This technology includes photons, ions, and free radicals and is advantageous, as it is low cost and environmentally friendly. It was reported that 50% aflatoxin reduction occurred as result of low pressure cold plasma treatment (Basaran et al., 2008). Another study claims that a 66% reduction of mycotoxin was observed in maize after a 10-min exposure to cold plasma. A 90% reduction in trichothecenes was observed when cold atmospheric pressure was applied for 8 min in foods. Moreover, 5-second plasma treatment has reported to degrade various mycotoxins (Smith et al., 2016).

Mycotoxin binders such as activated carbon, aluminosilicates, non-digestible carbohydrates, and cholesterol help absorb mycotoxins by inhibiting them from entering the gut and then the bloodstream (Pierron et al., 2016). This technique is an alternative to other physical methods used to degrade microorganisms. Lactone ring cleavage is the target for microbial enzymes for reducing mycotoxins. Activated carbon was used to eliminate patulin from infected cider and milk. Although the mycotoxin content was reduced, safety still needs to be ensured for this technique (Karlovsky et al., 2016).

4.8.2 Chemical Control

Seed treatment with ammonia has been reported to reduce mycotoxin content and also inhibits the growth of fungi. In the European Union, the use of bases is restricted for human consumption. The collective approach of glycerol and calcium hydroxide is reported to have shown a detoxification effect on mycotoxins (Luo et al., 2018). Sodium hydroxide and potassium hydroxide are frequently used for decontaminating oil, but these chemicals may cause various types of secondary contamination and also degrade quality attributes (Ji & Xie, 2020).

Chitosan, the second most abundant polysaccharide, is reported to inhibit fungi, bacteria, and viruses. Chitosan has biocompatibility and antimicrobial potential, due to which it has emerged as an important preservative (Zachetti et al., 2019). A study has revealed that in maize and wheat, chitosan can reduce *Fusarium* contamination, thereby helping reduce the mycotoxin content. Irradiation treatment along with low-molecular-weight chitosan decreased mycotoxins. The addition of chitosan with essential oil reduced levels of mycotoxins in marine algae (Gunupuru et al., 2019).

Ozone treatment is a simple technique without any harmful residue and has been reported to degrade many mycotoxins (Piemontese et al., 2018). Disinfection of various cereals, vegetables, and fruits was reported to degrade several mycotoxins. This treatment under optimized conditions showed a decrease of mycotoxin in wheat. Microbial inhibition was observed in wheat in another study without any chemical and rheological changes (Agriopoulou et al., 2016).

Researchers over the last 20 years have been trying to control mycotoxins with the help of biological agents (Wang et al., 2019). Antagonistic use of bacteria, fungi, and yeast has been reported for the inhibition of mycotoxins in food and feed. Biological treatments are a safe alternative, as they have no toxic effects on food materials. An in vitro study reported that decontamination or detoxification with pure cultures of microbes is very effective. It is also reported that fermentation helps in reducing mycotoxin content (Sarrocco & Vannacci, 2018). *Flavobacterium aurantiacum* is the only bacteria among 1000 that has been proven effective. *Enterococcus faecium* was used for detoxifying mycotoxins by binding to cell walls of bacteria. Peptidoglycans and polysaccharides, along with microorganisms, are reported to be responsible for binding mycotoxins (Cooper et al., 2007). It has been reported that lactic acid bacteria are effective against mycotoxins in aqueous solution. Yeasts such as *Saccharomyces cerevisiae* eliminate patulin in fermented foods by increasing time and temperature. Fermentation is undoubtably a desirable method for reducing mycotoxin content. Different fungi such *Aspergillus, Rhizopus, Trichoderma, Clonostachys*, and *Penicillium* spp. are used for detoxifying mycotoxins. A huge number of various non-toxic strains of *A. flavus* and *A parasiticus* bind with soil and fight against toxic strains (Adebo et al., 2019).

Enzymatic detoxification of mycotoxins includes chemical and biological processes. This technique is highly precise and specialized and needs mild conditions without any toxicity. Some strains of *Aspergillus* species produce enzymes that have the inbuilt capacity to decontaminate fumonisins, including *Fusarium* (Lyagin & Efremenko, 2019). Fruit spoilage fungi are degraded by using β-1, 3-glucanase, and chitinase. Spraying of 50% β-glucanase, 50% chitinase, and 40% concentrations. Thus, β-glucanase and chitinase can be regarded as good alternatives for fermented sausage to control fungi contamination (Shanakhat et al., 2018).

Some emerging novel technology such as adsorbents of nanoparticles have been used to remove mycotoxins. Magnetic carbon nanoparticles, chitosan-coated Fe_3O_4 nanoparticles, and silver nanoparticles were reported to degrade mycotoxins (Tarazona et al., 2019). A novel photocatalyst nanoparticle, UCNP@ TiO_2, is very efficient, and products were found to be non-toxic after treatment. Activated carbon, bentonite, and aluminum oxide were reported to be effective against mycotoxins (González-Jartín et al., 2019). Degradation of mycotoxins was reported to be done by essential oil extracted from plant sources by inhibiting fungal and mycotoxigenic activity. The use of plant origin sources is preferred for removal of mycotoxins, as it is regarded as safe for humans. Aspergillus growth was inhibited by clove oil, as it contains eugenol as a major compound. The addition of 2–3% Spanish paprika smoker in meat was proven effective against mycotoxins (Kollia et al., 2019).

4.9 CONCLUSION

Rapid detection technology for the detection of mycotoxins has been introduced in recent years. This chapter discusses detection methods for trichothecenes and also current techniques, including quantitative and non-invasive methods. Pre-treatment of samples for mycotoxins in food matrices has been a challenging task. Noteworthy changes and advancements have been made by introducing IAC and SPE clean-up process of various samples. Novel adsorbent nanomaterials have been introduced. Different analytical methods for trichothecene detection have advanced for cereal and cereal-based products. Methods like TLC and HPLC are used conventionally. Significant advancement has been recorded regarding simple to multiple sample analysis for detection of mycotoxins. Advances in chemistry and immunochemistry have led to more sensitive, specific, and fast quantitative assays for quantification for mycotoxin analysis. Prevention is a vital step for fighting against mycotoxins. Thus, monitoring mycotoxins in food matrices can address various issues regarding health, marketing, distribution, and consumption.

REFERENCES

Adebiyi, J. A., Kayitesi, E., Adebo, O. A., Changwa, R., & Njobeh, P. B. (2019). Food fermentation and mycotoxin detoxification: An African perspective. *Food Control*, 106(May), 106731. https://doi.org/10.1016/j.foodcont.2019.106731

Adebo, O. A., Kayitesi, E., & Njobeh, P. B. (2019). Reduction of mycotoxins during fermentation of whole grain sorghum to whole grain ting (a southern African food). *Toxins*, 11(3). https://doi.org/10.3390/toxins11030180

Agriopoulou, S., Koliadima, A., Karaiskakis, G., & Kapolos, J. (2016). Kinetic study of aflatoxins' degradation in the presence of ozone. *Food Control*, 61, 221–226. https://doi.org/10.1016/j.foodcont.2015.09.013

Agriopoulou, S., Stamatelopoulou, E., & Varzakas, T. (2020). Advances in occurrence, importance, and mycotoxin control strategies: Prevention and detoxification in foods. *Foods*, 9, 137.

Alberts, J. F., Lilly, M., Rheeder, J. P., Burger, H. M., Shephard, G. S., & Gelderblom, W. C. A. (2017). Technological and community-based methods to reduce mycotoxin exposure. *Food Control*, 73, 101–109. https://doi.org/10.1016/j.foodcont.2016.05.029

Alexander, N. J., Hohn, T. M., & McCormick, S. P. (1998). The TRI11 gene of *Fusarium sporotrichioides* encodes a cytochrome P450 mono-oxygenase required for C-15 hydroxylation in trichothecene biosynthesis. *Applied and Environmental Microbiology*, 64(1), 221–225.

Alexander, N. J., Hohn, T. M., & McCormick, S. P. (1999a). TRI 12, a trichothecene efflux pump from *Fusarium sporotrichioides*: Gene isolation and expression in yeast. *Molecular and General Genetics*, 261, 977–984.

Andrade, G. C. R. M., Pimpinato, R. F., Francisco, J. G., Monteiro, S. H., Calori-Domingues, M. A., & Tornisielo, V. L. (2018). Evaluation of mycotoxins and their estimated daily intake in popcorn and cornflakes using LC-MS techniques. *LWT*, 95, 240–246. https://doi.org/10.1016/j.lwt.2018.04.073

Andrade, P. D., Dantas, R. R., Moura-Alves, T. L. da S. de, & Caldas, E. D. (2017). Determination of multi-mycotoxins in cereals and of total fumonisins in maize products using isotope labeled internal standard and liquid chromatography/tandem mass spectrometry with positive ionization. *Journal of Chromatography A*, 1490, 138–147. https://doi.org/10.1016/j.chroma.2017.02.027

Anfossi, L., Giovannoli, C., & Baggiani, C. (2016). Mycotoxin detection. *Current Opinion in Biotechnology*, 37, 120–126. https://doi.org/10.1016/j.copbio.2015.11.005

Ayofemi Olalekan Adeyeye, S. (2020). Aflatoxigenic fungi and mycotoxins in food: A review. *Critical Reviews in Food Science and Nutrition*, 60(5), 709–721. https://doi.org/10.1080/10408398.2018.1548429

Bankole, S. A., Schollenberger, M., & Drochner, W. (2010).Survey of ergosterol, zearalenone and trichothecene contamination in maize from Nigeria. *Journal of Food Composition and Analysis*, 23(8), 837–842.

Barthel, J., Gottschalk, C., Rapp, M., Berger, M., Bauer, J., & Meyer, K. (2012).Occurrence of type A, B and D trichothecenes in barley and barley products from the Bavarian market. *Mycotoxin Research*, 28(2), 97–106.

Basaran, P., Basaran-Akgul, N., & Oksuz, L. (2008). Elimination of *Aspergillus parasiticus* from nut surface with low pressure cold plasma (LPCP) treatment. *Food Microbiology*, 25(4), 626–632. https://doi.org/10.1016/j.fm.2007.12.005

Bata, A., Harrach, B., Ujszaszi, K., Kis-Tamas, A., & Lasztity, R. (1985). Macrocyclic trichothecene toxins produced by *Stachybotrysatra* strains isolated in Middle Europe. *Applied Environmental Microbiology*, 49, 678–681.

Berthiller, F., Crews, C., Dall'Asta, C., Saeger, S. D., Haesaert, G., Karlovsky, P., Oswald, I. P., Seefelder, W., Speijers, G., & Stroka, J. (2013). Masked mycotoxins. *Molecular Nutrition and Food Research*, 57, 165–186.

Berthiller, F., Schuhmacher, R., Adam, G., & Krska, R. (2009). Formation, determination and significance of masked and other conjugated mycotoxins. *Analytical and Bioanalytical Chemistry*, 395, 1243–1252.

Bertuzzi, T., Leggieri, M. C., Battilani, P., Pietri, A. (2014). Co-occurrence of type A and B trichothecenes and zearalenone in wheat grown in northern Italy over the years 2009–2011. *Food Additives and Contaminants: Part B Surveillance*, 7, 273–281.

Bhat, R., Rai, R. V., & Karim, A. (2010). Mycotoxins in food and feed: Present status and future concerns. *Comprehensive Review Food Science Food Safety*, 9, 57–81.

Bhat, R., & Reddy, K. R. N. (2017). Challenges and issues concerning mycotoxins contamination in oil seeds and their edible oils: Updates from last decade. *Food Chemistry*, 215, 425–437. https://doi.org/10.1016/j.foodchem.2016.07.161

Birzele, B., Prange, A., & Krämer, J. (2000). Deoxynivalenol and ochratoxin A in German wheat and changes of level in relation to storage parameters. *Food Additives and Contaminants*, 17, 1027–1035.

Brown, D. W., McCormiek, S. P., Alexander, N. J., Proctor, R. H., & Desjardins, A. E. (2001). A Genetic and biochemical approach to study trichothecene diversity in *Fusarium sporotrichioides* and *Fusarium graminearum*. *Fungal Genetics and Biology*, 32, 121–133.

Bryła, M., Ksieniewicz-Woźniak, E., Waśkiewicz, A., Szymczyk, K., & Jędrzejczak, R. (2018). Natural occurrence of nivalenol, deoxynivalenol, and deoxynivalenol-3-glucoside in polish winter wheat. *Toxins*, 10(2). https://doi.org/10.3390/toxins10020081

Calori-Domingues, M. A., Bernardi, C. M. G., Nardin, M. S., de Souza, G. V., dos Santos, F. G. R., de Abreu Stein, M., da Gloria, E. M., dos Santos Dias, C. T., & de Camargo, A. C. (2016). Co-occurrence and distribution of deoxynivalenol, nivalenol and zearalenone in wheat from Brazil. *Food Additives and Contaminants: Part B Surveillance*, 9, 142–151.

Cano-Sancho, G., Valle-Algarra, F. M., Jiménez, M., Burdaspal, P., Legarda, T. M., Ramos, A. J., Sanchis, V., & Marín, S. (2011). Presence of trichothecenes and co-occurrence in cereal-based food from Catalonia (Spain). *Food Control*, 22(3–4), 490–495. https://doi.org/10.1016/j.foodcont.2010.09.033

Chauhan, R., Singh, J., Sachdev, T., Basu, T., & Malhotra, B. D. (2016). Recent advances in mycotoxins detection. *Biosensors and Bioelectronics*, 81, 532–545. https://doi.org/10.1016/j.bios.2016.03.004

Chen, P., Xiang, B., Shi, H., Yu, P., Song, Y., & Li, S. (2020). Recent advances on type A trichothecenes in food and feed: Analysis, prevalence, toxicity, and decontamination techniques. *Food Control*, 118, 107–371.

Chilaka, C., De Boevre, M., Atanda, O., & De Saeger, S. (2016). Occurrence of *Fusarium* mycotoxins in cereal crops and processed products (ogi) from Nigeria. *Toxins*, 8(11), 342.

Cole, R. J., & Cox, R. H. (1981). *Handbook of Toxic Fungal Metabolites*. Academic Press, New York.

Cooper, D., Doucet, L., & Pratt, M. (2007). Understanding in multinational organizations. *Journal of Organizational Behavior*, 28(3), 303–325. https://doi.org/10.1002/j

Coppa, C. F. S. C., Khaneghah, A. M., Alvito, P., Assunção, R., Martins, C. E. S. I., Gonçalves, B. L., Valganon de Neeff, D., Santana, A. S., & Corassini, C. H. (2019). The occurrence of mycotoxins in breast milk, fruit products and cereal-based infant formula: A review. *Trends in Food Science and Technology*, 92, 81–93.

Darsanaki, R. K., Issazadeh, K., Aliabadi, M. A., & Chakoosari, M. M. D. (2015). Occurrence of deoxynivalenol (DON) in wheat flours in Guilan Province, northern Iran. *Annals of Agricultural and Environmental Medicine*, 22, 35–37.

De Almeida, A. P., Lamardo, L. C. A., Shundo, L., da Silva, S. A., Navas, S. A., Alaburda, J., Ruvieri, V., & Sabino, M. (2016). Occurrence of deoxynivalenol in wheat flour, instant noodle and biscuits commercialised in Brazil. *Food Additives and Contaminants: Part B Surveillance*, 9, 251–255.

De Lima Rocha, D. F., dos Santos Oliveira, M., Furlong, E. B., Junges, A., Paroul, N., Valduga, E., Toniazzo, G. B., Zeni, J., Cansian, R. L. (2017). Evaluation of the TLC quantification method and occurrence of deoxynivalenol in wheat flour of southern Brazil. *Food Additives and Contaminants—Part A Chemistry, Analysis, Control, Exposure and Risk Assessment*, 34, 2220–2229.

Dellafiora, L., & Dall'Asta, C. (2016). Masked mycotoxins: An emerging issue that makes renegotiable what is ordinary. *Food Chemistry*, 213(June), 534–535. https://doi.org/10.1016/j.foodchem.2016.06.112

De Santis, B., Debegnach, F., Gregori, E., Russo, S., Marchegiani, F., Moracci, G., & Brera, C. (2017). Development of a LC-MS/MS method for the multi-mycotoxin determination in composite cereal-based samples. *Toxins*, 9(5). https://doi.org/10.3390/toxins9050169

Desjardins, A. E., Hohn, T. M., & McCormick, S. P. (1993).Trichothecene biosynthesis in *Fusarium* species: Chemistry, genetics, and significance. *Microbiological Reviews*, 57, 595–604.

Desjardins, A. E., & Proctor, R. H. (2007). Molecular biology of *Fusarium* mycotoxins. *International Journal of Food Microbiology*, 119, 47–50.

Dzantiev, B. B., Byzova, N. A., Urusov, A. E., & Zherdev, A. V. (2014). Immunochromatographic methods in food analysis. *TrAC—Trends in Analytical Chemistry*, 55, 81–93. https://doi.org/10.1016/j.trac.2013.11.007

Edwards, S. G. (2009a). *Fusarium* mycotoxin content of UK organic and conventional oats. *Food Additives & Contaminants*, 26, 1063–1069.

Edwards, S. G. (2009b). *Fusarium* mycotoxin content of UK organic and conventional barley. *Food Additives & Contaminants*, 26, 1185–1190.

Edwards, S. G. (2009c). *Fusarium* mycotoxin content of UK organic and conventional wheat. *Food Additives & Contaminants*, 26, 496–506.

Edwards, S. G. (2017). Impact of agronomic and climatic factors on the mycotoxin content of harvested oats in the United Kingdom. *Food Additives and Contaminants—Part A Chemistry, Analysis, Control, Exposure and Risk Assessment*, 34(12), 2230–2241. https://doi.org/10.1080/19440049.2017.1372639

EFSA (European Food Safety Authority). (2011). Scientific opinion on the risks for animal and public health related to the presence of T-2 and HT-2 toxin in food and feed. *EFSA Journal*, 2481–2668.

EFSA (European Food Safety Authority). (2014). Scientific opinion on the risks for human and animal health related to the presence of modified forms of certain mycotoxins in food and feed. *EFSA Journal*, 12, 3916–4023.

EFSA CONTAM Panel (EFSA Panel on Contaminants in the Food Chain). (2014). Scientific opinion on the risks to human and animal health related to the presence of beauvericin and enniatins in food and feed. *EFSA Journal*, 12, 3802.

Elaridi, J., Yamani, O., al Matari, A., Dakroub, S., & Attieh, Z. (2019). Determination of ochratoxin A (OTA), ochratoxin B (OTB), T-2, and HT-2 toxins in wheat grains, wheat flour, and bread in Lebanon by LC-MS/MS. *Toxins*, 11(8). https://doi.org/10.3390/toxins11080471

Feltrin, A. C. P., Sibaja, K. V. M., Tusnski, C., Caldas, S. S., Primel, E. G., & Garda-Buffon, J. (2018). Evaluation of the suitability of analytical methods in trichothecene A and B degradation. *Journal of the Brazilian Chemical Society*, 29(10), 2117–2126. https://doi.org/10.21577/0103-5053.20180086

Fink-Gremmels, J., & Van der Merwe, D. (2019). Mycotoxins in the food chain: Contamination of foods of animal origin. In *Chemical Hazards in Foods of Animal Origin. Food Safety Assurance and Veterinary Public Health*. Eds. F. J. M. Smulders, I. M. C. M. Rietjens, M. D. Rose. Wageningen Academic Publishers, Wageningen, The Netherlands, 241–261.

Freire, L., & Sant'Ana, A. S. (2018). Modified mycotoxins: An updated review on their formation, detection, occurrence, and toxic effects. *Food and Chemical Toxicology*, 111(November 2017), 189–205. https://doi.org/10.1016/j.fct.2017.11.021

Girolamo, A. De, Ciasca, B., Pascale, M., & Lattanzio, V. M. T. (2020). Determination of zearalenone and trichothecenes, including deoxynivalenol and its acetylated derivatives, nivalenol, T-2 and HT-2 toxins, in wheat and wheat products by LC-MS/MS: A collaborative study. *Toxins*, 12(12), 1–17. https://doi.org/10.3390/toxins12120786

Gonçalves, A., Gkrillas, A., Dorne, J. L., Dall'Asta, C., Palumbo, R., Lima, N., Battilani, P., Venâncio, A., & Giorni, P. (2019). Pre- and postharvest strategies to minimize mycotoxin contamination in the rice food chain. *Comprehensive Reviews in Food Science and Food Safety*, 18(2), 441–454. https://doi.org/10.1111/1541-4337.12420

Gonçalves, C., Mischke, C., & Stroka, J. (2020). Determination of deoxynivalenol and its major conjugates in cereals using an organic solvent-free extraction and IAC cleanup coupled in-line with HPLC-PCD-FLD. *Food Additives and Contaminants—Part A Chemistry, Analysis, Control, Exposure and Risk Assessment*, 37(10), 1765–1776. https://doi.org/10.1080/19440049.2020.1800829

González-Jartín, J. M., de Castro Alves, L., Alfonso, A., Piñeiro, Y., Vilar, S. Y., Gomez, M. G., Osorio, Z. V., Sainz, M. J., Vieytes, M. R., Rivas, J., & Botana, L. M. (2019). Detoxification agents based on magnetic nanostructured particles as a novel strategy for mycotoxin mitigation in food. *Food Chemistry*, 294(April), 60–66. https://doi.org/10.1016/j.foodchem.2019.05.013

Gottschalk, C., Barthel, J., Engelhardt, G., Bauer, J., & Meyer, K. (2007). Occurrence of type A trichothecenes in conventionally and organically produced oats and oat products. *Molecular Nutrition & Food Research*, 51(12), 1547–1553

Gummadidala, P. M., Omebeyinje, M. H., Burch, J. A., Biswas, P. K., Banerjee, K., Wang, Q., Jesmin, R., Mitra, C., R Moeller, P. D., Scott, G. I., & Chanda, A. (2018). Complementary feeding may pose a risk of simultaneous exposures to aflatoxin 1 M1 and deoxynivalenol in Indian infants and toddlers: Lessons from a mini-2 survey of food samples obtained from Kolkata, India. *Food and Chemical Toxicology*, 123, 9–15.

Gunupuru, L. R., Patel, J. S., Sumarah, M. W., Renaud, J. B., Mantin, E. G., & Prithiviraj, B. (2019). A plant biostimulant made from the marine brown algae *Ascophyllum nodosum* and chitosan reduce *Fusarium* head blight and mycotoxin contamination in wheat. *PLoS ONE*, 14(9), 1–19. https://doi.org/10.1371/journal.pone.0220562.

Gupta, R. C. (Ed.). (2012). *Veterinary Toxicology: Basic and Clinical Principles*. Academic Press.

Habler, K., Gotthardt, M., Schüler, J., & Rychlik, M. (2017). Multi-mycotoxin stable isotope dilution LC–MS/MS method for *Fusarium* toxins in beer. *Food Chemistry*, 218, 447–454. https://doi.org/10.1016/j.foodchem.2016.09.100

Han, Z., Nie, D., Ediage, E. N., Yang, X., Wang, J., Chen, B., Li, S., On, S. L. W., De Saeger, S., & Wu, A. (2014). Cumulative health risk assessment of co-occurring mycotoxins of deoxynivalenol and its acetyl derivatives in wheat and maize: Case study, Shanghai, China. *Food and Chemical Toxicology*, 74, 334–342.

Haque, M. D. A., Wang, Y., Shen, Z., Li, X., Saleemi, M. K. (2020). Mycotoxin contamination and control strategy in human, domestic animal and poultry: A review. *Microbiology and Pathology*, 142, 104095.

He, J., Zhou, T., Young, J. C., Boland, G. J., & Scott, P. M. (2010). Chemical and biological transformations for detoxification of trichothecene mycotoxins in human and animal food chains: A review. *Trends Food Science and Technology*, 21, 67–76.

Hietaniemi, V., Rämö, S., Yli-Mattila, T., Jestoi, M., Peltonen, S., Kartio, M., Sieviläinen, E., Koivisto, T., & Parikka, P. (2016). Updated survey of *Fusarium* species and toxins in Finnish cereal grains. *Food Additives and Contaminants—Part A Chemistry, Analysis, Control, Exposure and Risk Assessment*, 33(5), 831–848. https://doi.org/10.1080/19440049.2016.1162112

Hohn, T. M., & Beremand, P. (1989). Isolation and nucleotide sequence of a sesquiterpene cyclase gene from the trichothecene-producing fungus *Fusarium sporotrichioides*. *Gene*, 79, 131–138.

Hohn, T. M., Desjardins, A. E., & McCormick, S. P. (1995). The TRI4 gene of *Fusarium sporotrichioides* encodes a cytochrome P450 mo-nooxygenase involved in trichothecene biosynthesis. *Molecular Genetics and Genomics*, 248, 95–102.

Hohn, T. M., Krishna, R., & Proctor, R. H. (1999). Characterization of a transcriptional activator controlling trichothecene toxin biosynthesis. *Fungal Genetics and Biology*, 26, 224.

Hohn, T. M., McCormick, S. P., & Desjardins, A. E. (1993). Evidence for a gene cluster involving trichothecene-pathway biosynthetic genes in *Fusarium sporotrichioides*. *Current Genetics*, 24, 291.

Homdork, S., Fehrmann, H., & Beck, R. (2000). Influence of different storage conditions on the mycotoxin production and quality of *Fusarium* infected wheat grain. *Journal of Phytopathology*, 148, 7–15.

Hove, M., De Boevre, M., Lachat, C., Jacxsens, L., Nyanga, L. K., & De Saeger, S. (2016). Occurrence and risk assessment of mycotoxins in subsistence farmed maize from Zimbabwe. *Food Control*, 69, 36–44.

Ji, F., He, D., Olaniran, A. O., Mokoena, M. P., Xu, J., & Shi, J. (2019). Occurrence, toxicity, production and detection of *Fusarium* mycotoxin: A review. *Food Production Processing and Nutrition*, 1, 1–14.

Ji, F., Xu, J., Liu, X., Yin, X., & Shi, J. (2014).Natural occurrence of deoxynivalenol and zearalenone in wheat from Jiangsu province, China. *Food Chemistry*, 157, 393–397.

Ji, J., & Xie, W. (2020). Detoxification of aflatoxin B1 by magnetic graphene composite adsorbents from contaminated oils. *Journal of Hazardous Materials*, 381(July 2019), 120915. https://doi.org/10.1016/j.jhazmat.2019.120915

Jodra, A., López, M. Á., & Escarpa, A. (2015). Disposable and reliable electrochemical magnetoimmunosensor for fumonisins simplified determination in maize-based foodstuffs. *Biosensors and Bioelectronics*, 64, 633–638. https://doi.org/10.1016/j.bios.2014.09.054

Juan, C., Ritieni, A., & Mañes, J. (2013).Occurrence of *Fusarium* mycotoxins in Italian cereal and cereal products from organic farming. *Food Chemistry*, 141(3), 1747–1755.

Kachuei, R., Rezaie, S., Yadegari, M. H., Safaie, N., Allameh, A. A., Aref-Poor, M. A., Fooladi, A. A. I., Riazipour, M., & Abadi, H. M. M. (2014). Determination of T-2 mycotoxin in *Fusarium* strains by HPLC with fluorescence detector. *Journal of Applied Biotechnology Reports*, 1(1), 38–43.

Karlovsky, P., Suman, M., Berthiller, F., De Meester, J., Eisenbrand, G., Perrin, I., Oswald, I. P., Speijers, G., Chiodini, A., Recker, T., & Dussort, P. (2016). Impact of food processing and detoxification treatments on mycotoxin contamination. *Mycotoxin Research*, 32(4), 179–205. https://doi.org/10.1007/s12550-016-0257-7

Keller, N. P., & Hohn, T. M. (1997). Metabolic pathway gene clusters in filamentous fungi. *Fungal Genetics and Biology*, 21, 17–29.

Khaneghah, A. M., Farhadi, A., Nematollahi, A., Vasseghian, Y., & Fakhri, Y. A. (2020). Systematic review and meta-analysis to investigate the concentration and prevalence of trichothecenes in the cereal-based food. *Trends Food Science and Technology*, 102, 193–202.

Kimura, M., Matsumoto, G., Shingu, Y., Yoneyama, K., & Yamaguchi, I. (1998). The mystery of the trichothecene 3-Oacetyltransferase gene. Analysis of the region around Tri 101 and characterisation of its homologue from *Fusarium sporotrichioides*. *FEBS Letters*, 435, 163–168.

Kimura, M., Tokai, T., Takahashi-Ando, N., Ohsato, S., & Fujimura, M. (2007). Molecular and genetic studies of *Fusarium* trichothecene biosynthesis: Pathways, genes, and evolution. *BiosciBiotechnolBiochem*, 71, 2105–2123.

Kollia, E., Proestos, C., Zoumpoulakis, P., & Markaki, P. (2019). Capsaicin, an inhibitor of ochratoxin A production by *Aspergillus* section Nigri strains in grapes (*Vitis vinifera* L.). *Food Additives and Contaminants—Part A Chemistry, Analysis, Control, Exposure and Risk Assessment*, 36(11), 1709–1721. https://doi.org/10.1080/19440049.2019.1652771.

Krska, R., Malachova, A., Berthiller, F., & Egmond, H. P. V. (2014). Determination of T-2 and HT-2 toxins in food and feed: An update. *World Mycotoxin Journal*, 18, 131–142.

Liu, L., Chao, Y., Cao, W., Wang, Y., Luo, C., Pang, X., Fan, D., & Wei, Q. (2014). A label-free amperometric immunosensor for detection of zearalenone based on trimetallic Au-core/AgPt-shell nanorattles and mesoporous carbon. *Analytica Chimica Acta*, 847, 29–36. https://doi.org/10.1016/j.aca.2014.07.026

Liu, Y., Lu, Y., Wang, L., Chang, F., & Yang, L. (2015). Survey of 11 mycotoxins in wheat flour in Hebei province, China. *Food Additives and Contaminants: Part B Surveillance*, 8, 250–254.

Llorens, A., Mateo, R., Hinojo, M. J., Valle-Algarra, F. M., & Jiménez, M. (2004). Influence of environmental factors on the biosynthesis of type B trichothecenes by isolates of *Fusarium* spp. from Spanish crops. *International Journal of Food Microbiology*, 94(1), 43–54.

Luo, Y., Liu, X., & Li, J. (2018). Updating techniques on controlling mycotoxins—A review. *Food Control*, 89, 123–132. https://doi.org/10.1016/j.foodcont.2018.01.016

Lyagin, I., & Efremenko, E. (2019). Enzymes for detoxification of various mycotoxins: Origins and mechanisms of catalytic action. *Molecules*, 1, 1–39.

Majeed, S., de Boevre, M., de Saeger, S., Rauf, W., Tawab, A., Fazal-e-Habib, Rahman, M., & Iqbal, M. (2018). Multiple mycotoxins in rice: Occurrence and health risk assessment in children and adults of Punjab, Pakistan. *Toxins*, 10(2). https://doi.org/10.3390/toxins10020077

Mantle, P. G. (1991). Miscellaneous toxigenic fungi. In *Mycotoxins in Animal Foods*. Eds. J. E. Smith, R. S. Henderson. CRC Press, Inc., Boca Raton, FL, 141–152.

McCormick, S. P., Hohn, T. M., & Desjardins, A. E. (1996). Isolation and characterization of Tri3, a gene encoding 15-O-acetyltransferase from *Fusarium sporotrichioides*. *Applied and Environmental Microbiology*, 62, 353–359.

McCormick, S. P., Stanley, A. M., Stover, N. A., & Alexander, N. J. (2011). Trichothecenes: From simple to complex mycotoxins. *Toxins*, 3, 802–814.

Meneely, J. P., Ricci, F., van Egmond, H. P., & Elliott, C. T. (2011). Current methods of analysis for the determination of trichothecene mycotoxins in food. *TrAC—Trends in Analytical Chemistry*, 30(2), 192–203. https://doi.org/10.1016/j.trac.2010.06.012

Merhej, J., Richard-Forget, F., & Barreau, C. (2011). Regulation of trichothecene biosynthesis in *Fusarium*: Recent advances and new insights. *Applied Microbiology and Biotechnology*, 91, 519–528.

Miller, J. D. (2002). Aspects of the ecology of *Fusarium* toxins in cereals. In *Mycotoxins and Food Safety*. Eds. J. W. DeVries, M. W. Trucksess, L. S. Jackson. AdvExp Med Biol 54. Kluwer Academic/Plenum Publishers, New York, 19–27.

Morcia, C., Tumino, G., Ghizzoni, R., Badeck, F., Lattanzio, V., Pascale, M., et al. (2016). Occurrence of *Fusarium langsethiae* and T-2 and HT-2 toxins in Italian malting barley. *Toxins*, 8(8), 247.

Moretti, A., Susca, A., Mulè, G., Logrieco, A. F., & Proctor, R. H. (2013). Molecular diversity of mycotoxigenic fungi that threaten food safety. *International Journal of Food Microbiology*, 167, 57–66.

Nathanail, A. V., Syvähuoko, J., Malachová, A., Jestoi, M., Varga, E., Michlmayr, H., Adam, G., Sieviläinen, E., Berthiller, F., Peltonen, K. (2015). Simultaneous determination of major type A and B trichothecenes, zearalenone and certain modified metabolites in Finnish cereal grains with a novel liquid chromatography-tandem mass spectrometric method. *Analytical and Bioanalytical Chemistry*, 407, 4745–4755.

Neme, K., & Mohammed, A. (2017). Mycotoxin occurrence in grains and the role of postharvest management as a mitigation strategies. A review. *Food Control*, 78, 412–425. https://doi.org/10.1016/j.foodcont.2017.03.012

Ojuri, O. T., Ezekiel, C. N., Eskola, M. K., Šarkanj, B., Babalola, A. D., Sulyok, M., Hajšlová, J., Elliott, C. T., & Krska, R. (2019).Mycotoxin co-exposures in infants and young children consuming household- and industrially-processed complementary foods in Nigeria and risk management advice. *Food Control*, 98, 312–322.

Oswald, S., Karsunke, X. Y. Z., Dietrich, R., Märtlbauer, E., Niessner, R., & Knopp, D. (2013). Automated regenerable microarray-based immunoassay for rapid parallel quantification of mycotoxins in cereals. *Analytical and Bioanalytical Chemistry*, 405(20), 6405–6415. https://doi.org/10.1007/s00216-013-6920-3

Palacios, S. A., Erazo, J. G., Ciasca, B., Lattanzio, V. M. T., Reynoso, M. M., Farnochi, M. C., & Torres, A. M. (2017). Occurrence of deoxynivalenol and deoxynivalenol-3-glucoside in durum wheat from Argentina. *Food Chemistry*, 230, 728–734.

Pallarés, N., Font, G., Mañes, J., & Ferrer, E. (2017). Multimycotoxin LC-MS/MS analysis in tea beverages after dispersive liquid-liquid microextraction (DLLME). *Journal of Agricultural and Food Chemistry*, 65(47), 10282–10289. https://doi.org/10.1021/acs.jafc.7b03507

Paterson, R. R. M., & Lima, N. (2010). How will climate change affect mycotoxins in food? *Food Research International*, 43, 1902–1914.

Piec, J., Pallez, M., Beyer, M., Vogelgsang, S., Hoffmann, L., & Pasquali, M. (2016). The Luxembourg database of trichothecene type B *F. graminearum* and *F. culmorum* producers. *Bioinformation*, 12, 1–3.

Piemontese, L., Messia, M. C., Marconi, E., Falasca, L., Zivoli, R., Gambacorta, L., Perrone, G., & Solfrizzo, M. (2018). Effect of gaseous ozone treatments on DON, microbial contaminants and technological parameters of wheat and semolina. *Food Additives and Contaminants—Part A Chemistry, Analysis, Control, Exposure and Risk Assessment*, 35(4), 760–771. https://doi.org/10.1080/19440049.2017.1419285

Pierron, A., Alassane-Kpembi, I., Oswald, I. P., Kamle M., Mahato D. K., Devi S., Lee K. E., Kang, S. G., Kumar, P., Martins, H. M. L., Almeida, I. F. M., Camacho, C. R. L., Santos, S. M. O., Costa, J. M. G., & Bernardo, F. M. A. (2016). Fumonisins: Impact on agriculture, food, and human health and their management strategies Madhu. *Porcine Health Management*, 29(3), 1–8. http://dx.doi.org/10.1186/s40813-016-0041-2

Pinton, P., & Oswald, I. P. (2014). Effect of deoxynivalenol and other Type B trichothecenes on the intestine: A review. *Toxins*, 6, 1615–1643

Proctor, R. H., Hohn, T. M., McCormick, S. P., & Desjardins, A. E. (1996). Tri 6 encodes an unusual zinc finger protein involved in regulation of trichothecene biosynthesis in *Fusarium sporotrichioides*. *Applied and Environmental Microbiology*, 61, 1923–1930.

Proctor, R. H., McCormick, S. P., Alexnader, N. J., & Desjardins, A. E. (2009). Evidence that a secondary metabolic biosynthetic gene cluster has grown by gene relocation during evolution of the filamentous fungus *Fusarium*. *MolMicrobiol*, 74, 1128–1142.

Righetti, L., Bergmann, A., Galaverna, G., Rolfsson, O., Paglia, G., & Dall'Asta, C. (2018). Ion mobility-derived collision cross section database: Application to mycotoxin analysis. *Analytica Chimica Acta*, 1014, 50–57. https://doi.org/10.1016/j.aca.2018.01.047

Rocha, O., Ansari, K., & Doohan, F. M. (2005). Effects of trichothecene mycotoxins on eukaryotic cells: A review. *Food Additive and Contaminants*, 22, 369–378.

Rodríguez-Carrasco, Y., Berrada, H., Font, G., & Mañes, J. (2012). Multi-mycotoxin analysis in wheat semolina using an acetonitrile-based extraction procedure and gas chromatography-tandem mass spectrometry. *Journal of Chromatography A*, 1270, 28–40. https://doi.org/10.1016/j.chroma.2012.10.061

Rybecky, A. I., Chulze, S. N., & Chiotta, M. L. (2018). Effect of water activity and temperature on growth and trichothecene production by *Fusarium meridionale*. *International Journal of Food Microbiology*, 285, 69–73.

Sarrocco, S., Mauro, A., & Battilani, P. (2019). Use of competitive filamentous fungi as an alternative approach for mycotoxin risk reduction in staple cereals: State of art and future perspectives. *Toxins*, 11(12), 1–18. https://doi.org/10.3390/toxins11120701

Sarrocco, S., & Vannacci, G. (2018). Preharvest application of beneficial fungi as a strategy to prevent postharvest mycotoxin contamination: A review. *Crop Protection*, 110(April), 160–170. https://doi.org/10.1016/j.cropro.2017.11.013

Savi, G. D., Piacentini, K. C., Tibola, C. S., & Scussel, V. M. (2014). Mycoflora and deoxynivalenol in whole wheat grains (*Triticum aestivum* L.) from Southern. *Food Additives & Contaminants: Part B*, 7, 232–237.

Schwabe, M., & Krämer, J. (1995). Influence of water activity on the production of T-2 toxin by *Fusarium sporotrichioides*. *Mycotoxin Research*, 11, 35–39.

Scott, P. M. (1989). The natural occurrence of trichothecenes. In *Trichothecene Mycotoxicosis: Pathophysiologic Effects*, vol. I. Ed. V. R. Beasley. CRC Press, Inc., Boca Raton, FL, 1–26.

Scudamore, K. A., Patel, S., & Edwards, S. G. (2009). HT-2 toxin and T-2 toxin in commercial cereal processing in the United Kingdom, 2004–2007. *World Mycotoxin Journal*, 2, 357–365.

Shanakhat, H., Sorrentino, A., Raiola, A., Romano, A., Masi, P., & Cavella, S. (2018). Current methods for mycotoxins analysis and innovative strategies for their reduction in cereals: an overview. *Journal of the Science of Food and Agriculture*, 98(11), 4003–4013. https://doi.org/10.1002/jsfa.8933

Shi, H., Li, S., Bai, Y., Prates, L. L., Lei, Y., & Yu, P. (2018). Mycotoxin contamination of food and feed in China: Occurrence, detection techniques, toxicological effects and advances in mitigation technologies. *Food Control*, 91, 202–215. https://doi.org/10.1016/j.foodcont.2018.03.036

Silva, M. V., Pante, G. C., Romoli, J. C. Z., de Souza, A. P. M., Rocha, G. H. O. da, Ferreira, F. D., Feijó, A. L. R., Moscardi, S. M. P., de Paula, K. R., Bando, E., Nerilo, S. B., & Machinski, M. (2018). Occurrence and risk assessment of population exposed to deoxynivalenol in foods derived from wheat flour in Brazil. *Food Additives and Contaminants—Part A Chemistry, Analysis, Control, Exposure and Risk Assessment*, 35(3), 546–554. https://doi.org/10.1080/19440049.2017.1411613

Singh, J., & Mehta, A. (2020). Rapid and sensitive detection of mycotoxins by advanced and emerging analytical methods: A review. *Food Science and Nutrition*, 8(5), 2183–2204. https://doi.org/10.1002/fsn3.1474

Smith, M. C., Madec, S., Coton, E., & Hymery, N. (2016). Natural co-occurrence of mycotoxins in foods and feeds and their in vitro combined toxicological effects. *Toxins*, 8(4). https://doi.org/10.3390/toxins8040094

Song, S., Liu, N., Zhao, Z., Njumbe Ediage, E., Wu, S., Sun, C., De Saeger, S., & Wu, A. (2014). Multiplex lateral flow immunoassay for mycotoxin determination. *Analytical Chemistry*, 86(10), 4995–5001. https://doi.org/10.1021/ac500540z

Streit, E., Schatzmayr, G., Tassis, P., Tzika, E., Marin, D., Taranu, I., Tabuc, C., Nicolau, A., Aprodu, I., & Puel, O. (2012). Current situation of mycotoxin contamination and co-occurrence in animal feed-focus on Europe. *Toxins*, 4, 788–809.

Tanaka, T., Hasegawa, A., Yamamoto, S., Lee, U. S., Sugiura, Y., & Ueno, Y. (1988). World-wide contamination of cereals by the *Fusarium* mycotoxins nivalenol, deoxyniva-lenol, and zearalenone: 1. Survey of 19 countries. *Journal of Agricultural and Food Chemistry*, 36, 979–983.

Tarazona, A., Gómez, J. V., Mateo, E. M., Jiménez, M., & Mateo, F. (2019). Antifungal effect of engineered silver nanoparticles on phytopathogenic and toxigenic *Fusarium* spp. and their impact on mycotoxin accumulation. *International Journal of Food Microbiology*, 306(March), 108259. https://doi.org/10.1016/j.ijfoodmicro.2019.108259

Tericiolo, C., Maresca, M., Pinton, P., & Oswald, I. P. (2018,).Role of satiety hormones in anorexia induction by trichothecene mycotoxins. *Food Chemistry and Toxicology*, 121, 701–714.

Thrane, U., Adler, A., Clasen, P. E., Galvano, F., Langseth, W., Lew, H., Logrieco, A., Nielsen, K. F., & Ritieni, A. (2004). Diversity in metabolite production by *Fusarium langs-ethiae*, *Fusarium poae*, and *Fusarium sporotrichioides*. *International Journal of Food Microbiology*, 95, 257–266.

Tima, H., Brückner, A., Mohácsi-Farkas, C., & Kiskó, G. (2016). *Fusarium* mycotoxins in cereals harvested from Hungarian fields. *Food Additives and Contaminants: Part B Surveillance*, 9(2), 127–131. https://doi.org/10.1080/19393210.2016.1151948

Toregeani-Mendes, K. A., Arroteia, C. C., Kemmelmeier, C., Dalpasquale, V. A., Bando, É., Alves, A. F., Marques, O. J., Nishiyama, P., Mossini, S. A., & Machinski, M. Jr. (2011). Application of hazard analysis critical control points system for the control of afla-toxins in the Brazilian groundnut-based food industry. *International Journal on Food Science and Technology*, 46, 2611–2618.

Torres-Chavolla, E., & Alocilja, E. C. (2009). Aptasensors for detection of microbial and viral pathogens. *Biosensors and Bioelectronics*, 24(11), 3175–3182. https://doi.org/10.1016/j.bios.2008.11.010

Tralamazza, S. M., Bemvenuti, R. H., Zorzete, P., De Souza Garcia, F., & Corrêa, B. (2016). Fungal diversity and natural occurrence of deoxynivalenol and zearalenone in freshly harvested wheat grains from Brazil. *Food Chemistry*, 196, 445–450.

Trapp, S. C., Hohn, T. M., McCormick, S., & Jarvis, B. B. (1998). Characterisation of the gene cluster for biosynthesis of macrocyclic trichothecenes in *Myrothecium rori-dum*. *Molecular Genetics and Genomics*, 257, 421–432.

Van der Fels-Klerx, H. J. (2010). Occurrence data of trichothecene mycotoxins T-2 toxin and HT-2 toxin in food and feed. *EFSA Supporting Publications*, 7(7)

Vogelgsang, S., Musa, T., Bänziger, I., Kägi, A., Bucheli, T. D., Wettstein, F. E., Pasquali, M., & Forrer, H. R. (2017). *Fusarium* mycotoxins in Swiss wheat: A survey of growers' samples between 2007 and 2014 shows strong year and minor geographic effects. *Toxins*, 9(8). https://doi.org/10.3390/toxins9080246

Wang, J. H., Li, H. P., Qu, B., Zhang, J. B., Huang, T., Chen, F. F., & Liao, Y. C. (2008). Development of a generic PCR detection of 3-acetyldeoxy-nivalenol-, 15-acetyldeoxynivalenol- and nivalenol-chemotypes of *Fusarium graminearum* clade. *International Journal of Molecular Sciences*, 9(12), 2495–2504. https://doi.org/10.3390/ijms9122495

Wang, L., Wu, J., Liu, Z., Shi, Y., Liu, J., Xu, X., Hao, S., Mu, P., Deng, F., & Deng, Y. (2019). Aflatoxin B1 degradation and detoxification by *Escherichia coli* CG1061 isolated from chicken cecum. *Frontiers in Pharmacology*, 9(January), 1–9. https://doi.org/10.3389/fphar.2018.01548

Xing, F., Liu, X., Wang, L., Selvaraj, J. N., Jin, N., Wang, Y., Zhao, Y., Liu, Y. (2017). Distribution and variation of fungi and major mycotoxins in pre- and post-nature drying maize in North China Plain. *Food Control*, 80, 244–251.

Zachetti, V. G. L., Cendoya, E., Nichea, M. J., Chulze, S. N., & Ramirez, M. L. (2019). Preliminary study on the use of chitosan as an eco-friendly alternative to control fusarium growth and mycotoxin production on maize and wheat. *Pathogens*, 8(1). https://doi.org/10.3390/pathogens8010029

Zaki, M. (2012). Mycotoxins in animals: Occurrence, effects, prevention and management. *Journal of Toxicology and Environmental Health Sciences*, 4(1), 13–28. https://doi.org/10.5897/jtehs11.072

FURTHER READINGS

Adegoke, G. O., & Letuma, P. (2013). Strategies for the prevention and reduction of mycotoxins in developing countries. In *Mycotoxin and Food Safety in Developing Countries*. Ed. H. A. Makun. InTech Croat, 123–136.

Adhikari, M., Negi, B., Kaushik, N., Adhikari, A., Al-Khedhairy, A. A., Nagendra Kumar Kaushik, N. K., & Ha Choi, E. (2017). T-2 mycotoxin: Toxicological effects and decontamination strategies. *Oncotarget*, 8, 33933–33952.

Alexander, N. J., McCormick, S. P., & Hohn, T. M. (1999c). Tri12, a trichothecene efflux pump from *Fusarium sporotrichioides*: Gene isolation and expression in yeast. *Molecular Genetics and Genomics*, 261, 977.

Alexander, N. J., McCormick, S. P., & Ziegenhorn, S. L. (1999b). Phytotoxicity of selected trichothecenes using *Chlamydomonas reinhardtii* as a model system. *Natural Toxins*, 7, 265–269.

Amirahmadi, M., Shoeibi, S., Rastegar, H., Elmi, M., & Khangeghah, A. M. (2018). Simultaneous analysis of mycotoxins in corn flour using LC/MS-MS combined with a modified QuEChERS procedure. *Toxin Review*, 37, 187–195.

Aoki, T., O'Donnell, K., & Geiser, D. M. (2014). Systematics of key phytopathogenic *Fusarium* species: Current status and future challenges. *Journal of General Plant Pathology*, 80, 189–201.

Creppy, E. E. (2002). Update of survey, regulation and toxic effects of mycotoxins in Europe. *Toxicology Letter*, 127, 19–28.

Gil-Serna, J., Vázquez, C., González-Jaén, M. T., & Patiño, B. (2014). Mycotoxins: Toxicology. In *Encyclopedia of Food Microbiology*. Eds. C. Batt, M. L. Tortorello, 2nd edn. Elsevier Ltd. Academic Press, Amsterdam, 1539–1547.

Hussein, H. S., & Brasel, J. M. (2001). Toxicity, metabolism, and impact of mycotoxins on humans and animals. *Toxicology*, 167, 101–134

Magan, N. V., & Aldred, D. (2004). Role of spoilage fungi in seed deterioration. In *Fungal Biotechnology in Agricultural, Food and Environmental Applications*. Ed. D. K. Aurora. Marcell Dekker, New York, 311–323.

McCormick, S. P., Alexander, N. J., Trapp, S. E., & Hohn, T. M. (1999). Disruption of Tri101, the gene encoding trichothecene 3-O-acetyltransferase, from *Fusarium sporotrichioides*. *Applied and Environmental Microbiology*, 65, 5252.

Nathanail, A. V., Gibson, B., Han, L., Peltonen, K., Ollilainen, V., Jestoi, M., et al. (2016). The lager yeast Saccharomyces pastorianus removes and transforms *Fusarium* trichothecene mycotoxins during fermentation of brewer's wort. *Food Chemistry*, 203, 448–455

Oueslati, S., Berrada, H., Mañes, J., & Juan, C. (2018). Presence of mycotoxins in Tunisian infant foods samples and subsequent risk assessment. *Food Control*, 84, 362–369.

Proctor, R. H., Desjardins, A. E., McCormick, S. P., Plattner, R. D., Alexander, N. J., & Brown, D. W. (2002). Genetic analysis of the role of trichothecene and fumonisin mycotoxins in the virulence of *Fusarium. European Journal of Plant Pathology*, 108, 691–698.

Proctor, R. H., McCormick, S. P., Kim, H. S., Cardoza, R. E., Stanley, A. M., Lindo, L., et al. (2018). Evolution of structural diversity of trichothecenes, a family of toxins produced by plant pathogenic and entomopathogenic fungi. *PLoS Pathog*, 14, e1006946.

Restuccia, C., Giusino, F., Licciardello, F., Randazzo, C., Caggia, C., & Muratore, G. (2006). Biological control of peach fungal pathogens by commercial products and indigenous yeasts. *Journal of Food Protection*, 69, 2465–2470.

Sudakin, D. L. (1998). Toxigenic fungi in a water-damaged building: An intervention study. *American Journal of Industrial Medicine*, 34, 183–190.

Swanson, S. P., Nicoletti, J., Rood, H. D., Buck, W. B., Côte, L. M., & Yoshizawa, T. (1987). Metabolism of three trichothecene mycotoxins, T-2 toxin, diacetoxyscirpenol, and deoxynivalenol, by bovine rumen microorganisms. *Journal on Chromatography*, 414, 335–342.

Tolosa, J., Graziani, G., Gaspari, A., Chianese, D., Ferrer, E., Manes, J., & Ritieni, A. (2017). Multi-mycotoxin analysis in durum wheat pasta by liquid chromatography coupled to quadrupole orbitrap mass spectrometry. *Toxins*, 9, 59.

Van Egmond, H. P., Schothorst, R. C., & Jonker, M. A. (2007). Regulation relating to mycotoxins in food. Perspectives in a global and European context. *Analytical and Bioanalytical Chemistry*, 389, 147–157.

Zhu, M., Cen, Y., Ye, W., Li, S., & Zhang, W. (2020). Recent advances on macrocyclic trichothecenes, their bioactivities and biosynthetic pathway. *Toxins*, 12, 417.

Chapter 5

Detection and Management Strategies for Deoxynivalenol in Food and Feed

An Overview

Raman Selvakumar, Dalasanuru Chandregowda
Manjunathagowda, Arun Kumar Pandey, Dipendra
Kumar Mahato, Akansha Gupta, Shikha Pandhi,
Raveena Kargwal, Madhu Kamle, Pradeep Kumar

CONTENTS

5.1	Introduction	120
5.2	Major Sources of Deoxynivalenol	121
5.3	Chemistry and Biosynthesis of Deoxynivalenol	122
5.4	Genes Responsible for Production	124
5.5	Occurrence in Food and Feed	124
5.6	Effects on Agricultural Food and Feed	128
5.7	Mechanism of Toxicity and Health Effects of Deoxynivalenol	129
5.8	Effects of Processing on Deoxynivalenol	130
5.9	Effects of Environmental Factors on Deoxynivalenol Production	132
5.10	Detection Techniques	133
5.11	Masked Mycotoxins as a Major Concern in Detection	134
5.12	Degradation Kinetics	135
5.13	Management and Control Strategies	137
5.14	Conclusion	138

DOI: 10.1201/9781003242208-5

5.1 INTRODUCTION

Natural foods and feeds include toxins that are a significant source of worry for human and animal health (Kumar et al., 2017; Pleadin et al., 2019; Mahato et al., 2021). It is mycotoxins, which are toxic fungal metabolites that develop in food or feed, that pose a significant health danger to both humans and livestock (Kamle et al., 2019; Mahato et al., 2019; Kumar et al., 2020; Xu et al., 2021). When mycotoxins enter the bodies of humans and animals, they are known to cause major health problems in a number of ways. Mycotoxins have been reported to be carcinogenic, mutagenic, teratogenic, and immunosuppressive when consumed in food or animal feed (Kebede et al., 2020). *Aspergillus*, *Fusarium*, and *Penicillium* are the most prevalent fungal genera that produce mycotoxins. Mycotoxin contamination is a global issue, but it is aggravated in warm, humid environments that encourage fungus growth and the creation of mycotoxins. Microbial contamination in agricultural and food production has a negative economic effect on the industry. Apart from the negative effects on people's health, other major mycotoxin problems include reduced agricultural productivity, the removal and destruction of mycotoxin-contaminated products, and the economic challenges associated with mycotoxin contamination of crops and food items (Afsah-Hejri et al., 2020).

One of the most prevalent mycotoxins identified in grain harvests across the world is deoxynivalenol (DON). DON is a type B trichothecene that *Fusarium graminearum* and *Fusarium culmorum* produce frequently (Nagl et al., 2015; Yuan et al., 2017). It's also known as vomitoxin (Zhou et al., 2020) because of its emetic properties in pigs and gastrointestinal symptoms in people. It's found in wheat, maize, barley, rye, oat, and safflower seed. Africa, Asia (especially China), America, Europe, and the Middle East are all affected by DON pollution (Guo et al., 2020). DON is a colorless powder that dissolves in water, methanol, ethanol, acetonitrile, and ethyl acetate, among other solvents. It is stable throughout storage, grinding, and processing, and it is heat resistant in food and feed (Pleadin et al., 2019). DON is a mycotoxin that occurs naturally in food and is readily created as a consequence of environmental changes. It's often found in grains during preharvest, processing, drying, and storage (e.g., temperature, humidity). DON is also a prevalent food contaminant in mycotoxin pollution because it can survive temperatures ranging from 170 to 350°C (no substantial decrease after 30 minutes of treatment at 170°C) (Zhou et al., 2020).

DON's acute and chronic toxicity, including teratogenicity, cytotoxicity, genotoxicity, and immunotoxicity, as well as gastrointestinal symptoms such nausea, vomiting, and food rejection, have been shown in several investigations (Zhou et al., 2020). According to the International Agency for Research on Cancer, DON is a category 3 mycotoxin (Khaneghah et al., 2018a). Because of its ubiquitous incidence and harmful consequences for human and animal health, DON has attracted worldwide attention. As a result, techniques for eliminating, excluding, or inactivating DON from meals and feeds are needed. DON contamination may occur at any time throughout the manufacturing process, from preharvest to postharvest. DON management strategies have been developed to (a) avoid DON formation or contamination prior to harvesting; (b) degrade or remove DON from contaminated meals and feeds; and (c) limit DON bioavailability by reducing gastrointestinal absorption (Guo et al., 2020). As a consequence, regulatory criteria for DON residue limits in food and feed have been set by specific nations and food safety organizations to restrict daily DON consumption. In Canada, the maximum permissible concentration of DON in cereal goods is 2 parts per million (ppm) for wheat and 1.2 parts per million (ppm) for wheat flour; however, in China and the United States, the maximum allowable concentration of DON in cereal products is 1 part per million (ppm) (Yuan et al., 2017). DON contamination may occur in conjunction with other mycotoxins, as well as in disguised forms in food and feeds, in a real-world situation involving an eco-environmental system, posing significant challenges for government organizations charged with setting scientific regulations and standards (Zhou et al., 2020).

Given the plethora of publications and research papers on mycotoxins, this chapter aims to offer an updated systemic overview of the key sources, biosynthetic process, and genes responsible for DON in food and feed, as well as to analyze its health impacts and toxicity mechanism. A emphasis on masked mycotoxins is one of several topics covered, including the impact of industrial and environmental factors on growth, degradation, mitigation, and detection.

5.2 MAJOR SOURCES OF DEOXYNIVALENOL

DON is a naturally occurring metabolite produced by the *Fusarium* fungus, namely *Fusarium graminearum*, *Fusarium crookwellense*, and *Fusarium culmorum*, all of which contaminate food and feed over the world (Holanda et al., 2021; Bretz et al., 2006). Warm temperatures and excessive humidity encourage fungal development at all phases of blossoming and maturity (Gruber-Dorninger et al., 2019). Water activity (*a*w), pH, and nutrient content are further environmental factors that influence growth and toxin accumulation (Femenias et al., 2020). Cereals such as wheat (Petcu et al., 2019), maize (Juan et al., 2014), barley (Hietaniemi et al., 2016), rice (Juan et al., 2014), and oats (Lindblad et al., 2012), as well as their derivatives, such as breakfast cereals (Oueslati et al., 2018), infant cereals, meals, feed, and baby mix, are the most often contaminated food categories. Fusarium head blight (FHB) infection and DON contamination in cereal crops are affected by crop rotation, tillage, fungicide usage, cultivars resistant to FHB, and climatic circumstances such as spring rains and high temperatures (Pascari et al., 2019). The most common *Fusarium graminearum* disease in cereal crops such as wheat and barley is FHB, commonly known as scab, which not only causes economic losses but also raises safety concerns due to the production of mycotoxin (DON) in commodities due to its stability (Nakagawa et al., 2017). Mycelia mature into perithecia, which generate ascospores, which are then carried by wind or rain to crop plants (Xia et al., 2021). It is possible to distinguish two strains of *F. graminearum* by their production of 3-ADON or 15-ADON, which vary in the location of the acetyl group in the peptidyl transferase protein RPL3 of the 60S ribosomal subunit, which suppresses protein synthesis (Brauer et al., 2020). FHB is caused by *Fusarium culmorum*, which infects cereal crops such as durum wheat, triticale, rye, and bread wheat with DON (Gaikpa et al., 2020). DON is produced by *Fusarium crookwellense*, which was identified in infected oat kernels (Mielniczuk et al., 2020). The growth of FHB and DON in durum wheat has been connected to *Fusarium cerealis* (Palacios et al., 2021). Another strain that accumulates FHB and DON is *Fusarium equiseti*, which has been tested for infection in wheat in South Africa (Minnaar-Ontong et al., 2017). Only a few species have lately been identified to induce infection, including *F. verticillioides, F. poae, F. proliferatum, F. subglutinans,* and *F. temperatum* (Pfordt et al., 2020).

5.3 CHEMISTRY AND BIOSYNTHESIS OF DEOXYNIVALENOL

DON (trichothecene mycotoxin 3,7,15-trihydroxy-12,13-epoxytrichothecine-8) is a toxin found mostly in cereals and cereal-based products (cereals 14 and 34). Because of the existence of a double bond between carbon 8 and oxygen, DON is categorized as a type B tri-chothecene (Park et al., 2018). In higher eukaryotic cells, this DON protein synthesis inhibitor has been found to be effective. It poses a threat to the health of both humans and animals, and it also plays a vital role in the spread of plant disease. By binding to peptidyl transferase, an enzyme that limits protein creation, oxidative stress causes DNA damage and death. DON poisoning may cause anorexia, malnutrition, gastroenteritis, and even shock-like death in humans and cattle (Deng et al., 2021).

Trans-farnesyl pyrophosphate (FPP)-derived mevalonate units are joined to form tricodiene, a precursor to tri-chothecene, which is then used to make DON. DON synthesis enzymes and regulatory proteins are encoded by 15 genes distributed across three chromosomes (Khaneghah et al., 2018a). It contains *TRI1-Tri16* as well as one locus containing a total of 12 genes. An enzyme encoded by the *TRI5* gene catalyzes a step that produces trichothecene, which is subsequently followed by nine further processes that produce various DON precursors, each catalyzed by an enzyme encoded by another *TRI4* family gene. An acidic pH, which is a frequent environmental situation, facilitates DON biosynthesis. Ammonium levels

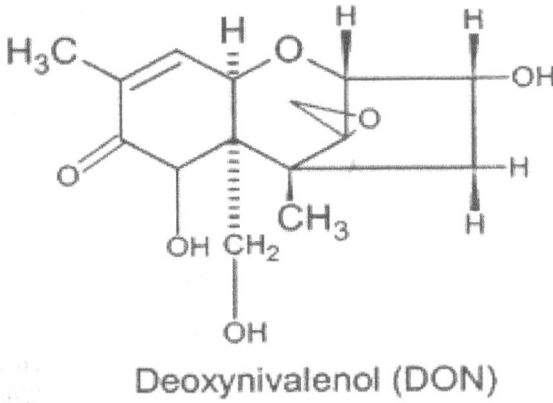

Deoxynivalenol (DON)

Figure 5.1   Structure of deoxynivalenol.

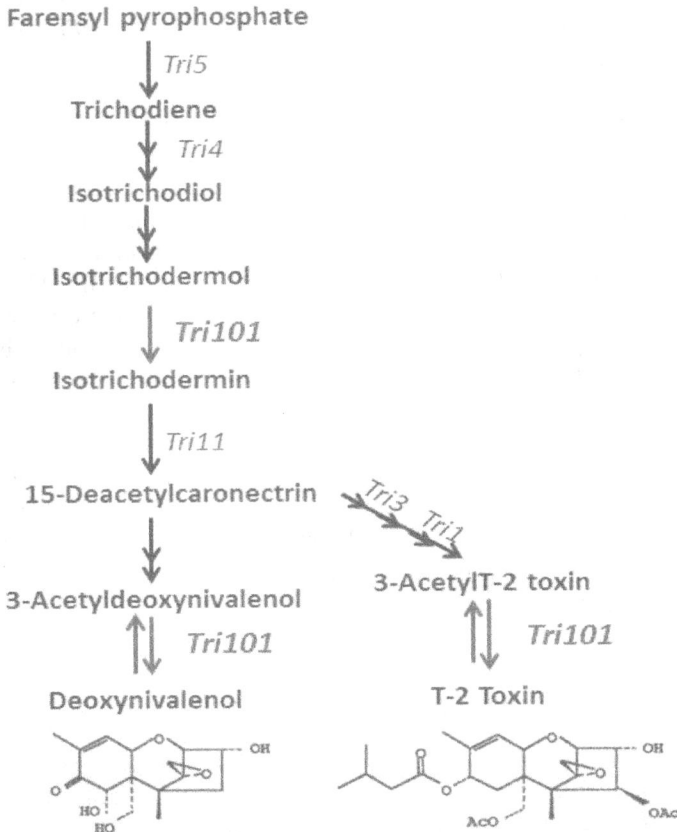

Figure 5.2   Biosynthetic pathway of deoxynivalenol.

may increase as a result of the absorption of nitrogen-containing substances from the food or growing medium. The *TRI5* gene product, which is generated when this situation occurs, enables tricodiene, an early precursor to DON, and FPP cyclization. In addition, in reaction to fungal invasion, plant defense mechanisms stimulate DON biosynthesis. Fungi infected with *F. graminearum* create hyphae, which spread to healthy fungi, which then produce DON.

5.4 GENES RESPONSIBLE FOR PRODUCTION

All of the *TRI* genes involved in trichothecene synthesis in *Fusarium graminearum* and *Fusarium sporotrichioides* have been identified in these two fungi. The remaining TRI genes, with the exception of *TRI1, TRI16,* and *TRI101,* are included in the largest *TRI* gene cluster (Hallen-Adams et al., 2011), which has the largest number of TRI genes. Multiple host and environmental conditions govern the production of the mycotoxin DON in *Fusarium graminearum*, which is controlled by two pathway-specific transcription factors, *TRI6* and *TRI10*, which are found in the fungus. Previously, the *TRI6* binding site in the promoters of *TRI* genes has been discovered and described in detail. In *F. graminearum*, deletion of the *TRI10* gene causes a reduction in *TRI* gene expression, despite the fact that its precise function is uncertain. Nevertheless, it is still unclear how the transcription factors Tri6 and TRI10 are designed to regulate the expression of other *TRI* genes (Jiang et al., 2016). According to the findings of this study, DON production in *F. graminearum* is dependent on the activity of the cyclic adenosine monophosphate (cAMP) signaling pathway as well as the three MAP kinase pathways (Zheng et al., 2012; Hu et al., 2014). External cues, sent via any of these critical signaling pathways, are likely to activate the transcription factors TRI6 and TRI10, which regulate TRI expression. The *TRI6* and *TRI10* transcription factors, which regulate TRI gene expression, are likely to be activated by any of these key signaling pathways (Jiang et al., 2016). Furthermore, the sequenced *F. graminearum* strain PH-1 is defective in functional TRI13 and TRI7, which prevents the creation of DON and the production of nivalenol (NIV) in the presence of these factors. It is known as the TRI5 cluster gene because it is involved in the first step of trichothecene biosynthesis, which is the creation of trichodiene from farnesyl pyrophosphate (Hallen-Adams et al., 2011).

5.5 OCCURRENCE IN FOOD AND FEED

Cereals such as wheat, maize, barley, rice, and oats, as well as their products such as breakfast cereals, infant cereals, meals, feed, and baby mix, are the most often contaminated foodstuffs. DON contamination is caused by fungicide use and FHB-resistant genotypes. Agronomic techniques such as crop rotation and tillage, as well as environmental conditions such as spring rains and high temperatures, all contribute to DON contamination. The presence of DON contamination in wheat and its by-products represents a substantial threat to human health. The presence of DON in foods and feeds (Table 5.1) is a significant source of worry for the food business all over the globe, particularly in developing countries.

TABLE 5.1   OCCURRENCE OF DEOXYNIVALENOL IN FOOD AND FEED AROUND THE WORLD

Food/Feed Matrix	Country	Range (µg/kg)	Detection Technique	Reference
Food				
Barley/ bakery products	Argentina	2360	HPLC-UV	Nogueira et al., 2018
	Brazil	310–15,500	LC-MS/MS	Piacentini et al., 2018
	Hungary	97–3065	HPLC	Ambrus et al., 2011
	Romania	0–4,000	ELISA	Tabuc et al., 2009
	Tunisia	500–3600	HPLC	Bensassi et al., 2011
Corn	South Korea	3.3–232.56	HPLC	Kim et al., 2016
Corn/corn germ meal	China	100–4320.9/ 100–4402.7	HPLC	Wu et al., 2016
Corn flour/ cornflakes	Serbia	931/878	HPLC	Torović, 2018
Maize	China	100–19,811.0	HPLC-UV	Liu et al., 2016
	Egypt	26–807	LC-MS/MS	Abdallah et al., 2017
	Hungary	225–2963	ELISA	Tima et al., 2016a
	Nepal	>1	HPLC	Desjardins et al., 2000
	Poland	1.0–6688	HPLC	Kosicki et al., 2016
	Serbia	260.4–9050	ELISA	Kos et al., 2017
	South Africa	9176	LC-MS/MS	Gruber-Dorninger et al., 2018
Oats	Canada	50–2340	HPLC-PDA	Tittlemier et al., 2013
	Finland	21,608	GC-MS	Hietaniemi et al., 2016
	Portugal	17,900	HPLC	Marques et al., 2008
	Russia	50–1030	HPLC-MS	Tutelyan et al., 2013
	Sweden	99–5544	HPLC/ESI-MS/MS	Fredlund et al., 2013
	UK	1866	LC-MS/MS	Edwards et al., 2017

(Continued)

TABLE 5.1   OCCURRENCE OF DEOXYNIVALENOL IN FOOD AND FEED AROUND THE WORLD   (CONTINUED)

Food/Feed Matrix	Country	Range (µg/kg)	Detection Technique	Reference
Noodles and pasta	Italy	35–450	LC-MS/MS	Raiola et al., 2012
Rice	Pakistan	6.99	LC-MS/MS	Majeed et al., 2018
Wheat	Argentina	9480	LC-MS/MS	Palacios et al., 2017
	Albania	1916	LCMS	Topi et al., 2021
	Brazil	73–2794	HPLC	Silva et al., 2018
	Canada	4700	HPLC-PDA	Tittlemier et al., 2013
	China	33–3030	HPLC	Zhao et al., 2018
	Finland	5510	LC-MS/MS	Nathanail et al., 2015
	Hungary	1880	ELISA	Tima et al., 2016a
	India	70–4730	HPLC	Mishra et al., 2013
	Italy	56–27,088	GC-MS	Bertuzzi et al., 2014
	Iran	23–1270	ELISA	Darsanaki et al., 2015
	Israel	1.2–1746	LC-MS/MS	Vogelgsang et al., 2017
	Iran	23–1270	ELISA	Darsanaki et al., 2015
	Norway	5–94	HPLC	Bernhoft et al., 2016
	Nigeria	119–2560	LC-MS	Egbontan et al., 2017
	Netherlands	100–11,000	LC-MS/MS	Hoogenboom et al., 2008
	Poland	10–1265	HPLC	Bryła et al., 2016
	Romania	110–1787	LC-MS/MS	Urusov et al., 2018
	Slovakia	788	ELISA	Šliková et al., 2013
	Spain	6178	HPLC	Vidal et al., 2013
	Serbia	64–4808	HPLC/ELISA	Jajić et al., 2014
	Serbia	154–16,528	ELISA	Jakšić et al., 2012
	Serbia	630–1840	HPLC	Jajić et al., 2008
	Sweden	1189	HPLC/ESI-MS/MS	Lindblad et al., 2013
	Switzerland	10,600	LC-MS/MS	Vogelgsang et al., 2017
	Sweden	1189	HPLC/ESI-MS/MS	Lindblad et al., 2013
	Uruguay	1400–3400	HPLC/UV	Pan et al., 2009

Food/Feed Matrix	Country	Range (µg/kg)	Detection Technique	Reference
Spring wheat	Lithuania	100–10,644.0	UPLC/MS	Supronienė et al., 2016
Wheat dust	Belgium	607–14,043	UPLC/MS	Sanders et al., 2016
Winter wheat	Lithuania	100–1393.0	UPLC/MS	Supronienė et al., 2016
Winter wheat	Slovak Republic	20–2651.79	HPLC-DAD	Remža et al., 2021
Wheat flour	Spain	501	HPLC	Femenias et al., 2021
Wheat flour and bread	Iran	0.78	ELISA	Gholampour Azizi et al., 2020
Infant food	United States	10–224	HPLC-UV	Dombrink-Kurtzman et al., 2010
Infant food	India	5–228	ELISA	Gummadidala et al., 2019
Barley/ pasta	Romania	21.52–721.88/ 28.23–173.55	ELISA and HPLC	Petcu et al., 2019
Flour and breakfast cereals	Romania	31.56–172.37	ELISA and HPLC	Petcu et al., 2019
Feed				
Broiler feeds	Thailand	33.58–60.81	LC-MS	Kongkapan et al., 2015
Cattle compound feed	Spain	289.9	UPLC–MS/ MS and UPLC– QTOF–MS	Romera et al., 2018
Cattle/ chicken/pig feed	South Korea	91.65–950.25/ 3.3–603.10/ 32.38–932.48	HPLC	Kim et al., 2016
Concentrated feed/formula feed/premixed feed	China	11.6– 277.6/47.1– 864.5/97.4– 776.3	UPHLC-MS	Fan et al., 2016
Dairy concentrate feed	Kenya	18.53–179.89	ELISA	Makau et al., 2016
Duck complete feed	China	100–2613.7	HPLC-UV	Liu et al., 2016

(Continued)

TABLE 5.1   OCCURRENCE OF DEOXYNIVALENOL IN FOOD AND FEED AROUND THE WORLD   (CONTINUED)

Food/Feed Matrix	Country	Range (µg/kg)	Detection Technique	Reference
Finished feed	South Africa	9805	LC-MS	Gruber-Dorninger et al., 2018
Forage maize	Northern Germany	2237–3038	LC-HRMS	Birr et al., 2021
Compound feeds	South Africa	3.22–56.52	UHPLC-MS/MS	Changwa et al., 2018
Feed	Egypt	1516	LC-MS/MS	Abdallah et al., 2017
Pig complete feed (powder)/ (pellet)	China	100–2767.6/ 100–3346.0	HPLC	Wu et al., 2016
Silage	Brazil	300	HPLC	Carvalho et al., 2016
	Spain	43.1–6685.6	LC-MS	Dagnac et al., 2016
	England	10.0–7111	UPLC	Cogan et al., 2017
	Poland	1.0–7860	HPLC	Kosicki et al., 2016
Poultry/sheep/ swine compound feed	Spain	250/250/254.9	UPLC–MS/MS and UPLC–QTOF–MS	Romera et al., 2018
Swine feed	Hungary	137–997	ELISA	Tima et al., 2016b

5.6 EFFECTS ON AGRICULTURAL FOOD AND FEED

Important contamination in food and feed has been observed since DON is found in toxicologically significant amounts in food and feed all over the globe. Humans and animals also face significant health risks from DON-contaminated agricultural food and feed (Awad et al., 2010). Wheat, maize, oats, and barley are among the grain groups that might include DON-contaminated food and feed. The sensitivity of DON varies based on the species, age range, and kind of animals being studied. DON pollution has been shown to be particularly harmful to pigs and other ruminants, as well as poultry, cats, dogs, and rodents (Mishra et al., 2020; Yao et al., 2020). DON contamination has a negative impact on egg, meat, and milk products in terms of both quality and quantity. Due to cytotoxicity, DON concentrations in the range of 0.1–2 g/ml may impact protein synthesis in lymphocytes and fibroblasts (Feizollahi et al., 2021), which are linked to the immune system. DON contamination also results in poor offspring, dead births, stiff pigs, female pig abortions, worse quality and quantity of eggs produced in poultry, and lower cow production performance (Bracarense et al., 2012). DON pollution has also been associated with an increased incidence of embryo deformities and chromosomal abnormalities in chickens (Ostry et al., 2017). The amount of DON administered to cows rose from 1.5 to 6.4 mg/kg (Knutsen et al., 2017), which was associated with a decrease in grain intake. It was discovered that pigs began vomiting at 12 mg/kg of DON and refused to eat at 20 mg/kg of DON. When the DON feed concentration was raised to 16–20 mg/kg (Eriksen et al., 2004), the chickens' lower weight growth and feed rejection were less noticeable than in pigs.

The typical dietary quantity of DON and its derivatives in nursing cows and meat sheep, according to current estimates, ranges from 64.2 to 996 g/kg. Dinitrophenol is less toxic to ruminants than it is to fowl. It has been established that the DON was transmitted from feed to foodstuffs in cows, chickens, and pigs. Chickens have been shown to be harmed by DON concentrations in feed of 9 mg/kg of feed (Knutsen et al., 2017). Several nations have imposed limits on the amount of DON that may be found in food and feed. The European Commission established the DON limits for wheat, maize, and oats at 1750 g/kg for all three cereals. DON levels in cattle and poultry feed are regulated by the United States Food and Drug Administration (USFDA), with the lowest limit being 1,000 g/kg of DON for wheat and wheat products and the highest limit being 5000 g/kg of DON for beef and chicken feed (Zhou et al., 2021).

5.7 MECHANISM OF TOXICITY AND HEALTH EFFECTS OF DEOXYNIVALENOL

Consumption of DON-contaminated food and feed may jeopardize human and animal health. DON is the most prevalent mycotoxin identified in cereals and has been linked to gastroenteritis and immune system dysfunction (Vidal et al., 2014a; Vidal et al., 2014b). The toxin may cause a range of problems, including digestive problems, unwillingness to eat, diarrhea, reproductive problems, nutritional malabsorption, increased disease incidence, and endocrine disruption (Berthiller et al., 2011; Li et al., 2014; Urbanek et al., 2018). DON promotes oxidative stress as a result of the formation of free radicals, which damages DNA and the cell membrane. Additionally, it increases ribosome disintegration, which results in ribotoxic stress, reduction of protein synthesis, and finally death (Berthiller et al., 2011; Payros et al., 2016; Schmeits et al., 2014). DON has been associated with an increase in reactive oxygen species, which results in lipid peroxidation and hepatotoxicity, since the liver is the first organ to experience oxidative stress (Li et al., 2014; Mikami et al., 2010; Strasser et al., 2013; Wu et al., 2014).

Even short-term exposure to a high dose of DON causes gastrointestinal problems in humans and animals. Even prolonged exposure to a negligible quantity increases the risk of cancer. Huang et al. (2021) used oral gavage to provide different doses of the toxin to pregnant female rats (F0 generation) and their pups (F_1 generation) up to 27 days postnatally (0, 0.03, 0.1, 0.3, 1, and 3 mg/ kg/day). The results suggest that a dose of up to 3 mg/kg/day has no effect on the body weight or survival of the F0 generation but that a dose of 3 mg/kg/day causes offspring to lose weight (F_1 generation). The decrease in body weight of F_1 rats may be attributed directly to DON, given there was no indication of maternal harm during gestation or lactation.

Additionally, DON has been related to neurotoxicity. Wang fed pigs a DON-based diet at two different levels (1.3 and 2.2 mg/kg) for 60 days and examined the hippocampus, cerebral cortex, and cerebellum. The growing concentration of the toxin was associated with oxidative damage in many brain locations, with scanning electron microscopy demonstrating damage to the hippocampus's cell structure. Additionally, an increase in calcium and calcium-dependent kinase II (CaMKII) concentrations was observed. Additionally, another study has shown a link between calcium ions and neurotransmitter release and cell proliferation, suggesting that interrupting Ca2+ homeostasis may have a detrimental effect on the hippocampus's neuronal circuits (Adelsberger et al., 2005). According to Wang et al. (2020c), DON toxicity is associated with the Ca2+/ CaM/CaMKII signaling pathway involved in neurotransmission and lipid peroxidation. Additionally, the cytotoxicity of DON has been studied in mammary epithelial cells. According to a study, cows fed contaminated feed produced less milk (Mayer et al., 2017; Seeling et al., 2006). Lee et al. (2019) investigated the effects of DON on the epithelial cells of the cow mammary gland (MAC-T). On MAC-T cells, the impact of various toxin dosages (1–10 M) was evaluated. The increase in toxin concentration resulted in a significant reduction in cell proliferative activity, with the greatest effect at 10 M. The phosphorylation of the signaling molecules phosphoinositide-3-kinase (PI3K) and mitogen-activated protein kinase (MAPK) increased in MAC-T cells, indicating that DON controls these signaling pathways.

5.8 EFFECTS OF PROCESSING ON DEOXYNIVALENOL

Wheat and wheat-derived products emerge significantly in DON's marketing materials. Wheat is used to make a variety of products, including biscuits, cereals, bread, and pasta, in addition to being one of the most widely consumed grains. Wheat's susceptibility to fungal infection, on the other hand, presents a threat to both the economy and human nutrition (Webb et al., 2003; Jones, 2005; Campagnollo et al., 2016; Amirahmadi et al., 2018). It is feasible to reduce the quantity of DON present in wheat by using simple procedures such as washing, sorting, and cleaning wheat grains, as well as removing immature and damaged grains. These methods are thought to be successful in removing more than 70% of the poison from the body. Furthermore, fusarium-infected grains become pink and must be removed from the field. Granules are classified into three types depending on size, shape, and thickness (Cheli et al., 2013; Milani et al., 2014; Hazel et al., 2004). DON contamination of cereal-based items has the potential to cause substantial quality issues throughout the cereal manufacturing process (Prange et al., 2005). The concentration of DON in wheat varies throughout the grain, with the outside layer, or bran, typically with a greater concentration than the internal layer, or germ. As a consequence, the flour generated by milling is far less toxic than the bran and husks that are often used as animal feed (Vidal et al., 2016). Wang et al. (2016) studied the impact of ozone gas on toxin-containing wheat grains. The results showed that 90 minutes of ozone treatment (75 mg/L) lowered DON levels by more than half without impairing their nutritional value. The researchers determined that DON contamination might occur in a wide range of items, including cereal-based snacks, which are especially popular among children. Because commodities such as wheat, barley, and maize have been shown to have higher DON concentrations (Juan et al., 2014; Lombaert et al., 2003), the European Union has imposed a maximum DON concentration limit for such foods (not to exceed 200 g/kg) since the threat was discovered. During numerous phases of grain processing, including milling and drying, changes in the quantity of DON present have been reported. Cereal flour may be flavored by roasting it at 120°C for around 30 minutes. Because of the toxin's high thermostability (Wolf et al., 1998, Fernández-Artigas et al., 1999), there is minimal reduction in the flour after roasting. Furthermore, long-term storage at 120°C causes the formation of acrylamide, a carcinogenic toxin that is especially hazardous to neonates (Erkekoğlu et al., 2010; Israel-Roming et al., 2010). However, although some studies demonstrate that raising the temperature to 220°C reduces DON concentrations, it is necessary to consider the probability of acrylamide formation (Yumbe-Guevara et al., 2003).

The fermentation of the dough is an important step in the bread-making process. According to Khaneghah, the present study on the effect of fermentation on DON concentrations has yielded ambiguous findings. Furthermore, it is difficult to quantify because to the generation of veiled mycotoxin (Pereira et al., 2014). The presence of concealed DON has also been linked to the observed increase in DON concentrations during fermentation. According to Young et al. (1984), during fermentation, the yeast converts the DON precursors already present in the flour to DON, increasing the concentration of DON in the final product. Zhang et al. (2014) studied the effects of fermentation on DON and deoxynivalenol-3-glucoside, one of its masked forms, using a Chinese steamed bed culture (D3G). According to the study, when fermented dough was turned into bread, the DON concentration rose by twice, while the D3G level reduced by around 50%. Zhang and Wang (2015) discovered that cooking noodles reduced DON and D3G levels by up to 50% and 20%, respectively, when compared to raw noodles. It is likely that poison spilled into the water during the heating process, producing the issue. As a result, food processing may lower the amount of DON in foods; nevertheless, additional research is needed to create more effective and efficient processing processes for detecting and eradicating the hidden mycotoxin.

5.9 EFFECTS OF ENVIRONMENTAL FACTORS ON DEOXYNIVALENOL PRODUCTION

The most significant environmental factors that impact mycotoxin production are the presence of fungal spores; physical damage; competition; and an appropriate temperature, moisture, and water activity (a_w) on the substrate (Pleadin et al., 2019; Kai et al., 2019). Raindrops transport macroconidia from the lower parts of the plant to the upper parts of the plant under high-intensity warm and wet conditions and heavy precipitation. The optimal temperature range for blooming is 10–25°C with humidity values of 85–90% (Kotowicz et al., 2014). Throughout the growth and harvesting phases, high to excessive humidity promotes mold and mycotoxin formation (Pleadin et al., 2019). Warm temperatures and precipitation promote toxin generation in cereals, as shown by Pascari, Marin, Ramos, Molino, and Sanchis (2015). Toxin production is also influenced by a_w and incubation time, which Han et al. (2018)] discovered to have an impact on DON production through TRI5 gene expression at ideal settings of 20–30°C, 0.95–0.98 in, and 7–28 days. Because temperature and humidity are critical for development and flowering, rainfall aided mold growth and DON generation (Czembor et al., 2015). Ramirez and colleagues (2004) discovered that growth was best at a_w values of 0.99 and 25°C, and that it was negatively influenced when the a_w of the medium declined. Martins and Martins (2002) discovered that DON production increased after 35 days of incubation at 22°C.

Environmental circumstances influence not only mold formation but also the production of hydrolytic enzymes, which play an important role in fungal establishment on a substrate (Mannaa and Kim, 2017). At 30°C, the maximum DON production occurred between 400 and 800 ppm CO_2 and 0.98 a_w (Peter Mshelia et al., 2020) Furthermore, the DON concentration was found to be greatest at 0.97 a_w and 30°C (Belizán et al., 2019). The optimal conditions for

DON production, according to Rybecky et al. (2018), were 25°C and 0.98–0.99 a_w. Furthermore, when a_w was lowered, fungus growth peaked at 0.99, then fell from 0.97 to 0.94 (2019).

5.10 DETECTION TECHNIQUES

For the detection of DON, the most often employed analytical methods are gas chromatography mass spectrometry (GCMS), high-performance liquid chromatography (HPLC) (Ok et al., 2018; Ji et al., 2017), and liquid chromatography tandem mass spectrometry (LC MS/MS). Despite the fact that these therapies have a high degree of sensitivity and specificity, they are time consuming and costly and need highly skilled professionals to complete (Ok et al., 2018). The enzyme-linked immunosorbent assay (ELISA), lateral flow immunochromatographic test, lateral flow immunoassay, fluorescence, surface-enhanced Raman scattering (SERS), and electrochemical detection techniques are all simple, low-cost, and high-sensitivity methods for detecting pathogens.

Methods such as lateral flow immunoassays (LFA) (Xu et al., 2019; Goud et al., 2018) and biosensing assays (Hossain et al., 2018; Machado et al., 2018) were developed in order to address the need for real-time monitoring of mycotoxins. Mycotoxins are considered the cheapest, quickest, and most straightforward LFA approach (Raeisossadati et al., 2016). Jin et al. (2021) established that detection of DON in corn was possible by means of a double near-infrared fluorescence-based LFA that was capable of detecting DON in both grains. For the detection of DON residues in various agricultural products, Zhao and colleagues (2021) developed an ultra-rapid lateral flow fluorescent microsphere immunoassay test strip (FM-ICTS). In addition to their high transmission, ease of use, and cheap cost (Seo et al., 2019; Zhang et al., 2020a), electrochemical and biosensors are very effective devices. Ong et al. (2020) used iron nanoflorets graphene nickel (INFGN) to build a selective biosensing device for the detection of DON, which was later validated. In their study, Li et al. (2021) demonstrated that a sensor based on molecularly imprinted poly(L-arginine) (P-Arg-MIP) on carboxylic acid functionalized carbon nanotubes (COOH-MWCNTs) can be used to detect dinitrophenol in agri-food items with high specificity and sensitivity, as well as high selectivity.

Multiplex immunochromatographic assays (MICAs) are another technique that may be utilized to detect mycotoxins (Hristov et al., 2019). Recent research has shown that AuNPs and quantum dots (QDs) may be employed as signal tags to detect a range of mycotoxins in foods (Huang et al., 2020a; Shao et al., 2019). To achieve high sensitivity by Forster resonance energy transfer, semiconductor QDs with a characteristic light property are used in the development of biosensors (FRET). DON may be detected using the fluorescent immunoassay developed by Goryacheva et al. (2021), which is based on FRET. Due to the low sensitivity of existing immunochromatographic assays based on AuNPs, Li et al. (2020b) developed an immunochromatographic test for visual detection of DON utilizing polydopamine-coated zirconium metal-organic framework-labeled antibodies (ZrPA-Ab). According to Chang et al. (2020) and Cheng et al. (2019), it is not known if AuNP or QD-based micro-incubators are more sensitive than QD-based microincubators. A multiplex immunochromatographic assay for the simultaneous detection of DON and aflatoxins B1 was developed by Zhao et al. (2021) by utilizing innovative Fe_2O_3 nanocubes (FNCs) as signal tags. The work of Huang et al. (2020) resulted in the development of an immunochromatographic test strip for the simultaneous detection of fumonisin B1 and DON in grains, and the work of Subak et al. (2021) resulted in the development of an aptasensor for the detection of DON in food and feed samples.

5.11 MASKED MYCOTOXINS AS A MAJOR CONCERN IN DETECTION

When DON interacts with proteins or carbohydrates, it generates a variety of structures known as "modified mycotoxins," which include both "biologically and chemically altered" forms (Berthiller et al., 2005; De Angelis et al., 2013; Rychlik et al., 2014). A "biologically modified" mycotoxin is one that has been conjugated by plants and is known as a "masked mycotoxin" (Rychlik et al., 2014). D3G is produced by the enzymatic interaction of DON with glucose, while 3Ac-DON and 15Ac-DON are produced by DON deacetylation (Freire et al., 2018; Tian et al., 2016). D3G, 3Ac-DON, and 15Ac-DON, among other forms, have been found in cereal and cereal-based goods (Khaneghah et al., 2018a). Despite the fact that toxicodynamics research on these forms is scarce, absorption, bioavailability, and toxicity have all been investigated (Nagl et al., 2014; Broekaert et al., 2015).

These modified or camouflaged variants of mycotoxins cannot be identified using standard methods, resulting in underreporting of their presence. As a consequence, utilizing DON that has been changed or concealed poses substantial health risks. This raises even greater concerns about the toxicity of these compounds to animals (Mahato et al., 2021; Kamle et al., 2019; Berthiller et al., 2009; Kumar et al., 2020; Mahato et al., 2019; Berthiller et al., 2009), since they may revert to their original condition. Modified forms are often found in close proximity to one another in food and feed samples. Thus, Fan and colleagues (2016) devised and validated a UHPLC-MS method for simultaneous detection of masked DON in feed samples utilizing an ultra-high performance liquid chromatography-tandem mass spectrometry (UHPLC-MS) system.

Measurement of DON and its masked forms, namely 3Ac and 15Ac, was carried out by Olopade and colleagues using an LC-MS/MS approach. Understanding the cumulative influence of these molecules will need toxicokinetic experiments using masked copies of the original substances (Lorenz et al., 2019; Lu et al., 2020). Additional measures, such as hydrolytic processes utilizing alkaline, acidic, or enzymatic methods (Beloglazova et al., 2013; Vidal et al., 2018) and a comprehensive approach to the detection of modified mycotoxins, such as that proposed by Lu et al. (2020) for altered DON, could be designed to resolve such concerns and secure the security of food or feed.

5.12 DEGRADATION KINETICS

When pre- and post-harvesting measures fail to limit DON contamination in food and feed, the development and deployment of effective degrading technologies is required (Awad et al., 2010). These detoxification processes are classified as physical, chemical, enzymatic, and biological (Feizollahi et al., 2021). Washing; cleaning; density separation; sieving; dehulling; adsorption; heat treatments; and the application of gamma, UV, and visible light radiations are examples of physical methods (Zhang et al., 2016). Thermal treatment in food processing procedures is classified as dry methods (baking, roasting, and frying) and wet methods (steaming and cooking) (Schaarschmidt et al., 2021); however, the degradation products created might have more toxic and hazardous effects than the primary mycotoxin (Stadler et al., 2020). Stadler et al. (2019) concluded that baking may produce a partial degradation of DON since increasing baking time and temperature resulted in a significant decrease in DON and the degradation by-products were less hazardous. The decrease during baking is attributable to either binding to matrix components or transformation into additional toxins but not to true toxin annihilation (Vidal et al., 2015). An effective decrease in DON was seen after roasting between 180°C and 220°C for 30 minutes, and the reduction increased with increasing temperature (Pleadin et al., 2019). However, owing to the great thermal stability and release of bound forms, frying had no significant influence on DON levels (Schaarschmidt et al., 2021). Furthermore, Kalagatur et al. (2018) discovered that heating at high pressure (1000–5500 bar) between 30°C and 60°C for 10–30 minutes dramatically lowered DON levels. The leaching of DON into the broth resulted in a more than 40% decrease in DON in the cooked product (Vidal et al., 2016). Furthermore, gamma rays have effectively decreased DON levels in wheat (Awad et al., 2010). Several studies have also documented DON breakdown utilizing UV (Jajić et al., 2016; Popović et al., 2018), where substantial results were obtained with no changes in the color or protein content of the treated samples (Yi-gang et al., 2016).

The biochemical technique employs compounds such as sodium bisulfite and calcium hydroxide monomethylamine (Sun et al., 2020; Santos Alexandre et al., 2018; Rempe et al., 2013), wet and dry ozone (Sun et al., 2020), ascorbic acid and ammonium hydroxide, and acid and alkaline electrolyzed water. However, physical and chemical methods have several drawbacks, including limited efficacy, high costs, harmful chemical residues, loss of nutritional value, sophisticated equipment requirements, and safety concerns, so environmentally friendly techniques utilizing microorganisms and enzymes have been developed (Ji et al., 2016).

Biological detoxification entails using fungi, bacteria, and actinomycetes to decrease or fully eliminate DON from goods by adsorption or enzymatic breakdown (Wu et al., 2020; Hathout et al., 2014). Li et al. (2020a) discovered that *Bacillus subtilis* had the highest detoxification rate out of 16 bacteria strains and demonstrated a synergistic effect with *Lacto-bacillus plantarum*, *B. velezensis* RC 218, and *Streptomyces albidoflavus* RC 87B, resulting in a 51% DON reduction in durum wheat, making it a potential biocontrol agent (Palazzini et al., 2018). Because of the breakdown capacity inherent in extracellular enzymes or proteins, *B. subtilis* ASAG 216 digested 81.1% DON at optimal temperature (35–50°C), time period (8 hours), and pH (6.5–9.0) conditions (Jia et al., 2020). *Bacillus licheniformis* strain YB9 decreased DON by 82.6% (Wang et al., 2020a). Furthermore, at 30°C and pH 8, the bacterial consortium C20 was able to decompose DON (Wang et al., 2020b). *Devosia insulae* A16 could likewise decompose 88% DON in 48 hours at 35°C and pH 7 (Wang et al., 2019b). *Slackia* sp. D-G6 reduced DON by deep oxidation at 37–47°C and pH 6–10, producing non-toxic DOM-1 as a byproduct (Gao et al., 2020). Furthermore, soil-derived bacteria *Pseudomonas* sp. Y1 and *Lysobacter* sp. S1 demonstrated considerable DON-degrading capacity through an enzymatic transformation of DON into less harmful 3-epi-DON (Zhai et al., 2019). At 40°C and pH 8, *Pelagibacterium halo-tolerans* ANSP101 degraded DON by converting it to the less hazardous 3-keto-deoxynivalenol (Zhang et al., 2020a). DON was entirely removed from wheat by *Devosia* strain D6–9, which catabolized it into 3-keto-DON and 3-epi-DON (He et al., 2020). Several studies, including those cited by Gao et al. (2020), have shown the biodegradation of DON by fungi. Enzymatic processes such as deep oxidation, oxidation, epimerization, and glycosylation may also be used to detoxify DON, resulting in the generation of by-products such as DOM-1, 3-keto DON, 3-epi DON, and DON-3-glucoside (D3G) (Tian et al., 2016).

However, due to limitations in physical, chemical, and biological methods such as low efficacy, the need for expensive chemicals, sophisticated equipment, and the formation of harmful chemical residues, a novel technology such as photocatalytic degradation has been developed with advantages such as ease of operation, low cost, environmental friendliness, reusability, and high stability (Li et al., 2020). The target toxin is degraded utilizing a photocatalyst such as TiON@PdO nanoparticle (Zhang et al., 2013), carbon-supported TiO_2 (Sun et al., 2019), ZnO@graphene hybrids (Bai et al., 2017), upconversion nanoparticles @ TiO_2 composite (Wu et al., 2020), and dendritic-like Fe_2O_3 (Wang et al., 2019a). Another approach, electrochemical oxidation (ECO), has evolved as a novel oxidation technology that destroys DON via many processes, and DON was greatly decreased in an experiment utilizing graphite as an electrode owing to the technique's high potential and acidic circumstances (Xiong et al., 2019). Another unique non-thermal technique for DON detoxification is atmospheric cold plasma technology (ACP), which employs plasma under low air pressure and has the benefit of stronger chemical reactivity and effectiveness than ozone or UV treatments, resulting in DON decontamination in seconds (Hojnik et al., 2017; Feizollahi et al., 2020).

5.13 MANAGEMENT AND CONTROL STRATEGIES

It is imperative that the control and prevention of DON accumulation and its entry into the food supply chain be given significant attention in light of developing worldwide food safety concerns. Before and after harvest, agricultural interventions may be used to improve yields (Murugesan et al., 2021). Good agricultural practices (GAPs), good storage practices (GSPs), and good management practices (GMPs) are among the interventions being implemented (Agriopoulou et al., 2020).

Identifying effective pre- and post-agronomic strategies to minimize grain contamination is crucial given that mycotoxin contamination occurs throughout the crop life cycle (Gromadzka et al., 2016). Pre-harvest strategies include the use of resistant varieties, weed control, damaged kernel removal, soil analysis, herbicide, insecticide, and fungicide application to eradicate pest and fungus, seed treatment, seed decontamination, crop rotation, soil management and ploughing, and fertilizers for primary productivity (Adebiyi et al., 2019; Mandappa et al., 2018; Degraeve et al., 2016). Genetically modified plants can also be used to suppress mycotoxin productions. Genetically resistant cultivars, pedigree selection, and the discovery of important infection paths, such as silk channels, during flowering time are the most effective approaches to cure this illness (Czembor et al., 2015). To prevent fungal development or to minimize mycotoxin production in plants, gene editing may be used to target an aflC gene in the mycotoxin biosynthesis pathway using a kernel-specific RNA interference (RNAi) gene cassette. As a disease control strategy, crop rotation or tillage has been suggested (Hofer et al., 2016). As a result, agricultural residue, which is where the majority of the fungus flourishes, is reduced and regulated. The threat of mycotoxin exposure increases as a result of the use of fertilizers, especially nitrogen, which encourages the development of *Fusarium* while concurrently enhancing plant growth in the field.

As a greater percentage of overripe or immature seeds may raise mycotoxin levels in final products, harvesting timing is critical when gathering seeds. Prevent early harvest, gather damaged kernels, mechanical damage and soil contact to prevent crop stress during harvest (Munkvold, 2014). Containers and vehicles must be free of insects and fungal development while transporting harvested grain (WHO, 2012). Before grains are stored, they must be dried to remove excess moisture, which promotes the development of fungi. Mycotoxin concentrations may be reduced by using solar dryers instead of sun drying (Pitt et al., 2013). Preventing the growth of fungi and the production of mycotoxins in stored grains requires careful management of their moisture, temperature, and relative humidity (Neme et al., 2017). It is best to store your food at temperatures between 1–4°C and 10–15°C in winter and less than 70% relative humidity in the summer. It is also important that the storage chambers be kept clean and well ventilated in order to protect them from rain and drainage. To keep them at a consistent temperature, they should be aerated with moving air (WHO, 2012). Effective grain storage management relies heavily on the concept of sanitation, loading, aeration, and monitoring (SLAM).

In order to lower mycotoxin levels after harvest, processes such as drying and packing, sorting and cleaning and drying, and insect control and pesticide use in storage rooms must be used in tandem. In order to encourage the adoption of best practices, these therapies are likely to incorporate instructional and awareness-raising efforts as well. Cleanliness and the application of natural and chemical agents after harvest are critical throughout the length of the crop's storage period. Hazard analysis critical control point (HACCP) postharvest planning should include methods to avoid contamination from field to table (Agriopoulou et al., 2020). Physical, chemical, or biological approaches may be used to remove mycotoxin contamination from agricultural products (Agriopoulou et al., 2020).

5.14 CONCLUSION

The mycotoxin DON has been found in agricultural goods across the globe, posing a threat to the health of both people and animals that consume these products. It is necessary to develop swift and effective therapies because of the extensive occurrence of DON in small-grain cereals around the globe. This is especially important in developing nations, since DON and its modified variants are devalued in these areas. Priority should be given to measures aimed at ensuring safe food handling, with particular attention being paid to the unknown health consequences of these myco-toxins co-occurring in vulnerable groups, such as the elderly and children. The most efficient method of determining the presence of total mycotox-ins in cereals and cereal-based products is to prevent their accumulation before and after harvest. Efforts in the future should be directed toward elucidating previously unknown pathways underlying both individual and cumulative toxic effects, as well as toward establishing a more objective risk level for chronic DON toxicity in human and animal populations. These results will be important in formulating more precise laws and regulations to safeguard the worldwide protection and reliability of food, feed, and feed additives.

REFERENCES

Abdallah, M.F.; Girgin, G.; Baydar, T.; Krska, R.; Sulyok, M. Occurrence of multiple myco-toxins and other fungal metabolites in animal feed and maize samples from Egypt using LC-MS/MS. *J. Sci. Food Agric.* **2017**, 97, 4419–4428.

Adebiyi, J.A.; Kayitesi, E.; Adebo, O.A.; Changwa, R.; Njobeh, P.B. Food fermentation and mycotoxin detoxification: An African perspective. *Food Control.* **2019**, 106, 106731.

Adegoke, G.O.; Letuma, P. Strategies for the prevention and reduction of mycotoxins in developing countries. In *Mycotoxin and Food Safety in Developing Countries*, Makun, H.A., Ed. IntechOpen. **2013**; pp. 123–136.

Afsah-Hejri, L.; Hajeb, P.; Ehsani, R.J. Application of ozone for degradation of mycotoxins in food: A review. *Compr. Rev. Food Sci. Food Saf.* **2020**, 19, 1777–1808.

Agriopoulou, S.; Stamatelopoulou, E.; Varzakas, T. Advances in occurrence, importance, and mycotoxin control strategies: Prevention and detoxification in foods. *Foods.* **2020**, 9, 137.

Ambrus, Á.; Szeitznè-Szabó, M.; Zentai, A.; Sali, J.; Szabo, I.J. Exposure of consumers to deoxynivalenol from consumption of white bread in Hungary. *Food Addit. Contam.* **2011**, 28, 209–217.

Amirahmadi, M.; Shoeibi, S.; Rastegar, H.; Elmi, M.; Mousavi Khaneghah, A. Simultaneous analysis of mycotoxins in corn flour using LC/MS-MS combined with a modified QuEChERS procedure. *Toxin Rev.* **2018**, 37, 187–195.

Awad, W.A.; Ghareeb, K.; Böhm, J.; Zentek, J. Decontamination and detoxification strategies for the *Fusarium* mycotoxin deoxynivalenol in animal feed and the effectiveness of microbial biodegradation. *Food Addit. Contam.* **2010**, 27, 510–520.

Bai, X.; Sun, C.; Liu, D.; Luo, X.; Li, D.; Wang, J.; Wang, N.; Chang, X.; Zong, R.; Zhu, Y. Photocatalytic degradation of deoxynivalenol using graphene/ZnO hybrids in aqueous suspension. *Appl. Catal. B Environ.* **2017**, 204, 11–20.

Belizán, M.M.E.; Gomez, A.D.L.A.; Baptista, Z.P.T.; Jimenez, C.M.; Matías, M.D.H.S.; Catalán, C.A.N.; Sampietro, D.A. Influence of water activity and temperature on growth and production of trichothecenes by *Fusarium graminearum* sensu stricto and related species in maize grains. *Int. J. Food Microbiol.* **2019**, 305, 108242.

Beloglazova, N.V.; De Boevre, M.; Goryacheva, I.Y.; Werbrouck, S.; Guo, Y.; De Saeger, S. Immunochemical approach for zearalenone-4-glucoside determination. *Talanta.* **2013**, 106, 422–430.

Bensassi, F.; Rjiba, I.; Zarrouk, A.; Rhouma, A.; Hajlaoui, M.R.; Bacha, H. Deoxynivalenol contamination in Tunisian barley in the 2009 harvest. *Food Addit. Contam. Part B.* **2011**, 4, 205–211.

Bernhoft, A.; Christensen, E.; Sandvik, M. *The Surveillance Programme for Mycotoxins and Fungi in Feed Materials, and Complete and Complementary Feed in Norway 2015.* Norwegian Veterinary Institute: Oslo, Norway. **2016**.

Berthiller, F.; Dall'Asta, C.; Schuhmacher, R.; Lemmens, M.; Adam, G.; Krska, R. Masked mycotoxins: Determination of a deoxynivalenol glucoside in artificially and naturally contaminated wheat by liquid chromatography–tandem mass spectrometry. *J. Agric. Food Chem.* **2005**, 53, 3421–3425.

Berthiller, F.; Krska, R.; Domig, K.J.; Kneifel, W.; Juge, N.; Schuhmacher, R.; Adam, G. Hydrolytic fate of deoxynivalenol-3-glucoside during digestion. *Toxicol. Lett.* **2011**, 206, 264–267.

Berthiller, F.; Schuhmacher, R.; Adam, G.; Krska, R. Formation, determination and significance of masked and other conjugated mycotoxins. *Anal. Bioanal. Chem.* **2009**, 395, 1243–1252.

Bertuzzi, T.; Camardo Leggieri, M.; Battilani, P.; Pietri, A. Co-occurrence of type A and B trichothecenes and zearalenone in wheat grown in northern Italy over the years 2009–2011. *Food Addit. Contam. Part B.* **2014**, 7, 273–281.

Birr, T.; Jensen, T.; Preußke, N.; Sönnichsen, F.D.; De Boevre, M.; De Saeger, S.; Hasler, M.; Verreet, J.-A.; Klink, H. Occurrence of *Fusarium* mycotoxins and their modified forms in forage maize cultivars. *Toxins.* **2021**, 13, 110.

Bracarense, A.-P.F.; Lucioli, J.; Grenier, B.; Pacheco, G.D.; Moll, W.-D.; Schatzmayr, G.; Oswald, I.P. Chronic ingestion of deoxynivalenol and fumonisin, alone or in interaction, induces morphological and immunological changes in the intestine of piglets. *Br. J. Nutr.* **2012**, 107, 1776–1786.

Brauer, E.K.; Balcerzak, M.; Rocheleau, H.; Leung, W.; Schernthaner, J.; Subramaniam, R.; Ouellet, T. Genome editing of a deoxynivalenol-induced transcription factor confers resistance to *Fusarium graminearum* in wheat. *Mol. Plant Microbe Interact.* **2020**, 33, 553–560.

Bretz, M.; Beyer, M.; Cramer, B.; Knecht, A.; Humpf, H.-U. Thermal degradation of the *Fusarium* mycotoxin deoxynivalenol. *J. Agric. Food Chem.* **2006**, 54, 6445–6451.

Broekaert, N.; Devreese, M.; De Mil, T.; Fraeyman, S.; Antonissen, G.; De Baere, S.; De Backer, P.; Vermeulen, A.; Croubels, S. Oral bioavailability, hydrolysis, and comparative toxicokinetics of 3-acetyldeoxynivalenol and 15-acetyldeoxynivalenol in broiler chickens and pigs. *J. Agric. Food Chem.* **2015**, 63, 8734–8742.

Bryła, M.; Waśkiewicz, A.; Podolska, G.; Szymczyk, K.; Jędrzejczak, R.; Damaziak, K.; Sułek, A. Occurrence of 26 mycotoxins in the grain of cereals cultivated in Poland. *Toxins.* **2016**, 8, 160.

Campagnollo, F.B.; Ganev, K.C.; Khaneghah, A.M.; Portela, J.B.; Cruz, A.G.; Granato, D.; Corassin, C.H.; Oliveira, C.A.F.; Sant'Ana, A.S. The occurrence and effect of unit operations for dairy products processing on the fate of aflatoxin M1: A review. *Food Control.* **2016**, 68, 310–329.

Carvalho, B.F.; Ávila, C.L.S.; Krempser, P.M.; Batista, L.R.; Pereira, M.N.; Schwan, R.F. Occurrence of mycotoxins and yeasts and moulds identification in corn silages in tropical climate. *J. Appl. Microbiol.* **2016**, 120, 1181–1192.

Chang, X.; Zhang, Y.; Liu, H.; Tao, X. A quadruple-label time-resolved fluorescence immunochromatographic assay for simultaneous quantitative determination of three mycotoxins in grains. *Anal. Methods.* **2020**, 12, 247–254.

Changwa, R.; Abia, W.; Msagati, T.; Nyoni, H.; Ndleve, K.; Njobeh, P. Multi-mycotoxin occurrence in dairy cattle feeds from the Gauteng province of South Africa: A pilot study using UHPLC-QTOF-MS/MS. *Toxins.* **2018**, 10, 294.

Channaiah, L.H.; Maier, D.E. Best stored maize management practices for the prevention of mycotoxin contamination. In *Mycotoxin Reduction in Grain Chains*, Leslie, J.F., Logrieco, A.F., Eds. Wiley & Sons, Ltd. **2014**; p. 78. http://doi.org/10.1002/9781118832790.ch6.

Cheng, N.; Shi, Q.; Zhu, C.; Li, S.; Lin, Y.; Du, D. Pt–Ni (OH) 2 nanosheets amplified two-way lateral flow immunoassays with smartphone readout for quantification of pesticides. *Biosens. Bioelectron.* **2019**, 142, 111498.

Cogan, T.; Hawkey, R.; Higgie, E.; Lee, M.R.F.; Mee, E.; Parfitt, D.; Raj, J.; Roderick, S.; Walker, N.; Ward, P. Silage and total mixed ration hygienic quality on commercial farms: Implications for animal production. *Grass Forage Sci.* **2017**, 72, 601–613.

Czembor, E.; Stępień, Ł.; Waśkiewicz, A. Effect of environmental factors on *Fusarium* species and associated mycotoxins in maize grain grown in Poland. *PLoS ONE.* **2015**, 10, e0133644.

Dagnac, T.; Latorre, A.; Fernández Lorenzo, B.; Llompart, M. Validation and application of a liquid chromatography-tandem mass spectrometry based method for the assessment of the co-occurrence of mycotoxins in maize silages from dairy farms in NW Spain. *Food Addit. Contam. Part A.* **2016**, 33, 1850–1863.

Darsanaki, R.K.; Issazadeh, K.; Aliabadi, M.A.; Chakoosari, M.M.D. Occurrence of deoxynivalenol (DON) in wheat flours in Guilan Province, northern Iran. *Ann. Agric. Environ. Med.* **2015**, 22, 35–37.

De Angelis, E.; Monaci, L.; Pascale, M.; Visconti, A. Fate of deoxynivalenol, T-2 and HT-2 toxins and their glucoside conjugates from flour to bread: An investigation by high-performance liquid chromatography high-resolution mass spectrometry. *Food Addit. Contam. Part A.* **2013**, 30, 345–355.

Degraeve, S.; Madege, R.; Audenaert, K.; Kamala, A.; Ortiz, J.; Kimanya, M.; Tiisekwa, B.; De Meulenaer, B.; Haesaert, G. Impact of local pre-harvest management practices in maize on the occurrence of *Fusarium* species and associated mycotoxins in two agro-ecosystems in Tanzania. *Food Control.* **2016**, 59, 225–233.

Deng, Y.; You, L.; Nepovimova, E.; Wang, X.; Musilek, K.; Wu, Q.; Wu, W.; Kuca, K. Biomarkers of deoxynivalenol (DON) and its modified form DON-3-glucoside (DON-3G) in humans. *Trends Food Sci. Technol.* **2021**.

Desjardins, A.E.; Manandhar, G.; Plattner, R.D.; Maragos, C.M.; Shrestha, K.; McCormick, S.P. Occurrence of *Fusarium* species and mycotoxins in Nepalese maize and wheat and the effect of traditional processing methods on mycotoxin levels. *J. Agric. Food Chem.* **2000**, 48, 1377–1383.

Dombrink-Kurtzman, M.A.; Poling, S.M.; Kendra, D.F. Determination of deoxynivalenol in infant cereal by immunoaffinity column cleanup and high-pressure liquid chromatography–UV detection. *J. Food Prot.* **2010**, 73, 1073–1076.

Edwards, S.G. Impact of agronomic and climatic factors on the mycotoxin content of harvested oats in the United Kingdom. *Food Addit. Contam. Part A.* **2017**, 34, 2230–2241.

Egbontan, A.O.; Afolabi, C.G.; Kehinde, I.A.; Enikuomehin, O.A.; Ezekiel, C.N.; Sulyok, M.; Warth, B.; Krska, R. A mini-survey of moulds and mycotoxins in locally grown and imported wheat grains in Nigeria. *Mycotoxin Res.* **2017**, 33, 59–64.

Eriksen, G.S.; Pettersson, H. Toxicological evaluation of trichothecenes in animal feed. *Anim. Feed Sci. Technol.* **2004**, 114, 205–239.

Erkekoğlu, P.; Baydar, T. Toxicity of acrylamide and evaluation of its exposure in baby foods. *Nutr. Res. Rev.* **2010**, 23, 323–333.

Fan, Z.; Bai, B.; Jin, P.; Fan, K.; Guo, W.; Zhao, Z.; Han, Z. Development and validation of an ultra-high performance liquid chromatography-tandem mass spectrometry method for simultaneous determination of four type B trichothecenes and masked deoxynivalenol in various feed products. *Molecules.* **2016**, 21, 747.

Feizollahi, E.; Iqdiam, B.; Vasanthan, T.; Thilakarathna, M.S.; Roopesh, M.S. Effects of atmospheric-pressure cold plasma treatment on deoxynivalenol degradation, quality parameters, and germination of barley grains. *Appl. Sci.* **2020**, 10, 3530.

Feizollahi, E.; Roopesh, M.S. Mechanisms of deoxynivalenol (DON) degradation during different treatments: A review. *Crit. Rev. Food Sci. Nutr.* **2021**, 1–22.

Femenias, A.; Gatius, F.; Ramos, A.J.; Sanchis, V.; Marín, S. Near-infrared hyperspectral imaging for deoxynivalenol and ergosterol estimation in wheat samples. *Food Chem.* **2021**, 341, 128206.

Femenias, A.; Gatius, F.; Ramos, A.J.; Sanchis, V.; Marín, S. Use of hyperspectral imaging as a tool for *Fusarium* and deoxynivalenol risk management in cereals: A review. *Food Control.* **2020**, 108, 106819.

Fernández-Artigas, P.; Guerra-Hernández, E.; García-Villanova, B. Browning indicators in model systems and baby cereals. *J. Agric. Food Chem.* **1999**, 47, 2872–2878.

Fredlund, E.; Gidlund, A.; Sulyok, M.; Börjesson, T.; Krska, R.; Olsen, M.; Lindblad, M. Deoxynivalenol and other selected *Fusarium* toxins in Swedish oats—Occurrence and correlation to specific *Fusarium* species. *Int. J. Food Microbiol.* **2013**, 167, 276–283.

Freire, L.; Sant'Ana, A.S. Modified mycotoxins: An updated review on their formation, detection, occurrence, and toxic effects. *Food Chem. Toxicol.* **2018**, 111, 189–205.

Gaikpa, D.S.; Lieberherr, B.; Maurer, H.P.; Longin, C.F.H.; Miedaner, T. Comparison of rye, triticale, durum wheat and bread wheat genotypes for *Fusarium* head blight resistance and deoxynivalenol contamination. *Plant Breed.* **2020**, 139, 251–262.

Gao, X.; Mu, P.; Zhu, X.; Chen, X.; Tang, S.; Wu, Y.; Miao, X.; Wang, X.; Wen, J.; Deng, Y. Dual function of a novel bacterium, slackia sp. D-G6: Detoxifying deoxynivalenol and producing the natural estrogen analogue, equol. *Toxins.* **2020**, 12, 85.

Gholampour Azizi, I.; Arjmandi, J.; Ahmadi, S.; Rouhi, S. A comparative study on deoxynivalenol mycotoxin level in wheat flour and bread samples. *J. Inflamm. Dis.* **2020**, 24, 366–373.

Goryacheva, O.A.; Beloglazova, N.V.; Goryacheva, I.Y.; De Saeger, S. Homogenous FRET-based fluorescent immunoassay for deoxynivalenol detection by controlling the distance of donor-acceptor couple. *Talanta.* **2021**, 225, 121973.

Goud, K.Y.; Kailasa, S.K.; Kumar, V.; Tsang, Y.F.; Gobi, K.V.; Kim, K.-H. Progress on nano-structured electrochemical sensors and their recognition elements for detection of mycotoxins: A review. *Biosens. Bioelectron.* **2018**, 121, 205–222.

Gromadzka, K.; Górna, K.; Chełkowski, J.; Waśkiewicz, A. Mycotoxins and related *Fusarium* species in preharvest maize ear rot in Poland. *Plant Soil Environ.* **2016**, 62, 348–354.

Gruber-Dorninger, C.; Jenkins, T.; Schatzmayr, G. Global mycotoxin occurrence in feed: A ten-year survey. *Toxins.* **2019**, 11, 375.

Gruber-Dorninger, C.; Jenkins, T.; Schatzmayr, G. Multi-mycotoxin screening of feed and feed raw materials from Africa. *World Mycotoxin J.* **2018**, 11, 369–383.

Gummadidala, P.M.; Omebeyinje, M.H.; Burch, J.A.; Chakraborty, P.; Biswas, P.K.; Banerjee, K.; Wang, Q.; Jesmin, R.; Mitra, C.; Moeller, P.D.R. Complementary feeding may pose a risk of simultaneous exposures to aflatoxin M1 and deoxynivalenol in Indian infants and toddlers: Lessons from a mini-survey of food samples obtained from Kolkata, India. *Food Chem. Toxicol.* **2019**, 123, 9–15.

Guo, H.; Ji, J.; Wang, J.s.; Sun, X. Deoxynivalenol: Masked forms, fate during food processing, and potential biological remedies. *Compr. Rev. Food Sci. Food Saf.* **2020**, 19, 895–926.

Hallen-Adams, H.E.; Wenner, N.; Kuldau, G.A.; Trail, F. Deoxynivalenol biosynthesis-related gene expression during wheat kernel colonization by *Fusarium graminearum. Phytopathology.* **2011**, 101, 1091–1096.

Han, Z.; Shen, Y.; Di Mavungu, J.D.; Zhang, D.; Nie, D.; Jiang, K.; De Saeger, S.; Zhao, Z. Relationship between environmental conditions, TRI5 gene expression and deoxynivalenol production in stored Lentinula edodes infected with *Fusarium graminearum. World Mycotoxin J.* **2018**, 11, 177–186.

Hathout, A.S.; Aly, S.E. Biological detoxification of mycotoxins: A review. *Ann. Microbiol.* **2014**, 64, 905–919.

Hazel, C.M.; Patel, S. Influence of processing on trichothecene levels. *Toxicol. Lett.* **2004**, 153, 51–59.

He, W.-J.; Shi, M.-M.; Yang, P.; Huang, T.; Zhao, Y.; Wu, A.-B.; Dong, W.-B.; Li, H.-P.; Zhang, J.-B.; Liao, Y.-C. A quinone-dependent dehydrogenase and two NADPH-dependent aldo/keto reductases detoxify deoxynivalenol in wheat via epimerization in a Devosia strain. *Food Chem.* **2020**, 321, 126703.

Hietaniemi, V.; Rämö, S.; Yli-Mattila, T.; Jestoi, M.; Peltonen, S.; Kartio, M.; Siev-Iäinen, E.; Koivisto, T.; Parikka, P. Updated survey of *Fusarium* species and toxins in Finnish cereal grains. *Food Addit. Contam. Part A.* **2016**, 33, 831–848.

Hofer, K.; Barmeier, G.; Schmidhalter, U.; Habler, K.; Rychlik, M.; Hückelhoven, R.; Hess, M. Effect of nitrogen fertilization on *Fusarium* head blight in spring barley. *Crop Prot.* **2016**, 88, 18–27.

Hojnik, N.; Cvelbar, U.; Tavčar-Kalcher, G.; Walsh, J.L.; Križaj, I. Mycotoxin decontamina-tion of food: Cold atmospheric pressure plasma versus "classic" decontamination. *Toxins.* **2017**, 9, 151.

Holanda, D.M.; Kim, S.W. Mycotoxin occurrence, toxicity, and detoxifying agents in pig production with an emphasis on deoxynivalenol. *Toxins.* **2021**, 13, 171.

Hoogenboom, L.A.P.; Bokhorst, J.G.; Northolt, M.D.; Van de Vijver, L.P.L.; Broex, N.J.G.; Mevius, D.J.; Meijs, J.A.C.; Van der Roest, J. Contaminants and microorganisms in Dutch organic food products: A comparison with conventional products. *Food Addit. Contam.* **2008**, 25, 1195–1207.

Hossain, M.Z.; Maragos, C.M. Gold nanoparticle-enhanced multiplexed imaging surface plasmon resonance (iSPR) detection of *Fusarium* mycotoxins in wheat. *Biosens. Bioelectron.* **2018**, 101, 245–252.

Hristov, D.R.; Rodriguez-Quijada, C.; Gomez-Marquez, J.; Hamad-Schifferli, K. Designing paper-based immunoassays for biomedical applications. *Sensors.* **2019**, 19, 554.

Hu, S.; Zhou, X.; Gu, X.; Cao, S.; Wang, C.; Xu, J.-R. The cAMP-PKA pathway regulates growth, sexual and asexual differentiation, and pathogenesis in *Fusarium graminearum. Mol. Plant Microbe Interact.* **2014**, 27, 557–566.

Huang, M.C.; Furr, J.R.; Robinson, V.G.; Betz, L.; Shockley, K.; Cunny, H.; Witt, K.; Waidyanatha, S.; Germolec, D. Oral deoxynivalenol toxicity in Harlan Sprague Dawley (Hsd: Sprague Dawley˙ SD˙) rat dams and their offspring. *Food Chem. Toxicol.* **2021**, 148, 111963.

Huang, X.; Huang, T.; Li, X.; Huang, Z. Flower-like gold nanoparticles-based immuno-chromatographic test strip for rapid simultaneous detection of fumonisin B1 and deoxynivalenol in Chinese traditional medicine. *J. Pharm. Biomed. Anal.* **2020a**, 177, 112895.

Huang, X.; Huang, X.; Xie, J.; Li, X.; Huang, Z. Rapid simultaneous detection of fumoni-sin B1 and deoxynivalenol in grain by immunochromatographic test strip. *Anal. Biochem.* **2020b**, 606, 113878.

Israel-Roming, F.; Avram, M. Deoxynivalenol stability during wheat processing. *Rom. Biotechnol. Lett.* **2010**, 15, 48.

Jajić, I.; Jakšić, S.; Krstović, S.; Abramović, B. Preliminary results on deoxynivalenol degradation in maize by UVA and UVC irradiation. *Contemp. Agric.* **2016**, 65, 7–12.

Jajić, I.; Jurić, V.; Glamočić, D.; Abramović, B. Occurrence of deoxynivalenol in maize and wheat in Serbia. *Int. J. Mol. Sci.* **2008**, 9, 2114–2126.

Jajić, I.; Krstović, S.; Kos, J.; Abramović, B. Incidence of deoxynivalenol in Serbian wheat and barley. *J. Food Prot.* **2014**, 77, 853–858.

Jakšić, S.; Abramović, B.; Jajić, I.; Baloš, M.Ž.; Mihaljev, Ž.; Despotović, V.; Šojić, D. Co-occurrence of fumonisins and deoxynivalenol in wheat and maize harvested in Serbia. *Bull. Environ. Contam. Toxicol.* **2012**, 89, 615–619.

Ji, C.; Fan, Y.; Zhao, L. Review on biological degradation of mycotoxins. *Anim. Nutr.* **2016**, 2, 127–133.

Ji, J.; Zhu, P.; Cui, F.; Pi, F.; Zhang, Y.; Sun, X. The disorder metabolic profiling in kidney and spleen of mice induced by mycotoxins deoxynivalenol through gas chromatogra-phy mass spectrometry. *Chemosphere.* **2017**, 180, 267–274.

Jia, R.; Cao, L.; Liu, W.; Shen, Z. Detoxification of deoxynivalenol by *Bacillus subtilis* ASAG 216 and characterization the degradation process. *Eur. Food Res. Technol.* **2020**, 247, 67–76.

Jiang, C.; Zhang, C.; Wu, C.; Sun, P.; Hou, R.; Liu, H.; Wang, C.; Xu, J.R. TRI6 and TRI10 play different roles in the regulation of deoxynivalenol (DON) production by cAMP sig-nalling in *Fusarium graminearum. Environ. Microbiol.* **2016**, 18, 3689–3701.

Jin, Y.; Chen, Q.; Luo, S.; He, L.; Fan, R.; Zhang, S.; Yang, C.; Chen, Y. Dual near-infrared fluorescence-based lateral flow immunosensor for the detection of zearalenone and deoxynivalenol in maize. *Food Chem.* **2021**, 336, 127718.

Jones, H.D. Wheat transformation: Current technology and applications to grain devel-opment and composition. *J. Cereal Sci.* **2005**, 41, 137–147.

Juan, C.; Raiola, A.; Mañes, J.; Ritieni, A. Presence of mycotoxin in commercial infant formulas and baby foods from Italian market. *Food Control.* **2014**, 39, 227–236.

Kai, F.; Xing, L.; WenBo, G.; ZhiQi, Z.; JiaJia, M.; Ying, D.; DongXia, N. Production of free and modified forms of deoxynivalenol in wheat grain. *J. Food Saf. Qual.* **2019**, 10, 2545–2554.

Kalagatur, N.K.; Kamasani, J.R.; Mudili, V.; Krishna, K.; Chauhan, O.P.; Sreepathi, M.H. Effect of high pressure processing on growth and mycotoxin production of *Fusarium graminearum* in maize. *Food Biosci.* **2018**, 21, 53–59.

Kamle, M.; Mahato, D.K.; Devi, S.; Lee, K.E.; Kang, S.G.; Kumar, P. Fumonisins: Impact on agriculture, food, and human health and their management strategies. *Toxins.* **2019**, 11, 328.

Kebede, H.; Liu, X.; Jin, J.; Xing, F. Current status of major mycotoxins contamination in food and feed in Africa. *Food Control.* **2020**, 110, 106975.

Khaneghah, A.M.; Fakhri, Y.; Raeisi, S.; Armoon, B.; Sant'Ana, A.S. Prevalence and concentration of ochratoxin A, zearalenone, deoxynivalenol and total aflatoxin in cereal-based products: A systematic review and meta-analysis. *Food Chem. Toxicol.* **2018a**, 118, 830–848.

Khaneghah, A.M.; Fakhri, Y.; Sant'Ana, A.S. Impact of unit operations during processing of cereal-based products on the levels of deoxynivalenol, total aflatoxin, ochratoxin A, and zearalenone: A systematic review and meta-analysis. *Food Chem.* **2018b**, 268, 611–624.

Khaneghah, A.M.; Martins, L.M.; von Hertwig, A.M.; Bertoldo, R.; Sant'Ana, A.S. Deoxynivalenol and its masked forms: Characteristics, incidence, control and fate during wheat and wheat based products processing: A review. *Trends Food Sci. Technol.* **2018c**, 71, 13–24.

Kim, D.-H.; Hong, S.-Y.; Jeon, M.-H.; An, J.-M.; Kim, S.-Y.; Kim, H.-Y.; Yoon, B.R.; Chung, S.H. Simultaneous determination of the levels of deoxynivalenol, 3-acetyldeoxynivalenol, and nivalenol in grain and feed samples from South Korea using a high-performance liquid chromatography–photodiode array detector. *Appl. Biol. Chem.* **2016**, 59, 881–887.

Knutsen, H.K.; Alexander, J.; Barregård, L.; Bignami, M.; Brüschweiler, B.; Ceccatelli, S.; Cottrill, B.; Dinovi, M.; Grasl-Kraupp, B. EFSA panel on contaminants in the food chain. Risks to human and animal health related to the presence of deoxynivalenol and its acetylated and modified forms in food and feed. *EFSA J.* **2017**, 15, e04718.

Kongkapan, J.; Poapolathep, S.; Isariyodom, S.; Kumagai, S.; Poapolathep, A. Simultaneous detection of multiple mycotoxins in broiler feeds using a liquid chromatography tandem-mass spectrometry. *J. Vet. Med. Sci.* **2015**, 78, 259–264.

Kos, J.; Hajnal, E.J.; Šarić, B.; Jovanov, P.; Nedeljković, N.; Milovanović, I.; Krulj, J. The influence of climate conditions on the occurrence of deoxynivalenol in maize harvested in Serbia during 2013–2015. *Food Control.* **2017**, 73, 734–740.

Kosicki, R.; Błajet-Kosicka, A.; Grajewski, J.; Twarużek, M. Multiannual mycotoxin survey in feed materials and feedingstuffs. *Anim. Feed Sci. Technol.* **2016**, 215, 165–180.

Kotowicz, N.K.; Frąc, M.; Lipiec, J. The importance of *Fusarium* fungi in wheat cultivation–pathogenicity and mycotoxins production: A review. *J. Anim. Plant Sci.* **2014**, 21, 3326–3243.

Kumar, P.; Mahato, D.K.; Kamle, M.; Mohanta, T.K.; Kang, S.G. Aflatoxins: A global concern for food safety, human health and their management. *Front. Microbiol.* **2017**, 7, 2170.

Kumar, P.; Mahato, D.K.; Sharma, B.; Borah, R.; Haque, S.; Mahmud, M.M.C.; Shah, A.K.; Rawal, D.; Bora, H.; Bui, S. Ochratoxins in food and feed: Occurrence and its impact on human health and management strategies. *Toxicon.* **2020**.

Lee, J.-Y.; Lim, W.; Park, S.; Kim, J.; You, S.; Song, G. Deoxynivalenol induces apoptosis and disrupts cellular homeostasis through MAPK signaling pathways in bovine mammary epithelial cells. *Environ. Pollut.* **2019**, 252, 879–887.

Li, D.; Ye, Y.; Lin, S.; Deng, L.; Fan, X.; Zhang, Y.; Deng, X.; Li, Y.; Yan, H.; Ma, Y. Evaluation of deoxynivalenol-induced toxic effects on DF-1 cells in vitro: Cell-cycle arrest, oxidative stress, and apoptosis. *Environ. Toxicol. Pharmacol.* **2014**, 37, 141–149.

Li, J.; Yan, H.; Tan, X.; Lu, Z.; Han, H. Cauliflower-inspired 3D SERS substrate for multiple mycotoxins detection. *Anal. Chem.* **2019**, 91, 3885–3892.

Li, Q.; Deng, Y.; Dai, S.; Wu, Y.; Li, W.; Zhuo, S.; Jiao, S.; Wang, S.; Jin, Y.; Li, J. Microfluidic assembly synthesis of magnetic TiO 2@ SiO 2 hybrid photonic crystal microspheres for photocatalytic degradation of deoxynivalenol. *J. Inorg. Organomet. Polym. Mater.* **2020a**, 1–8.

Li, R.; Bu, T.; Zhao, Y.; Sun, X.; Wang, Q.; Tian, Y.; Bai, F.; Wang, L. Polydopamine coated zirconium metal-organic frameworks-based immunochromatographic assay for highly sensitive detection of deoxynivalenol. *Anal. Chim. Acta.* **2020b**, 1131, 109–117.

Li, W.; Diao, K.; Qiu, D.; Zeng, Y.; Tang, K.; Zhu, Y.; Sheng, Y.; Wen, Y.; Li, M. A highly-sensitive and selective antibody-like sensor based on molecularly imprinted poly (L-arginine) on COOH-MWCNTs for electrochemical recognition and detection of deoxynivalenol. *Food Chem.* **2021**, 350, 129229.

Li, W.; Yu, S.Y.; Cheng, P.; Zhang, J.B. Screening, identification, and application of deoxynivalenol-detoxifying bacteria from mildewed alfalfa silage. *Int. Food Res. J.* **2020c**, 27, 350–356.

Lindblad, M.; Börjesson, T.; Hietaniemi, V.; Elen, O. Statistical analysis of agronomical factors and weather conditions influencing deoxynivalenol levels in oats in Scandinavia. *Food Addit. Contam. Part A.* **2012**, 29, 1566–1571.

Lindblad, M.; Gidlund, A.; Sulyok, M.; Börjesson, T.; Krska, R.; Olsen, M.; Fredlund, E. Deoxynivalenol and other selected *Fusarium* toxins in Swedish wheat—Occurrence and correlation to specific *Fusarium* species. *Int. J. Food Microbiol.* **2013**, 167, 284–291.

Liu, J.; Sun, L.; Zhang, J.; Guo, J.; Chen, L.; Qi, D.; Zhang, N. Aflatoxin B1, zearalenone and deoxynivalenol in feed ingredients and complete feed from central China. *Food Addit. Contam. Part B.* **2016**, 9, 91–97.

Lombaert, G.A.; Pellaers, P.; Roscoe, V.; Mankotia, M.; Neil, R.; Scott, P.M. Mycotoxins in infant cereal foods from the Canadian retail market. *Food Addit. Contam.* **2003**, 20, 494–504.

Lorenz, N.; Dänicke, S.; Edler, L.; Gottschalk, C.; Lassek, E.; Marko, D.; Rychlik, M.; Mally, A. A critical evaluation of health risk assessment of modified mycotoxins with a special focus on zearalenone. *Mycotoxin Res.* **2019**, 35, 27–46.

Lu, Q.; Qin, J.-A.; Fu, Y.-W.; Luo, J.-Y.; Lu, J.-H.; Logrieco, A.F.; Yang, M.-H. Modified mycotoxins in foodstuffs, animal feed, and herbal medicine: A systematic review on global occurrence, transformation mechanism and analysis methods. *Trends Analyt. Chem.* **2020**, 133, 116088.

Machado, J.M.D.; Soares, R.R.G.; Chu, V.; Conde, J.P. Multiplexed capillary microfluidic immunoassay with smartphone data acquisition for parallel mycotoxin detection. Biosens. *Bioelectron.* **2018**, 99, 40–46.

Mahato, D.K.; Devi, S.; Pandhi, S.; Sharma, B.; Maurya, K.K.; Mishra, S.; Dhawan, K.; Selvakumar, R.; Kamle, M.; Mishra, A.K. Occurrence, impact on agriculture, human health, and management strategies of zearalenone in food and feed: A review. *Toxins.* **2021**, 13, 92.

Mahato, D.K.; Lee, K.E.; Kamle, M.; Devi, S.; Dewangan, K.N.; Kumar, P.; Kang, S.G. Aflatoxins in food and feed: An overview on prevalence, detection and control strategies. *Front. Microbiol.* **2019**, 10, 2266.

Majeed, S.; De Boevre, M.; De Saeger, S.; Rauf, W.; Tawab, A.; Rahman, M.; Iqbal, M. Multiple mycotoxins in rice: Occurrence and health risk assessment in children and adults of Punjab, Pakistan. *Toxins.* **2018**, 10, 77.

Makau, C.M.; Matofari, J.W.; Muliro, P.S.; Bebe, B.O. Aflatoxin B 1 and Deoxynivalenol contamination of dairy feeds and presence of aflatoxin M 1 contamination in milk from smallholder dairy systems in Nakuru, Kenya. *Int. J. Food Contam.* **2016**, 3, 1–10.

Mandappa, I.M.; Basavaraj, K.; Manonmani, H.K. Analysis of mycotoxins in fruit juices. In *Fruit Juices.* Elsevier. **2018**; pp. 763–777.

Mannaa, M.; Kim, K.D. Influence of temperature and water activity on deleterious fungi and mycotoxin production during grain storage. *Mycobiology.* **2017**, 45, 240–254.

Marques, M.F.; Martins, H.M.; Costa, J.M.; Bernardo, F. Co-occurrence of deoxynivalenol and zearalenone in crops marketed in Portugal. *Food Addit. Contam.* **2008**, 1, 130–133.

Martins, M.L.G.; Martins, H.M. Influence of water activity, temperature and incubation time on the simultaneous production of deoxynivalenol and zearalenone in corn (*Zea mays*) by *Fusarium graminearum. Food Chem.* **2002**, 79, 315–318.

Mayer, E.; Novak, B.; Springler, A.; Schwartz-Zimmermann, H.E.; Nagl, V.; Reisinger, N.; Hessenberger, S.; Schatzmayr, G. Effects of deoxynivalenol (DON) and its microbial biotransformation product deepoxy-deoxynivalenol (DOM-1) on a trout, pig, mouse, and human cell line. *Mycotoxin Res.* **2017**, 33, 297–308.

Mielniczuk, E.; Cegiełko, M.; Kiecan, I.; Perkowski, J.; Pastucha, A. The pathogenicity and toxigenic properties of *Fusarium* crookwellense LW Burgess, PE Nelson & Toussoun depending on weather conditions. *Acta Sci. Pol. Hortorum Cultus.* **2020**, 19, 17–24.

Mikami, O.; Yamaguchi, H.; Murata, H.; Nakajima, Y.; Miyazaki, S. Induction of apoptotic lesions in liver and lymphoid tissues and modulation of cytokine mRNA expression by acute exposure to deoxynivalenol in piglets. *J. Vet. Sci.* **2010**, 11, 107.

Milani, J.; Maleki, G. Effects of processing on mycotoxin stability in cereals. *J. Sci. Food Agric.* **2014**, 94, 2372–2375.

Minnaar-Ontong, A.; Herselman, L.; Kriel, W.-M.; Leslie, J.F. Morphological characterization and trichothecene genotype analysis of a *Fusarium* head blight population in South Africa. *Eur. J. Plant Pathol.* **2017**, 148, 261–269.

Mishra, S.; Ansari, K.M.; Dwivedi, P.D.; Pandey, H.P.; Das, M. Occurrence of deoxynivalenol in cereals and exposure risk assessment in Indian population. *Food Control.* **2013**, 30, 549–555.

Mishra, S.; Srivastava, S.; Dewangan, J.; Divakar, A.; Kumar Rath, S. Global occurrence of deoxynivalenol in food commodities and exposure risk assessment in humans in the last decade: A survey. *Crit. Rev. Food Sci. Nutr.* **2020**, 60, 1346–1374.

Munkvold, G.P. Crop management practices to minimize the risk of mycotoxins contamination in temperate-zone maize. In *Mycotoxin Reduction in Grain Chains*, Leslie, J.F., Logrieco, A.F., Eds. Wiley. **2014**; pp. 59–75.

Murugesan, P.; Brunda, D.K.; Moses, J.; Anandharamakrishnan, C. Photolytic and photocatalytic detoxification of mycotoxins in foods. *Food Control.* **2021**, 123, 107748.

Nagl, V.; Schatzmayr, G. Deoxynivalenol and its masked forms in food and feed. *Curr. Opin. Food Sci.* **2015**, 5, 43–49.

Nagl, V.; Woechtl, B.; Schwartz-Zimmermann, H.E.; Hennig-Pauka, I.; Moll, W.-D.; Adam, G.; Berthiller, F. Metabolism of the masked mycotoxin deoxynivalenol-3-glucoside in pigs. *Toxicol. Lett.* **2014**, 229, 190–197.

Nakagawa, H.; He, X.; Matsuo, Y.; Singh, P.K.; Kushiro, M. Analysis of the masked metabolite of deoxynivalenol and *Fusarium* resistance in CIMMYT wheat germplasm. *Toxins.* **2017**, 9, 238.

Nathanail, A.V.; Syvähuoko, J.; Malachová, A.; Jestoi, M.; Varga, E.; Michlmayr, H.; Adam, G.; Sieviläinen, E.; Berthiller, F.; Peltonen, K. Simultaneous determination of major type A and B trichothecenes, zearalenone and certain modified metabolites in Finnish cereal grains with a novel liquid chromatography-tandem mass spectrometric method. *Anal. Bioanal. Chem.* **2015**, 407, 4745–4755.

Neme, K.; Mohammed, A. Mycotoxin occurrence in grains and the role of postharvest management as a mitigation strategies. A review. *Food Control.* **2017**, 78, 412–425.

Nogueira, M.S.; Decundo, J.; Martinez, M.; Dieguez, S.N.; Moreyra, F.; Moreno, M.V.; Stenglein, S.A. Natural contamination with mycotoxins produced by *Fusarium graminearum* and *Fusarium poae* in malting barley in Argentina. *Toxins.* **2018**, 10, 78.

Ok, H.E.; Lee, S.Y.; Chun, H.S. Occurrence and simultaneous determination of nivalenol and deoxynivalenol in rice and bran by HPLC-UV detection and immunoaffinity cleanup. *Food Control.* **2018**, 87, 53–59.

Ong, C.C.; Sangu, S.S.; Illias, N.M.; Gopinath, S.C.B.; Saheed, M.S.M. Iron nanoflorets on 3D-graphene-nickel: A 'dandelion' nanostructure for selective deoxynivalenol detection. *Biosens. Bioelectron.* **2020**, 154, 112088.

Ostry, V.; Malir, F.; Toman, J.; Grosse, Y. Mycotoxins as human carcinogens—The IARC monographs classification. *Mycotoxin Res.* **2017**, 33, 65–73.

Oueslati, S.; Berrada, H.; Mañes, J.; Juan, C. Presence of mycotoxins in Tunisian infant foods samples and subsequent risk assessment. *Food Control.* **2018**, 84, 362–369.

Palacios, S.A.; Del Canto, A.; Erazo, J.; Torres, A.M. *Fusarium* cerealis causing *Fusarium* head blight of durum wheat and its associated mycotoxins. *Int. J. Food Microbiol.* **2021**, 346, 109161.

Palacios, S.A.; Erazo, J.G.; Ciasca, B.; Lattanzio, V.M.T.; Reynoso, M.M.; Farnochi, M.C.; Torres, A.M. Occurrence of deoxynivalenol and deoxynivalenol-3-glucoside in durum wheat from Argentina. *Food Chem.* **2017**, 230, 728–734.

Palazzini, J.; Roncallo, P.; Cantoro, R.; Chiotta, M.; Yerkovich, N.; Palacios, S.; Echenique, V.; Torres, A.; Ramirez, M.; Karlovsky, P. Biocontrol of *Fusarium graminearum* sensu stricto, reduction of deoxynivalenol accumulation and phytohormone induction by two selected antagonists. *Toxins.* **2018**, 10, 88.

Pan, D.; Graneri, J.; Bettucci, L. Correlation of rainfall and levels of deoxynivalenol in wheat from Uruguay, 1997–2003. *Food Addit. Contam.* **2009**, 2, 162–165.

Park, J.; Chang, H.; Kim, D.; Chung, S.; Lee, C. Long-term occurrence of deoxynivalenol in feed and feed raw materials with a special focus on South Korea. *Toxins.* **2018**, 10, 127.

Pascari, X.; Marín, S.; Ramos, A.J.; Molino, F.; Sanchis, V. Deoxynivalenol in cereal-based baby food production process. A review. *Food Control.* **2019**, 99, 11–20.

Payros, D.; Alassane-Kpembi, I.; Pierron, A.; Loiseau, N.; Pinton, P.; Oswald, I.P. Toxicology of deoxynivalenol and its acetylated and modified forms. *Arch. Toxicol.* **2016**, 90, 2931–2957

Pereira, V.L.; Fernandes, J.O.; Cunha, S.C. Mycotoxins in cereals and related foodstuffs: A review on occurrence and recent methods of analysis. *Trends Food Sci. Technol.* **2014**, 36, 96–136.

Petcu, C.D.; Georgescu, I.M.; Zvorişteanu, O.V.; Negreanu, C.N. Study referring to the appearance of contamination with deoxynivalenol in grains, grain flour and bakery products on the Romanian market. *Sci. Papers Ser. D. Anim. Sci.* **2019**, 62, 214–245.

Peter Mshelia, L.; Selamat, J.; Iskandar Putra Samsudin, N.; Rafii, M.Y.; Abdul Mutalib, N.-A.; Nordin, N.; Berthiller, F. Effect of temperature, water activity and carbon dioxide on fungal growth and mycotoxin production of acclimatised isolates of *Fusarium* verticillioides and F. graminearum. *Toxins.* **2020**, 12, 478.

Pfordt, A.; Schiwek, S.; Rathgeb, A.; Rodemann, C.; Bollmann, N.; Buchholz, M.; Karlovsky, P.; von Tiedemann, A. Occurrence, pathogenicity, and mycotoxin production of fusarium temperatum in relation to other fusarium species on maize in Germany. *Pathogens.* **2020**, 9, 864.

Piacentini, K.C.; Rocha, L.O.; Savi, G.D.; Carnielli-Queiroz, L.; Almeida, F.G.; Minella, E.; Corrêa, B. Occurrence of deoxynivalenol and zearalenone in brewing barley grains from Brazil. *Mycotoxin Res.* **2018**, 34, 173–178.

Pitt, J.I.; Taniwaki, M.H.; Cole, M.B. Mycotoxin production in major crops as influenced by growing, harvesting, storage and processing, with emphasis on the achievement of Food Safety Objectives. *Food Control.* **2013**, 32, 205–215.

Pleadin, J.; Babić, J.; Vulić, A.; Kudumija, N.; Aladić, K.; Kiš, M.; Jaki Tkalec, V.; Škrivanko, M.; Lolić, M.; Šubarić, D. The effect of thermal processing on the reduction of deoxynivalenol and zearalenone cereal content. *Croat. J. Food Sci. Technol.* **2019**, 11, 44–51.

Pleadin, J.; Frece, J.; Markov, K. Mycotoxins in food and feed. *Adv. Food Nutr. Res.* **2019**, 89, 297–345.

Popović, V.; Fairbanks, N.; Pierscianowski, J.; Biancaniello, M.; Zhou, T.; Koutchma, T. Feasibility of 3D UV-C treatment to reduce fungal growth and mycotoxin loads on maize and wheat kernels. *Mycotoxin Res.* **2018**, 34, 211–221.

Prange, A.; Birzele, B.; Krämer, J.; Meier, A.; Modrow, H.; Köhler, P. *Fusarium*-inoculated wheat: Deoxynivalenol contents and baking quality in relation to infection time. *Food Control.* **2005**, 16, 739–745.

Raeisossadati, M.J.; Danesh, N.M.; Borna, F.; Gholamzad, M.; Ramezani, M.; Abnous, K.; Taghdisi, S.M. Lateral flow based immunobiosensors for detection of food contaminants. *Biosens. Bioelectron.* **2016**, 86, 235–246.

Raiola, A.; Meca, G.; Mañes, J.; Ritieni, A. Bioaccessibility of deoxynivalenol and its natural co-occurrence with ochratoxin A and aflatoxin B1 in Italian commercial pasta. *Food Chem. Toxicol.* **2012**, 50, 280–287.

Ramirez, M.L.; Chulze, S.; Magan, N. Impact of environmental factors and fungicides on growth and deoxinivalenol production by *Fusarium graminearum* isolates from Argentinian wheat. *Crop Prot.* **2004**, 23, 117–125.

Rempe, I.; Kersten, S.; Valenta, H.; Dänicke, S. Hydrothermal treatment of naturally contaminated maize in the presence of sodium metabisulfite, methylamine and calcium hydroxide; effects on the concentration of zearalenone and deoxynivalenol. *Mycotoxin Res.* **2013**, 29, 169–175.

Remža, J.; Lacko-Bartošová, M.; Kosík, T. Official control of wheat mycotoxins con-tamination in the Slovak Republic. *J. Microbiol. Biotechnol. Food Sci.* **2021**, 2021, 270–272.

Romera, D.; Mateo, E.M.; Mateo-Castro, R.; Gomez, J.V.; Gimeno-Adelantado, J.V.; Jimenez, M. Determination of multiple mycotoxins in feedstuffs by combined use of UPLC–MS/MS and UPLC–QTOF–MS. *Food Chem.* **2018**, 267, 140–148.

Rybecky, A.I.; Chulze, S.N.; Chiotta, M.L. Effect of water activity and temperature on growth and trichothecene production by *Fusarium* meridionale. *Int. J. Food Microbiol.* **2018**, 285, 69–73.

Rychlik, M.; Humpf, H.-U.; Marko, D.; Dänicke, S.; Mally, A.; Berthiller, F.; Klaffke, H.; Lorenz, N. Proposal of a comprehensive definition of modified and other forms of mycotoxins including "masked" mycotoxins. *Mycotoxin Res.* **2014**, 30, 197–205.

Sanders, M.; McPartlin, D.; Moran, K.; Guo, Y.; Eeckhout, M.; O'Kennedy, R.; De Saeger, S.; Maragos, C. Comparison of enzyme-linked immunosorbent assay, surface plasmon resonance and biolayer interferometry for screening of deoxynivalenol in wheat and wheat dust. *Toxins.* **2016**, 8, 103.

Santos Alexandre, A.P.; Vela-Paredes, R.S.; Santos, A.S.; Costa, N.S.; Canniatti-Brazaca, S.G.; Calori-Domingues, M.A.; Augusto, P.E.D. Ozone treatment to reduce deoxyni-valenol (DON) and zearalenone (ZEN) contamination in wheat bran and its impact on nutritional quality. *Food Addit. Contam. Part A.* **2018**, 35, 1189–1199.

Schaarschmidt, S.; Fauhl-Hassek, C. The fate of mycotoxins during secondary food pro-cessing of maize for human consumption. *Compr. Rev. Food Sci. Food Saf.* **2021**, 20, 91–148.

Schmeits, P.C.J.; Katika, M.R.; Peijnenburg, A.A.C.M.; van Loveren, H.; Hendriksen, P.J.M. DON shares a similar mode of action as the ribotoxic stress inducer anisomycin while TBTO shares ER stress patterns with the ER stress inducer thapsigargin based on compara-tive gene expression profiling in Jurkat T cells. *Toxicol. Lett.* **2014**, 224, 395–406.

Seeling, K.; Lebzien, P.; Dänicke, S.; Spilke, J.; Südekum, K.H.; Flachowsky, G. Effects of level of feed intake and *Fusarium* toxin-contaminated wheat on rumen fermentation as well as on blood and milk parameters in cows. *J. Anim. Physiol. Anim. Nutr.* **2006**, 90, 103–115.

Seo, S.E.; Tabei, F.; Park, S.J.; Askarian, B.; Kim, K.H.; Moallem, G.; Chong, J.W.; Kwon, O.S. Smartphone with optical, physical, and electrochemical nanobiosensors. *J. Ind. Eng. Chem.* **2019**, 77, 1–11.

Shao, Y.; Duan, H.; Zhou, S.; Ma, T.; Guo, L.; Huang, X.; Xiong, Y. Biotin–streptavidin sys-tem-mediated ratiometric multiplex immunochromatographic assay for simulta-neous and accurate quantification of three mycotoxins. *J. Agric. Food Chem.* **2019**, 67, 9022–9031.

Silva, M.V.; Pante, G.C.; Romoli, J.C.Z.; de Souza, A.P.M.; Rocha, G.H.O.D.; Ferreira, F.D.; Feijó, A.L.R.; Moscardi, S.M.P.; de Paula, K.R.; Bando, E. Occurrence and risk assess-ment of population exposed to deoxynivalenol in foods derived from wheat flour in Brazil. *Food Addit. Contam. Part A.* **2018**, 35, 546–554.

Šliková, S.; Gavurníková, S.; Šudyová, V.; Gregová, E. Occurrence of deoxynivalenol in wheat in Slovakia during 2010 and 2011. *Toxins.* **2013**, 5, 1353–1361.

Stadler, D.; Berthiller, F.; Suman, M.; Schuhmacher, R.; Krska, R. Novel analytical meth-ods to study the fate of mycotoxins during thermal food processing. *Anal. Bioanal. Chem.* **2020**, 412, 9–16.

Stadler, D.; Lambertini, F.; Woelflingseder, L.; Schwartz-Zimmermann, H.; Marko, D.; Suman, M.; Berthiller, F.; Krska, R. The Influence of processing parameters on the mitigation of deoxynivalenol during industrial baking. *Toxins.* **2019**, 11, 317.

Strasser, A.; Carra, M.; Ghareeb, K.; Awad, W.; Böhm, J. Protective effects of antioxidants on deoxynivalenol-induced damage in murine lymphoma cells. *Mycotoxin Res.* **2013**, 29, 203–208.

Subak, H.; Selvolini, G.; Macchiagodena, M.; Ozkan-Ariksoysal, D.; Pagliai, M.; Procacci, P.; Marrazza, G. Mycotoxins aptasensing: From molecular docking to electrochemical detection of deoxynivalenol. *Bioelectrochemistry.* **2021**, 138, 107691.

Sun, S.; Zhao, R.; Xie, Y.; Liu, Y. Photocatalytic degradation of aflatoxin B1 by activated carbon supported TiO2 catalyst. *Food Control.* **2019**, 100, 183–188.

Sun, X.; Ji, J.; Gao, Y.; Zhang, Y.; Zhao, G.; Sun, C. Fate of deoxynivalenol and degradation products degraded by aqueous ozone in contaminated wheat. *Food Res. Int.* **2020**, 137, 109357.

Supronienė, S.; Sakalauskas, S.; Mankevičienė, A.; Barčauskaitė, K.; Jonavičienė, A. Distribution of B type trichothecene producing *Fusarium* species in wheat grain and relation to mycotoxins DON and NIV concentrations. *Zemdirbyste-Agriculture.* **2016**, 103.

Tabuc, C.; Marin, D.; Guerre, P.; Sesan, T.; Bailly, J.D. Molds and mycotoxin content of cereals in southeastern Romania. *J. Food Prot.* **2009**, 72, 662–665.

Tian, Y.; Tan, Y.; Liu, N.; Yan, Z.; Liao, Y.; Chen, J.; De Saeger, S.; Yang, H.; Zhang, Q.; Wu, A. Detoxification of deoxynivalenol via glycosylation represents novel insights on antagonistic activities of *Trichoderma* when confronted with *Fusarium graminearum*. *Toxins.* **2016**, 8, 335.

Tima, H.; Brückner, A.; Mohácsi-Farkas, C.; Kiskó, G. *Fusarium* mycotoxins in cereals harvested from Hungarian fields. *Food Addit. Contam. Part B.* **2016a**, 9, 127–131.

Tima, H.; Rácz, A.; Guld, Z.; Mohácsi-Farkas, C.; Kiskó, G. Deoxynivalenol, zearalenone and T-2 in grain based swine feed in Hungary *Food Addit. Contam. Part B.* **2016b**, 9, 275–280.

Tittlemier, S.A.; Roscoe, M.; Trelka, R.; Gaba, D.; Chan, J.M.; Patrick, S.K.; Sulyok, M.; Krska, R.; McKendry, T.; Gräfenhan, T. *Fusarium* damage in small cereal grains from Western Canada. 2. Occurrence of *Fusarium* toxins and their source organisms in durum wheat harvested in 2010. *J. Agric. Food Chem.* **2013**, 61, 5438–5448.

Topi, D.; Babič, J.; Pavšič-Vrtač, K.; Tavčar-Kalcher, G.; Jakovac-Strajn, B. Incidence of *Fusarium* mycotoxins in wheat and maize from Albania. *Molecules.* **2021**, 26, 172.

Torović, L. *Fusarium* toxins in corn food products: A survey of the Serbian retail market. *Food Addit. Contam. Part A.* **2018**, 35, 1596–1609.

Tutelyan, V.A.; Zakharova, L.P.; Sedova, I.B.; Perederyaev, O.I.; Aristarkhova, T.V.; Eller, K.I. Fusariotoxins in Russian federation 2005–2010 grain harvests. *Food Addit. Contam. Part B.* **2013**, 6, 139–145.

Urbanek, K.A.; Habrowska-Górczyńska, D.E.; Kowalska, K.; Stańczyk, A.; Domińska, K.; Piastowska-Ciesielska, A.W. Deoxynivalenol as potential modulator of human steroidogenesis. *J. Appl. Toxicol.* **2018**, 38, 1450–1459.

Urusov, A.E.; Gubaidullina, M.K.; Petrakova, A.V.; Zherdev, A.V.; Dzantiev, B.B. A new kind of highly sensitive competitive lateral flow immunoassay displaying direct analyte-signal dependence. Application to the determination of the mycotoxin deoxynivalenol. *Microchim. Acta.* **2018**, 185, 1–7.

Vidal, A.; Bendicho, J.; Sanchis, V.; Ramos, A.J.; Marín, S. Stability and kinetics of leaching of deoxynivalenol, deoxynivalenol-3-glucoside and ochratoxin A during boiling of wheat spaghettis. *Food Res. Int.* **2016a**, 85, 182–190.

Vidal, A.; Marín, S.; Morales, H.; Ramos, A.J.; Sanchis, V. The fate of deoxynivalenol and ochratoxin A during the breadmaking process, effects of sourdough use and bran content. *Food Chem. Toxicol.* **2014a**, 68, 53–60.

Vidal, A.; Marín, S.; Ramos, A.J.; Cano-Sancho, G.; Sanchis, V. Determination of aflatoxins, deoxynivalenol, ochratoxin A and zearalenone in wheat and oat based bran supplements sold in the Spanish market. *Food Chem. Toxicol.* **2013**, 53, 133–138.

Vidal, A.; Marín, S.; Sanchis, V.; De Saeger, S.; De Boevre, M. Hydrolysers of modified mycotoxins in maize: α-Amylase and cellulase induce an underestimation of the total aflatoxin content. *Food Chem.* **2018**, 248, 86–92.

Vidal, A.; Morales, H.; Sanchis, V.; Ramos, A.J.; Marín, S. Stability of DON and OTA during the breadmaking process and determination of process and performance criteria. *Food Control.* **2014b**, 40, 234–242.

Vidal, A.; Sanchis, V.; Ramos, A.J.; Marín, S. The fate of deoxynivalenol through wheat processing to food products. *Curr. Opin. Food Sci.* **2016**, 11, 34–39.

Vidal, A.; Sanchis, V.; Ramos, A.J.; Marín, S. Thermal stability and kinetics of degradation of deoxynivalenol, deoxynivalenol conjugates and ochratoxin A during baking of wheat bakery products. *Food Chem.* **2015**, 178, 276–286.

Vogelgsang, S.; Musa, T.; Bänziger, I.; Kägi, A.; Bucheli, T.D.; Wettstein, F.E.; Pasquali, M.; Forrer, H.-R. *Fusarium* mycotoxins in Swiss wheat: A survey of growers' samples between 2007 and 2014 shows strong year and minor geographic effects. *Toxins.* **2017**, 9, 246.

Wang, G.; Wang, Y.; Ji, F.; Xu, L.; Yu, M.; Shi, J.; Xu, J. Biodegradation of deoxynivalenol and its derivatives by Devosia insulae A16. *Food Chem.* **2019a**, 276, 436–442.

Wang, H.; Mao, J.; Zhang, Z.; Zhang, Q.; Zhang, L.; Zhang, W.; Li, P. Photocatalytic degradation of deoxynivalenol over dendritic-like α-Fe2O3 under visible light irradiation. *Toxins.* **2019b**, 11, 105.

Wang, L.; Shao, H.; Luo, X.; Wang, R.; Li, Y.; Li, Y.; Luo, Y.; Chen, Z. Effect of ozone treatment on deoxynivalenol and wheat quality. *PLoS ONE.* **2016**, 11, e0147613.

Wang, S.; Hou, Q.; Guo, Q.; Zhang, J.; Sun, Y.; Wei, H.; Shen, L. Isolation and characterization of a deoxynivalenol-degrading bacterium *Bacillus licheniformis* YB9 with the capability of modulating intestinal microbial flora of mice. *Toxins.* **2020a**, 12, 184.

Wang, X.; Chen, X.; Cao, L.; Zhu, L.; Zhang, Y.; Chu, X.; Zhu, D.; ur Rahman, S.; Peng, C.; Feng, S. Mechanism of deoxynivalenol-induced neurotoxicity in weaned piglets is linked to lipid peroxidation, dampened neurotransmitter levels, and interference with calcium signaling. *Ecotoxicol. Environ. Saf.* **2020b**, 194, 110382.

Wang, Y.; Wang, G.; Dai, Y.; Wang, Y.; Lee, Y.-W.; Shi, J.; Xu, J. Biodegradation of deoxynivalenol by a novel microbial consortium. *Front. Microbiol.* **2020c**, 10, 2964.

Webb, C.; Owens, G.W. Milling and flour quality. In *Bread Making: Improving Quality*, Stan, C., Ed. Woodhead Publishing Series in Food Science, Technology and Nutrition. **2003**; pp. 200–219.

WHO. *World Health Organization. Prevention and Reduction of Food and Feed Contamination.* **2012**. www.fao.org/home/en/.

Wolf, C.E.; Bullerman, L.B. Heat and pH alter the concentration of deoxynivalenol in an aqueous environment. *J. Food Prot.* **1998**, 61, 365–367.

Wu, L.; Li, J.; Li, Y.; Li, T.; He, Q.; Tang, Y.; Liu, H.; Su, Y.; Yin, Y.; Liao, P. Aflatoxin B 1, zeara-lenone and deoxynivalenol in feed ingredients and complete feed from different Province in China. *J. Anim. Sci. Biotechnol.* **2016**, 7, 1–10.

Wu, Q.-H.; Wang, X.; Yang, W.; Nüssler, A.K.; Xiong, L.-Y.; Kuča, K.; Dohnal, V.; Zhang, X.-J.; Yuan, Z.-H. Oxidative stress-mediated cytotoxicity and metabolism of T-2 toxin and deoxynivalenol in animals and humans: An update. *Arch. Toxicol.* **2014**, 88, 1309–1326.

Wu, S.; Wang, F.; Li, Q.; Wang, J.; Zhou, Y.; Duan, N.; Niazi, S.; Wang, Z. Photocatalysis and degradation products identification of deoxynivalenol in wheat using upconver-sion nanoparticles@ TiO2 composite. *Food Chem.* **2020**, 323, 126823.

Xia, R.; Schaafsma, A.W.; Wu, F.; Hooker, D.C. The change in winter wheat response to deoxynivalenol and *Fusarium* head blight through technological and agronomic progress. *Plant Dis.* **2021**, 105, 840–850.

Xiong, S.; Li, X.; Zhao, C.; Gao, J.; Yuan, W.; Zhang, J. The degradation of deoxynivalenol by using electrochemical oxidation with graphite electrodes and the toxicity assess-ment of degradation products. *Toxins.* **2019**, 11, 478.

Xu, H.; Wang, L.; Sun, J.; Wang, L.; Guo, H.; Ye, Y.; Sun, X. Microbial detoxification of myco-toxins in food and feed. *Crit. Rev. Food Sci. Nutr.* **2021**, http://doi.org/10.1080/1040 8398.10402021.11879730

Xu, S.; Zhang, G.; Fang, B.; Xiong, Q.; Duan, H.; Lai, W. Lateral flow immunoassay based on polydopamine-coated gold nanoparticles for the sensitive detection of zearale-none in maize. *ACS Appl. Mater. Interfaces.* **2019**, 11, 31283–31290.

Yao, Y.; Long, M. The biological detoxification of deoxynivalenol: A review. *Food Chem. Toxicol.* **2020**, 145, 111649.

Yi-gang, Y.; Han-ruo, M.; Rui, H.; Yu-qian, T.; Xing-long, X. Degradation of deoxynivalenol in flour by ozone and ultraviolet light and their effects on flour quality. *Mod. Food Sci. Technol.* **2016**, 32, 196–202.

Young, J.C.; Fulcher, R.G.; Hayhoe, J.H.; Scott, P.M.; Dexter, J.E. Effect of milling and baking on deoxynivalenol (vomitoxin) content of eastern Canadian wheats. *J. Agric. Food Chem.* **1984**, 32, 659–664.

Yuan, J.; Sun, C.; Guo, X.; Yang, T.; Wang, H.; Fu, S.; Li, C.; Yang, H. A rapid Raman detection of deoxynivalenol in agricultural products. *Food Chem.* **2017**, 221, 797–802.

Yumbe-Guevara, B.E.; Imoto, T.; Yoshizawa, T. Effects of heating procedures on deoxyni-valenol, nivalenol and zearalenone levels in naturally contaminated barley and wheat. *Food Addit. Contam.* **2003**, 20, 1132–1140.

Zhai, Y.; Zhong, L.; Gao, H.; Lu, Z.; Bie, X.; Zhao, H.; Zhang, C.; Lu, F. Detoxification of deoxynivalenol by a mixed culture of soil bacteria with 3-epi-deoxynivalenol as the main intermediate. *Front. Microbiol.* **2019**, 10, 2172.

Zhang, H.; Dong, M.; Yang, Q.; Apaliya, M.T.; Li, J.; Zhang, X. Biodegradation of zearale-none by Saccharomyces cerevisiae: Possible involvement of ZEN responsive pro-teins of the yeast. *J. Proteom.* **2016**, 143, 416–423.

Zhang, H.; Wang, B. Fate of deoxynivalenol and deoxynivalenol-3-glucoside during wheat milling and Chinese steamed bread processing. *Food Control.* **2014**, 44, 86–91.

Zhang, H.; Wang, B. Fates of deoxynivalenol and deoxynivalenol-3-glucoside during bread and noodle processing. *Food Control.* **2015**, 50, 754–757.

Zhang, J.; Liu, Y.; Li, Q.; Zhang, X.; Shang, J.K. Antifungal activity and mechanism of palla-dium-modified nitrogen-doped titanium oxide photocatalyst on agricultural patho-genic fungi *Fusarium graminearum. ACS Appl. Mater. Interfaces.* **2013**, 5, 10953–10959.

Zhang, J.; Qin, X.; Guo, Y.; Zhang, Q.; Ma, Q.; Ji, C.; Zhao, L. Enzymatic degradation of deoxynivalenol by a novel bacterium, Pelagibacterium halotolerans ANSP101. *Food Chem. Toxicol.* **2020a**, 140, 111276.

Zhang, K.; Ge, Y.; He, S.; Ge, F.; Huang, Q.; Huang, Z.; Wang, X.; Wen, Y.; Wang, B. Development of new electrochemical sensor based on kudzu vine biochar modified flexible carbon electrode for portable wireless intelligent analysis of clenbuterol. *Int. J. Electrochem. Sci.* **2020b**, 15, 7326–7336.

Zhang, W.; Tang, S.; Jin, Y.; Yang, C.; He, L.; Wang, J.; Chen, Y. Multiplex SERS-based lateral flow immunosensor for the detection of major mycotoxins in maize utilizing dual Raman labels and triple test lines. *J. Hazard. Mater.* **2020c**, 393, 122348.

Zhao, S.; Bu, T.; He, K.; Bai, F.; Zhang, M.; Tian, Y.; Sun, X.; Wang, X.; Zhangsun, H.; Wang, L. A novel α-Fe2O3 nanocubes-based multiplex immunochromatographic assay for simultaneous detection of deoxynivalenol and aflatoxin B1 in food samples. *Food Control.* **2021**, 123, 107811.

Zhao, Y.; Guan, X.; Zong, Y.; Hua, X.; Xing, F.; Wang, Y.; Wang, F.; Liu, Y. Deoxynivalenol in wheat from the Northwestern region in China. *Food Addit. Contam. Part B.* **2018**, 11, 281–285.

Zheng, D.; Zhang, S.; Zhou, X.; Wang, C.; Xiang, P.; Zheng, Q.; Xu, J.-R. The FgHOG1 pathway regulates hyphal growth, stress responses, and plant infection in *Fusarium graminearum*. *PLoS ONE.* **2012**, 7, e49495.

Zhou, H.; Guog, T.; Dai, H.; Yu, Y.; Zhang, Y.; Ma, L. Deoxynivalenol: Toxicological profiles and perspective views for future research. *World Mycotoxin J.* **2020**, 13, 179–188.

Zhou, S.; Xu, L.; Kuang, H.; Xiao, J.; Xu, C. Fluorescent microsphere immunochromatographic sensor for ultrasensitive monitoring deoxynivalenol in agricultural products. *Microchem. J.* **2021**, 164, 106024.

FURTHER READINGS

Adelsberger, H.; Garaschuk, O.; Konnerth, A. Cortical calcium waves in resting newborn mice. *Nat. Neurosci.* **2005**, 8, 988–990.

Berthiller, F.; Crews, C.; Dall'Asta, C.; Saeger, S.D.; Haesaert, G.; Karlovsky, P.; Oswald, I.P.; Seefelder, W.; Speijers, G.; Stroka, J. Masked mycotoxins: A review. *Mol. Nutr. Food Res.* **2013**, 57, 165–186.

Borràs-Vallverdú, B.; Ramos, A.J.; Marín, S.; Sanchis, V.; Rodríguez-Bencomo, J.J. Deoxynivalenol degradation in wheat kernels by exposition to ammonia vapours: A tentative strategy for detoxification. *Food Control.* **2020**, 118, 107444.

Cambaza, E.; Koseki, S.; Kawamura, S. *Fusarium graminearum* colors and Deoxynivalenol synthesis at different water activity. *Foods.* **2019**, 8, 7.

Cheli, F.; Pinotti, L.; Rossi, L.; Dell'Orto, V. Effect of milling procedures on mycotoxin distribution in wheat fractions: A review. *LWT-Food Sci. Technol.* **2013**, 54, 307–314.

Chen, Y.; Yang, Y.; Wang, Y.; Peng, Y.; Nie, J.; Gao, G.; Zhi, J. Development of an *Escherichia coli*-based electrochemical biosensor for mycotoxin toxicity detection. *Bioelectrochemistry.* **2020**, 133, 107453.

European Commission. Commission directive 2006/125/EC on processed cereal-based foods and baby foods for infants and young children. *OJEU.* **2006a**, 396, 16–35.

Flores-Flores, M.E.; González-Peñas, E. An LC–MS/MS method for multi-mycotoxin quantification in cow milk. *Food Chem.* **2017**, 218, 378–385.

Guo, L.; Shao, Y.; Duan, H.; Ma, W.; Leng, Y.; Huang, X.; Xiong, Y. Magnetic quantum dot nanobead-based fluorescent immunochromatographic assay for the highly sensitive detection of aflatoxin B1 in dark soy sauce. *Anal. Chem.* **2019**, 91, 4727–4734.

Kong, D.; Wu, X.; Li, Y.; Liu, L.; Song, S.; Zheng, Q.; Kuang, H.; Xu, C. Ultrasensitive and eco-friendly immunoassays based monoclonal antibody for detection of deoxynivalenol in cereal and feed samples. *Food Chem.* **2019**, 270, 130–137.

Lyu, F.; Gao, F.; Zhou, X.; Zhang, J.; Ding, Y. Using acid and alkaline electrolyzed water to reduce deoxynivalenol and mycological contaminations in wheat grains. *Food Control.* **2018**, 88, 98–104.

Munkvold, G.P.; Arias, S.; Taschl, I.; Gruber-Dorninger, C. Mycotoxins in corn: Occurrence, impacts, and management. In *Corn*, Serna-Saldivar, S.O., Ed. Elsevier. **2019**; pp. 235–287.

Olopade, B.K.; Oranusi, S.U.; Nwinyi, O.C.; Gbashi, S.; Njobeh, P.B. Occurrences of deoxynivalenol, zearalenone and some of their masked forms in selected cereals from Southwest Nigeria. *NFS J.* **2021**, 23, 24–29.

Ran, L.; Ze-Jia, L.; Jin-Yi, Y.; Zhen-Lin, X.; Hong, W.; Hong-Tao, L.; Yuan-Ming, S.; Yu-Dong, S. An indirect competitive enzyme-linked immunosorbent assay for simultaneous determination of florfenicol and thiamphenicol in animal meat and urine. *Chinese J. Anal. Chem.* **2018**, 46, 1321–1328.

Schlegel, K.M.; Elsinghorst, P.W. Myco-DES: Enabling remote extraction of mycotoxins for robust and reliable quantification by stable isotope dilution LC–MS/MS. *Anal. Chem.* **2020**, 92, 5387–5395.

Thakare, D.; Zhang, J.; Wing, R.A.; Cotty, P.J.; Schmidt, M.A. Aflatoxin-free transgenic maize using host-induced gene silencing. *Sci. Adv.* **2017**, 3, e1602382.

Torres, A.M.; Barros, G.G.; Palacios, S.A.; Chulze, S.N.; Battilani, P. Review on pre-and post-harvest management of peanuts to minimize aflatoxin contamination. *Food Res. Int.* **2014**, 62, 11–19.

Tsien, R.W.; Tsien, R.Y. Calcium channels, stores, and oscillations. *Annu. Rev. Cell Biol.* **1990**, 6, 715–760.

Urusov, A.E.; Petrakova, A.V.; Zherdev, A.V.; Dzantiev, B.B. "Multistage in one touch" design with a universal labelling conjugate for high-sensitive lateral flow immunoassays. *Biosens. Bioelectron.* **2016**, 86, 575–579.

Chapter 6

Occurrence, Detection, and Management of T-2 Toxin and HT-2 Toxin in Food and Feed

Pooja Yadav, Nabendu Debnath,
Shalini Arora, Ashok Kumar Yadav

CONTENTS

6.1	Introduction	158
6.2	Properties of T-2 Toxin	160
6.3	Ecological Ubiquity and Factors Triggering Toxin Production	160
6.4	Manifestations of T-2 Toxin and Metabolite Exposure	161
6.5	Metabolism of T-2 and HT-2 Toxins	161
	6.5.1 Hydrolysis	162
	6.5.2 Hydroxylation	162
	6.5.3 De-Epoxidation	163
6.6	Effect on Metabolic Pathways	163
6.7	Genes Responsible for T-2 Toxin Synthesis Pathway	164
6.8	T-2 Toxin and Metabolite Detection	167
	6.8.1 Chromatography	167
	6.8.1.1 TLC Method	167
	6.8.1.2 Gas Chromatography	168
	6.8.1.3 Liquid Chromatography	168
	6.8.1.3.1 Limitations of HPLC	169
	6.8.2 Ion Mobility Spectrometry	169
	6.8.3 Rapid Methods of Detection	170
	6.8.3.1 Immunological Assays	170
	6.8.3.2 Biosensors	170
	6.8.3.3 Proteomic Methods	170

DOI: 10.1201/9781003242208-6

6.9   Control Strategies   170
    6.9.1   Pre-Harvest Control Plan   174
    6.9.2   Harvest Control Plan   174
    6.9.3   Post-Harvest Control Plan   174
        6.9.3.1   Physical Post-Harvest Control Methods   175
            6.9.3.1.1   Separation Methods   175
            6.9.3.1.2   Radiation   175
            6.9.3.1.3   Mycotoxin Adsorbers   176
        6.9.3.2   Chemical Methods   176
            6.9.3.2.1   Alkalinizing Agents   176
            6.9.3.2.2   Oxidizing Agents   177
            6.9.3.2.3   Miscellaneous   177
        6.9.3.3   Biological Methods   178
6.10 Summary   178

6.1 INTRODUCTION

Many fungal mold species produce numerous metabolites in their later phase known as secondary metabolites, which somehow help in the survival of fungi in adverse conditions but may harm others. Fungal species release several secondary metabolites, including plant growth promoters, antibiotics, plant pigments, and several types of toxins (McCormick et al. 2011). For example, many endophytic fungi such as *Aspergillus niger, Aspergillus flavus, Fusarium oxysporum*, and so on, release gibberellins and indole acetic acid (IAA) that act as plant growth promoters. Fungal species release several antibiotics; for example, *Penicillium chrysogenum* produces penicillin, and *Cephalosporium acremonium* produces cephalosporin. Similarly, some toxic compounds are also released by fungi in later stages of their life, such as trichothecenes, aflatoxins, fumonisins, zearalenone, citrinin, and ergot alkaloids. These fungal metabolites are known as mycotoxins. Mycotoxins are compounds capable of inducing a series of toxic effects when consumed in specific quantities. Some major fungal genera that produce mycotoxins are *Aspergillus, Fusarium, Penicillium, Claviceps,* and *Alternaria.* There are approximately more than 500 mycotoxins reported from 10,000 fungi identified to date with the toxic potency of posing health concerns to both humans and animals. Such toxins are aflatoxins (AFTs), ochratoxins (OTs), trichothecenes (TCTs), fumonisins (FUMs), zearalenone (ZEN), patulin (PAT), citrinin (CT), and ergot alkaloids (EAs) (Grenier and Oswald 2011; Ukwuru and Ohaegbu 2018).

The first documented instance of human mycotoxicosis can be traced back to the Middle Ages: "St. Anthony's Fire" related to ergot alkaloid mycotoxin induced by *Claviceps purpura* in rye. Mycotoxins have lethal reverberations with carcinogenic, tumorigenic, mutagenic, immunosuppressant qualities and can cause disruption of endocrine effects in humans and animals (Haque et al. 2020). When consumed in sufficient quantities, they can also cause liver and kidney damage, affecting the proper functioning of both and hence leading to death.

In 1940, the term stachybotryotoxicosis was coined by Soviet scientists to explain a syndrome with symptoms such as pharyngitis, rhinorrhea, hyperpnea, croak, and pyrexia resulting from expiration of the "*Stachybotrys* myco-toxin." During World War II, it was found that T-2 mycotoxin could be used as a biological weapon in an incident where civilians in Orenburg, Russia, accidentally consumed wheat contaminated with *Fusarium* fungi. Civilians developed symptoms similar to alimentary toxic aleukia (ATA). Twenty years after the occurrence the primary toxin responsible for ATA was found to be 4β,15-diacetoxy-3α-hydroxy-8α-[3-methylbutyryloxy]-12,13-epoxytrichothec-9-ene, or T-2 toxin, which causes symptoms such as skin necrosis, dermal exasperation, gastric exasperation, immunosuppression, and increased death rate in afflicted humans and animals (Haque et al. 2020). T-2 toxin is a type-A trichothecene synthesized by different fungal species as a secondary metabolite. These toxins are derived from ring structured trichothecenes. There are more than 200 varieties of trichothecenes and their derivatives. Trichothecenes possess composite and diversified structures for two reasons: 1) varied 15-carbon assembled backbone of cyclic sesquiterpenoids and 2) varied and distinct functional groups attached to the structural backbone in a specified and specific stereoisomeric pattern. Hence, trichothecenes are categorized into four (A, B, C and D) groups based on the position of tricyclic 12, 13-epoxytrichothec-9-ene (EPT) (McCormick et al. 2011). In terms of chemical structure, type A trichothecenes characteristically have T-2 toxin and HT-2 toxin varying at four carbon

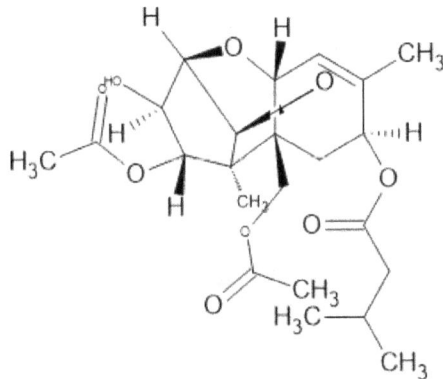

Figure 6.1   Chemical structure of T-2 toxin.

sites (C-3, C-4, C-57, C-15), and the carbonyl group is not present, whereas type B (e.g., deoxynivalenol and nivalenol) has a carbonyl group at C-8. In type C, another epoxy group is present between C-7 and C-8 or C-8 and C-9. Type D has a "macrocyclic ring" between C-4 and C-15 (Wu et al. 2013). Satratoxin and roridin are representatives of type D. It is reported that type A trichothecenes, particularly HT-2 and T-2 toxin, are the most toxic, with higher toxic potency as compared to type B trichothecenes (Morcia et al. 2016).

6.2 PROPERTIES OF T-2 TOXIN

T-2 toxin is non-volatile and resistant to unfavorable surrounding conditions, ideal varying temperature, and light levels, but harsh acidic and alkaline conditions degrade it (Adhikari et al. 2017). T-2 toxin does not solubilize in water but solubilizes in other less polar solvents (Duffy and Reid 1993). These toxins are not easily deactivated, not even in an autoclave; for that, the temperature should be raised to 900°F or 500°F for 10 and 30 min, respectively (Kachuei et al. 2014). T-2 toxin is a part of a broad family of chemically similar toxins secreted by fungal genera such as *Fusarium*, particularly by *F. sporotrichioides*, *F. poae*, *F. equiseti*, and *F. acuminatum*; *Myrothecium*; and *Stachybotrys*.

T-2 toxin affects cereals such as maize, wheat, rice, barley, oats, and soybeans and their by-products and has a deleterious effect on human and animal well-being. T-2 toxin and its metabolites show acute symptoms to humans once ingested. The results include vomiting, diarrhea, nausea, skin pain, irritation, redness, necrosis, weight loss, bloody stools, cartilage tissue damage, reduced blood glucose levels, epidermal sloughing pruritus, redness, and blisters. In contrast, other deleterious symptoms in chronic poisoning are also reported such as prostration, weakness, ataxia, pathological changes in liver failure, heart failure, shock, and death (Hussein 2001; Joint FAO/WHO Expert Committee on Food Additives 2001). Ingestion of type A trichothecenes in animals results in lower feed appetite, vomiting, and suppressed immune system. In severe poisoning, they are explicitly responsible for reduced blood cell count and immune activity, and thus can cause several other diseases (Eriksen and Pettersson 2004; Gholampour Azizi et al. 2014; Glenn 2007; Joint FAO/WHO Expert Committee on Food Additives 2001).

6.3 ECOLOGICAL UBIQUITY AND FACTORS TRIGGERING TOXIN PRODUCTION

T-2 toxin and other metabolite production is dependent on environmental factors such as weather conditions, monsoons, untimely rains in harvesting season, floods, grain defects, moisture content present in grains (13 to 22%), and temperature range (0 to 50°C, dependent on fungal species) (Adhikari et al. 2017). The humidity percentage of the substrate is between 10 and 20%, whereas relative humidity (greater than or equal to 70%) and oxygen availability play a crucial role in the stimulation T-2 and HT-2 toxins. Reports support that growth of *Fusarium* species is more prominent in cereals grown in countries such as Vietnam, China, Thailand, and South Korea with humid climatic conditions (Ukwuru and Ohaegbu 2018). During pre- and post-harvest, some intrinsic and extraneous factors of agricultural by-products that promote toxin production should be monitored: 1) intrinsic factors—moisture content, substrate type, nutrient composition, plant type, and water availability; 2) extraneous factors—climatic conditions, temperature, and oxygen level; 3) post-harvest processes such as blending, drying, grain handling, preservative addition, and storage; and 4) indirect factors—fungal strains, insect interactions, and other microbes (Adhikari et al. 2017).

In 2011, for safety evaluation standards, the Scientific Committee for Food (SCF) of the European Commission set the legitimate limit temporary tolerable daily intake (TDI) at 0.02 µg/kg bw/day, which was further revised and reduced in 2017 to 0.01 µg/kg bw/day (Harcz et al. 2007). In 2017, the acute reference dose (ARfD) was set as 0.03 µg/kg bw/day (Committee 2011). The established temporary tolerable daily intake is described as either the T-2 and HT-2 toxin sum or the individual toxicity lying in the same range. This value is set to protect people from both toxins' hazardous effects (Lavinia et al. 2011).

6.4 MANIFESTATIONS OF T-2 TOXIN AND METABOLITE EXPOSURE

There are several proposed mechanisms for T-2 toxin and HT-2 toxin action in humans and other animals. Per the *Missouri Department of Health and Senior Services Communicable Disease Investigation Reference Manual*, T-2 toxin and its metabolites can enter the body via three routes: 1) absorption through the skin, 2) inhalation of toxins in the environment, and 3) ingestion of mycotoxins by food or feed contaminants. After exposure, symptoms can arise within minutes or a few hours, depending on the amount, duration of exposure, and route of toxin taken. Symptoms can be mild or severe and may include any of the following listed in Table 6.1.

6.5 METABOLISM OF T-2 AND HT-2 TOXINS

Once T-2 and HT-2 toxin enters the body, it is rapidly absorbed by the gastrointestinal tract or via respiratory mucosal membranes and metabolized in the liver. Immediate distribution occurs in organs with or without accumulation.

TABLE 6.1   SYMPTOMS OF T-2 TOXIN AND METABOLITE EXPOSURE IN THE HUMAN BODY

Route of Toxin	Symptoms
Skin contact	Burning sensation, redness, blisters, tissue damage visible in the form of patches on the skin, itching
Eye contact	Pain, redness, tears, impaired vision, itching
Inhalation	Nasal, mouth and throat pain, nasal bleeding, coughing, runny nose, shortness of breath, blood in sputum and saliva
Ingestion	Loss of appetite, nausea, vomiting, abdominal pain, watery or blood-imbued diarrhea
Severe exposure by any route	Weakness, drowsiness, ataxia, lack of coordination, fast heartbeat, low body temperature, shock, decreased cardiac output, reduced red and white blood cells

The half-lives of the toxin and its breakdown compounds are usually very short, and they are removed from the body within 48 hours, but it depends on time, amount, and route of exposure. Maximum plasma concentration was reported in rodents. Steps involved in the metabolomic route of the toxin are hydrolysis (hydrolyzed at C-4, C-8, and C-15 positions), hydroxylation (hydroxylated at C-7, C-3′, and C-4′ positions), acetylation conjugation, and deoxidation (Duffy and Reid 1993). We will discuss each step and the resultant metabolites one by one.

6.5.1 Hydrolysis

There are three ester bonds present in T-2 toxin at carbon positions 4, 8, and 15; they are hydrolyzed to provide various metabolites. It has been found that HT-2 toxin is produced by hydrolytic cleavage at the fourth positioned carbon, which further hydrolyses to give neosaxitoxin and 4-deacetylneosolaniol (Yang et al. 2013). Other toxic compounds such as T-2 triol and T-2 tetraol are also observed in meager quantities (Wu et al. 2011).

6.5.2 Hydroxylation

Once the toxin enters the body, it is more susceptible to oxidation by oxidases present inside. Hydroxylation occurs at C-7, and C-9 is an essential pathway of mycotoxin metabolism. Hydroxylation at C-7 results in several metabolites such as 7-OH-HT-2, 3′, 7-dihydroxy-T-2, 3′, 7-dihydroxy-HT-2, isomeric forms, and de-epoxy3′,7-dihydroxy-HT-2. Hydroxylation at the C-9 position results in

9-OH-T-2 and 3′, 7-dihydroxy-T-2 (Yang et al. 2013). The C-3 and C-4 positions of isovaleryl T-2 toxin are also hydroxylased and result in various metabolites such as 3′-hydroxy-T-2, 3′-OH-T-2, 3′-OH-HT-2, 3′-OH-15-deacetyl-T-2, 3′-OH-T-2 triol, and 4′-OH-T-2 (Duffy and Reid 1993).

6.5.3 De-Epoxidation

De-epoxidation is crucial for detoxifying T-2 toxin in the intestine by anaerobic microbes present there. After this step, numerous metabolites are formed, including deepoxy-HT-2, de-epoxy-3′-OH-HT-2, de-epoxy-3′-OH-15-de-acetyl-T-2, and de-epoxy-3′, 7-dihydroxy-HT-2 (Yang et al. 2013).

6.6 EFFECT ON METABOLIC PATHWAYS

T-2 toxin has a thiol group, which facilitates inhibiting the crucial pathway of protein production by binding with peptidyl transferase enzyme and targets the 60s ribosomal unit, thus hindering translation process.

It also acts as a DNA inhibitor, as it causes fragmentation of DNA and lesion formation. The non-ribosomal impact of T-2 toxin is the main root of trouble in animals. It crosses the plasma membrane and forms reactive oxygen species (ROS), which cause oxidative stress on cells and cell organelles. Intensive damage to the cells is caused by ROS exemplary hydrogen peroxide, hydroxyl radicals, and superoxide molecules (Mackei et al. 2020). Oxidative stress causes endoplasmic reticulum stress and enhances lipid peroxidation and methylation of DNA, resulting in damaged mitochondrial DNA. It also initiates inflammatory pathways and disturbs signaling pathways needed for the normal functioning of life (Chaudhary and Lakshmana Rao 2010; Mackei et al. 2020; Pestka 2010). T-2 toxin also displays the oxidative cellular damage property that targets essential biomolecules such as nucleic acids, proteins, and lipids. T-2 toxin causes harmful effects from apoptosis in various cell organs such as skin, kidney, brain, bone marrow, spleen, and thymus lymphocytes. Once animals are exposed to T-2 toxin, it leads to cell death in these cell types by either mitochondrial or non-mitochondrial mechanisms.

Other consequences of T-2 toxin exposure are inhibition of antibody production, impaired dendritic cell formation, decreased proliferation of lymphocytes, and delayed-type hypersensitivity and thus hindered proper functioning of the membrane (Li et al. 2011). It has been reported in *in vitro* experiments that mycotoxin induces inflammatory responses by activating IL-1β and IL-18. They trigger inflammatory responses (Kankkunen et al. 2009; Seeboth et al. 2012).

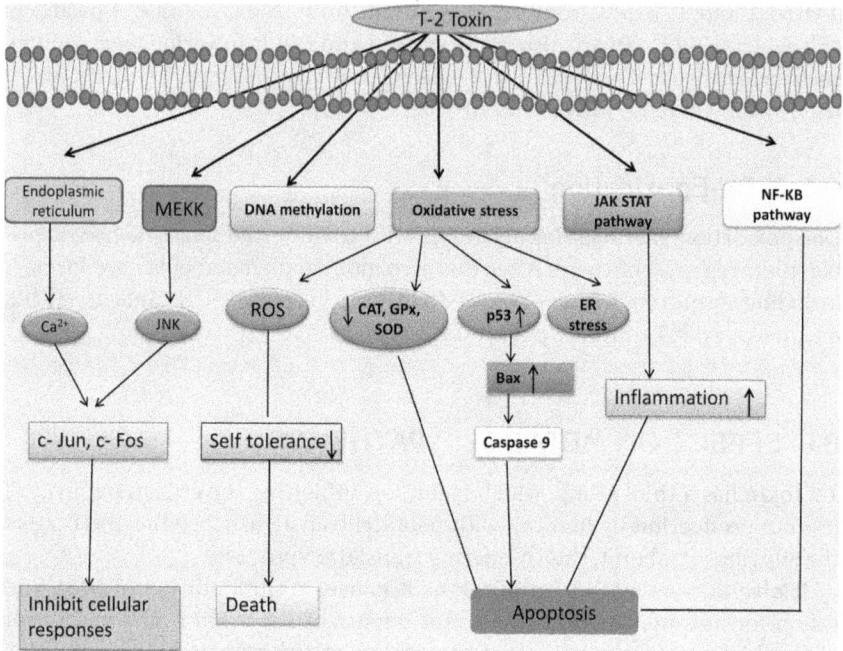

Figure 6.2   Molecular mechanism of T-2 toxin on various functions of the cell.

6.7 GENES RESPONSIBLE FOR T-2 TOXIN SYNTHESIS PATHWAY

Mycotoxin-producing genes are arranged in the form of a cluster known as the biosynthetic gene cluster. These are clusters of two or more genes placed at a particular location in the genome. The genes of the cluster are encoders of various enzymes, including polyketide synthases (PKS), non-ribosomal peptide synthetases (NRPS), hybrids (PKS-NRPS), terpene cyclases (TCs), and prenyl-transferases (PTs) needed for toxin production. They also regulate the expression and transportation of metabolites (Inglis et al. 2013; Lazarus et al. 2014).

The gene cluster that induces secondary metabolites production in fungal species is the trichothecene biosynthetic gene (TRI) (Villafana et al. 2019). The conventional method of naming this gene cluster consists of the three uppercase letters TRI and a number representing the gene (Proctor et al. 2018; Villafana et al. 2019). The TRI cluster gene reported in *F. graminearum* and *F. sporotrichi-oides* is a 26 kb DNA fragment consisting three loci: a single-gene TRI101 locus, a 2-gene TRI1-TRI16 locus, and a 12-gene core TRI cluster (Brown et al. 2004; Villafana et al. 2019). These genes are responsible for encoding enzymes needed

TABLE 6.2   GENES OF TRI GENE CLUSTERS WITH THE ENZYMES THEY ENCODE AND THE COMPOUNDS FORMED BY CONVERTING VARIOUS SUBSTRATES TO PRODUCTS (PROCTOR ET AL. 2018)

Gene	Protein Annotation	Substrate	Function	Product
TRI8	Esterase	3-acetyl-T-2 toxin	Deacetylation at C3 Or C15	T-2 toxin
TRI7	Acetyl transferase	3,15 diacetoxyscerpinol	—	3,4, 15 triacetoxyscerpinol
TRI3	Acetyl transferase	15-decalonectrin	Acetylation at C15 (*Fusarium*)	Calonectrin
TRI4	Cytochrome P450 monooxygenase	Trichodiene 2-hydroxytrichodiene 12,13 epoxy-9,10 trichoene-2ol Isotrichodiol	Oxygenation at C2, C3, C11, and C13 (*Fusarium*)	2-hydroxytrichodiene 12,13 epoxy-9,10 trichoene-2 ol Isotrichotriol
TRI6	Zn_2His_2 transcription factor	—	TRI gene expression transcript regulator (*Fusarium*)	—
TRI5	Terpene synthase	(2e,6e) farnesyl diphosphate	Farnesyl pyrophosphate cyclized to trichodiene (*Fusarium, Trichoderma*)	Trichodiene
TRI10	Transcriptional regulator	—	Tri gene expression transcript regulator (*Fusarium*)	—
TRI9	Unknown	—	Unknown	—
TRI11	Cytochrome P450 monooxygenase	Isotrichodermin	Hydroxylation at C15 (*Fusarium*)	15-decalonectrin
TRI12	Major facilitator superfamily transporter	—	Trichothecene efflux pump (*Fusarium*)	—
TRI13	Cytochrome P450 monooxygenase	Calonectrin	Hydroxylation at C4 (*Fusarium*)	3,15 diacetoxyscerpinol
TRI14	Unknown	—	Unknown	—
TRI16	Acyl transferase	3-acetylneosolaniol	Acylation at C8 (*Fusarium*)	3-acetyl T-2 toxin
TRI101	Acetyl transferase	Isotrichodermol	Acetylation at C3 (*Fusarium*)	Isotrichodermin

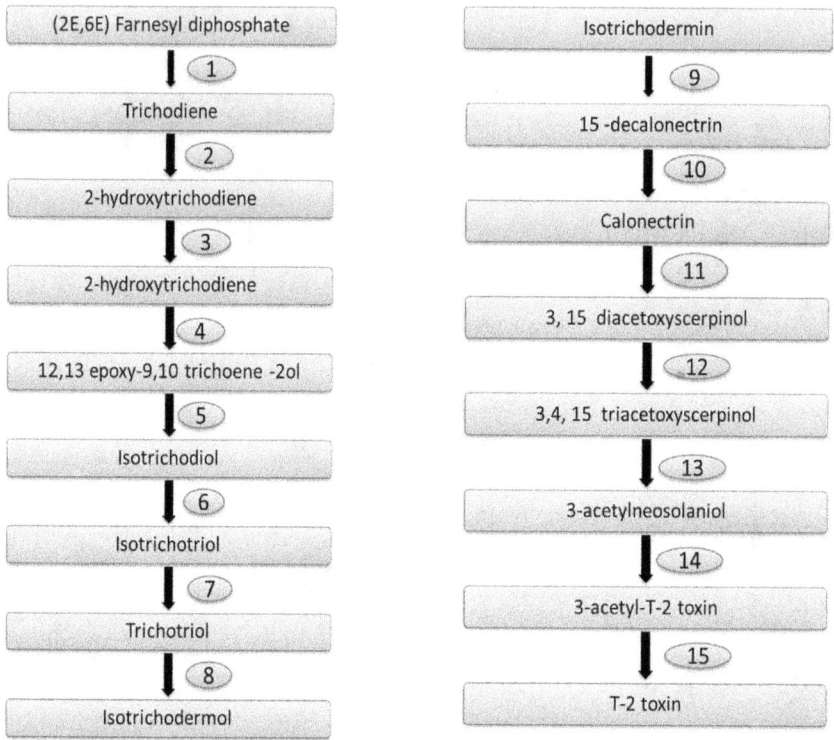

Figure 6.3 Schematic representation of T-2 toxin biosynthesis pathway: Enzymes involves in this pathway are represented by numerals.

to synthesize intermediates in trichothecene formation and help transportation across the plasma membrane. At the time of evolution, other TRI genes were also identified in some *Fusarium* species; for example, TR101 was found inserted in the core gene cluster of *F. equiseti*.

Steps in the synthesis of T-2 toxin and other trichothecenes involve the biosynthesis of calonectrin. The synthesis pathway of calonectrin is the same in all trichothecene biosynthesis, followed by respective pathways depending upon their chemical nature. Steps involved in calonectrin synthesis are as follow:

TRI5 trichodiene synthase; **2**. TRI4 trichodiene oxygenase; 3: TRI4 2-hydroxydiene 12,13 epoxidase; 4: TRI4 12,13 epoxy-9,10 trichoene-2ol –11-hydroxylase; 5: TRI4 isotrichodiol 3-hydroxylase; 6: spontaneous; 7: spontaneous; 8: TRI101 isotrichodermol 3-actyltransferase; 9: TRI11 isotrichodermin 15 hydroxylase; 10: TRI 3 trichothecene 15—O acetylytransferase; 11: TRI13 trichothecene 4-hydroxylase; 12: TRI7 trichothecene C-4-acetyltransferase; 13:

TRI1 trichothecene-8-hydroxylase; 14: TRI16 trichothecene C-8 acetyltransferase; 15: TRI8 trichothecene C-3-esterase.

6.8 T-2 TOXIN AND METABOLITE DETECTION

Techniques available to determine T-2 toxin and its metabolites are many and have advanced with time. Accurate, sensitive, and faster methods have always been essential to understand the global impact of contamination caused by mycotoxins and find a mitigation strategy to control them.

Many countries have established legitimate limits for food and feed. Highly sensitive and dependable techniques are required to determine and separate toxins from food (Pereira et al. 2014a).

Traditional practice methods to detect toxins are thin-layer chromatography (TLC), gas chromatography (GC), and high-performance liquid chromatography (HPLC). But nowadays, several other methods are also utilized, such as immunoassay-based methods, biosensors, electrochemical biosensors, piezoelectric biosensors, and proteomic and genomic methods for T-2 and metabolite determination.

Steps involved in toxin determination are 1) suitable extraction solvent, 2) clean-up method to remove impurities, and 3) instrumentation to detect toxins. The solvent to be chosen is basically based on the analyte to be determined; on a general basis, organic solvents are employed for food and feed contaminants. All components except the analyte have a combined impact on the selection of the method to be used. Anything that hinders accuracy has to be removed in the second step. The third step compels good instrumentation that will detect even small limit of detection (LOD) of the toxin.

6.8.1 Chromatography

6.8.1.1 TLC Method

This is a technique with a stationary phase supported by a backing system that separates components in a group (Omurtag and Yazicioğlu 2000). The stationary phase in TLC is polar, and the mobile phase is non-polar in nature. There is thin glass, plastic, or aluminum foil coated by an adsorbent material such as silica gel, aluminum, or cellulose, termed the stationary phase. Once the sample is applied to the plate, the mobile phase moves up due to capillary action. Separation is analyzed based on the different rate at which another analyte will move. TLC plates are treated by sulfuric acid or chromotropic salts (disodium 4,5-dihydroxynaphthalene-2,7-disulfonate dihydrate) (Zain 2011). Several strains of *Fusarium culmorum* from maize grain samples from south India were checked and analyzed for trichothecenes (Venkataramana et al.

2013). Other examples of this technique used in separating and quantifying T-2 toxin are many, including T-2 toxin and its metabolite identification in Wistar rats (Chandratre et al. 2014) and separation of trichothecenes in *Fusarium* species (Eldebaiky 2018; Nagaraja et al. 2016). The modern version of this method, high-performance thin layer chromatography, is an analytical, qualitative, and quantitative method employed for T-2 toxin and metabolite detection (Andola and Purohit 2010; Li et al. 2014). Sodium methoxide was sprayed on a TLC plate and fluorescent spots under UV light indicate diphenylindenone sulfonyl (Dis) esters of trichothecenes (Zain 2011). It is reported that in the TLC method, each spot carries 20–25 ng of T-2 toxin (Betina 1989) (Adhikari et al. 2017).

6.8.1.2 Gas Chromatography

This is a method that depends upon the volatile nature of the analyte and its affinity to the stationary phase. On a general basis, mycotoxins are volatile and if pre-processed can be analyzed by gas chromatography. Specific pre-processing and detection examples are reported where trichothecene hydroxyl groups are converted to other derivatives such as trimethylsilyl, trifluoroacetyl, and heptafluorobutyryl (Jyotirmayee and Sarangi 2013). The mobile phase in gas chromatography is a carrier gas (helium or nitrogen), whereas the stationary phase can be a liquid or polymer supported by inert material inside a column. Once the gaseous analyte is applied inside the column, it interacts with the stationary phase and elutes out at different intervals according to affinity. Reports support that T-2 toxin and other type A toxins were detected at a concentration of 200 g kg⁻¹ (single clean-up) and 50–100 g kg⁻¹ (dual chromatographic clean-ups). Researchers developed a highly sensitive and specific GC-MS technique to analyze mycotoxins in grains and human urine (Rodríguez-Carrasco et al. 2014; Rodríguez-Carrasco et al. 2013; Rodríguez-Carrasco et al. 2012). Gas chromatography is often combined with flame ionization (FID), electron capture (ECD), or mass spectrometer.

6.8.1.3 Liquid Chromatography

This method overcomes the barriers of TLC like restricted plate length, temperature, and dampness, as TLC occurs in an open system. It is a separation technique in which liquid acts as a mobile phase and analyses are separate based on an affinity in the stationary and mobile phase. The more advanced version is called high-performance liquid chromatography, where a high constraint is applied to force liquid out of the column. HPLC is based on the polarity of both mobile and stationary phases. The main advantage of HPLC is it can be employed to separate polar, non-polar, non-volatile, and thermolabile substances (Ngundi et al. 2005). LC is usually combined with UV absorption, amperometric detectors, a fluorescence detection stage (FLD), or a mass spectrometer (Cirlini et al. 2012; Núñez et al. 2012; Singh and Mehta 2020). One efficient labelling reagent is 1-anthroylnitrile (1-AN) used for T-2 toxin detection by HPLC-fluorescence detection. HPLC is a sensitized and accurate technique to identify T-2 toxins from cereals such as barley, rice, oats, corn, and sorghum (Li et al. 2014). The setup consists of a column with specified antibodies to bind with T-2 extract and clean the column, 1-AN derivatized on precolumn, and HPLC with fluorophore for the detection. The sample is generally ground cereal extracted in methanol:water (80:20, v/v), refined by an affinity column, and quantified excitation and emission wavelengths of 381 and 470 nm, respectively. The detection limit of this process is decided on a signal:noise ratio (3:1), 0.005 microg/g (Jyotirmayee and Sarangi 2013). To determine the toxin with more accuracy and visual detection, fluorescent labelers are used such as 1-naphthoyl chloride (1-NC), 2-naphthoyl chloride (2-NC), and pyrene-1-carbonyl cyanide (PCC) (Lippolis et al. 2008).

6.8.1.3.1 Limitations of HPLC

The detection of T-2 toxin and other metabolites by HPLC needs additional steps of clean-up and derivative formation, limiting its usage. Other derivatizations of the column are also available, which are T-2 toxin esters, including p-nitrobenzoate (10 ng T-2), and diphenylindenone sulfonyl (30 ng T-2), which involves ultraviolet detection (Zain 2011).

A more advanced technique known as ultrafast liquid chromatography coupled to tandem mass spectrometry combines two-liquid chromatography and mass spectrometer. This technique was used by Xing et al. (2016) to detect mycotoxin. They analyzed 21 mycotoxins qualitatively and quantitatively in Radix Paeoniae Alba using ultrafast liquid chromatography coupled with Quick, easy, cheap, effective, rugged, and safe (QuEChERS).

6.8.2 Ion Mobility Spectrometry

The principle of this technique is based on ion movement at the different rates used to determine distinct shapes and confirmation of protein in ion mobility spectrometry (IMS) drift. This instrument consists of an electron spray ionization (ESI) source, a mass-collecting quadrupole, IMS drift tube, and mass-measuring quadrupole. Different reports have identified a procedure: first the protein solution is ionized by ESI; then this state is collected by the first quadrupole, which is further forced into the drift tube. Based on different shapes and confirmation, it is separated and enters the second quadrupole. Drift time gives information about the confirmation of protein. In 2018, researchers used this technique to determine T-2 and other mycotoxins in cereals like wheat, malt, maize, and rye. It is better than other techniques because it produces a low noise ratio and provides more data on retention time.

6.8.3 Rapid Methods of Detection

6.8.3.1 Immunological Assays

These are fast and accurate methods for the diagnosis of mycotoxins, include enzyme-linked sorbent assay (ELISA), dipsticks, and immunodipsticks (LFDs) (Alshannaq and Yu 2017). ELISA is used for the rapid and accurate detection of mycotoxins. The basic principle behind ELISA is antigen–antibody interaction linked with the chromogenic substrate, which results in visible and quantifiable results (Agriopoulou et al. 2020). A few researchers have used this method for the detection of T-2/HT-2 toxins (Nolan et al. 2019)

Dipsticks work on the same principle as ELISA, but more applications of this technique in T-2 detection are still to be explored.

LFDs, also known as immune-strips or immunodipsticks, are a rapid, portable technique with the same principle as ELISA, but they work competitively. There is limited application of LFDs in T-2 toxin detection; some examples are given in Table 6.3.

6.8.3.2 Biosensors

Various biosensors have been developed in the past few years and have widely been accepted due to their rapid analysis, sensitivity, accuracy, and stability. Different biosensors are available to detect mycotoxins based on other optical principles. Techniques like surface plasmon resonance (SPR; fluorescence), piezoelectricity (quartz crystal microbalance; QCM), and electrochemical biosensors have been utilized to detect toxins released by fungi. Research was published in 2019 based on the usage of biosensors to detect toxin production at the initial stage (Oliveira et al. 2019; Pereira et al. 2014b). Examples of toxins detected by different biosensors are shown in Table 6.3.

6.8.3.3 Proteomic Methods

This is an emerging technology to be used for toxin detection. It involves extracted proteins/peptides from fungi being analyzed by matrix-assisted laser desorption or ionization-time of flight mass spectrometry (MALDI-TOF MS). It is one of the most accurate and rapid techniques researchers use for T-2 toxin and mycotoxin detection (Hsba 2017).

6.9 CONTROL STRATEGIES

The steps to be followed from choosing the crop until its usage are crucial for controlling mycotoxins in crops. Before planning control strategies to remove or reduce mycotoxin, it is essential to know the source of the toxin and its needs

TABLE 6.3   METHODS EMPLOYED FOR THE DETECTION OF T-2 TOXIN AND ITS METABOLITES

Toxin	Method	Sensitivity	Sample	References
T-2/H-T-2-toxin	Lateral flow detectors	80 µg/kg	Barley	Foubert et al. 2017
T-2 toxin (T-2)	Lateral flow detectors	13 µg/kg	Maize	Li et al. 2019
T-2 toxin	Lateral flow time resolved Fluro-immunoassay	0.09 ng g^{-1} and 0.17 ng g^{-1}	Rice, maize, and feed, respectively	Zhang et al. 2015
HT-2/T-2 toxin	Biosensor (amperometric)	0.4 ng/ml–1 ng/ml	Human urine	Schulz et al. 2019
T-2/HT-2	Optical biosensor	25 microg/kg for baby food and breakfast cereal and 26 microg/kg for wheat	Cereals and maize-based baby food	Meneely et al. 2010
T-2 toxin antibody wheat	Optical biosensor (SPR)	12 µg/kg	Wheat	Hossain and Maragos 2018
T-2 toxin, T-2 toxin-3-glucoside (T-2-G)	Optical biosensor (SPR)	1.2 ng/ml	Wheat	Hossain et al. 2018
T-2	Optical biosensor (SPR)	26 µg/kg	Barley	Joshi et al. 2016
T-2/T-2 triol	Liquid chromatography/mass spectrometry (LC/MS)	0.4 µg/kg, 0.2 µg/kg	Corn	Aniołowska and Steininger 2013
HT-2/T-2	Gas chromatography/mass spectrometry (GC/MS)	2–12 µg/kg	Cereals, food, feed	Schollenberger et al. 1998
T-2	Gas chromatography/mass spectrometry	5–10 g/kg	Corn, Wheat	Tanaka et al. 2000

(Continued)

TABLE 6.3   METHODS EMPLOYED FOR THE DETECTION OF T-2 TOXIN AND ITS METABOLITES   (CONTINUED)

Toxin	Method	Sensitivity	Sample	References
T-2/HT-2	GC-MS/MS	10 µg/kg	Oats	Edwards 2009
T-2/HT-2	GC-MS	1.88 and 0.47ng/g	Chinese herbal medicine	Kong et al. 2012
T-2	LC-MS	0.41–0.5 µg/kg	Barley	Barthel et al. 2012
T-2/HT-2	HPLC	5 microg/kg for T-2 toxin and 3 microg/kg for HT-2 toxin	Cereal grains	Visconti et al. 2005
T-2/HT-2	HPLC	0.005 microg/g	Cereal grains	Pascale et al. 2003
T-2/HT-2	HPLC	10–50 g/kg	Maize	Mateo et al. 2002
HT-2	LC-MS/MS	76–904 µg/kg	Oat	Pettersson et al. 2011
T-2/HT-2	UHPLC/MS	16.6–47.2 µg/kg 18.4–36.7 µg/kg	Barley, maize, rice, and wheat grains	Mahdjoubi et al. 2020
T-2/HT-2	LC-MS/MS	103 µg/kg	Oats	De Boevre et al. 2012
T-2/HT-2	LC-MS/MS	0.7–13 µg/kg 3–32 µg/kg	Wheat	Nathanail et al. 2015
HT-2	LC-HRMS	163 µg/kg	Barley	Lattanzio et al. 2015
HT-2	LC-MS	6.5–31 µg/kg	Rice products	Chilaka et al. 2016
T-2 toxin HT-2 toxin	LC-MS/MS	4–8 µg/kg 20 µg/kg	Cereal	De Santis et al. 2017
T-2 toxin HT-2 toxin	LC-MS/MS	0.02 and 10.14 ng/mL	Cow milk	Flores-Flores and González-Peñas 2017
T-2 toxin	LC-MS/MS	30.1–37.4	Barley	Pleadin 2018
T-2 toxin	LC-MS/MS	0.1–1	Cereals	Yoshinari et al. 2018

Toxin	Method	Sensitivity	Sample	References
T-2 toxin	LC-MS/MS	4.5 µg/kg	Barley	Di Marco Pisciottano et al. 2020
T-2/HT-2	LC-MS/MS	2.5–17.6 µg/kg	Biscuits	Ostry et al. 2020 Mahato et al. 2022
T-2 toxin	EC voltametric	0.15 µg/g	Cereals and human samples	Gao et al. 2014
T-2 toxin	Lateral flow device	100 µg kg^{-1}	Wheat and oats	Molinelli et al. 2008
T-2/HT-2 toxin	Surface plasmon resonance	25 µg kg^{-1} 26 µg kg^{-1}	Baby food and breakfast cereal wheat	Meneely et al. 2010
T-2 toxin	AlphaLISA	0.03–500 ng/mL	Food and feed	Zhang et al. 2021
T-2/HT-2 toxin	HPLC-MS	—	Barley	Pernica et al. 2022
T-2 toxin	ELISA	146 µg/kg	Maize	Tima et al. 2016 Mahato et al. 2022
T-2/HT-2	ELISA	12.5 µg/kg and 7.5 µg/kg	Cereals and baby food	Oplatowska-Stachowiak et al. 2017
T-2 toxin	ELISA	6.7–15.9 µg/kg	Barley	Kiš et al. 2021

to grow. The main reason for toxin production is fungus growth in crops and their products. Several factors can cause and promote mycotoxins, including plants inclined to fungal invasion, climatic conditions, temperature, moisture content, damaged seeds, improper storage methods, and availability of fungal substrate.

Control strategies can be studied in two major forms: pre-harvest and post-harvest approaches (Karimi and Mehri 2014). Both methods are discussed in this section.

6.9.1 Pre-Harvest Control Plan

Yearly varying weather conditions play a significant role in a pre-harvest fungal infestation. But the seriousness of this issue is also dependent on the state of the seed; it should not be damaged or insect infected. A few methods can be used to reduce fungal infection, but those are very reliable: application of good agricultural practices (GAPs), good environmental conditions, good manufacturing practices (GMPs), use of fungicides and insecticides, and good storage practices (Luo et al. 123–132).

All methods that will benefit crop productivity come under good agricultural practices. These include crop rotation methods, use of registered and certified chemicals (herbicides, fungicides, insecticides, and weedicides), proper soil treatment, soil analysis, good genetically modified fungus resistant seeds, seedbed formation, use of dry cow dung compost, and so on. Food processing is also necessary at the productivity stage to implement GAPs to have better resistance towards fungal infection. Other environmental factors that promote mycotoxin fungi should be checked every day. Essential measures should be taken in cleaning storage places and maintaining unfavorable conditions for mycotoxic fungi and mycotoxin production.

While choosing a field for seed planting, we should consider the biodiversity around it to ensure the type and sources of infection such as pests, insects, mammals, and diseases in the local diversity of plants and animals. The soil profile should be well known, like what kind of minerals are present and their deficiencies and organic matter deficiencies.

6.9.2 Harvest Control Plan

Appropriate methods need to be chosen for harvesting crops to have minor mechanical damage to the seeds. The chances of contamination are very high in damaged seeds, so one can reduce fungal infection by checking at the initial stage. To minimize moisture, crops should be harvested at a correct time interval (Choudhary and Kumari 2010).

6.9.3 Post-Harvest Control Plan

Once harvesting is done, waste material should be discarded soon, and storage places should be cleaned and dried with low or no moisture content, insect–pest controlled, and temperature stable. The decontamination method must have key precautionary steps followed: kill, remove, or deactivate the mycotoxin; discard food and feed by-products immediately; keep the nutritional value of food; and kill fungal spores, which can be agents responsible for new toxin production. Above all, the method should be readily available, easy to implement, and cost effective.

Post-harvest control comprises physical, natural, biological, or chemical treatment to make grains and feed material free from mycotoxin contamination. In comparison to physical and chemical methods, biological and natural processes are more reliable.

6.9.3.1 Physical Post-Harvest Control Methods

The physical method of treatment of seeds is removing affected parts or removing damaged seeds from the lot. The basic principle behind this method is physically identifying rotted or damaged grains depending upon smell, size, color, and shape. It includes separation methods, sorting, milling, boiling, washing, processing, cleaning, heating, differentiating, and so on.

6.9.3.1.1 Separation Methods

Categorizing and separating constitutes the initial step of the physical method. It is the easiest method of fungal removal and stops its propagation in all food materials with the most negligible side effects. According to one study, it is reported that color sorting helps remove infected oats, which can bring down the mycotoxin quantity in the final product of oat flakes. The study also reported that removing the outer skin of oat reduces toxin concentration to lower than 65 and 55 µg/kg (Scudamore et al. 2009). Rapid drying is a frequently used method for mycotoxin decontamination as it reduces the moisture and creates unfavorable conditions for fungal growth and proliferation. It is reported that drying of maize seeds to 15.5% moisture content immediately after harvesting for 24–48 hrs reduces the risk of fungal growth (Conte et al. 2020). While storing the grains, the moisture content should also be properly monitored in the storage room. It is reported that post-harvest methods such as drying, winnowing, sorting, and washing are very effective in mycotoxin removal from grains. The principle of these methods is selection of good seeds from the sample so that they can lower the overall contamination. Simple washing followed by drying can also be utilized to lower the concentration of some mycotoxins in corn seeds (Munkvold 2003). However, heating is not as suitable for this because most mycotoxins are heat stable.

6.9.3.1.2 Radiation

Radiation is a method of detoxification in which radiation (ionizing or non-ionizing) is used. It can reduce pathogens from food by using energy in a series of reactions that change food material's structure. In 2011, Medina and Magan found a radiation method or increase in temperature can lower the water content, which reduces T-2 and HT-2 contamination. In this method, contaminated grains are exposed to UV light, and physical separation is possible in green UV light. But the grains that are internally infected by mycotoxins will not be visible in UV light; hence, this method cannot be used for those. Other basic cooking methods, such as boiling and baking, do not have a perfect impact on T-2 and HT-2 toxin decontamination, as they are heat stable (Olopade et al. 2019).

Another new method, pulse electric field, which uses short pulses of electric field from 80 kV/cm to 100 V/cm affects cell membrane porosity. These pulses are very useful in killing microbes and mycotoxin degradation. The efficiency of the method depends on the exposure time and the food matrix (Hassan and Hussein 2017). Internally infected grains are treated with gamma radiation for specific time intervals, which can result in 94.5% mycotoxin reduction (Serra et al. 2018). Meanwhile, direct sunlight is also helpful in reducing contamination if applied evenly for 30 hours or more.

6.9.3.1.3 Mycotoxin Adsorbers

There are different types of adsorbents or binders available that give protection against mycotoxins. For example, montmorillonite, activated carbon, aluminosilicates, complex non-digestible carbohydrates, and cholesterol are such adsorbers. Montmorillonite is a tested adsorber that, when applied at 8%, is claimed to reduce 66% of T-2 contamination in maize (Olopade et al. 2019). Though several types of montmorillonites are available on the market, sodium montmorillonite is more effective than unmodified types. The neutral charge of clay due to sodium ions enhances T-2 toxin adsorption. The second most efficient is lemongrass mixed with montmorillonite, which makes the clay more hydrophobic as compared to unmodified. Lemongrass with montmorillonites at 12% is claimed to reduce 56% of toxins in maize. The ability of sodium calcium aluminosilicate to decontaminate T-2 toxin has also been evaluated in broilers (Wei et al. 2019).

6.9.3.2 Chemical Methods

These include chemical compounds such as bases, organic compounds, ozone, and oxidizers to decontaminate grains from mycotoxins (Luo et al. 2018). Chemicals employed for the decontamination of toxins could be categorized as electron donors (sodium bi-sulfite, D-glucose and D-fructose), electron acceptors (ozone, hydrogen peroxide), miscellaneous agents (chlorinating agents, bentonite, etc.), acids (acetic acid, phosphoric acid, formic acid, propionic acid), and bases (sodium hydroxide, calcium hydroxide, ammonia gas) (Janik et al. 2021). However, not all of these are utilized for the decontamination of T-2 toxin; reports suggests that oxidizers and basic and acidic reagents are not that sufficient for decontamination of T-2 toxin.

6.9.3.2.1 Alkalinizing Agents

Sodium hydroxide and sodium hypochlorite are combined and have been proved to stop T-2 toxin activity (Zhang et al. 2018). A combined mixture of glycerol and calcium hydroxide was also applied for mycotoxin decontamination (Čolović et al. 2019). Calcium hydroxide is used in decontamination of foodstuff with T-2 toxin because it creates alkaline conditions that change the structure of mycotoxins. An approach of treating seeds with 0.25% NaOCl-0.025 mol/L NaOH for four hours was proved to deactivate T-2 toxin and other trichothecenes (Adhikari et al. 2017).

6.9.3.2.2 Oxidizing Agents

Ozone is a strong oxidizer and is used for mycotoxin decontamination in raw and compound feed. It is used in storage places to stop mold growth (Čolović et al. 2019). Researchers have proposed that ozone basically oxidizes the trichothecenes at C-9 and C-10 double bonds by the addition of two oxygen atoms, whereas other molecules remain unchanged. The process of oxidation by ozone is pH dependent; at lower pH, the degree of reactivity is higher, whereas at pH 7, it is constrained by state of carbon 8, and at pH 8 or more, very little or no reaction is observed. This method is beneficial in the case of vegetables, cereals, and fruits. Ozone has the ability to reduce contamination and improve the microbial status of the grains or treated material. It has been used over the years with a high success rate in decontamination of food products, but it cannot be overlooked that these chemicals could hamper the nutritional and physical properties of treated food and feed. Hydrogen peroxide has also been used for years for this purpose (Oner and Demirci 2016).

6.9.3.2.3 Miscellaneous

Bentonite is utilized to inhibit the toxicity of T-2 toxin by modulating assimilation and increasing fecal defecation of the toxin. Bentonite potentially binds to T-2 and other mycotoxins (Eya et al. 2008). However, it has been reported that to efficiently decontaminate T-2 toxin with bentonite, more than 10 g/kg of it is required (Janik et al. 2021).

Sodium bisulfite, a reducing agent, is also used; it reacts with trichothecenes and forms less toxic sulfonate derivatives. These derivatives are known to minimize the detrimental effects of these toxins on health when combined with D-glucose or D-fructose sugar (Hathout and Aly 2014).

Ammoniation, treatment with ammonia gas, has gained attention in decontamination of mycotoxins. However, the efficiency of this method varies with the type of mycotoxin. This method can also lower the quality of food due to excessive ammonia usage for the decontamination of food products (Čolović et al. 2019). Conjugating toxins by glycosidation to produce glucuronides and glucosides is also used to reduce the risk from toxins. Glycosidation is the most crucial biochemical pathway for T-2 and HT-2 toxin breakdown inside the biological system. Glucuronidase, the enzyme required for the process, is derived from *Rattus* liver (Wu et al. 2007). All these methods are available, but the choice of chemicals should be based on the type of toxin, sample size, exposure time, condition of samples, and so on, and the harmful effects of excessive chemical use should also be considered.

6.9.3.3 Biological Methods

These approaches are more dependable but they need more investigation of biological agents that may be employed for mycotoxin detoxification. It is well known that yeast, bacteria, and fungi kill mycotoxins in food and feed. Some microbes have been proved to degrade mycotoxins or can reduce the effect of toxins. Some researchers claim that ruminants have something in their body that makes them less susceptible to mycotoxins. It may be possible that microbes in the ruminant could lessen or degrade toxins (Ogunade et al. 2018). Six strains of *Saccharomyces cerevisiae* and 12 strains of *Lactobacillus* sp. bacteria have been proved to detoxify mycotoxins. Yeast was tested to reduce T-2, zearalenone, and aflatoxin concentration by 69%, 52%, and 60%, respectively, whereas *Lactobacillus* reduced T-2, zearalenone, and aflatoxin by 61%, 57%, and 60%, respectively. There are a few reports of detoxification of trichothecenes, such as de-acylation and de-epoxidation. The preliminary step of detoxification is de-acylation, in which T-2 toxin converts to HT-2 and then T-2 triol as the last product, by *Curtobacterium* sp. strain 114–2. This T-2 triol is 23 and 13 times less harmful than T-2 and HT-2, respectively (Russell et al. 2012). De-epoxification is the next target to reduce toxicity; some studies have proved microflora of pig or rat can convert trichothecenes to their less-toxic metabolites (Gratz et al. 2018; Ueno et al. 1983; Vanhoutte et al. 2016). A product is available in the market for detoxifying mycotoxins in animal feed known as Biomin BBSH 797 (He et al. 2010). This was isolated from eubacterium BBSH 797, which can degrade trichothecenes taken from ruminal fluid (He et al. 2016; Yu et al. 2010). There are species of yeast that help lessen toxins; their cell wall or cell extract is especially useful in this regard. Nathanail et al. in 2016 stated that if trichothecenes are fermented with *Saccharomyces cerevisiae*, it can reduce the combined concentration of T-2 and DON by 53%. Though there are examples of biological degradation, they are limited to only laboratories; to realize field benefits, much research needs to be done.

6.10 SUMMARY

Mycotoxins are well known as food contaminating agents; their acceptable limit has already been fixed, so a priority should be developing highly sensitive and reliable methods for detecting these toxins. Type A trichothecenes, specifically T-2 toxin and HT-2 toxin, are frequent contaminants in agricultural produce. Their presence and concentration can be varied, as they are dependent on various elements such as geographic location, atmospheric conditions, storage techniques, and so on. Many methods have been evolved and are in progress to detect these toxins in a more specific and sensitive way. The first step is extraction and purification protocols before using any techniques employed to determine T-2 and HT-2 toxins in contaminated samples. Techniques are differentiated based on sample preparation steps, sample and solvent volumes, analysis time, detection sensitivity, extraction techniques, and their scale of extraction. There are numerous techniques available that are being optimized and analyzed, and many other methods are in progress. Physical methods are cost effective and used for large samples, but if fast detection is needed, rapid immunoassays and biosensors should be used. Recent microbiological and biotechnological methods such as recombinant DNA techniques and engineered proteins are biological approaches that could be further promoted to detect toxins.

REFERENCES

1. Adhikari M, Negi B, Kaushik N, Adhikari A, Al-Khedhairy AA, Kaushik NK, et al. T-2 mycotoxin: toxicological effects and decontamination strategies. *Oncotarget*. 2017 May;8(20):33933–33952.
2. Agriopoulou S, Stamatelopoulou E, Varzakas T. Advances in analysis and detection of major mycotoxins in foods. *Foods*. 2020 Apr;9:518.
3. Alshannaq A, Yu J-H. Occurrence, toxicity, and analysis of major mycotoxins in food. *Int. J. Environ. Res. Public Health*. 2017 Jun;14(6):632.
4. Andola H, Purohit VK. High performance thin layer chromatography (HPTLC): a modern analytical tool for biological analysis. *Nat. Sci.* 2010 Jan 1;8:58–61.
5. Aniołowska M, Steininger M. Determination of trichothecenes and zearalenone in different corn (*Zea mays*) cultivars for human consumption in Poland. *J. Food Compos. Anal.* 2013 Jan;33.
6. Barthel J, Gottschalk C, Rapp M, Berger M, Bauer J, Meyer K. Occurrence of type A, B and D trichothecenes in barley and barley products from the Bavarian market. *Mycotoxin Res.* [Internet]. 2012;28(2):97–106. https://doi.org/10.1007/s12550-012-0123-1
7. Betina V. Chromatographic methods as tools in the field of mycotoxins. *J. Chromatogr.* 1989;477(2):187–233.
8. Brown DW, Dyer RB, McCormick SP, Kendra DF, Plattner RD. Functional demarcation of the *Fusarium* core trichothecene gene cluster. *Fungal Genet. Biol.* 2004;41(4):454–462.
9. Chandratre G, Telang A, Badgujar P, Raut S, Sharma A. Toxicopathological alterations induced by high dose dietary T-2 mycotoxin and its residue detection in Wistar rats. *Arch. Environ. Contam. Toxicol.* 2014 Feb;67.
10. Chaudhary M, Lakshmana Rao PV. Brain oxidative stress after dermal and subcutaneous exposure of T-2 toxin in mice. *Food Chem. Toxicol.* 2010;48(12):3436–3442.

11. Chilaka CA, De Boevre M, Atanda OO, De Saeger S. Occurrence of fusarium myco-toxins in cereal crops and processed products (Ogi) from Nigeria. *Toxins (Basel)*. [Internet]. 2016 Nov 18;8(11):342. https://pubmed.ncbi.nlm.nih.gov/27869703

12. Choudhary AK, Kumari P. Management of mycotoxin contamination in prehar-vest and post harvest crops: Present status and future prospects. *J. Phytol.* 2010 Nov;2:37–52.

13. Cirlini M, Dall'Asta C, Galaverna G. Hyphenated chromatographic techniques for structural characterization and determination of masked mycotoxins. *J. Chromatogr. A.* 2012;1255:145–152.

14. Čolović R, Puvača N, Cheli F, Avantaggiato G, Greco D, Đuragić O, et al. Decontamination of mycotoxin-contaminated feedstuffs and compound feed. *Toxins (Basel)*. 2019 Oct;11(11):617.

15. Committee, EFSA Scientific. 2011. Scientific opinion on genotoxicity testing strat-egies applicable to food and feed safety assessment. *EFSA Journal* 2011;9(9):2379. https://doi.org/10.2903/j.efsa.2011.2379

16. Conte G, Fontanelli M, Galli F, Cotrozzi L, Pagni L, Pellegrini E. Mycotoxins in feed and food and the role of ozone in their detoxification and degradation: An update. *Toxins*. 2020;12(8):486.

17. De Boevre M, Di Mavungu JD, Maene P, Audenaert K, Deforce D, Haesaert G, et al. Development and validation of an LC-MS/MS method for the simultaneous deter-mination of deoxynivalenol, zearalenone, T-2-toxin and some masked metabolites in different cereals and cereal-derived food. *Food Addit. Contam. Part A* [Internet]. 2012 May 1;29(5):819–835. https://doi.org/10.1080/19440049.2012.656707

18. De Santis B, Debegnach F, Gregori E, Russo S, Marchegiani F, Moracci G, et al. Development of a LC-MS/MS method for the multi-mycotoxin determination in composite cereal-based samples. *Toxins (Basel)*. 2017 May;9(5):169.

19. Di Marco Pisciottano I, Imperato C, Urbani V, Guadagnuolo G, Imbimbo S, De Crescenzo M, et al. T-2 and HT-2 toxins in feed and food from southern Italy, determined by LC-MS/MS after immunoaffinity clean-up. *Food Addit. Contam. Part B* [Internet]. 2020 Oct 1;13(4):275–283. https://doi.org/10.1080/19393210.2020 .1771776

20. Duffy MJ, Reid RS. Measurement of the stability of T-2 toxin in aqueous solution. *Chem. Res. Toxicol.* 1993 Jul;6(4):524–529.

21. Edwards SG. *Fusarium* mycotoxin content of UK organic and conventional oats. *Food Addit. Contam. Part A* [Internet]. 2009 Jul 1;26(7):1063–1069. https://doi. org/10.1080/02652030902788953

22. Eldebaiky S. Detection of T-2 toxin from some Egyptian *Fusarium* strains with special reference to its cytotoxicity. *Fusarium*. 2018 Jul;6.

    Eriksen, GS, Pettersson H. Toxicological evaluation of trichothecenes in animal feed. *Animal Feed Science and Technology* 2004;114(1–4):205–239. doi: 10.1016/j. anifeedsci.2003.08.008

23. Eya JC, Parsons A, Haile I, Jagidi P. Effects of dietary zeolites (bentonite and mor-denite) on the performance juvenile rainbow trout *Onchorhynchus myskis*. *Aust. J. Basic Appl. Sci.* 2008;2(4):961–967.

24. Flores-Flores ME, González-Peñas E. An LC-MS/MS method for multi-mycotoxin quantification in cow milk. *Food Chem.* [Internet]. 2017;218:378–385. Available from: www.sciencedirect.com/science/article/pii/S0308814616314960

25. Foubert A, Beloglazova NV, Gordienko A, Tessier MD, Drijvers E, Hens Z, et al. Development of a rainbow lateral flow immunoassay for the simultaneous detection of four mycotoxins. *J. Agric. Food Chem.* 2017 Aug;65(33):7121–7130.

26. Gao X, Cao W, Chen M, Xiong H, Zhang X, Wang S. A high sensitivity electrochemical sensor based on Fe3+-ion molecularly imprinted film for the detection of T-2 toxin. *Electroanalysis.* 2014;26(12):2739–2746.

27. Gholampour Azizi I, Azarmi M, Danesh Pouya N, Rouhi S. T-2 toxin analysis in poultry and cattle feedstuff. *Jundishapur J. Nat. Pharm. Prod.* 2014 Apr;9(2):e13734–e13734.

28. Glenn AE. Mycotoxigenic *Fusarium* species in animal feed. *Anim. Feed Sci. Technol.* 2007;137(3):213–240.

29. Gratz SW, Currie V, Richardson AJ, Duncan G, Holtrop G, Farquharson F, et al. Porcine small and large intestinal microbiota rapidly hydrolyze the masked mycotoxin deoxynivalenol-3-glucoside and release deoxynivalenol in spiked batch cultures in vitro. *Appl. Environ. Microbiol.* 2018 Jan;84(2):e02106–e02117.

30. Grenier B, Oswald I. Mycotoxin co-contamination of food and feed: Meta-analysis of publications describing toxicological interactions. *World Mycotoxin J.* 2011 Aug;4:285–313.

31. Haque M, Wang Y, Shen Z, Li X, Saleemi M, He C. Mycotoxin contamination and control strategy in human, domestic animal and poultry: A review. *Microb. Pathog.* 2020 Feb;142:104095.

32. Harcz P, Temmerman L, Voghel S, Waegeneers N, Wilmart O, Vromman V, et al. Contaminants in organically and conventionally produced winter wheat (*Triticum aestivum*) in Belgium. *Food Addit. Contam.* 2007 Aug;24:713–720.

33. Hassan FF, Hussein HZ. Detection of aflatoxin M1 in pasteurized canned milk and using of UV radiation for detoxification. *Int. J. Adv. Chem. Eng. Biol. Sci.* 2017;4:130–133.

34. Hathout AS, Aly SE. Biological detoxification of mycotoxins: a review. *Ann. Microbiol. BioMed Central.* 2014;64(3):905–919.

35. He JW, Hassan YI, Perilla N, Li X-Z, Boland GJ, Zhou T. Bacterial epimerization as a route for deoxynivalenol detoxification: The influence of growth and environmental conditions. *Front. Microbiol.* 2016;572.

36. He JW, Zhou T, Young JC, Boland G, Scott P. Chemical and biological transformations for detoxification of trichothecene mycotoxins in human and animal food chains: A review. *Trends Food Sci. & Technol.* 2010;21(2):67–76.

37. Hossain MZ, Maragos CM. Gold nanoparticle-enhanced multiplexed imaging surface plasmon resonance (iSPR) detection of *Fusarium* mycotoxins in wheat. *Biosens. Bioelectron.* 2018;101:245–252.

38. Hossain MZ, McCormick SP, Maragos CM. An imaging surface plasmon resonance biosensor assay for the detection of T-2 toxin and masked T-2 toxin-3-glucoside in wheat. *Toxins.* 2018;10(3):119.

39. Hsba L, Císarová M, Shariati MA, Tancinová D. Detection of mycotoxins using MALDI-TOF mass spectrometry. *J. Microbiol. Biotechnol. Food Sci.* 2017;7:181–185.

40. Hussein SH, Brasel JM. Toxicity, metabolism, and impact of mycotoxins on humans and animals. *Toxicology.* 2001;167:101–134.

41. Inglis DO, Binkley J, Skrzypek MS, Arnaud MB, Cerqueira GC, Shah P, et al. Comprehensive annotation of secondary metabolite biosynthetic genes and gene clusters of *Aspergillus nidulans, A. fumigatus, A. niger* and *A. oryzae. BMC Microbiol.* 2013;13(1):91.

42. Janik E, Niemcewicz M, Podogrocki M, Ceremuga M, Stela M, Bijak M. T-2 toxin—The most toxic trichothecene mycotoxin: Metabolism, toxicity, and decontamination strategies. *Molecules* [Internet]. 2021 Nov 14;26(22):6868. https://pubmed.ncbi.nlm.nih.gov/34833960

43. Joint FAO/WHO Expert Committee on Food Additives (2000: Geneva, Switzerland), World Health Organization, and Food and Agriculture Organization of the United Nations. 2001. Evaluation of certain food additives and contaminants: Fifty-fifth report of the Joint FAO/WHO Expert Committee on Food Additives. https://apps.who.int/iris/handle/10665/42388

44. Joshi S, Segarra-Fas A, Peters J, Zuilhof H, van Beek TA, Nielen MWF. Multiplex surface plasmon resonance biosensing and its transferability towards imaging nanoplasmonics for detection of mycotoxins in barley. *Analyst.* 2016;141(4):1307–1318.

45. Jyotirmayee K, Sarangi M. Thin layer chromatography: A tool of biotechnology for isolation of bioactive compounds from medicinal plants. *Int. J. Pharm. Sci. Rev. Res.* 2013 Jan;18:126–132.

46. Kachuei R, Rezaie S, Yadegari MH, Safaie N, Allameh AA, Aref-poor MA, Fooladi AAI, Riazipour MAH. Determination of T-2 mycotoxin in fusarium strains by HPLC with fluorescence detector. *J. Appl. Biotech. Rep.* 2014;1:38–43.

47. Kankkunen P, Rintahaka J, Aalto A, Leino M, Majuri M-L, Alenius H, et al. Trichothecene mycotoxins activate inflammatory response in human macrophages. *J. Immunol.* 2009 May;182(10):6418–6425.

48. Karimi G, Mehri D. Mycotoxins. In: Gopalakrishnakone P, editor. *Toxinology.* Dordrecht: Springer; 2014, pp. 1–15. 10.1007/978-94-007-6645-7_10-1.

49. Kiš M, Vulić A, Kudumija N, Šarkanj B, Jaki Tkalec V, Aladić K, et al. A two-year occurrence of *Fusarium* T-2 and HT-2 toxin in Croatian cereals relative of the regional weather. *Toxins.* 2021;13(1):39.

50. Kong W, Zhang X, Shen H, Ou-Yang Z, Yang M. Validation of a gas chromatography-electron capture detection of T-2 and HT-2 toxins in Chinese herbal medicines and related products after immunoaffinity column clean-up and precolumn derivatization. *Food Chem.* [Internet]. 2012;132(1):574–581. Available from: www.sciencedirect.com/science/article/pii/S0308814611015263

51. Lattanzio VMT, Ciasca B, Terzi V, Ghizzoni R, McCormick SP, Pascale M. Study of the natural occurrence of T-2 and HT-2 toxins and their glucosyl derivatives from field barley to malt by high-resolution Orbitrap mass spectrometry. *Food Addit. Contam. Part A* [Internet]. 2015 Oct 3;32(10):1647–1655. https://doi.org/10.1080/19440049.2015.1048750

52. Lavinia P, Alexandra T, Damiescu L, Simion G. T-2 Toxin occurrence in cereals and cereal-based foods. *Bulletin of University of Agricultural Sciences and Veterinary Medicine Cluj-Napoca Agriculture* 2011 Jan;68:1843–5386.

53. Lazarus CM, Williams K, Bailey AM. Reconstructing fungal natural product biosynthetic pathways. *Nat. Prod. Rep.* 2014;31(10):1339–1347.

54. Li R, Meng C, Wen Y, Fu W, He P. Fluorometric lateral flow immunoassay for simultaneous determination of three mycotoxins (aflatoxin B1, zearalenone and deoxynivalenol) using quantum dot microbeads. *Microchim. Acta.* 2019 Nov;186.

55. Li Y, Luo X, Yang S, Cao X, Wang Z, Shi W, et al. High specific monoclonal antibody production and development of an ELISA method for monitoring T-2 toxin in rice. *J. Agric. Food Chem.* 2014 Feb;62(7):1492–1497.

56. Li Y, Wang Z, Beier RC, Shen J, Smet D De, De Saeger S, et al. T-2 toxin, a trichothecene mycotoxin: Review of toxicity, metabolism, and analytical methods. *J. Agric. Food Chem.* 2011 Apr;59(8):3441–3453.

57. Lippolis V, Pascale M, Maragos CM, Visconti A. Improvement of detection sensitivity of T-2 and HT-2 toxins using different fluorescent labeling reagents by high-performance liquid chromatography. *Talanta.* 2008;74(5):1476–1483.

58. Luo, Y, Liu, X, Li J. Updating techniques on controlling mycotoxins—A review. *Food Control.* 2018;89:123–132.

59. Mackei M, Orbán K, Molnár A, Pál L, Dublecz K, Husvéth F, et al. Cellular effects of T-2 toxin on primary hepatic cell culture models of chickens. *Toxins.* 2020;12(1):46.

60. Mahato DK, Pandhi S, Kamle M, Gupta A, Sharma B, Panda BK, et al. Trichothecenes in food and feed: Occurrence, impact on human health and their detection and management strategies. *Toxicon* [Internet]. 2022;208:62–77. Available from: www.sciencedirect.com/science/article/pii/S0041010122 000253

61. Mahdjoubi CK, Arroyo-Manzanares N, Hamini-Kadar N, García-Campaña AM, Mebrouk K, Gámiz-Gracia L. Multi-mycotoxin occurrence and exposure assessment approach in foodstuffs from Algeria. *Toxins.* 2020;12(3):194.

62. Mateo JJ, Mateo R, Jiménez M. Accumulation of type A trichothecenes in maize, wheat and rice by *Fusarium sporotrichioides* isolates under diverse culture conditions. *Int. J. Food Microbiol.* 2002;72(1):115–123.

63. McCormick SP, Stanley AM, Stover NA, Alexander NJ. Trichothecenes: From simple to complex mycotoxins. *Toxins (Basel).* 2011 Jul;3(7):802–814.

64. Medina A, Magan N. Temperature and water activity effects on production of T-2 and HT-2 by *Fusarium* langsethiae strains from north European countries. *Food Microbiol.* 2011;28(3):392–398.

65. Meneely JP, Sulyok M, Baumgartner S, Krska R, Elliott CT. A rapid optical immunoassay for the screening of T-2 and HT-2 toxin in cereals and maize-based baby food. *Talanta* [Internet]. 2010;81(1):630–636. Available from: www.sciencedirect.com/science/article/pii/S0039914010000159

66. Molinelli A, Grossalber K, Führer M, Baumgartner S, Sulyok M, Krska R. Development of qualitative and semiquantitative immunoassay-based rapid strip tests for the detection of T-2 toxin in wheat and oat. *J. Agric. Food Chem.* [Internet]. 2008 Apr 1;56(8):2589–2594. https://doi.org/10.1021/jf800393j

67. Morcia C, Tumino G, Ghizzoni R, Badeck FW, Lattanzio VMT, Pascale M, et al. Occurrence of *Fusarium* langsethiae and T-2 and HT-2 toxins in Italian malting barley. *Toxins* (Basel). MDPI; 2016 Aug;8(8):247.

68. Munkvold GP. Cultural and genetic approaches to managing mycotoxins in maize. *Annu. Rev. Phytopathol* [Internet]. 2003 Sep 1;41(1):99–116. https://doi.org/10.1146/annurev.phyto.41.052002.095510

69. Nagaraja H, Chennappa G, Poorna Chandra Rao K, Mahadev Prasad G, Sreenivasa MY. Diversity of toxic and phytopathogenic *Fusarium* species occurring on cereals grown in Karnataka state, India. *3 Biotech.* 2016 Jun;6(1):57.

70. Nathanail AV, Gibson B, Han L, Peltonen K, Ollilainen V, Jestoi M, et al. The lager yeast *Saccharomyces pastorianus* removes and transforms *Fusarium* trichothecene mycotoxins during fermentation of brewer's wort. *Food Chem.* 2016;203:448–455.

71. Nathanail AV, Syvähuoko J, Malachová A, Jestoi M, Varga E, Michlmayr H, et al. Simultaneous determination of major type A and B trichothecenes, zearalenone and certain modified metabolites in Finnish cereal grains with a novel liquid chromatography-tandem mass spectrometric method. *Anal. Bioanal. Chem.* 2015 Jun;407(16):4745–4755.

72. Ngundi MM, Shriver-Lake LC, Moore MH, Lassman ME, Ligler FS, Taitt CR. Array biosensor for detection of ochratoxin A in cereals and beverages. *Anal. Chem.* 2005 Jan;77(1):148–154.

73. Nolan P, Auer S, Spehar A, Elliott CT, Campbell K. Current trends in rapid tests for mycotoxins. *Food Addit. Contam. Part A Chem. Anal. Control. Expo. Risk Assess.* 2019;36(5):800–814.

74. Núñez O, Gallart-Ayala H, Martins CPB, Lucci P. New trends in fast liquid chromatography for food and environmental analysis. *J. Chromatogr. A.* 2012;1228:298–323.

75. Ogunade IM, Martinez-Tuppia C, Queiroz OCM, Jiang Y, Drouin P, Wu F, et al. Silage review: Mycotoxins in silage: Occurrence, effects, prevention, and mitigation. *J. Dairy Sci.* 2018;101(5):4034–4059.

76. Oliveira IS, Junior AGDS, De Andrade CAS, Oliveira MDL. Biosensors for early detection of fungi spoilage and toxigenic and mycotoxins in food. *Curr. Opin. Food Sci.* 2019;29:64–79.

77. Olopade BK, Oranusi SU, Nwinyi OC, Lawal IA, Gbashi S, Njobeh PB. Decontamination of T-2 toxin in maize by modified montmorillonite clay. *Toxins (Basel).* 2019 Oct;11(11):616.

78. Omurtag G, Yazicioğlu D. Determination of T-2 toxin in grain and grain products by HPLC and TLC. *J. Environ. Sci. Health. B.* 2000 Dec;35:797–807.

79. M. E. Oner and A. Demirci, "Chapter 33 - Ozone for Food Decontamination: Theory and Applications**This chapter is dedicated to the Late Professor Louise Fielding, the chapter author of the first edition, who passed away in 2013.," in *Woodhead Publishing Series in Food Science, Technology and Nutrition*, H. Lelieveld, J. Holah, and D. B. T.-H. of H. C. in the F. I. (Second E. Gabrić, Eds. San Diego: Woodhead Publishing, 2016), pp. 491–501. doi: https://doi.org/10.1016/B978-0-08-100155-4.00033-9.

80. Oplatowska-Stachowiak M, Kleintjens T, Sajic N, Haasnoot W, Campbell K, Elliott CT, et al. *T-2 Toxin/HT-2 Toxin and Ochratoxin A ELISAs Development and In-House Validation in Food in Accordance with the Commission Regulation (EU) No 519/2014* [Internet]. Toxins (Basel). EuroProxima B.V., Arnhem 6827 BN, The Netherlands. michalina.oplatowska@europroxima.com.; 2017. Available from: http://europepmc.org/abstract/MED/29189752

81. Ostry V, Dofkova M, Blahova J, Malir F, Kavrik R, Rehurkova I, et al. Dietary exposure assessment of sum deoxynivalenol forms, sum T-2/HT-2 toxins and zearalenone from cereal-based foods and beer. *Food Chem. Toxicol.* 2020 Mar 1;139:111280.

82. Pascale M, Haidukowski M, Visconti A. Determination of T-2 toxin in cereal grains by liquid chromatography with fluorescence detection after immunoaffinity column clean-up and derivatization with 1-anthroylnitrile. *J. Chromatogr. A* [Internet]. 2003;989(2):257–264. Available from: www.sciencedirect.com/science/article/pii/S0021967303000815

83. Pereira VL, Fernandes JO, Cunha SC. Mycotoxins in cereals and related foodstuffs: A review on occurrence and recent methods of analysis. *Trends Food Sci. & Technol.* 2014b;36(2):96–136.

84. Pereira VL, Fernandes JO, Cunha SC. Review. *Trends Food Sci. Technol.* 2014a;36(2):96–136.

85. Pernica M, Kyralová B, Svoboda Z, Boško R, Brožková I, Česlová L, et al. Levels of T-2 toxin and its metabolites, and the occurrence of *Fusarium* fungi in spring barley in the Czech Republic. *Food Microbiol* [Internet]. 2022;102:103875. Available from: www.sciencedirect.com/science/article/pii/S0740002021001404

86. Pestka JJ. Deoxynivalenol-induced proinflammatory gene expression: Mechanisms and pathological sequelae. *Toxins (Basel).* 2010 Jun;2(6):1300–1317.

87. Pettersson H, Brown C, Hauk J, Hoth S, Meyer J, Wessels D. Survey of T-2 and HT-2 toxins by LC–MS/MS in oats and oat products from European oat mills in 2005–2009. *Food Addit. Contam. Part B.* 2011 Jun 1;4:110–115.

88. Pleadin J. The Incidence of T-2 and HT-2 toxins in cereals and methods of their reduction practice by the food industry. In: Askun T, editor. *Fusarium— Plant Diseases, Pathogen Diversity, Genetic Diversity, Resistance and Molecular Markers*, London: IntechOpen; 2018. p. Ch. 4. https://doi.org/10.5772/intechopen. 71550.

89. Proctor RH, McCormick SP, Kim H-S, Cardoza RE, Stanley AM, Lindo L, et al. Evolution of structural diversity of trichothecenes, a family of toxins produced by plant pathogenic and entomopathogenic fungi. *PLOS Pathog.* 2018 Apr;14(4):e1006946.

90. Rodríguez-Carrasco Y, Berrada H, Font G, Mañes J. Multi-mycotoxin analysis in wheat semolina using an acetonitrile-based extraction procedure and gas chromatography-tandem mass spectrometry. *J. Chromatogr. A.* 2012 Nov;1270.

91. Rodríguez-Carrasco Y, Font G, Manes J, Berrada H. Determination of mycotoxins in bee pollen by gas chromatography-tandem mass spectrometry. *J. Agric. Food Chem.* 2013 Feb;61.

92. Rodríguez-Carrasco Y, Moltó JC, Mañes J, Berrada H. Exposure assessment approach through mycotoxin/creatinine ratio evaluation in urine by GC–MS/MS. *Food Chem. Toxicol.* 2014;72:69–75.

93. Russell SL, Gold MJ, Hartmann M, Willing BP, Thorson L, Wlodarska M, et al. Early life antibiotic-driven changes in microbiota enhance susceptibility to allergic asthma. *EMBO Rep.* 2012 May;13(5):440–447.

94. Schollenberger M, Lauber U, Jara HT, Suchy S, Drochner W, Müller H-M. Determination of eight trichothecenes by gas chromatography–mass spectrometry after sample clean-up by a two-stage solid-phase extraction. *J. Chromatogr. A.* 1998;815(1):123–132.

95. Schulz K, Pöhlmann C, Dietrich R, Märtlbauer E, Elßner T. Electrochemical biochip assays based on anti-idiotypic antibodies for rapid and automated on-site detection of low molecular weight toxins. *Front. Chem.* 2019 Feb;7:31.

96. Scudamore K, Patel S, Edwards S. HT-2 toxin and T-2 toxin in commercial cereal processing in the United Kingdom, 2004–2007. *World Mycotoxin J.* 2009;2(3):357–365.

97. Seeboth J, Solinhac R, Oswald IP, Guzylack-Piriou L. The fungal T-2 toxin alters the activation of primary macrophages induced by TLR-agonists resulting in a decrease of the inflammatory response in the pig. *Vet. Res.* 2012 Apr;43(1):35.

98. Serra MS, Pulles MB, Mayanquer FT, Vallejo MC, Rosero MI, Ortega JM, et al. Evaluation of the use of gamma radiation for reduction of aflatoxin B1 in corn (*Zea mays*) used in the production of feed for broiler chickens. *J. Agric. Chem. Environ.* 2018;7(1):21–33.

99. Singh J, Mehta A. Rapid and sensitive detection of mycotoxins by advanced and emerging analytical methods: A review. *Food Sci. Nutr.* 2020 May;8(5):2183–2204.

100. Tanaka T, Yoneda A, Inoue S, Sugiura Y, Ueno Y. Simultaneous determination of trichothecene mycotoxins and zearalenone in cereals by gas chromatography-mass spectrometry. *J. Chromatogr. A.* 2000;882(1):23–28.

101. Tima H, Brückner A, Mohácsi-Farkas C, Kiskó G. *Fusarium* mycotoxins in cereals harvested from Hungarian fields. *Food Addit. Contam. Part B* [Internet]. 2016 Apr 2;9(2):127–131. https://doi.org/10.1080/19393210.2016.1151948

102. Ueno Y, Nakayama K, Ishii K, Tashiro F, Minoda Y, Omori T, et al. Metabolism of T-2 toxin in *Curtobacterium* sp. strain 114–2. *Appl. Environ. Microbiol.* 1983 Jul;46(1):120–127.

103. Ukwuru MU, Ohaegbu CG, Muritala A. An overview of mycotoxin contamination of foods and feeds. *J Biochem Microb Toxicol.* 2018;1:101.

104. Vanhoutte I, Audenaert K, De Gelder L. Biodegradation of mycotoxins: Tales from known and unexplored worlds. *Front. Microbiol.* 2016 Apr;7:561.

105. Venkataramana M, Shilpa P, Balakrishna K, Murali HS, Batra HV. Incidence and multiplex PCR based detection of trichothecene chemotypes of *Fusarium* culmorum isolates collected from freshly harvested maize kernels in southern India. *Braz. J. Microbiol.* 2013 Oct;44(2):401–406.

106. Villafana RT, Ramdass AC, Rampersad SN. Selection of fusarium trichothecene toxin genes for molecular detection depends on TRI gene cluster organization and gene function. *Toxins.* 2019;11(1):36.

107. Visconti A, Lattanzio VMT, Pascale M, Haidukowski M. Analysis of T-2 and HT-2 toxins in cereal grains by immunoaffinity clean-up and liquid chromatography with fluorescence detection. *J. Chromatogr. A* [Internet]. 2005;1075(1):151–158. Available from: www.sciencedirect.com/science/article/pii/S0021967305007387

108. Wei J-T, Wu K-T, Sun H, Khalil MM, Dai J-F, Liu Y, et al. A novel modified hydrated sodium calcium aluminosilicate (HSCAS) adsorbent can effectively reduce T-2 toxin-induced toxicity in growth performance, nutrient digestibility, serum biochemistry, and small intestinal morphology in chicks. *Toxins (Basel).* 2019;11(4):199.

109. Wu Q, Dohnal V, Yuan KK and Z. Trichothecenes: Structure-toxic activity relationships. *Curr. Drug Metab.* 2013;641–660.

110. Wu Q, Huang L, Liu Z et al. A comparison of hepatic in vitro metabolism of T-2 toxin in rats, pigs, chickens, and carp. *Xenobiotica.* 2011;41(10):863–873.

111. Wu X, Murphy P, Cunnick J, Hendrich S. Synthesis and characterization of deoxynivalenol glucuronide: Its comparative immunotoxicity with deoxynivalenol. *Food Chem. Toxicol.* 2007;45(10):1846–1855.

112. Xing Y, Meng W, Sun W, Li D, Yu Z, Tong L, et al. Simultaneous qualitative and quantitative analysis of 21 mycotoxins in Radix Paeoniae Alba by ultra-high performance liquid chromatography quadrupole linear ion trap mass spectrometry and QuEChERS for sample preparation. *J. Chromatogr. B.* 2016;1031:202–213.

113. Yang S, Li Y, Cao X, Hu D, Wang Z, Wang Y, et al. Metabolic pathways of T-2 toxin in vivo and in vitro systems of Wistar rats. *J. Agric. Food Chem.* 2013 Oct;61(40):9734–9743.
114. Yoshinari T, Takeda N, Watanabe M, Sugita-Konishi Y. Development of an analytical method for simultaneous determination of the modified forms of 4,15-diacetoxyscirpenol and their occurrence in Japanese retail food. *Toxins.* 2018;10(5):178.
115. Yu H, Zhou T, Gong J, Young C, Su X, Li X-Z, et al. Isolation of deoxynivalenol-transforming bacteria from the chicken intestines using the approach of PCR-DGGE guided microbial selection. *BMC Microbiol.* 2010;10(1):182.
116. Zain ME. Impact of mycotoxins on humans and animals. *J. Saudi Chem. Soc.* 2011;15(2):129–144.
117. Zhang L, Dou X-W, Zhang C, Logrieco AF, Yang M-H. A review of current methods for analysis of mycotoxins in herbal medicines. *Toxins (Basel).* 2018 Feb;10(2):65.
118. Zhang L, Lv Q, Zheng Y, Chen X, Kong D, Huang W, et al. A rapid and accurate method for screening T-2 toxin in food and feed using competitive AlphaLISA. *FEMS Microbiol. Lett* [Internet]. 2021 Mar 1;368(6):fnab029. https://doi.org/10.1093/femsle/fnab029
119. Zhang Z, Du W, Li J, Zhang Q, Li PW. Monoclonal antibody-Europium conjugate-based lateral flow time-resolved fluoroimmunoassay for quantitative determination of T-2 toxin in cereals and feed. *Anal. Methods.* 2015 Feb 17;7.

Chapter 7

Detection and Management of Ergot Alkaloids and Their Therapeutic Applications

Nabendu Debnath, Pooja Yadav,
Shalini Arora, Ashok Kumar Yadav

CONTENTS

7.1	Introduction	190
7.2	Biosynthetic Pathway of Ergot Alkaloid Formation	192
	7.2.1 D-Lysergic Acid Formation	193
	7.2.2 Ergoamide and Ergopeptine Formation	195
7.3	Occurrence and Health Impacts of Ergot Alkaloids in Humans and Animals	195
7.4	Modern Methods Used in Ergot Alkaloid Detection	197
	7.4.1 Capillary Electrophoresis	198
	7.4.2 Liquid Chromatography with Fluorescence Detection	198
	7.4.3 Liquid Chromatography with Mass Spectrometry	198
	7.4.4 Immunological Techniques	199
7.5	Control Strategies of Ergot Alkaloids	199
	7.5.1 Pre-Harvest Management	199
	7.5.1.1 Use of Chemical and Biological Agents	199
	7.5.1.2 Field Management	200
	7.5.2 Post-Harvest Management	201
	7.5.2.1 Improving of Drying and Storage Conditions	201
	7.5.2.2 Use of Chemical and Natural Agents	201
	7.5.2.3 Physical Separation	202
	7.5.2.4 Chemical Methods	202
7.6	Pharmacological Properties of Ergot Alkaloids	203
7.7	Summary	205

DOI: 10.1201/9781003242208-7

7.1 INTRODUCTION

Ergot alkaloids (EAs) are nitrogen-based chemicals that are classified as indole alkaloids. These secondary metabolites are generated by a variety of fungi that infect plant seed heads throughout the flowering season, particularly during periods of heavy rainfall. This pathogenic fungus, which typically infects rye and triticale, is mostly from the Clavicipitaceae (e.g., Claviceps) and Trichocomaceae (containing *Aspergillus* and *Penicillium*) families. *C. purpurea* infects about 400 *Poaceae* species, including weedy grasses and cereals including rye, wheat, triticale, barley, millet, and oats (Arroyo-Manzanares et al. 2017). Ergot is derived from the old French term "argot," which means "cock's spur." During infections, the fungi's hyphae infect the host plant's ovule, engulfing the whole ovary, and eventually replace the growing grain or seed with sclerotia, a black, hardened form of hyphae that protrudes from the ordinary grains. The last stage of the illness is represented by this stage. Sclerotia grow in abundance during periods of high rainfall and damp soils. Sclerotia contamination of grains occurs during the harvesting process. Even after eliminating ergot bodies, EAs might be found in the grains. The sclerotia are also broken and combined with flour during milling. Sometimes sclerotia of similar size to the grain are developed in dry climates that are even more difficult to remove (Alexander et al. 2012). Some cereals are more prone to this infection than others; for example, rye is more prone than wheat. The contamination of plants with this endotoxin is unpredictable, as it depends on several factors, including climatic conditions, growing, harvesting, fungal growth, weather, and crop management. Historically, extensive poisoning incidents have been documented in the Middle Ages, first Crusade (1095), and Russian campaign (1720–22). In modern ages, upgraded agricultural and milling techniques along with a better understanding of diseases have impacted progression and outbreaks. Milling techniques such as grading, sieving, and sorting, along with several grain processing mechanisms like dockage removing, separators, air screens, density separators, and color sorting, have successfully reduced EA contamination in grains. However, EA contamination can be observed in crops at any stage including harvesting, transportation, and storage due to the fact that when ergot sclerotia break into smaller fragments, it becomes difficult to separate them with conventional techniques. Thus, control strategies are focused on two main stages, pre-harvest and post-harvest, which will be discussed in this chapter in later sections.

EAs are made up of indole alkaloids and have a complex chemical structure. They feature a tetracyclic ergoline ring structure and are 3,4-substituted indole derivatives (Figure 7.1). EAs are split into two categories based on their structural complexity: D-lysergic acid derivatives (ergotamine) and clavine alkaloids (agrocalvine). A carboxy group at the 8-position is used to bind a simple amino alcohol or a short peptide chain to the ergoline nucleus in amide linkage in D-lysergic acid derivatives. The clavine alkaloids are the most basic forms of EA, with the carboxyl group replaced by methyl or hydroxymethyl and no additional side group attachment allowed (Tudzynski et al. 2002). Ergot poisoning in humans and animals is known as ergotism, and these chemicals have pharmaceutical properties. Hallucinations, itching and blistering skin, gangrene, loss of hands and feet, and even death are all symptoms of ergotism, one of the oldest known disorders. With the exception of 6,7-secoergolenes and 6,7-secoergolines, the EA metabolites constitute a broad and structurally varied family that shares the initial metabolic step, the production of a tetracyclic ergoline ring system. Beginning in the 1950s, the biosynthetic route of EAs was elucidated over many years. Some processes, however, are mostly unknown. (Gerhards et al. 2014). Different fungi produce a different EAs; for example, *Clavicipitaceae* typically has either lysergic acid-derived EAs or dihydroergot alkaloids, and *Trichocomaceae* produce alkaloids derived from festuclavine (Jakubczyk et al. 2014). In addition to EAs, ergopeptines also occur in nature. Ergopeptines are the peptide EA consisting of (+)-lysergic acid and a tripeptide system containing L-proline. Natural ergoline-derived alkaloids have a double bond in ring D of the tetracyclic ring system, either as $\Delta^{9,10}$ or $\Delta^{8,9}$ (the ring system termed ergolene). The nitrogen in ring D is consistently methylated (Uhlig and Petersen 2008; Uhlig et al. 2021). Several studies have identified and reported genes responsible for synthesizing these metabolites. The ergoline ring skeleton is derived from L-tryptophan and dimethylallyl diphosphate (DMAPP). Gebler and Poulter in 1992 isolated the first enzyme in the biosynthesis of ergot alkaloids. This enzyme catalyzes the C4-prenylation of L-tryptophan and is encoded by the gene dmaW (Gebler and Poulter 1992). Later on, this gene was also isolated and sequenced. Through genomic walking, Tudzynksi et al. identified a dmaW gene cluster containing 14 genes in *C. purpurea* strain P1 (1999). All the genes are involved in the biosynthesis of ergopeptines, ergotamines, ergocrypine, ergoamide, and ergonovine. In comparison with the genome of *C. purpurea*, several fungal genomes have been sequenced to identify the biosynthetic genes for ergot alkaloids. In *C. fusiformis*, nine homologous genes have been identified lacking functional copies of non-ribosomal peptide synthetase (NRPS), whereas in the symbiotic endophyte Epichloë festucae, 12 homologues have been found. The end product, in the case of *Aspergillus fumigatus* (*A. fumigatus*), is fumigaclavine C. The gene cluster behind the biosynthesis comprises seven homologous genes and some additional genes encoding the modifying enzyme. Similarly, seven homologous gene clusters in *Penicillium commune* (*P. commune*) are identified via a cosmid library that produces fumigaclavine A (Martin et al. 2017). In species of the *Arthrodermataceae*, five homologous genes corresponding to those of *C. purpurea* has been found, signifying a different end product from that in *Aspergillus* or *Claviceps* spp. In all ergot-alkaloid synthesizing fungi, the chanoclavine-I aldehyde is formed as an end product. The quantitative mycotoxin analysis is an important means to check the amount of endotoxin present in the crops and is carried out via liquid chromatography-tandem mass spectrometry (LC-MS/MS). It is also useful to researchers in telling apart other mycotoxins present in food (Zöllner and Mayer-Helm 2006; Alshannaq and Yu 2017; Agriopoulou et al. 2020). In addition to the LC-MS/MS technique, liquid chromatography coupled with fluorescence detectors (LC-FLD), as well as QuEChERS techniques (quick, easy, cheap, effective, rugged, and safe) are also utilized (Storm et al. 2008). Decades of investigations have provided the distinctive functions of EAs. Due to their structural similarities to several neurotransmitters such as noradrenaline, dopamine, or serotonin, EAs became the point of research to evaluate their therapeutic potential. In this chapter, we will discuss the current perspectives of EAs as a therapeutic tool along with their control strategies.

7.2 BIOSYNTHETIC PATHWAY OF ERGOT ALKALOID FORMATION

Ergots are a group of metabolites produced by different fungal species. The bio-synthetic pathway of ergot formation involves different genes and proteins, although proteins involved in the ergot synthetic pathway vary according to different fungal species. Ergot ring formation is the shared process in all species. The precursors of ergot ring formation are L-tryptophan and dimethylallyl diphosphate. In 1992, Gebler and Poulter purified the first enzyme from a *Claviceps* sp. of the biosynthesis pathway of ergot alkaloids. A gene cluster of 14 genes was identified in *C. purpurea* strain P1 by Tudzynksi et al., which was involved in ergot formation (1999). The first step of the pathway is the conversion of L-tryptophan to 4-γ, γ-dimethylallyltryptophan (DMAT) catalyzed by an enzyme, 4-dimethylallyltryptophan synthase (DMATS) (Figure 7.2). This step involves prenylation of L-tryptophan at C-4 position by a prenyl donor dimethylallyl diphosphate, resulting in DMAT formation (Steffan et al. 2007; Metzger et al. 2009; Steffan and Li 2009). The second step of the pathway involves N-methylation of DMAT by 4-dimethylallyltryptophan *N*-methyltransferase in the presence of S-adenosylmethionine (SAM). At this step, DMAT is converted to 4-dimethylallyl-L-abrine (4-DMA-L-abrine). The third step of the pathway is the formation of chanoclavine-I, which involves at least one decarboxylation and two oxidation steps (Gröger and Floss 1998; Tudzynski et al. 2002; Schardl et al. 2006; Florea et al. 2017). Enzymes involved at this step are FAD-dependent oxidoreductase, EasE, and the catalase EasC (known as FgaOx1 and FgaCat in *A. fumigatus*, respectively) (Lorenz et al. 2010; Goetz et al. 2011). The next metabolite of the pathway is chanoclavine-I aldehyde, which is catalyzed by the

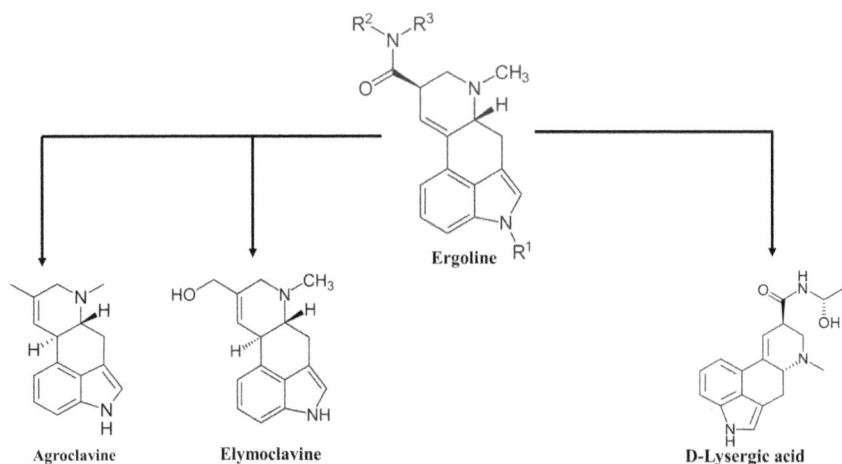

Figure 7.1   Ergoline ring structure, clavine alkaloids, and D-lysergic acid amides.

short-chain dehydrogenase/reductase (SDR) EasD. Chanoclavine-I aldehyde is the last common pathway intermediate and represents a branch point of pathways involved in several fungal species (Jakubczyk et al. 2014; Robinson and Panaccione 2015). In different fungal species, several products are formed; for example, festuclavine is formed in *A. fumigatus*, pyroclavine is formed in *P. commune*, and agroclavine is formed in *C. purpurea* (Matuschek et al. 2012) (Figure 7.2). In *A. fumigatus* and *P. commune,* the primary substrates for the formation of fumigaclavins are festuclavine and pyroclavine, respectively. 8R and 8S isomers of fumigaclavins B are the results of hydroxylation in *A. fumigatus* and *P. commune,* respectively. Monooxygenase FgaP450-2 catalyzes the reaction in *A. fumigatus,* whereas in *P. commune,* it is done by its analog FgaP450-2PC (Wallwey and Li 2011). Then, in the presence of acetyl-CoA, acetyltransferase FgaAT converts the substrates to fumigaclavine A (Liu et al. 2009). Finally, in *A. fumigatus,* prenyltransferase FgaPT1 catalyzes the reaction to form fumigaclavine C (Lorenz et al. 2007).

7.2.1 D-Lysergic Acid Formation

Agroclavine is converted to elymoclavine via 2-electron oxidation and then converted to paspalic acid via 4-electron oxidation (Schardl et al. 2006). These reactions are catalyzed by isomers of cytochrome P-450 monooxygenases (CloA). CloA isomer isolated from *C. fusiform* converts agroclavine to elymoclavine via 2-electron oxidation, whereas another isomer of CloA isolated from *C. purpurea* converts agroclavine to paspalic acid via 6-electron oxidation

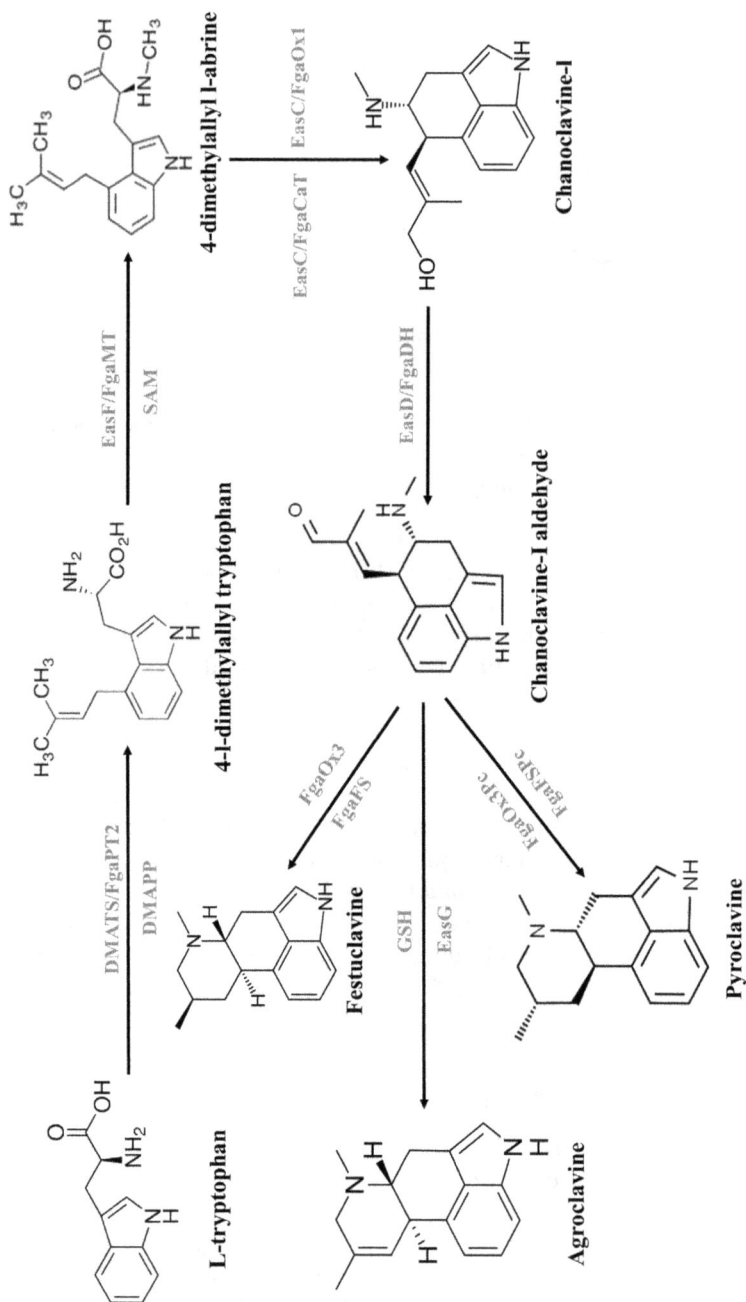

Figure 7.2 Formation of the ergoline ring structure in the biosynthetic pathway.

(Lorenz et al. 2007; Robinson and Panaccione 2015). Paspalic acid thus formed is converted to lysergic acid.

7.2.2 Ergoamide and Ergopeptine Formation

D-lysergic acid is the prime substrate for the formation of ergopeptines catalyzed by an enzymatic complex known as nonribosomal peptide synthetase. This complex contains d-lysergyl peptide synthetases 1 and 2 (LPS1 and LPS2) (Gerhards et al. 2014; Jakubczyk et al. 2014). First, LPS2 initiates the reaction by activating D-lysergic acid and then LPS1 catalyzes the reaction to form intermediates such as d-lysergyl mono-, di-, and tripeptide thioester that finally get converted to d-lysergyl tripeptide lactam or ergopeptam (Xie et al. 2011; Robinson and Panaccione 2015). The next step is catalyzed by mono-oxygenase easH that forms an intermediate, which when cyclized forms ergopeptines (Wallwey and Li 2011). Ergoamides, such as ergometrine, are also formed from D-lysergic acid. This is performed by the LPS2 and LPS3 protein of the complex (Ortel and Keller 2009).

7.3 OCCURRENCE AND HEALTH IMPACTS OF ERGOT ALKALOIDS IN HUMANS AND ANIMALS

Rapid Alert System for Food and Feed recently published a report showing that mycotoxins are still the leading cause of food hazards (Pigłowski 2020). In a different study, European Agency for Food Safety Authority (EFSA) analyzed food specimens during 2011 and 2016 in 15 different European countries for the evolution of EA infections in foods and feeds. This study identified several products that are highly contaminated, such as "rye milling products," "mixed wheat and rye bread and rolls," "rye bread and rolls," and "rye flakes" (Safety et al. 2017). In children and pregnant women, the toxic effects of EA poisoning are more severe than for the rest of the population. Additionally, EFSA has recommended the lowest value of EAs, 64 μg/kg for children in the age group of two to five (Müller et al. 2006; Debegnach et al. 2019). Moreover, it has been found that several cereal-based foods, specifically foods containing rye and wheat, which are consumed routinely in human foods, are contaminated regularly by *Claviceps* species. However, in northern Europe, it has been found that rye-containing foods have 1121 μg/kg, and wheat-containing foods were found to have 591 μg/kg (Malysheva et al. 2014). This mycotoxin affects not just European nations but also the Indian subcontinent. In previous investigations, *C. fusiformis* has been linked to toxic events in bajra (pearl millet, *Pennisetum typhoides*) in India. (Krishnamachari and Bhat 1976; Saraswat and Krishna 2019). Recent representative studies with the presence of EAs in different food products and different countries are presented in Table 7.1.

TABLE 7.1   OCCURRENCE OF EA TOXICITY IN DIFFERENT COUNTRIES

Sample Product	Origin Country	Positive Cases	Method	Reference
Barley	Switzerland	3–17%	LC-MS/MS	Drakopoulos et al. 2021
Cool-season barley grains	Western Canada	NA	NIR and ATR-FT/MIR spectroscopy	Shi et al. 2019
Cereals	Canada	NA	LC-MS/MS	Tittlemier et al. 2015
Rye-based products Wheat-based products	Italy	85	LC–MS/MS	Debegnach et al. 2019
Barley	Czech Republic	NA	UHPLC–MS	Rubert et al. 2012
Cereal products	China	NA	LC–MS/MS	Guo et al. 2016
Cereal grains	Slovenia	17	LC–MS/MS	Babič et al. 2020
Cereals and cereal products	European countries	59	LC–MS/MS	Malysheva et al. 2014
Organic rye flour Conventional rye flour	Denmark	88	HPLC–FLD	Storm et al. 2008
Rye product samples	Germany	92	HPLC–FLD	Müller et al. 2009
Cereal samples	Algeria	20	QuEChERS–UHPLC–MS/MS	Carbonell-Rozas et al. 2021
Cereal-based food	The Netherlands	54	LC–MS/MS	Mulder et al. 2015
Rye grain, rye flour, rye bran, rye flakes	Poland	83–100	LC–IT–MS/MS	Bryła et al. 2015
Winter rye	Germany Austria Poland	40	HPLC	Kodisch et al. 2020

Hallucinations, convulsions, agalactia, burning feeling, vasoconstriction, and gangrenous loss of limbs are the most common signs of ergotism disorders. (Tittlemier et al. 2015; Shi et al. 2019; Waret-Szkuta et al. 2019). Nausea, changes in endocrine function, vomiting, weakness, numbness, severe cardiovascular consequences, and even death have been recorded as additional symptoms (Coufal-Majewski et al. 2016; Arroyo-Manzanares et al. 2017; Maruo et al. 2018; Baldim et al. 2020).

In addition to the toxic effects on human populations, animals are equally affected due to EA contamination. The primary concern is that animals used in the breeding and meat industry such as pigs, cattle, sheep, rabbits, and poultry have been highly susceptible to EA toxins after ingesting EA-contaminated grains (Korn et al. 2014; Craig et al. 2015). Specifically, symptoms such as black beaks and combs of the legs have been observed in poultry animals (Tkachenko et al. 2021). Additionally, dyspnea, agalactia, and necrosis of the extremities have also been found in animals infected with EA toxins (Bryden 2012). EAs have been employed for biomedical applications owing to their antibacterial, antiproliferative, and antioxidant activities, despite the fact that they are poisonous to people and animals. They have the potential to be effective therapeutic agents.

7.4 MODERN METHODS USED IN ERGOT ALKALOID DETECTION

It is important to determine the presence of ergot alkaloids in foodstuffs because they cause poisoning in cattle and consumers and thus result in economic losses. The first attempt to analyze grain crops adulterated with ergots of *Claviceps* species was based on actual enumeration of the sclerotia in grains. The EFSA panel has also validated that the pattern of the presence of ergot in crops varies. Thus, analytical techniques should be developed to determine and quantify ergot alkaloids. It has always been essential to find an accurate, sensitive, and fast method to understand the global impact of contamination by toxins and form a strategy for controlling them.

Techniques used for ergot detection can be categorized broadly into immunological enzyme-linked immunosorbent assay (ELISA) and radioimmunoassay (RIA) and analytical techniques (capillary electrophoresis, liquid chromatography with fluorescence detection [LC-FLD], and high-performance liquid chromatography with mass spectrometry [LC-MS]). Alternatively, some methods are also used in conditions like gas chromatography with mass spectrometric detection (GC-MS) and thin-layer chromatography (TLC).

According to EFSA, the first and mandatory step in the determination of ergot is sample treatment to remove interference and to concentrate the sample. In the case of cereals and foodstuffs, sample size is fairly large because ergots are not evenly distributed. It should also be noted that while preparing the sample, it is complicated to distinguish between epimers of ergot alkaloids, though attempts have been made to avoid this situation.

7.4.1 Capillary Electrophoresis

Though this method is rarely used for ergot detection, it offers some advantages over other techniques such as lower assay time, better efficiency, less sample requirement, and use of fewer reagents and thus can be used for a wide range of chemicals. Still, sensitivity is one factor that should be improved to have wider applications. Fanali and group for the first time used capillary electrophoresis to separate a mixture of ergot alkaloid enantiomer derivatives (1992). It has been utilized to determine ergot alkaloids and their epimers by using fused silica capillary and chemicals such as urea, polyvinyl alcohol, and β- and γ-cyclodextrins. This technique is used to determine ergotamine and caffeine in tablet formulation (Aboul-Enein and Bakr 1997). The method was also combined with mass spectrophotometric time of flight (TOF) detection and UV to determine lysergic acid, *iso*-lysergic acid, and the related paspalic acid in reaction mixtures. It was also optimized by running electrolytes such as methanol with asparagine, sodium tetraborate, or ammonium acetate. The determined limit of detection (LOD) was below 0.5 mg L^{-1} and 0.1 mg L^{-1} for UV and TOF detection, respectively (Crews 2015).

7.4.2 Liquid Chromatography with Fluorescence Detection

Reverse phase-based chromatography is the first choice for separating ergot alkaloids by using C-8 and C-18 columns because normal phase procedures are limited in use for this purpose. The solvent systems used in this method are methanol-water or acetonitrile-water with added chemicals to provide alkaline pH such as ammonium hydroxide, ammonium carbonate, ammonium carbamate, or triethylamine. High-performance liquid chromatography was used to separate six ergot alkaloids with their epimers quickly. This method can also distinguish α- and β-epimers of ergocryptinine, which is usually not possible with the other techniques. This technique has been utilized to detect 12 ergot alkaloids in rye and rye products under a basic environment (Müller et al. 2006, 2009; Debegnach et al. 2019). In another procedure, a fescue seed and straw sample was treated with a QuEChERS-based extraction method, and then ergoyaline was detected by HPLC-FLD. Sample treatment was done in two separate steps; first was the partitioning step, followed by the clean-up step using dispersive SPE. Not just separation, HPLC-FLD has also been used to diversify ergot alkaloid distribution by Beaulieu et al. (2013, 2021). This method gives better resolution in detecting ergot alkaloids and thus is usually used for ergot alkaloids.

7.4.3 Liquid Chromatography with Mass Spectrometry

This technique is nowadays more relevant in determining various toxins with better sensitivity. This method has been used to detect significant ergot alkaloids and their epimers such as (ergometrine, ergosine, ergotamine, ergocornine, ergocryptine, and ergocristine) and their epimers in rye and wheat (Kokkonen and Jestoi 2010). In various foodstuffs, scientists have determined EAs by using a two-step sample preparation method (Malysheva et al. 2014). Additionally, mass spectrometry has also identified the lesser-known EAs, and for that, a simple approach was also devised by Arroyo-Manzanares et al. (2015, 2021) The approach was based on the utilization of MS for studying the fragmentation pattern of ergot alkaloids.

7.4.4 Immunological Techniques

Immunological techniques such as ELISA and RIA are based on the specific binding of antigens and antibodies. These methods are rapid and less costly than LC-MS or LC-FLD, but specificity and accuracy are a little bit compromised in comparison to both (Crews 2015). In both ELISA and RIA, ELISA has the upper hand because it does not use radioisotopes and provides the same sensitivity to the process. Nowadays, monoclonal antibodies are in use because they are more specific and can recognize a wide range of ergot alkaloids. Various commercial kits are available to provide a better detection limit of around 2 µg/kg (Alexander et al. 2012). However, these kits give the total concentration of ergots. Still, when we need quantification of individual ergot alkaloids, ELISA cannot be used; therefore, HPLC–FLD or LC-MS is mandatory.

7.5 CONTROL STRATEGIES OF ERGOT ALKALOIDS

The control of ergot alkaloid mycotoxins is complicated by various factors. First, the contamination of crops with fungi and their growth are not consistent and vary by the climatic conditions, crop management, and initial contamination level. The success of each control strategy is also influenced by the associated cost and safety concerns (Mavrommatis et al. 2021). Mycotoxin contamination can occur in any stage of crop production, including growth, harvesting, storage and transportation. Based on the aforementioned possible scenarios, various management strategies have been developed to combat against mycotoxigenic fungal infections, and it is apt to divide the control strategies into the steps of food processing: pre-harvest and post-harvest. These strategies are briefly described in the following sections.

7.5.1 Pre-Harvest Management

7.5.1.1 Use of Chemical and Biological Agents

Pre-harvest control strategies focus on adopting good agricultural practices (GAPs) and biological control through microorganisms (Pleadin et al. 2019). Pesticides and fungicides are undesirable, as they may develop resistance and are not approved for the environment. However, biological agents have a limited reach compared to chemicals (Peles et al. 2021). It has been observed that mycotoxin infections are often coupled with insect damage. It was found that various insects augment the entry of mycotoxin, producing fungi such as *Claviceps* to varieties of crops; therefore, insecticides can be useful by indirectly controlling mycotoxigenic fungi and mycotoxins (Dowd 2003). Microorganisms like yeast, bacteria, and fungi cause the degradation and adsorption of mycotoxins. Certain enzymes are also utilized to transform and absorb mycotoxins (Nešić et al. 2021). Non-toxicogenic microorganisms such as *A. flavus/A. parasiticus* can also be very useful in preventing mycotoxin contamination by competing against pathogens (dorner et al. 1999; Dorner et al. 2003). Even though the detoxification of food via biological agents is desirable, very little published work is available in this regard. *Pseudomonas aureofaciens, Trichoderma lignorum*, and *Fusarium roseum* have been tested against EA, but no significant effect on sclerotia germination was seen. *Rhodococcus erythropolis* is used to degrade EAs, but the mechanism is unclear (Battilani et al. 2009; Dabkevičius and Semaškienė 2017; Lyagin and Efremenko 2019). Further, to prevent the germination of sclerotia in the soil, it is wise to spray fungicide. The number of sclerotia could also be reduced by deep-burrowing them. Choosing a resistant variety of crops is also helpful in lowering sclerotia contamination (Battilani et al. 2009). The use of fungicides is a widespread practice that aims to reduce the growth of pathogens and prevent contamination by self-seeding. The ingredients and composition of fungicides vary, but all are projected towards eliminating mycotoxins (Thabit et al. 2021). After the ripening of the crop, the sclerotia become visibly distinct from the grain and can be removed easily by handpicking. To avoid the laborious work of handpicking, a late harvest strategy is adopted where the harvesting is delayed so that possible winds separate sclerotia. Finally, gravity and color sorters are used to separate sclerotia from food grains post-harvest (Karlovsky et al. 2016).

7.5.1.2 Field Management

Sclerotia is very important for the fungus, as it helps the organism to survive the extreme winter season. Exposure to a period of cold temperature is also essential for the fungi to germinate in the next season. However, by adopting good agricultural practices, one can control the growth and spread of EA. To avoid the germination of ergots during the spring season, calcium cyanamide as a fertilizer is very effective. The growth of EA can also be controlled indirectly by maintaining a robust crop. This can be achieved by crop rotation, weed control, and nutrient enrichment (Battilani et al. 2009; Uppala et al. 2016). Crop rotation helps reduce the amount of endotoxin because the sclerotia can survive from one to three years. For example, where maize is cultivated at a higher rate, maize stubble is prone to infection following the cereal crop (Landschoot et al.

2013; Janssen et al. 2019). Weed control can eliminate the chances of contamination through grass weeds, as they can also produce sclerotia (Menzies and Turkington 2015). However, the first and foremost prevention of EA endotoxin is the use of uninfected seeds. Additionally, ploughing, minimum tillage, and no-till soil cultivation techniques are very useful in minimizing the infection rate (Schaafsma et al. 2001).

7.5.2 Post-Harvest Management

The aforementioned pre-harvest management procedures are not fully capable of total suppression of EA contamination and require several downstream strategies, which are included in post-harvest management techniques. Better drying and storage ambiance along with chemical and natural agents are utilized in post-harvest management. Additionally, decontamination procedures are also used if previous approaches do not provide adequate protection.

7.5.2.1 Improving of Drying and Storage Conditions

Improper storage conditions such as higher moisture content in storage facilities facilitate mycotoxin infection and EA poisoning (Aldred and Magan 2004; Daou et al. 2021). Moreover, the ambient temperature of the storage facility also influences contamination risk. Therefore, temperature controlling systems should be used during the storage process. In contrast, modified atmospheric conditions, using alternative gases such as carbon dioxide, nitrogen, carbon monoxide, and sulfur dioxide, could reduce the chances of mycotoxin contamination (Cairns-Fuller et al. 2005; Magan and Aldred 2007; Taniwaki et al. 2010).

7.5.2.2 Use of Chemical and Natural Agents

Mycotoxin insecticides can successfully diminish the infestation and infection rates of EA contamination in stored crops (Felix D'Mello et al. 1998; Liu et al. 2016; Habschied et al. 2021). Apart from chemical insecticides, naturally occurring compounds such as phytochemicals could be used as an alternative approach to prevent mycotoxin contamination. Phytochemicals such as trans-cinnamic acid and ferulic acid have been shown to reduce fungal contamination successfully (Nesci et al. 2007). In addition, phenolic compounds such as vanillic and caffeic acid as well as antagonistic microorganisms can also be applied to reduce mycotoxin formation in storage facilities (Samapundo et al. 2007; Ren et al. 2020; Nešić et al. 2021). Several other strategies can also be applied in post-harvest management of EA contamination, but several technological and economical complications are present when natural and chemical agents are used to reduce mycotoxin contamination. Therefore, more studies are required to find potent new agents with fewer complications to reduce the occurrence of EA contamination.

7.5.2.3 Physical Separation

As mentioned, chemical and biological methods are not adequate to remove EA contamination from stored crops; therefore, physical separation techniques could be used and may provide a better alternative to suppress EA infections. Physical separation includes cleaning, sorting and segregation, washing, and milling processes. In addition, solvent adsorption techniques can also be used.

After storing crops, regular cleaning can reduce the incidence of EA contamination. For example, a study showed that if maize is cleaned at regular intervals, aflatoxin levels can be reduced up to 40–80% (Park 2002; Samapundo et al. 2007). Washing with distilled water and 1M sodium carbonate solution have been shown to reduce mycotoxin levels in food grains and are frequently used before wet milling and brewing steps (Jaukovic et al. 2017).

A sorting and segregation strategy is a very helpful physical separation method based on the size of the grain, handpicking, and using mechanical techniques. For example, mycotoxin infection damages grain shape and structure, which helps in the sorting step and later mechanical steps such as density segregation; fluorescence sorting is used to eliminate mycotoxin levels in grains (Hruska et al. 2014; Shi et al. 2014; Matumba et al. 2015; Karlovsky et al. 2016; Nasrollahzadeh et al. 2022).

Mycotoxin removal can also be observed by using variety of solvents such as ethanol, aqueous isopropanol, and methanol–water. Additionally, different types of adsorbents such as activated carbon and bentonite and microorganisms like yeast have also been shown to be equally effective for mycotoxin removal (Diaz et al. 2003; Avantaggiato et al. 2004; Cecchini et al. 2006; Var et al. 2008).

7.5.2.4 Chemical Methods

One of the factors that play an important role in the selection of strategy to remove mycotoxins is the type of food or feed. Several chemicals such as acids, bases, oxidizing compounds, reducing reagents, and chlorinating agents have been used simultaneously with other physical treatments to reduce mycotoxin levels. This combination of strategies is showing satisfactory results in recent times. Strong acids such as HCl (pH 2) inhibit the biological activities of mycotoxins (Gratz et al. 2018). On the other hand, several bases are also used in different techniques such as the ammoniation process, which includes either ammonium hydrochloride or gaseous ammonia (NH3), which has successfully reduced mycotoxin levels in food grains (Burgos-Hernández et al. 2002). Other widely used chemicals and techniques include sodium bicarbonate and hydrogen peroxide, calcium hydroxide, and nixtamalization. Hydrogen peroxide is used in varying concentrations such as 0.2%, 0.05%, and 10%, which have shown very promising results (ALTUG˘ et al. 1990; Fouler et al. 1994; Alla 1998). In a different approach, reducing and chlorinating agents such as sodium bisulfite and chlorine at a gas concentration of 11 mg g^{-1} were used, and mycotoxin level reduced significantly (Dänicke et al. 2012; Yu et al. 2021).

Although the aforementioned chemicals used to remove EAs and other mycotoxins have been successful over the years, most of them do not fulfill Food and Agriculture Organization (FAO) requirements due to their safety issues (Galvano et al. 2001; Varga et al. 2010). Therefore, safer chemicals as an alternative approach will be more favorable and accepted.

7.6 PHARMACOLOGICAL PROPERTIES OF ERGOT ALKALOIDS

The argot alkaloids are known to have pharmacological properties. They initiate a biochemical response by modulating the neurotransmitter receptor. These receptors such as dopamine, serotonin, and noradrenaline have structural similarities to the ergoline ring of EA (Dellafiora et al. 2015). The EA has a broad range of action determined by the affinity of individual agents for the receptors and the kind of receptor. The effect can be inhibitory or stimulatory. The interaction between EA and neurotransmitters influences various systems like gastrointestinal, cardiovascular, endocrine, and so on. It may also influence smooth muscle contraction, temperature regulation, and appetite. The pharmaceutical properties of EA make them an interesting target for drug development. FDA has approved many EA-based drugs. Vasoconstriction and hypoprolactinemia are the most well-known effects produced by EA-based drugs (Denef et al. 1980; Sibley and Creese 1983; Schillo et al. 1988; Sharma et al. 2016; Hwang et al. 2017). Other than this, uterine contraction is an important effect of EA. The first official use of EA in medical treatment was to treat postpartum hemorrhage and accelerate uterine involution in the puerperium (Komarova and Tolkachev 2001). EAs like lysergic acid diethylamide (LSD) are powerful hallucinogens. They are used as drugs in psychiatric treatment. They mimic the action of serotonin and crosses the blood–brain barrier to bind with serotonin and dopamine receptor (Sessa 2011; Baggott 2013). EAs show different vascular smooth muscle effects depending upon the agent and type of blood vessel. Ergotamine, a natural alkaloid, causes constriction of the blood vessels in both arteries and veins. Moreover, it acts as an antagonist to the α1-receptor, reversing the effects of adrenalin. Compounds like ergotamine, ergonovine, and methysergide relieve migraine or cluster headache symptoms by stimulating serotonin, decreasing inflammation, and reversing blood vessel dilation around the brain (Dahlöf and Brink 2012; Reddy et al. 2020). EA confers a stimulatory effect on the uterus muscles guided by hormonal status such as during pregnancy. The magnitude of response that EAs show on the uterine muscles also varies according to the dosage. A small dose brings about a rhythmic contraction in the muscles, whereas a large dose causes strong and prolonged contractions. Ergonovine is a widely used EA in obstetrics. Many smart drugs have been derived from EAs. Smart drugs are nootropic drugs that enhance cognitive function and memory. Bromocriptine, hydergine, nicergoline, ergoloid mesylates, co-dergocrine, and dihydroergotoxine are famous EA-derived smart drugs. Bromocriptine is used against ischemia, a condition arising due to restriction of blood supply to any tissue. Bromocriptine acts as an agonist of neurotransmitters like dopamine D2 and various serotonin receptors. It reverses the glutamate GLT1 transporter and inhibits the release of glutamine (Shirasaki et al. 2010; Siddique et al. 2016). Apart from this, bromocriptine also acts as an aphrodisiac agent by causing dopamine enrichment. Thus, it is used to treat "weak orgasm syndrome," attributable to prolactin over-secretion (Sharma et al. 2016). The EA drug hydergine is an effective antioxidant. It is used to treat memory loss associated with ageing. It mimics the effect of nerve growth factor (NGF). NGF promotes growth of dendrites, which form a communication link between nerve and brain cells and thus affect memory and learning. Based on its effect on nerves, this drug is also used to treat Alzheimer's disease, dementia, and psychological disorders. It also helps to maintain brain oxygen level, which prevents free radical accumulation and other age-related toxins (Schneider et al. 2000). Nicergoline is another EA-based drug popularly known by the trade name Sermion. It is used to treat vascular disorders. It possesses neurotrophic and antioxidant properties and is thus also used to treat cognitive, affective, and behavioral disorders. Besides, it also has an effect on neurotransmitters by acting as an antagonist of α1-adrenoceptor and enhancing cholinergic and catecholaminergic neurotransmitter action. It also restrains platelet aggregation and accelerates metabolism (Miccheli et al. 2003; Winblad et al. 2008). Ergot-derived drugs have a wide range of therapeutic applications. They are used to treat migraine-related headaches. They form the first drugs against migraine and carry side effects like ergotism and gangrene. Keeping the side effects in mind, they are only prescribed when no other drug is effective for the patient. Ergotamine (ET) and dihydroergotamine (DHE) are the two important EA drugs used to treat migraines. They have different pharmacokinetic and pharmacodynamic properties. They are quickly absorbed into the body due to the presence of caffeine. They are also being developed as orally inhaled drugs with 60% oral assimilation (Schiff 2006). Apart from the popular uterotonic function of EA-derived drugs, they also affect prolactin secretion. Prolactin induces lactation and weight gain in women during pregnancy. An ergot derivative, bromocriptine, subdues prolactin with minimum effect on pituitary hormones. It rather shows a general suppressive effect on the functioning of the pituitary. Cabergoline, another EA drug, under brand names Caberlin, Dostinex, and Cabaser, is used to treat prolactinomas with fewer side effects (Cao et al. 2016; Hollander et al. 2016).

7.7 SUMMARY

EA contamination in foods and feeds has been observed since the time of the Middle Ages. Previous outbreaks in Russia (1926–1927); in Manchester, England (1928); and in India (1958, 1973, 1974) have shown the impact of these infections. To prevent this, standard safety protocols need to be applied, which will be able to suppress the growth of fungi and the production of mycotoxins in agricultural products. As previously described, cereals are mainly contaminated with EAs in pre-harvest and post-harvest stages, hindering economic growth. As a result, special care needs to be provided in all stages of cereal production. Moreover, heterogeneity has been observed in the distribution of EAs, and it depends on the fungal strains, geographic regions, and host plants. Not only this, the sporadic nature of the EA contamination makes it a multifactorial incident that involves the weather. For example, humid conditions promote the germination of sclerotia. For this reason, presently, better grain cleaning machines based on photocells have been used. For all their harmful effects, however, EAs possess vast medicinal importance. These beneficial effects range from helping childbirth to reducing headaches to suppressing psychological problems. EAs are structurally similar to various neurotransmitters, allowing them to bind homologous receptors in the nervous system and making them essential pharmacological compounds. Future research in alkaloid chemistry could provide a new era of drug research where ergot-based pharmacotherapy will eventually offer better therapeutic options.

REFERENCES

Aboul-Enein HY, Bakr SA. Simultaneous determination of caffeine and ergotamine in pharmaceutical dosage formulation by capillary electrophoresis. *J Liq Chromatogr Relat Technol.* 1997 Jan 1;20(1):47–55.

Agriopoulou S, Stamatelopoulou E, Varzakas T. Advances in analysis and detection of major mycotoxins in foods. *Foods.* 2020 Apr 20;9:518.

Aldred D, Magan N. Prevention strategies for trichothecenes. *Toxicol Lett.* 2004;153(1): 165–171.

Alexander J, Benford D, Boobis A, Ceccatelli S, Cottrill B, Cravedi J-P, et al. Scientific opinion on ergot alkaloids in food and feed EFSA panel on contaminants in the food chain (CONTAM). *EFSA J.* 2012 Aug 1;10:1–158.

Alla E. Zearalenone: Incidence, toxigenic fungi and chemical decontamination in Egyptian cereals. *Nahrung.* 1998 Jan 1;41:362–365.

Alshannaq A, Yu J-H. Occurrence, toxicity, and analysis of major mycotoxins in food. *Int J Environ Res Public Health.* 2017 Jun 13;14(6):632.

Altug̀ T, Yousef A, Marth E. Degradation of aflatoxin B 1 in dried figs by sodium bisulfite with or without heat, ultraviolet energy or hydrogen peroxide. *J Food Prot.* 1990 Jul 1;53:581–582.

Arroyo-Manzanares N, Campillo N, López-García I, Hernández-Córdoba M, Viñas P. High-resolution mass spectrometry for the determination of mycotoxins in biological samples. A review. *Microchem J.* 2021;166:106197.

Arroyo-Manzanares N, Gámiz-Gracia L, García-Campaña AM, Diana Di Mavungu J, De Saeger S. Ergot alkaloids: Chemistry, biosynthesis, bioactivity, and methods of analysis. In *Fungal Metabolites*. Mérillon J-M, Ramawat KG, editors. Cham: Springer International Publishing; 2017, pp. 887–929.

Arroyo-Manzanares N, Huertas-Pérez J, Gámiz-Gracia L, García-Campaña AM. Simple and efficient methodology to determine mycotoxins in cereal syrups. *Food Chem.* 2015;177.

Avantaggiato G, Havenaar R, Visconti A. Evaluation of the intestinal absorption of deoxynivalenol and nivalenol by an in vitro gastrointestinal model, and the binding efficacy of activated carbon and other adsorbent materials. *Food Chem Toxicol.* 2004;42(5):817–824.

Babič J, Tavčar-Kalcher G, Celar FA, Kos K, Červek M, Jakovac-Strajn B. Ergot and ergot alkaloids in cereal grains intended for animal feeding collected in Slovenia: Occurrence, pattern and correlations. *Toxins.* 2020;12.

Baggott M. The pharmacology of LSD: A critical review by Annelie Hintzen & Torsten Passie. *Drug Alcohol Rev.* 2013 Sep 1;32.

Baldim I, Oliveira WP, Kadian V, Rao R, Yadav N, Mahant S, et al. Natural ergot alkaloids in ocular pharmacotherapy: Known molecules for novel nanoparticle-based delivery systems. *Biomolecules.* 2020;10.

Battilani P, Costa LG, Dossena A, Gullino ML, Marchelli R, Galaverna G, et al. Scientific information on mycotoxins and natural plant toxicants. *EFSA Support Publ.* 2009 Dec 1;6(9):24E.

Beaulieu WT, Panaccione DG, Hazekamp CS, Mckee MC, Ryan KL, Clay K. Differential allocation of seed-borne ergot alkaloids during early ontogeny of morning glories (*Convolvulaceae*). *J Chem Ecol.* 2013;39(7):919–930.

Beaulieu WT, Panaccione DG, Quach QN, Smoot KL, Clay K. Diversification of ergot alkaloids and heritable fungal symbionts in morning glories. *Commun Biol.* 2021;4(1):1362.

Bryden W. Mycotoxin contamination of the feed supply chain: Implication of animal productivity and feed security. *Anim Feed Sci Technol.* 2012 Apr 20;173:134–158.

Bryła M, Szymczyk K, Jędrzejczak R, Roszko M. Application of liquid chromatography/ ion trap mass spectrometry technique to determine ergot alkaloids in grain products. *Food Technol Biotechnol.* 2015 Mar;53(1):18–28.

Burgos-Hernández A, Price RL, Jorgensen-Kornman K, López-García R, Njapau H, Park DL. Decontamination of aflatoxin B1-contaminated corn by ammonium persulphate during fermentation. *J Sci Food Agric.* 2002 Apr 1;82(5):546–552.

Cairns-Fuller V, Aldred D, Magan N. Water, temperature and gas composition interactions affect growth and ochratoxin A production by isolates of *Penicillium verrucosum* on wheat grain. *J Appl Microbiol.* 2005 Nov 1;99(5):1215–1221.

Cao Y, Wang F, Liu Z, Jiao B. Effects of preoperative bromocriptine treatment on prolactin-secreting pituitary adenoma surgery. *Exp Ther Med.* 2016 May;11(5):1977–1982.

Carbonell-Rozas L, Mahdjoubi CK, Arroyo-Manzanares N, García-Campaña AM, Gámiz-Gracia L. Occurrence of ergot alkaloids in barley and wheat from Algeria. *Toxins.* 2021;13.

Cecchini F, Morassut M, Garcia Moruno E, Di Stefano R. Influence of yeast strain on ochratoxin A content during fermentation of white and red must. *Food Microbiol.* 2006;23(5):411–417.

Coufal-Majewski S, Stanford K, McAllister T, Blakley B, McKinnon J, Chaves AV, et al. Impacts of cereal ergot in food animal production. *Front Vet Sci.* 2016 Feb 25;3:15.

Craig AM, Klotz JL, Duringer JM. Cases of ergotism in livestock and associated ergot alkaloid concentrations in feed. *Front Chem.* 2015 Feb 18;3:8.

Crews C. Analysis of ergot alkaloids. *Toxins (Basel).* 2015 Jun 1;7:2024–2050.

Dabkevičius Z, Semaškienė R. Control of ergot (*Claviceps purpurea* (Fr.) Tul.) ascoscarpus formation under the impact of chemical, biological seed dressing. *Plant Prot Sci.* 2017 Dec 31;38:681–683.

Dahlöf C, Brink A. Dihydroergotamine, ergotamine, methysergide and sumatriptan— Basic science in relation to migraine treatment. *Headache.* 2012 Mar 22;52:707–714.

Dänicke S, Kersten S, Valenta H, Breves G. Inactivation of deoxynivalenol-contaminated cereal grains with sodium metabisulfite: A review of procedures and toxicological aspects. *Mycotoxin Res.* 2012 Nov 1;28:199–218.

Daou R, Joubrane K, Maroun RG, Khabbaz LR, Ismail A, Khoury A El. Mycotoxins: Factors influencing production and control strategies. *AIMS Agric Food.* 2021 Feb 20;6:416+.

Debegnach F, Patriarca S, Brera C, Gregori E, Sonego E, Moracci G, et al. Ergot alkaloids in wheat and rye derived products in Italy. *Foods (Basel, Switzerland).* 2019 May 1;8(5):150.

Dellafiora L, Dall'Asta C, Cozzini P. Ergot alkaloids: From witchcraft till in silico analysis. Multi-receptor analysis of ergotamine metabolites. *Toxicol Reports.* 2015;2:535–545.

Denef C, Manet D, Dewals R. Dopaminergic stimulation of prolactin release. *Nature.* 1980;285(5762):243–246.

Diaz DE, Hagler WM, Hopkins BA, Whitlow LW. Aflatoxin Binders I: In vitro binding assay for aflatoxin B1 by several potential sequestering agents. *Mycopathologia.* 2003;156(3):223–226.

Dorner JW, Cole RJ, Connick WJ, Daigle DJ, McGuire MR, Shasha BS. Evaluation of biological control formulations to reduce aflatoxin contamination in peanuts. *Biol Control.* 2003;26(3):318–324.

Dorner JW, Cole RJ, Wicklow DT. Aflatoxin reduction in corn through field application of competitive fungi. *J Food Prot.* 1999 Jun 1;62(6):650–656.

Dowd PF. Insect management to facilitate preharvest mycotoxin management. *J Toxicol Toxin Rev.* 2003 Jan 1;22(2–3):327–350.

Drakopoulos D, Sulyok M, Krska R, Logrieco AF, Vogelgsang S. Raised concerns about the safety of barley grains and straw: A Swiss survey reveals a high diversity of mycotoxins and other fungal metabolites. *Food Control.* 2021;125:107919.

Fanali S, Flieger M, Steinerova N, Nardi A. Use of cyclodextrins for the enantioselective separation of ergot alkaloids by capillary zone electrophoresis. *Electrophoresis.* 1992 Jan 1;13(1):39–43.

Felix D'Mello JP, Macdonald AMC, Postel D, Dijksma WTP, Dujardin A, Placinta CM. Pesticide use and mycotoxin production in *Fusarium* and *Aspergillus* phytopathogens. *Eur J Plant Pathol.* 1998;104(8):741–751.

Florea S, Panaccione D, Schardl C. Ergot alkaloids of the clavicipitaceae. *Phytopathology.* 2017 Feb 7;107.

Fouler S, Trivedi A, Kitabatake N. Detoxification of citrinin and ochratoxin A by hydrogen peroxide. *J AOAC Int.* 1994 May 1;77:631–637.

Galvano F, Piva A, Ritieni A, Galvano G. Dietary strategies to counteract the effects of mycotoxins: A review. *J Food Prot.* 2001 Jan 1;64(1):120–131.

Gebler JC, Poulter CD. Purification and characterization of dimethylallyl tryptophan synthase from *Claviceps purpurea. Arch Biochem Biophys.* 1992;296(1):308–313.

Gerhards N, Neubauer L, Tudzynski P, Li S-M. Biosynthetic pathways of ergot alkaloids. *Toxins (Basel).* 2014 Dec 1;6:3281–3295.

Goetz K, Coyle C, Cheng J, O'Connor S, Panaccione D. Ergot cluster-encoded catalase is required for synthesis of chanoclavine-I in *Aspergillus fumigatus. Curr Genet.* 2011 Mar 1;57:201–211.

Gratz SW, Currie V, Richardson AJ, Duncan G, Holtrop G, Farquharson F, et al. Porcine small and large intestinal microbiota rapidly hydrolyze the masked mycotoxin deoxynivalenol-3-glucoside and release deoxynivalenol in spiked batch cultures in Vitro. *Appl Environ Microbiol.* 2018 Jan 2;84(2):e02106–e02117.

Gröger D, Floss HG. Biochemistry of ergot alkaloids—Achievements and challenges. *Alkaloids.* 1998 Jan 1;50:171–218.

Guo Q, Shao B, Du Z, Zhang J. Simultaneous determination of 25 ergot alkaloids in cereal samples by ultraperformance liquid chromatography–tandem mass spectrometry. *J Agric Food Chem.* 2016 Sep 21;64(37):7033–7039.

Habschied K, Krstanović V, Zdunić Z, Babić J, Mastanjević K, Šarić GK. Mycotoxins biocontrol methods for healthier crops and stored products. *J Fungi (Basel, Switzerland).* 2021 Apr 29;7(5):348.

Hollander AB, Pastuszak AW, Hsieh T-C, Johnson WG, Scovell JM, Mai CK, et al. Cabergoline in the treatment of male orgasmic disorder—A retrospective pilot analysis. *Sex Med.* 2016 Mar;4(1):e28–e33.

Hruska Z, Yao H, Kincaid R, Brown R, Cleveland T, Bhatnagar D. Fluorescence excitation–emission features of aflatoxin and related secondary metabolites and their application for rapid detection of mycotoxins. *Food Bioprocess Technol.* 2014 Apr 1;7.

Hwang H-H, Yu M, Lai E-M. Agrobacterium-mediated plant transformation: Biology and applications. *Arab B.* 2017 Oct 20;15:e0186.

Jakubczyk D, Cheng JZ, O'Connor SE. Biosynthesis of the ergot alkaloids. *Nat Prod Rep.* 2014;31(10):1328–1338.

Janssen EM, Mourits MCM, van der Fels-Klerx HJ, Lansink AGJMO. Pre-harvest measures against *Fusarium* spp. infection and related mycotoxins implemented by Dutch wheat farmers. *Crop Prot.* 2019;122:9–18.

Jaukovic M, Zečević V, Stanković S, Krnjaja V, Nikić T, Bailović S, et al. Effect of dilute alkaline steeping on mold contamination, toxicity, and nutritive value of maize malt. *J Am Soc Brew Chem.* 2017 Sep 1;75:369–373.

Karlovsky P, Suman M, Berthiller F, De Meester J, Eisenbrand G, Perrin I, et al. Impact of food processing and detoxification treatments on mycotoxin contamination. *Mycotoxin Res.* 2016 Nov;32(4):179–205.

Kodisch A, Oberforster M, Raditschnig A, Rodemann B, Tratwal A, Danielewicz J, et al. Covariation of ergot severity and alkaloid content measured by HPLC and one ELISA Method in inoculated winter rye across three isolates and three European countries. *Toxins.* 2020;12.

Kokkonen M, Jestoi M. Determination of ergot alkaloids from grains with UPLC-MS/MS. *J Sep Sci.* 2010 Aug 1;33(15):2322–2327.

Komarova EL, Tolkachev ON. The chemistry of peptide ergot alkaloids. Part 1. Classification and chemistry of ergot peptides. *Pharm Chem J.* 2001;35(9):504–513.

Korn AK, Gross M, Usleber E, Thom N, Köhler K, Erhardt G. Dietary ergot alkaloids as a possible cause of tail necrosis in rabbits. *Mycotoxin Res.* 2014;30(4):241–250.

Krishnamachari KAVR, Bhat R. Poisoning by ergoty bajra (pearl millet) in man. *Indian J Med Res.* 1976 Dec 1;64:1624–1628.

Landschoot S, Audenaert K, Waegeman W, De Baets B, Haesaert G. Influence of maize–wheat rotation systems on *Fusarium* head blight infection and deoxynivalenol content in wheat under low versus high disease pressure. *Crop Prot.* 2013 Oct 1;52:14–21.

Liu X, Wang L, Steffan N, Yin W-B, Li S-M. Ergot alkaloid biosynthesis in *Aspergillus fumigatus*: FGAAT catalyses the acetylation of fumigaclavine B. *ChemBioChem.* 2009 Sep 21;10(14):2325–2328.

Liu Z, Zhang G, Zhang Y, Jin Q, Zhao J, Li J. Factors controlling mycotoxin contamination in maize and food in the Hebei province, China. *Agron Sustain Dev.* 2016;36(2):39.

Lorenz N, Olsovská J, Sulc M, Tudzynski P. Alkaloid cluster gene CCSA of the ergot fungus *Claviceps purpurea* encodes chanoclavine I synthase, a flavin adenine dinucleotide-containing oxidoreductase mediating the transformation of N-methyl-dimethylallyltryptophan to chanoclavine I. *Appl Environ Microbiol.* 2010 Mar;76(6):1822–1830.

Lorenz N, Wilson EV, Machado C, Schardl CL, Tudzynski P. Comparison of ergot alkaloid biosynthesis gene clusters in *Claviceps* species indicates loss of late pathway steps in evolution of *C. fusiformis*. *Appl Environ Microbiol.* 2007 Nov;73(22):7185–7191.

Lyagin I, Efremenko E. Enzymes for detoxification of various mycotoxins: Origins and mechanisms of catalytic action. *Molecules.* 2019 Jun 26;24(13):2362.

Magan N, Aldred D. Post-harvest control strategies: Minimizing mycotoxins in the food chain. *Int J Food Microbiol.* 2007 Nov 1;119:131–139.

Malysheva S, Larionova D, Diana Di Mavungu J, Saeger S. Pattern and distribution of ergot alkaloids in cereals and cereal products from European countries. *World Mycotoxin J.* 2014 Jan 1;7:217–230.

Martin JF, Álvarez-Álvarez R, Liras P. Clavine alkaloids gene clusters of penicillium and related fungi: Evolutionary combination of prenyltransferases, monooxygenases and dioxygenases. *Genes (Basel).* 2017 Nov 24;8(12):342.

Maruo VM, Bracarense AP, Metayer J-P, Vilarino M, Oswald IP, Pinton P. Ergot alkaloids at doses close to EU regulatory limits induce alterations of the liver and intestine. *Toxins.* 2018;10.

Matumba L, Van Poucke C, Njumbe Ediage E, Jacobs B, Saeger S. Effectiveness of hand sorting, flotation/washing, dehulling and combinations thereof on the decontamination of mycotoxin-contaminated white maize. *Food Addit Contam Part A Chem Anal Control Expo Risk Assess.* 2015 Mar 18;32.

Matuschek M, Wallwey C, Wollinsky B, Xie X, Li S-M. In vitro conversion of chanoclavine-I aldehyde to the stereoisomers festuclavine and pyroclavine controlled by the second reduction step. *RSC Adv.* 2012;2(9):3662–3669.

Mavrommatis A, Giamouri E, Tavrizelou S, Zacharioudaki M, Danezis G, Simitzis PE, et al. Impact of mycotoxins on animals' oxidative status. *Antioxidants (Basel, Switzerland).* 2021 Feb 1;10(2):214.

Menzies JG, Turkington TK. An overview of the ergot (*Claviceps purpurea*) issue in western Canada: Challenges and solutions. *Can J Plant Pathol.* 2015 Jan 2;37(1):40–51.

Metzger U, Schall C, Zocher G, Unsöld I, Stec E, Li S-M, et al. The structure of dimethylallyl tryptophan synthase reveals a common architecture of aromatic prenyltransferases in fungi and bacteria. *Proc Natl Acad Sci U S A*. 2009 Aug 25;106(34):14309–14314.

Miccheli A, Puccetti C, Capuani G, Di Cocco ME, Giardino L, Calzà L, et al. [1-13C]Glucose entry in neuronal and astrocytic intermediary metabolism of aged rats. A study of the effects of nicergoline treatment by 13C NMR spectroscopy. *Brain Res*. 2003 Apr 1;966:116–125.

Mulder PPJ, Pereboom-de Fauw DPKH, Hoogenboom RLAP, de Stoppelaar J, de Nijs M. Tropane and ergot alkaloids in grain-based products for infants and young children in the Netherlands in 2011–2014. *Food Addit Contam Part B*. 2015 Oct 2;8(4):284–90.

Müller C, Kemmlein S, Klaffke H, Krauthause W, Preiß-Weigert A, Wittkowski R. A basic tool for risk assessment: A new method for the analysis of ergot alkaloids in rye and selected rye products. *Mol Nutr Food Res*. 2009 Apr 1;53(4):500–507.

Müller C, Klaffke H, Krauthause W, Wittkowski R. Determination of ergot alkaloids in rye and rye flour. *Mycotoxin Res*. 2006 Dec 1;22:197–200.

Nasrollahzadeh A, Mokhtari S, Khomeiri M, Saris P. Mycotoxin detoxification of food by lactic acid bacteria. *Int J Food Contam*. 2022;9(1):1.

Nesci A, Gsponer N, Etcheverry M. Natural maize phenolic acids for control of aflatoxigenic fungi on maize. *J Food Sci*. 2007 Jun 1;72(5):M180–185.

Nešić K, Habschied K, Mastanjević K. Possibilities for the biological control of mycotoxins in food and feed. *Toxins (Basel)*. 2021 Mar 10;13(3):198.

Ortel I, Keller U. Combinatorial assembly of simple and complex D-lysergic acid alkaloid peptide classes in the ergot fungus *Claviceps purpurea*. *J Biol Chem*. 2009 Mar 13;284(11):6650–6660.

Park D. Effect of processing on aflatoxin. *Adv Exp Med Biol*. 2002 Feb 1;504:173–179.

Peles F, Sipos P, Kovács S, Győri Z, Pócsi I, Pusztahelyi T. Biological control and mitigation of aflatoxin contamination in commodities. *Toxins (Basel)*. 2021 Feb 1;13(2):104.

Pigłowski M. Food hazards on the European Union market: The data analysis of the Rapid Alert System for Food and Feed. *Food Sci Nutr*. 2020 Feb 11;8(3):1603–1627.

Pleadin J, Frece J, Markov K. Chapter eight—Mycotoxins in food and feed. In *Advances in Food and Nutrition Research*. FBT-AF and Toldrá NR, editor. Academic Press, USA; 2019, pp. 297–345.

Reddy P, Hemsworth J, Guthridge KM, Vinh A, Vassiliadis S, Ezernieks V, et al. Ergot alkaloid mycotoxins: Physiological effects, metabolism and distribution of the residual toxin in mice. *Sci Rep*. 2020;10(1):9714.

Ren X, Zhang Q, Zhang W, Mao J, Li P. Control of aflatoxigenic molds by antagonistic microorganisms: Inhibitory behaviors, bioactive compounds, related mechanisms, and influencing factors. *Toxins (Basel)*. 2020 Jan 1;12(1):24.

Robinson SL, Panaccione DG. Diversification of ergot alkaloids in natural and modified fungi. *Toxins (Basel)*. 2015 Jan 20;7(1):201–218.

Rubert J, Dzuman Z, Vaclavikova M, Zachariasova M, Soler C, Hajslova J. Analysis of mycotoxins in barley using ultra high liquid chromatography high resolution mass spectrometry: Comparison of efficiency and efficacy of different extraction procedures. *Talanta*. 2012;99:712–719.

Safety EF, Arcella D, Gómez Ruiz JÁ, Innocenti ML, Roldán R. Human and animal dietary exposure to ergot alkaloids. *EFSA J*. 2017 Jul 1;15(7):e04902.

Samapundo S, Meulenaer B, Osei-Nimoh D, Lamboni PhD LY, Debevere J, Devlieghere F. Can phenolic compounds be used for the protection of corn from fungal invasion and mycotoxin contamination during storage? *Food Microbiol.* 2007 Sep 1;24:465–473.

Saraswat S, Krishna S. Bajra (Pearl Millet). The millennium food. *Indian J Nutr Diet.* 2019 Jul 1;56:325.

Schaafsma AW, Ilinic LT-, Miller JD, Hooker DC. Agronomic considerations for reducing deoxynivalenol in wheat grain. *Can J Plant Pathol.* 2001 Sep 1;23(3):279–285.

Schardl CL, Panaccione DG, Tudzynski P. Chapter 2 Ergot alkaloids—Biology and molecular biology. In *Alkaloids: Chemistry and Biology.* Cordell GABT-TAC and B, editor. Academic Press; 2006, pp. 45–86.

Schiff PL. Ergot and its alkaloids. *Am J Pharm Educ.* 2006 Oct 15;70(5):98.

Schillo KK, Leshin LS, Boling JA, Gay N. Effects of endophyte-infected fescue on concentrations of prolactin in blood sera and the anterior pituitary and concentrations of dopamine and dopamine metabolites in brains of steers. *J Anim Sci.* 1988 Mar 1;66(3):713–718.

Schneider L, Olin JT, Novit A, Luczak S. Hydergine for dementia. *Cochrane Database Syst Rev.* 2000;3.

Sessa B. The pharmacology of LSD: A critical review. by Annelie Hintzen and Torsten Passie. *Br J Psychiatry.* 2011;199(3):258–259.

Sharma N, Sharma VK, Kumar Manikyam H, Bal Krishna A. Ergot alkaloids: A review on therapeutic applications. *EJMP.* 2016 Apr 21;14(3):1–7.

Shi H, Schwab W, Liu N, Yu P. Major ergot alkaloids in naturally contaminated cool-season barley grain grown under a cold climate condition in western Canada, explored with near-infrared (NIR) and Fourier transform mid-infrared (ATR-FT/MIR) spectroscopy. *Food Control.* 2019;102:221–230.

Shi H, Stroshine R, Ileleji K. Aflatoxin reduction in corn by cleaning and sorting. *Am Soc Agric Biol Eng Annu Int Meet 2014.* 2014 Jan 1;1:311–321.

Shirasaki Y, Sugimura M, Sato T. Bromocriptine, an ergot alkaloid, inhibits excitatory amino acid release mediated by glutamate transporter reversal. *Eur J Pharmacol.* 2010;643(1):48–57.

Sibley DR, Creese I. Interactions of ergot alkaloids with anterior pituitary D-2 dopamine receptors. *Mol Pharmacol.* 1983 May 1;23(3):585–593.

Siddique YH, Khan W, Fatima A, Jyoti S, Khanam S, Naz F, et al. Effect of bromocriptine alginate nanocomposite (BANC) on a transgenic *Drosophila* model of Parkinson's disease. *Dis Model Mech.* 2016 Jan;9(1):63–68.

Steffan N, Li S-M. Increasing structure diversity of prenylated diketopiperazine derivatives by using a 4-dimethylallyltryptophan synthase. *Arch Microbiol.* 2009 Apr 1;191:461–466.

Steffan N, Unsöld IA, Li S-M. Chemoenzymatic synthesis of prenylated indole derivatives by using a 4-dimethylallyltryptophan synthase from *Aspergillus fumigatus.* *ChemBioChem.* 2007 Jul 23;8(11):1298–1307.

Storm ID, Rasmussen PH, Strobel BW, Hansen HCB. Ergot alkaloids in rye flour determined by solid-phase cation-exchange and high-pressure liquid chromatography with fluorescence detection. *Food Addit Contam Part A.* 2008 Mar 1;25(3):338–346.

Taniwaki M, Hocking AD, Pitt J, Fleet GH. Growth and mycotoxin production by fungi in atmospheres containing 80% carbon dioxide and 20% oxygen. *Int J Food Microbiol.* 2010 Oct 15;143:218–225.

Thabit TMA, Abdelkareem EM, Bouqellah NA, Shokr SA. Triazole fungicide residues and their inhibitory effect on some trichothecenes mycotoxin excretion in wheat grains. *Molecules*. 2021;26.

Tittlemier SA, Drul D, Roscoe M, McKendry T. Occurrence of ergot and ergot alkaloids in western Canadian wheat and other cereals. *J Agric Food Chem*. 2015 Jul 29;63(29):6644–6650.

Tkachenko A, Benson K, Mostrom M, Guag J, Reimschuessel R, Webb B. Extensive evaluation via blinded testing of an UHPLC-MS/MS method for quantitation of ten ergot alkaloids in rye and wheat grains. *J AOAC Int*. 2021 Jun 1;104(3):546–554.

Tudzynski P, Correia T, Keller U. Biotechnology and genetics of ergot alkaloids. *Appl Microbiol Biotechnol*. 2002 Jan 1;57:593–605.

Tudzynski P, Hölter K, Correia T, Arntz C, Grammel N, Keller U. Evidence for an ergot alkaloid gene cluster in *Claviceps purpurea*. *Mol Gen Genet MGG*. 1999;261(1):133–141.

Uhlig S, Petersen D. Lactam ergot alkaloids (ergopeptams) as predominant alkaloids in sclerotia of *Claviceps purpurea* from Norwegian wild grasses. *Toxicon*. 2008;52(1):175–185.

Uhlig S, Rangel-Huerta OD, Divon HH, Rolén E, Pauchon K, Sumarah MW, et al. Unraveling the ergot alkaloid and indole diterpenoid metabolome in the *Claviceps purpurea* species complex using LC–HRMS/MS diagnostic fragmentation filtering. *J Agric Food Chem*. 2021 Jun 30;69(25):7137–7148.

Uppala SS, Wu BM, Alderman SC. Effects of temperature and duration of preconditioning cold treatment on sclerotial germination of *Claviceps purpurea*. *Plant Dis*. 2016 May 23;100(10):2080–2086.

Var I, Kabak B, Erginkaya Z. Reduction in ochratoxin A levels in white wine, following treatment with activated carbon and sodium bentonite. *Food Control*. 2008 Jun 1;19:592–598.

Varga J, Kocsubé S, Péteri Z, Vágvölgyi C, Tóth B. Chemical, physical and biological approaches to prevent ochratoxin induced toxicoses in humans and animals. *Toxins (Basel)*. 2010 Jul;2(7):1718–1750.

Wallwey C, Li S-M. Ergot alkaloids: structure diversity, biosynthetic gene clusters and functional proof of biosynthetic genes. *Nat Prod Rep*. 2011;28(3):496–510.

Waret-Szkuta A, Larraillet L, Oswald IP, Legrand X, Guerre P, Martineau G-P. Unusual acute neonatal mortality and sow agalactia linked with ergot alkaloid contamination of feed. *Porc Heal Manag*. 2019;5(1):24.

Winblad B, Fioravanti M, Dolezal T, Logina I, Milanov IG, Popescu DC, et al. Therapeutic use of nicergoline. *Clin Drug Investig*. 2008;28(9):533–552.

Xie X, Wallwey C, Matuschek M, Steinbach K, Li S-M. Formyl migration product of chanoclavine-I aldehyde in the presence of the old yellow enzyme FgaOx3 from *Aspergillus fumigatus*: A NMR structure elucidation. *Magn Reson Chem*. 2011 Oct 1;49(10):678–681.

Yu C, Lu P, Liu S, Li Q, Xu E, Gong J, et al. Efficiency of deoxynivalenol detoxification by microencapsulated sodium metabisulfite assessed via an in vitro bioassay based on intestinal porcine epithelial cells. *ACS Omega*. 2021 Mar 30;6(12):8382–8393.

Zöllner P, Mayer-Helm B. Trace mycotoxin analysis in complex biological and food matrices by liquid chromatography–atmospheric pressure ionisation mass spectrometry. *J Chromatogr A*. 2006;1136(2):123–169.

Chapter 8

Roquefortines in Food and Feed
Detection and Management Strategies

Saloni, Dinesh Chandra Rai, Himanshu Kumar Rai,
Arvind Kumar, Urvashi Vikranta, Shikha Pandhi,
Akansha Gupta, Dipendra Kumar Mahato

CONTENTS

8.1	Introduction	213
8.2	Occurrence of Roquefortines in Food and Feed	216
8.3	Major Sources of Roquefortines	218
8.4	Gene Responsible for Roquefortine Production	219
8.5	Biosynthesis of Roquefortines	220
8.6	Detection Methods	221
8.7	Management and Control Strategies	225
	8.7.1 Chemical Methods	231
	8.7.2 Biological Methods of Decontamination or Detoxification	231
8.8	Conclusion	233

8.1 INTRODUCTION

Secondary metabolites (SMs) are organic compounds produced mostly by bacteria, fungi, and plants. They are low-molecular-weight compounds with several biochemical structures and functions. The term "secondary metabolite" comes from the fact that its synthesis is not required for organism growth and reproduction; for example, antibiotics, anticancer and antiviral drugs, alkaloids, differentiation effectors, and so on are produced by filamentous fungi, making them an intriguing research matter (Kosalková et al., 2009). SMs are made for specific purposes like defense against harmful microbes, increasing tolerance to stresses, and repellence of unwanted feeders (Erb & Kliebenstein,

DOI: 10.1201/9781003242208-8

2020). Many secondary metabolites are formed during fungal growth, many of which can be toxic to vertebrates if present in high enough amounts. This class of compounds is known as mycotoxins. The most essential parameters influencing fungal growth and mycotoxin synthesis are the availability of nutrients, water activity, ambient temperature, and oxygen (Kokkonen et al., 2005). Secondary metabolites can be found in feed, raw materials, and processed foods. To safeguard human health, mycotoxin contamination evaluation in food focuses primarily on those major mycotoxins for which regulatory restrictions have been established (Fontaine et al., 2015a).

Among various mycotoxins, roquefortines are produced by a bacterium known as *Penicillium roqueforti*, as implied by the name. In the past, the term "roquefortines" was used to describe two distinct classes of secondary metabolites. The classes of roquefortines include roquefortine C (ROQC), D, F, L, M, and N, among which derivatives of ergoline are also called isofumigaclavine A and B, as shown in Figure 8.1. Roquefortine C seems to be modified diketopiperazine, neither an ergoline nor a cyclic diterpene, and it is frequently found in foods contaminated with *P. roqueforti*, whether intentionally or accidentally. In general, the fungus has been used to purposely ripen specific blue-veined cheeses. ROQC levels in those kinds of products can be considerable (mg kg^{-1}) (Maragos, 2020). Certain neurotoxins, such as cyclopiazonic acid (CPA) and ROQC, can be detected among the cheese mycotoxins. The substituted indole found in both mycotoxins is produced from tryptophan and is found in several neurotoxins, such as ergot alkaloids, paspalitrems, penitrems, and lolitrems (Reddy et al., 2019).

Many blue-veined kinds of cheese are available, including roquefort, Stilton, gorgonzola, and cambozola. Blue-veined cheeses, often known as "blue *Penicillium* cheeses," are usually inoculated with roqueforti (Maragos, 2021). Recently, its significance as a biosynthetic precursor of the triazaspirocyclic structure of oxaline, glandicoline B, and meleagrin has attracted attention. The triazaspirocyclic pattern, which is composed of three nitrogen atoms bonded to one quaternary carbon to form a spirocyclic structural framework, is a unique chemical functional group that has been found to have a wider range of biological activity, including antibacterial and antimutagenic activity against carcinoma cells, as well as anti-biofouling activity against sea life (Gober, 2017). *P. paneum, Aspergillus niger,* and *P. roqueforti* are commonly found in silage. The mycotoxin producers *P. roqueforti* and *P. paneum,* which primarily produce roquefortine C, were given special attention because of their potential impact on dairy cattle yield and health (ROC, neurotoxic) (Tangni et al., 2013). *Penicillium roqueforti* and *Penicillium camemberti* are two common cheese-ripening cultures that make a significant contribution to the texture, appearance, and flavor development of blue-veined and mold-ripened cheeses being in good health in terms of their potential to create a range of mycotoxins,

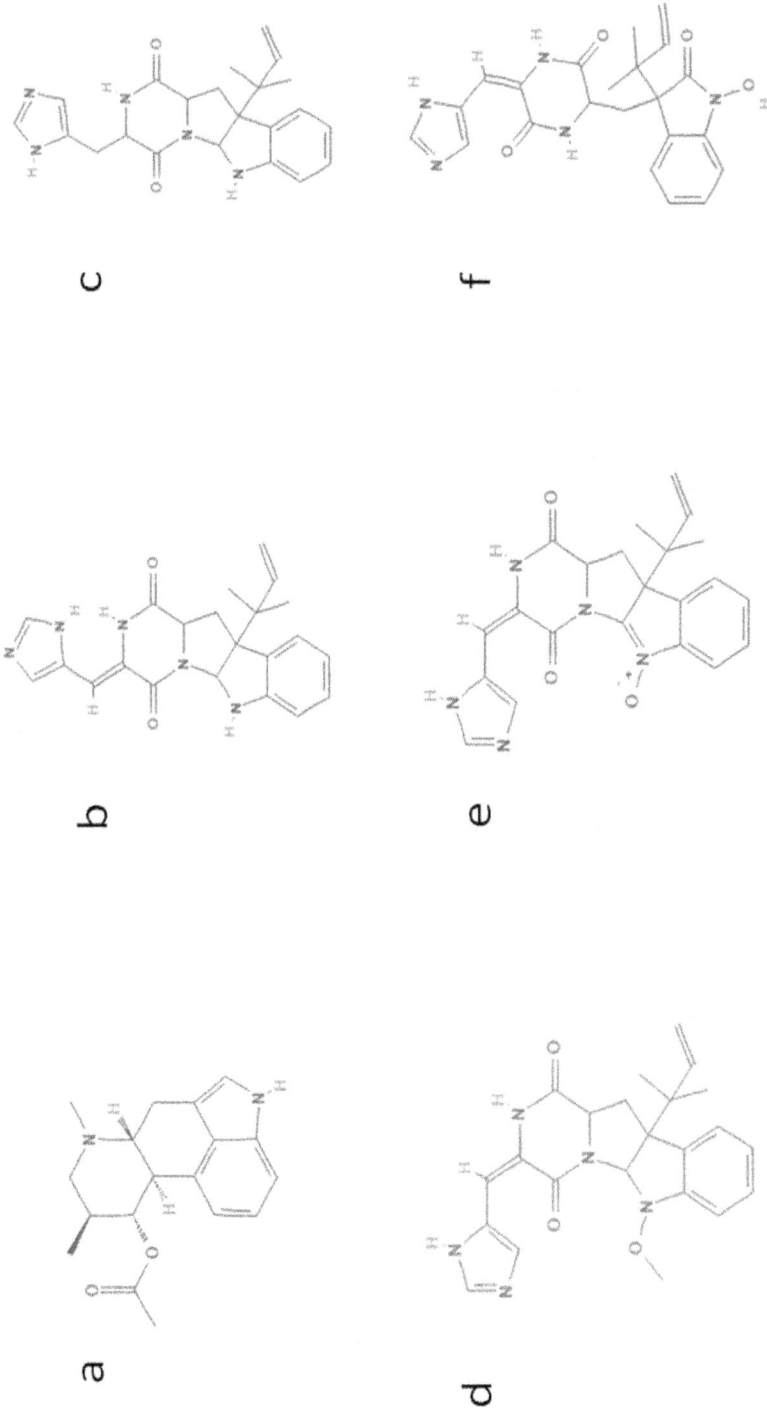

Figure 8.1 Structure of roquefortines: a) roquefortine A, b) roquefortine C, c) roquefortine D, d) roquefortine F, e) roquefortine L, f) roquefortine M.

such as *P. camemberti* synthesizes cyclopiazonic acid, whereas *P. roqueforti* produces mycophenolic acid (MPA) and roquefortine C. These metabolites, however, are less hazardous than those listed previously and can be found in small quantities in cheese. Because of their long history of safe use throughout the United States and European countries, the use of such fungal cultures in cheese-making is proven safe. The Food and Drug Administration (FDA) has also marked these microorganisms as generally recognized as safe (GRAS) (Hymery et al., 2014).

Several *Penicillium* species, including *Penicillium chrysogenum* and *Penicillium roqueforti*, as well as several saprophytic or plant-linked fungi, such as *Penicillium carneum, Penicillium expansive,* and *Penicillium glandicola*, have the ability to produce roquefortine-type alkaloids (Kosalková et al., 2015). *Penicillium* is a common fungus that may be grown locally in a wide range of areas. *Penicillium* is a saprophytic organism that is widely used in industrial mycology for the synthesis of antibiotics (like griseofulvin and penicillin) and enzymes (such as ribonuclease) and is also known to be a food spoilage agent. Another bacterium, *Penicillium roqueforti*, well known for its role in silage contamination, is also frequently found in spoiling different types of soft, semi-hard, and hard cheeses. However, several cultures of *P. roqueforti* have been used in the food industry and are important in the ripening of cheeses such as roquefort, stilton, gorgonzola, and danablu. To obtain these cheeses' typical flavors and fragrances, it is critical to inoculate the milk with the culture media of *P. roqueforti* due to the molds' strong lipolytic activity and the formation of short-chain fatty acids (caproic, butyric, caprylic, and acetic acid). When fatty acids are oxidized, methyl ketone is produced, which enhances the flavor (Vallone et al., 2014). As a result, the various sources, occurrences, and biosynthetic pathways of roquefortines in food and feed, as well as the numerous detection and management approaches to make sure of the safety and security of food and feed, are discussed in this chapter.

8.2 OCCURRENCE OF ROQUEFORTINES IN FOOD AND FEED

Mycotoxins, like toxic metabolic substances, are generally produced by certain molds and can be added to milk or milk products from two sources: (a) indirect contamination, which takes place when dairy cows eat mycotoxin-containing feed, and (b) direct contamination, which takes place when molds grow by purpose or by accident (Sengun et al., 2008). According to different experimental findings, various *Penicillium* species, which include *Penicillium chrysogenum, Penicillium roqueforti,* and various saprophytes such as *Penicillium glandicola, Penicillium carneum, Penicillium paneum, Penicillium expansum,* and

Penicillium griseofulvum, produce roquefortine-related alkaloids (Kosalková et al., 2015). Dimethylallyl-tryptophan (DMAT) is the basic substance that serves as a precursor for prenylated alkaloids, and DMAT is generated by using L-tryptophan and dimethylallylpyrophosphate (DMA-PP) as a substrate for the enzyme prenyltransferase. The prenylation reaction yields a DMA group, which is linked to the L-tryptophan indole ring at various points throughout the biosynthetic pathway, usually at the 3, 4 positions of carbon or 1 position of nitrogen. *Penicillium* spp. have been detected in over 200 species, making it the most frequent microscopic filamentous fungus in the food processing and formulation industry. *Penicillium* species, which are mesophiles, thrive in temperatures ranging from 5°C to 37°C, with a pH of 3 to 4.5 and water activity ranging from 0.78% to 0.88%. They can sometimes be present in decomposing plants and compost, dried foods, soil, spices, fresh fruits, vegetables, cereals, and also in the dust particles and air. They can also proliferate on the walls of buildings, particularly if the humidity of the construction materials is high. Antibiotics, antiviral agents, and mycotoxins are just a few of the many metabolites that *Penicillium* species are capable of producing (Demjanová et al., 2021). Roquefortine C is a diketopiperazine (DKP) alkaloid reported in *Penicillium roqueforti* cultures and later in cultures of other *Penicillium* species that grew on contaminated vegetables such as onions, cereal grain, beer, wine, and so on. The well-known neurotoxicity of roquefortine C's mycotoxin makes it a major focus when it comes to food and feed contamination (García-Estrada et al., 2011). It was reported that ROQC concentrations in 30 blue kinds of cheese varied from 50–1470 μg/kg (Fontaine et al., 2015a).

Penicillium roqueforti is the saprophytic fungus that shows its presence in the ecological niche, even if its natural reservoir is yet not known (Kosalková et al., 2015). It was found that the thickness of fungal colonies ranges from about 40–70 mm when grown on malt extract agar (MEA) and Czapek yeast extract agar (CYA) for 14 days, showing conidia (asexual spores) in blue-green color (Coton et al., 2020). *Penicillium roqueforti* has been divided into different species like *P. paneum, P. carneum,* and *P. roqueforti* (Nielsen et al., 2006). A variety of foods were used to isolate the active compound roquefortine C. This compound was found to be abundant in cheese, while roquefortines A and D were found to be rare. It was further described that the reverse condensation between 3' carbon of the DMA and C-3 of L-tryptophan occurs in roquefortine alkaloids (Li et al., 2009). A study showed that the cyclopiperazine nucleus may be found in roquefortines and other related compounds such as neoxaline, glandicoline, and meleagrin, which are all L-histidine and L-tryptophan derivatives. The condensation of these two amino acids occurs in the presence of roquefortine dipeptide synthetase RDS (RoqA), resulting in cyclodipeptide (cyclo-trp-his), afterward prenylated using roquefortines prenyltransferase (RPT), where the association of the isopentyl group with the tryptophan at C-3 in the presence of DMA-PP is observed (3, 12-dihydro-roquefortine C). Now roquefortine C is generated by oxidizing roquefortine D in the presence of roquefortine D dehydrogenase, which removes two hydrogen atoms and forms a double bond between 3 and 12 carbons. For the synthesis of roquefortine C, the presence of non-ribosomal cyclodipeptide synthetase and prenyltransferase is required (Kosalková et al., 2015). *Penicillium chrysogenum* has been shown to include genes that are involved in the production of meleagrin and roquefortine C (García-Estrada et al., 2011). Non-ribosomal cyclodipeptide synthetase (rds), and prenyltransferase (rpt) are responsible for encoding non-ribosomal cyclodipeptide synthetase and prenyltransferase, respectively, that function to synthesize roquefortine C (Häggblom, 1990). The existence of roquefortine C in feed grain was found by isolating the mycotoxin from feed grains that were heavily contaminated by *Penicillium roqueforti*. A strain was identified from maize silage that could manufacture indole alkaloids, indicating that it was related to *Penicillium roqueforti* (Ohmomo et al., 1994).

8.3 MAJOR SOURCES OF ROQUEFORTINES

Molds that can lead to contamination and other dairy products appear in a variety of genera and species, and they are extensively utilized in feed and food processing. Butter; yoghurt; blue-veined, semi-hard, hard, softer, mold-ripened, and semi-soft cheeses; and diverse milk-based derivative products are among the dairy products that thrive with *Penicillium* species (Dubey et al., 2018). Ripening in blue-veined cheeses (*Penicillium roqueforti*) has been achieved with the help of fungi which resulted in the formation of mycotoxin roquefortine C in blue-veined cheeses. A study reported that 99.34% of blue-veined slices of cheeses had roquefortine C, accounting for a maximum value of 6630 µg/kg (Maragos, 2021). In another study, it was also identified that 97.9% of blue-veined cheeses were found to have a quantifiable amount of roquefortine C with a mean value of 848 ± 1670 µg/kg, out of which 75% of cheeses had less than 792 µg/kg of roquefortine C (Fontaine et al., 2015a). When different strains of *Penicillium roqueforti* taken from blue cheese from three different countries (the United Kingdom, United States, and France) were analyzed, the complete 16.6-kb roquefortine gene cluster was found to be 98–99% equivalent in terms of organization and nucleotide sequence. The rds, rpt, rdh, and gmt genes encode roquefortine dipeptide synthetase, roquefortine prenyltransferase, roquefortine D-dehydrogenase, and methyltransferase, respectively, from the discovered gene cluster. As a result, this secondary metabolite is regarded as a blue cheese contaminant (Rathnayake et al., 2020). In 30, 14, 12, 5, 2, 1, and 12 sub-samples of bread/rolls, cheese, vegetables, nuts, jams, fruits, and others, respectively, the median concentration of 120 g/kg and maximum concentration of 84,000 g/kg of roquefortines were estimated, and the 247 subsamples of surface area and thickness of 1cm^2 and 0.5–1 cm, respectively, were made from 87 mold-affected foodstuffs. In non-moldy subsamples, a trace quantity of roquefortines was also found (Sulyok et al., 2010). Driehuis et al. (2008) demonstrated that there is a minimum probability of getting roquefortine C (81 g/kg) in grass silage, and if it is found, it may be linked to deterioration due to air infiltration during storage and feeding out, as 15–42% of grass silage and 8–30% of maize silage in Germany had shown the presence of roquefortine C. Only 0.8% of the 120 grass silage tested positive for roquefortine C at levels higher than the quantification level (50 g/kg). Different molds produced mycotoxins in silage, but *P. roqueforti* was the most prevalent, which can be explained by the fact that it can endure high concentrations of carbon dioxide and acetic acid, as well as survive and thrive on reduced oxygen levels (Driehuis, 2011). The presence of visible fungal growth is commonly detected in ensiled feed items, and they include toxigenic fungus species such as *P. roqueforti sensulato*, a combination of *Penicillium roqueforti sensustrictu* and *Penicillium paneum* (Wambacq, 2017). *Penicillium granulatum* MCCC3A00475, a deep-sea-derived fungus, yielded roquefortine J (unknown previously). Using column chromatography, fractionation of ethyl acetate (EtOAc) extract from fermented cultures of deep-sea-derived species was performed, leading to the identification of roquefortine alkaloid (roquefortine J) with ten other recognized chemicals (Niu et al., 2018). The presence of extremely virulent strains (X5 and PF1) of the species *P. expansum* was reported in dried chestnuts, chestnut granulates, chestnut flour, and other secondary metabolites such as patulin and chaetoglobosin (Prencipe et al., 2018). During research on natural bioactive compounds from the fermentation of ZSDS 1-F7, a sponge-derived fungus, three roquefortine derivatives, namely roquefortine C, (16S)-hydroxyroquefortine C, and (16R)-hydroxyroquefortine C, were discovered. Different spices were tested for the presence of mycotoxins, including paprika, coarse chili, ground chili, onion spices, vegetable spices, cheese spices, and fruit chutney spices. The presence of 57 µg/kg of roquefortine C was found in coarse chili (n = 14), while other spices had roquefortines below the limit of quantification (17 µg/kg) (Motloung et al., 2018).

8.4 GENE RESPONSIBLE FOR ROQUEFORTINE PRODUCTION

Secondary metabolite genes and/or groups were regularly passed from one organism to another during evolution. Diketopiperazines, a family of spontaneously produced secondary metabolites with pharmacological significance, have been found in a variety of fungus species. Roquefortine C, a diketopiperazine that is isolated in *Penicillium roqueforti*, has now been discovered in 25 distinct *Penicillium* species. Gram-positive bacteria are resistant to its antimicrobial effects. Even though its specific mode of action is uncertain, it is found to attach to cytochrome p450 and inhibit RNA synthesis. Roquefortine C is a contaminant in blue cheese that has been shown to have neurotoxic effects in mice. Meleagrin, a roquefortine C downstream product, is thought to be the precursor to neoxaline, an antibacterial molecule (Ali et al., 2013).

Roquefortine C; roquefortine D (3,12-dihidroroquefortine C); roquefortine F; glandicoline A; glandicoline B; roquefortines L, M, and N; meleagrin; and neoxaline are all members of the roquefortine family of secondary metabolites. Isofumigaclavines A and B, called roquefortines A and B, refer to the ergot alkaloids and hence are not included in this category. Roquefortine C and similar substances are prenylated diketopiperazine indole alkaloid molecules synthesized from L-histidine, L-tryptophan, and mevalonate produced by *Penicillium* species such as *P. roqueforti* and the *Penicillium chrysogenum* complex. It was found that indole alkaloids can be present in fermented foods, alcoholic beverages, and foods, particularly blue-veined cheeses. Based on the type and quantity of cheeses included in the observation, roquefortine C concentrations in this moldy food ranged from 50–1470 µg/kg, 800 to 12,000 g/kg, or 11 to 14,125 g/kg (García-Estrada and Martín, 2016).

Roquefortine C is an indole alkaloid derived from peptide chains condensed by the rds/roqA gene-encoded non-ribosomal peptide synthetase. The roquefortine C biosynthetic gene cluster also includes rdh/roqR, rpt/roqD, and the remaining pseudogene gmt/roqN. rdh/roqR and rpt/roqD are prenyltransferases and cytochrome P450 oxidoreductases, respectively. They later take part in the biosynthesis of roquefortine D and C. *P. roqueforti* possesses seven genes in the mycophenolic acid biosynthetic gene cluster (mpaA, mpaC, mpaB, mpaDE, mpaG, mpaF, and mpaH) (24.4 kbp). Mycophenolic acid is an organic acid, and the putative polyketide synthase encoded by MPA forms its backbone as the initial step in the metabolic route (Kirtil et al., 2021).

8.5 BIOSYNTHESIS OF ROQUEFORTINES

Six genes in the roquefortine/meleagrin pathway circumscribe a dimodular non-ribosomal peptide synthetase. Starting with histidyl-tryptophanyl diketopiperazine (HTD), which is created by the core synthetase enzyme RoqA deploying tryptophan and histidine as substrates, RoqD catalyzes the formation of roquefortine D by reversed prenylation of HTD at the C-3 position of its indole moiety. Simultaneously, HTD is converted to dehydro-histidyltryptophanyldiketo piperazine at its histidinyl moiety by RoqR, a cytochrome P450 oxidoreductase (DHTD). Both HTD reactions lead to a roquefortine/meleagrin pathway branch, with the first to DHTD either through RoqR oxidation and then to roquefortine C via RoqD dimethylallyl addition, and or the second to roquefortine C via an enzymatic order change. RoqD produces roquefortine D by dimethylallyl addition first, while RoqR produces roquefortine C by further oxidation. Although various silencing, labeling, and deletion experiments have been carried out, the subsequent biosynthetic events and genes involved remain unknown. Roquefortine C, for example, should be converted to glandicoline A, then to glandicoline B, with RoqM and RoqO catalyzing each step. It's not obvious yet which reaction they belong to. Additionally, neoxaline was hypothesized as a pathway end product derived from the hydrogenation of meleagrin, but no gene encoding the process could be located in the roq gene cluster (Ries et al., 2013). Exogenous roquefortine C concentrations influence roquefortine C production and excretion in fungus, according to studies. In addition, 14C-roquefortine C experiments showed that roquefortine C was not only taken up by developing mycelia but was also integrated into protein and mycelial residues. Roquefortine C was found to be transported over the cellular membrane through both energy-dependent and energy-independent pathways in in vitro experiments on *P. crustosum* strains. This suggests that roquefortine C may serve as an external nitrogen supply for the organism that produces it (Overy et al., 2005).

8.6 DETECTION METHODS

Penicillium roquefortine (PR) toxins can be detected by a variety of traditional and advanced techniques (Dubey et al., 2018) such as thin-layer chromatography (TLC), a chromatographic technique in which the mobile phase solvent has been preferred depending on the characteristics of the mixture's constituents, and the solid phase is a thin glass plate enclosed in either aluminum oxide or silica gel. The dispersion of a compound between a solid fixed phase and a liquid mobile phase (eluting solvent) moving over the solid phase is the basic principle of TLC, and a small amount of a compound or mixture is placed directly above the bottom of the TLC plate as just a preliminary step. After that, the plate is developed in a developing chamber with a solvent slightly lower than the sample's application level. By capillary action, the solvent is drawn up through the particles on the plate, and each component either remains solid or dissolves in the solvent and migrates up the plate as it passes over the mixture. Physical properties, particularly functional groups, which are influenced by molecular structure, determine whether a molecule moves up the plate or stays behind. The "like dissolves like" solubility rule is followed. Physical properties of a chemical that has been in the mobile phase for a long time should be similar to those of the mobile phase. In line with this, the most soluble compounds will be carried the furthest up the TLC plate by the mobile phase, and the compounds that have a stronger affinity for the TLC plate particles but are less soluble in the mobile phase will be left behind (Bele & Khale, 2011). Thin-layer chromatography was widely used as an early detection method for ROQC. As a result, TLC methods were used to detect ROQC in blue cheeses and fungi at significant concentrations. The absorption peaks of ROQC in the ultraviolet (UV) range have been reported in a variety of situations at around 209, 240, and 328 nm, respectively (Maragos, 2020).

Another technique like high-pressure liquid chromatography (HPLC) is an analytical chemistry technique that is mainly used for isolating, identifying, and analyzing each component in a mixture. In this technique, a pressurized liquid solvent containing the sample combination is conveyed through a solid adsorbent material-filled column by using pumps, and each component in the sample uniquely interacts with the adsorbent material, resulting in a wide range of flow velocity and separation of such constituents as they exit the column (Karger, 1997).

Blue cheese and its dressing, as well as silage, were studied using sophisticated computer technology like liquid chromatography with UV or photodiode array (PDA) detection or liquid chromatography with tandem mass spectrometry (LC-MS-MS), which integrates the separating power of liquid chromatography with triple quadrupole mass spectrometry's sensitivity and selectivity (Mezcua et al., 2006). To quantify ROQC in cheeses, silage, tree nuts, and fungal cultures, liquid chromatography with tandem mass spectrometry was commonly utilized. Ingesting moldy walnuts or almonds has been linked to at least one human infection and four dog infections, according to the available data (Maragos, 2020). The Ames test has never revealed any mutagenic activity for roquefortine C (Hymery et al., 2014). The retention duration (mean = 5%) and ion ratio calculated during the validation procedure were used to identify each mycotoxin in cheese samples. When the concentration exceeded the measurement range's top limit, the resulting suspension was diluted in blank extract before being re-injected appropriately (Fontaine et al., 2015b).

Immunochemical and chromatographic techniques are also used to detect mycotoxins. The detection of mycotoxins in cereals is done predominantly by the use of enzyme-linked immunoassay (ELISA), which is a plate-based assay technique used for detecting and quantifying soluble molecules such as antibodies, peptides, hormones, and proteins. In this technique, the antigen (target macromolecule) is immobilized on a solid surface (microplate) before being combined with an antibody linked to a reporting enzyme, and the reporter enzyme's activity is determined by incubating it with a suitable substrate to yield a quantifiable result. Thus, this technique is mainly explained by the interaction of a highly specific antibody with an antigen. When it came to different types of mycotoxins, ELISA was used to find out how much of each type of toxin was present in the samples. Among other mycotoxins, monoclonal antibodies

TABLE 8.1   ANALYTICAL METHODS FOR DETECTION OF ROQUEFORTINES

Source	Mold Species	Secondary Metabolites	Analysis	Mycotoxin Range (Roquefortine C)	References
Blue and blue-white mold cheese	ND	Roquefortine C	LC-MS/MS method	0.8 to 12 mg kg^{-1}	Kokkonen et al., 2005
Blue-veined cheeses	*Penicillium roqueforti*	Roquefortine C	LC-MS/MS	0.792 mg kg^{-1}	Fontaine et al., 2015a
Traditional mold-ripened Turkish civil cheese	*Penicillium roqueforti*	Roquefortine C	HPLC + diode array detection (DAD)	0.4–47.0 mg kg^{-1}	Cakmakci et al., 2015
Blue moldy tulum cheeses	*Penicillium roqueforti*	Roquefortines	Thin-layer chromatography	2.1–2.4 (at 5°C) 2.1–3.8 (at 12°C) mg kg^{-1}	Erdogan & Sert, 2004
Food waste Total Samples (97) = summer (48) and winter (49)	*Penicillium roqueforti*	Roquefortine C	HPLC system coupled + mass spectrometer (MS)	Summer: 0.040–0.920 mg kg^{-1} Winter: 190 mg kg^{-1}	Rundberget et al., 2004
Baled grass silage	*Penicillium roqueforti*	Roquefortine C; roquefortines A, B, and D; f mycophenolic acid, andrastin, estuclavine, marefortine A, and agroclavine	HPLC system coupled with a mass spectrometer	20 mg/kg	O'Brien et al., 2006

(Continued)

TABLE 8.1   ANALYTICAL METHODS FOR DETECTION OF ROQUEFORTINES   (CONTINUED)

Source	Mold Species	Secondary Metabolites	Analysis	Mycotoxin Range (Roquefortine C)	References
Feedstuffs of dairy cows in the Netherlands	*Penicillium roqueforti*	Deoxynivalenol, zearalenone, roquefortine C, and mycophenolic acid	Liquid chromatography-tandem mass spectrometry	0. 114 mg kg^{-1}	Driehuis et al., 2008
Maize silage	*Penicillium* sp.	Roquefortine C, patulin (PAT), cyclopiazonic acid (CPA), and mycophenolic acid (MPA)	HPLC system coupled with a mass spectrometer	0.38 μg/g	Mansfield et al., 2008
Media: CYA and PDA (lab conditions)	*P. roqueforti*	Roquefortine C	HPLC system	3.6 μg/ml	Taniwaki et al., 2009
Nut extracts (nut milk) and canine serum	*P. roqueforti*	Roquefortine C	LC-PDA-FLD-MS (mAb and ELISA)	Nut milk = 0.25 to 2 ng ml^{-1} Canine serum = 0.2 to 5 ng ml^{-1}	Maragos, 2020
Maize seedlings	*P. vulpinum* strain	Roquefortine C	TLC plates + 366-nm UV light + chemical spray reagents	6.50 μg ml^{-1}	Ismaiel & Papenbrock, 2014
Cheese	*Penicillium sp.*	Roquefortine C	(TLC) + (HPLC)	0.05 to 1.47 mg kg^{-1}	Finoli et al., 2001
Seed maize	*Penicillium sp.*	Roquefortine C	LC-MS/MS	2.9 μg·kg^{-1}	Betancourt et al., 2015

(*LC-MS/MS:* Liquid chromatographic mass spectrometry or tandem mass spectrometry, TLC: Thin-layer chromatography, mAb: Monoclonal antibodies, PDA: Potato dextrose agar, CYA: Czapek yeast extract agar, HPLC: High-performance liquid chromatography, ND: Not detected.)

have been produced for diacetoxyscirpentriol, 3- and 15-acetyl-DON, sporides-min A, and roridin (Engvall & Perlmann, 1971).

Modern mycotoxin testing methods, such as multi-mycotoxin liquid chro-matography with tandem mass spectrometry (MS/MS), use extremely precise, highly specific, and sensitive methods of mass spectrometry tandem liquid chromatography (Szulc et al., 2021), as described in Table 8.1. The LC-MS/MS method was utilized to develop a highly sensitive and specific method for detecting penitrem A and roquefortine C in a range of sample media, like urine and serum. Two probable cases of canine penitrem A infection were investi-gated using this technique. In case 1, strychnine and metaldehyde were found to be absent in the gastric lavage sample and urine analyses, with DLs of 1.0 and 0.1 mg/kg, respectively. The patient's gastric lavage sample (DL 5 25 ng/ml) revealed 0.6 g/ml of penitrem A and 1.3 g/ml of roquefortine C. The serum sample contained DL 5 1 ng/ml for both penitrem A and roquefortine C. Roquefortine C (DL 5 ng/ml) was found in the urine sample, despite the absence of penitrem A. Roquefortine C was discovered in case 2 at a concentration of 14.1 ng/ml. Penitrem A was detected; however, the concentration was well below the detec-tion threshold of 1 ng/ml. As a result, even at the DL level, penitrem A concen-tration can have clinical importance for diagnostic purposes. In such samples, roquefortine C is more likely to be a biomarker for penitrem A exposure than not (Tiwary et al., 2009).

8.7 MANAGEMENT AND CONTROL STRATEGIES

Nutritional parameters of foods and other derived products must be exam-ined to confirm the existence of different fungal toxins in food and feed. Furthermore, the cytotoxic response of ingested mycotoxins is crucial for determining harmful limits and exploring new techniques to reduce the haz-ardous effects of toxin-producing strains. Contamination with mycotoxins can occur in the field at any stage before harvesting the crops, during harvest-ing or storage of the crops, and during different processing conditions; hence approaches for preventing contamination with mycotoxins can be neatly categorized into three broad stages: pre-harvest, harvest, and post-harvest strategies. Different conventional approaches, such as good agricultural practices (GAPs); good manufacturing practices (GMPs); and a future supple-mentary control approach, the application of hazard analysis critical control points (HACCPs), have been used to inhibit mycotoxin development in cere-als (Kabak et al., 2006). Food processing operations such as trimming, manual selection or choosing, baking, polishing, cleaning or rinsing, roasting, fermen-tation, flaking, boiling, frying, canning, and nixtamalization all have consid-erable impacts on mycotoxin magnitude in agricultural goods. These include chemical treatments such as chlorine, ammonia, ozone, sodium or calcium hydroxide, and the execution of different adsorbents like activated charcoal and clay (Mohapatra et al., 2017). In recent years, a wide variety of different physicochemical and biological procedures, such as direct fluorimetry, fluorescence polarization ELISA, and other biosensors and strip methods, have been utilized to measure mycotoxin levels (Dubey et al., 2018). Generally, there are three stages to the intervention. Preventive measures should begin before any fungus infests plant material and continue during the fungal growth invasion and generation of mycotoxin, and the final stage is when significantly contaminated agricultural goods have been recognized. The HACCP food safety management system shares some similarities with this hazard analysis. To control mycotoxin contamination, HACCP is being employed in the production of safe animal feed and food. The second principle of HACCP, determination of critical control points, and the third principle, establishing critical limits, are the most commonly used. Since mycotoxins are difficult to remove practically once they are present, much of the work must be focused on the first two processes (Jouany, 2007). It was reported that, in northern Germany, calves given silage with 0.2 to 1.5 mg of roquefortine C/kg showed various symptoms like mastitis, infertility, and a lack of appetite, and there have also been reports of paralytic effects in cows given 4 to 8 mg/kg of roquefortine C. Meanwhile, there are not any clinical indications of infection found in sheep fed 25 mg/kg of roquefortine C for 18 days, showing that sheep are more resistant to roquefortine C as compared to cattle (Ogunade et al., 2018).

The most usual post-harvest fungi discovered in silage are *Penicillium* section *roquefortorum* species like *P. paneum* and *P. roqueforti*. To encourage fungus growth and the development of mycotoxins, various conditions like unfavorable weather or storage conditions play a vital role (Gallo et al., 2015). While several treatments have been identified to limit the creation of specific mycotoxins in various commodities, it is presently not realistic to eliminate all mycotoxin-contaminated commodities. Environmental conditions influence pre-harvest events; however, the farmer has a lot of control over post-harvest events. For an efficient mycotoxin management and control program, the entire silage production chain must be integrated and personalized for each farm. Field crop growing, silage manufacturing, and feeding out all need to be considered. Having an inexpensive, simple silage test is essential for quickly diagnosing the silage quality so that "in-field decision-making" may be done when determining if a given forage is acceptable to be used as animal feed. Routine sampling and testing of silage allow for the detection of any changes in mycotoxin contamination. There will be new issues in the future as more people turn to rapid approaches in the fieldwork. Managing mycotoxin-contaminated silage must also include a correct nutritional intervention (Dell'Orto et al., 2015). Nuts are contaminated throughout the supply chain by several *Penicillium* species from harvest to storage. Production of toxic substances by *Penicillium* species, such as patulin, penitrem A, and roquefortine C, is currently unregulated by European regulations, as well as cyclopenase inhibitors like cyclopenin and cyclopenol and a toxin called chaetoglobosin A. For the careful handling of fungal growth and mycotoxin contamination, it is consequently required to design suitable harvesting methods, processing procedures, storage conditions, and detoxification procedures (Spadaro et al., 2021), as shown in Table 8.2.

The most practical and cost-effective strategy for reducing mycotoxin risk is to avoid contamination by mycotoxigenic fungi. When prevention is no longer an option, mycotoxin detoxification may be considered (Spadaro et al., 2017). Acute and chronic mycotoxicoses are poisonings caused by mycotoxins that

TABLE 8.2   ALTERNATIVE TECHNOLOGIES FOR MANAGEMENT AND CONTROL OF MYCOTOXINS

Treatment	Source	Claims	References
Modified atmosphere 20–40% CO_2 + 1–5% O_2	Cheese	Reduction of fungal growth and mycotoxin production by 20–80%	Kabak et al., 2006
Modified atmosphere 40–60% CO_2 + N_2 + < 0.5% O_2	Growth media like potato dextrose agar + Czapek yeast extract agar	Growth of *P. roqueforti* was observed in 20% CO_2 but not in 40% and 60% CO_2	Taniwaki et al., 2009
8% of NaCl	Cheese	Significant decrease in ROQC	Fontaine et al., 2015b
Bacillus velezensis NRRL B-23189	Silage	Antagonistic effects against *Penicillium roqueforti* and *Penicillium paneum*	Wambacq et al., 2018
Bacillus subtilis YM 10–20	Plate count agar (PCA) media	Iturin-like compound permeation through fungal spores and inhibition of germination	Chitarra et al., 2003

can be lethal. Mycotoxins have been associated with various human and animal illnesses, including turkey's X disease, alimentary toxic aleukia, and yellow rain. Mycotoxin formation in foods can be influenced by temperature, food substrate, mold strain, and other environmental factors. To inhibit the growth of fungi, eliminate or reduce toxin levels, and degrade or detoxify toxins in meals and feed, physical, chemical, and biological approaches are used. Because degradation and detoxification have limitations, removing mold infestation is the most effective way to reduce mycotoxins in food and dairy products (Sengun et al., 2008). First and foremost, it is critical to understand the conditions that will allow mold to thrive and produce mycotoxins. This knowledge is required for the implementation of control measures that involve manipulating the environment that allows for growth and production of mycotoxins. Water activity, temperature, substrate, mold strain, gas composition, the presence of chemical preservatives, and microbial interactions are all elements that influence mycotoxin development. Different methods that are used to control roquefortine production are as follows.

Physical methods: To avoid and control *Penicillium roquefortine* (PR) toxin contaminations, a variety of conventional approaches or classic hurdle techniques are used alone or in combination (Dubey et al., 2018). Two of the most common approaches are manually sorting mold-infested grains based on their characteristics (mostly physical factors) or using fluorescence technologies. Other methods like extrusion, cleaning, milling, washing, winnowing, heat treatment, adsorption, solar, UV, X-rays, or microwave irradiation all work to remove PR toxins, but anaerobiosis and toxin extraction are the most effective. During most of these physical processes, PR toxin is somewhat stable and survives in final meals and silage. Heat and ultraviolet light have been reported to be effective ways of reducing PR toxin concentration in silage and cereal grains. However, because fungal infections and mycotoxin production can appear at any stage during the growing process, from pre-harvest throughout the field to harvest handling of food processing, complete eradication seems to be impossible. Physical methodologies include:

Sorting: As the major chances of mycotoxin contamination can be present in seeds or kernels, sorting and segregating damaged, discolored crops with obvious mold development can result in considerable amounts of mycotoxins being removed from these crops. Crop separation can be accomplished using manual, mechanical, or electronic methods. Manual selection is the simplest method for extracting tarnished grains, although it is time consuming, limiting its applicability in many situations. Fluorescence sorting of maize, cottonseed, and dried figs can partially remove mycotoxin, with contamination visible as bright greenish-yellow fluorescence after illumination with UV light, with a positive correlation reported between the observation of bright greenish-yellow fluorescence under long-wave (365 nm) UV light and the observation bright greenish-yellow fluorescence under long-wave (365 nm) UV light (Spadaro et al., 2017). Sorting has been demonstrated to be an effective post-harvest approach for preventing mycotoxin contamination. While electronic sorting systems are useful for removing infected seeds, they are unlikely to be deployed on a wide scale in the industrial sector due to cost concerns.

Density Segregation: Flotation and density segregation, which has been shown to mitigate mycotoxin levels in farms, is another effective technique for lowering mycotoxin levels. Because mold-damaged, mycotoxin-contaminated kernels differ in physical properties from unaffected kernels, they can be distinguished via density segregation through specialized liquids as well as fractionation using specific gravity tables.

Adsorption: The removal of mycotoxins from animal feeds is addressed as a tough problem. Montmorillonite (Mt) is a naturally occurring clay mineral that is currently gaining attention as a promising adsorbent of mycotoxins. It is popularly being used because of its low cost, high efficiency, and eco-friendly nature (Li et al., 2018). The major component of bentonite is Mt. Mt and its modified forms (raw Mt, Mt modified with organic molecules, pillared Mt, thermal-modified Mt) can be used in the adsorption of mycotoxins from feeds. Graphene oxide (GO) functionalized with amphiphilic didodecyl dimethyl ammonium bromide (DDAB) molecules can effectively adsorb mycotoxins. Graphene has a large surface area and a high capacity for binding mycotoxins. Organo-montmorillonites (OMts) treated with monomeric and di-alkyl positively charged surfactants can remove both polar and weakly polar hydrophobic mycotoxin simultaneously. Modified montmorillonites (altered by lauramidopropyl betaine and dodecyl dimethyl betaine) are two types of zwitterionic surfactants that improve mycotoxin detoxification efficacy. Organo-rectorites with a combination of different quaternary ammonium salts also aid in detoxification (Rogowska et al., 2019). Nanocomposites (diameters less than 15 μm) composed of aluminum oxide, activated carbon, and bentonite had the capability of removing nearly 87% of mycotoxins with a 450 g/g adsorption efficiency. Biopolymers, activated carbon, or graphene oxide spheres with a diameter of less than 3 mm, on the other hand, eliminated up to 70% of mycotoxins (adsorption of 598 ng/g). Beer detoxification was evaluated, and a circular structure made of alginate and activated carbon or pectin showed the potential to remove toxins from the beverage. As a result, this technology might be beneficial to the food business. Mycotoxins can also be reduced by using activated charcoal at a concentration of 3–5 g/L for 5 minutes (Gonzalez-Jartin et al., 2019).

Extraction with Solvents: Using several solvents, mycotoxins can be removed from contaminated food items such as oilseed peanuts and cottonseed. The most popularly utilized solvents include methanol-water, 95% ethanol, hexane-ethanol-water, 90% aqueous acetone, hexane-methanol, aqueous isopropanol, 80% isopropanol, acetonitrile-water, and acetone-hexane-water.

Whilst still solvent extraction can effectively remove mycotoxin from oilseed meals without the formation of hazardous residues or a loss of nutritive value or quality, its ubiquitous use has been inhibited by its significant expense and concerns about toxic extract disposal (Yang et al., 2020). In cottonseed and peanut meal, extraction with 80% isopropanol eliminated mycotoxins, but there was a loss of meal solids ranging from 8.7 to 9.5% (Kabak et al., 2006)

Dust removal is effective in the removal of mycotoxins. Per the reports of previous findings, mycotoxins are commonly deposited in grain, and consumption can be hazardous to animals and people. Segregation of mycotoxins into fine fractions that are removed from the raw materials or end products can be accomplished with efficient dust collection and separation technologies (Čolović et al., 2019).

In the feed mill, various dust separation methods could be used, as cold plasma, microwave heating, flotation, gamma radiation, and other physical therapies have all been demonstrated to be successful in mycotoxin decontamination. They are not widely employed in the feed business for a variety of reasons, including expensive investment costs, high operating expenses, and the lack of large-scale processing facilities required by feed manufacturers.

Storage conditions have an impact on overall fungal development and activity; hence they play an important role in mycotoxin control. The main storage factors that encourage fungi proliferation and mycotoxin secretions are high humidity and heating rate. Fungal growth and mycotoxin accumulation are reduced under controlled storage settings, which include optimum air humidity, ventilation, temperature management, and packing methods.

Thermal treatments: Even though a majority of mycotoxins are thermostable and can withstand temperatures ranging from 80–120°C, they are destroyed little or not at all by common cooking methods such as boiling and frying, or even after pasteurization. The following thermal processing factors are particularly important for the breakdown and reduction of mycotoxins in food and feed: pH, temperature, degree of heat penetration, initial mycotoxin concentration, time of exposure to high temperature, moisture content, and so on. Extrusion, frying, roasting, baking, canning, cooking, crumbling, flaking, pelleting, alkaline cooking, and other thermal food and feed treatment techniques can all have various effects on mycotoxins. The use of high-temperature techniques showed the highest promise for reducing mycotoxins among thermal treatments.

The use of non-thermal procedures has been observed because of the advantages such as short time requirements, low costs, and low food matrix–adverse effects. The newest and most promising technologies for reducing mycotoxins in grains include electron beam irradiation, cold plasma, and pulsed light (PL). Furthermore, the effects on food and plant quality may be minimal because the current procedures are non-thermal (Yousefi et al., 2021). Furthermore, the use of these strategies in combination could result in the safeguarding of food and feed.

8.7.1 Chemical Methods

Chemical procedures have long been considered the most effective way of inactivating or detoxifying PR poison from tainted goods. Ammoniation, ozonation, formaldehyde, deamination, oxidizing/reducing agents, acid/base treatment, chlorinated agents, and other chemical techniques have all been employed to degrade the PR toxin in infested commodities. Inorganic compounds, such as sulfite, ammonium sulfate, and nitrate; weak organic acids, such as lactic, acetic, sorbic, propionic, and benzoic acid; and antioxidants, such as beta-carotene, selenium, vitamins A and E, and ascorbic acid are the most widely used chemicals. Today, a variety of chemical agents for mycotoxin decontamination are available, including alkalines (sodium hydroxide, ammonia gas, calcium hydroxide) and acids (phosphoric acid, propionic acid, acetic acid, sorbic acid, sodium hypochlorite, formic acid). For the detoxification of mycotoxin-contaminated raw feed and compound feed, oxidizing chemicals like ozone and hydrogen peroxide are utilized. In addition to feed and compound feed, the utilization of ozone for the disinfection of food products has been considered with great success throughout the years. Many of the compounds listed previously could be used to effectively reduce mycotoxin levels in feed and complex foods, but they may also cause changes in the physical, nutritional, and sensory qualities of treated materials, which should be taken into account (Spadaro et al., 2017; Čolović et al., 2019). The proper application of these chemical preservatives throughout the pre-harvesting and packaging processes may help to prevent fungal infection and, as a result, mycotoxin contamination in a variety of food and feed products. Nanotechnology has made significant progress, especially in the field of encapsulation and its applications in a variety of food preservation fields, such as mycotoxin detoxification via adsorption methods or photocatalytic destruction (Prakash et al., 2018).

8.7.2 Biological Methods of Decontamination or Detoxification

Penicillium roqueforti infestations are common in a broad spectrum of consumable food, where they may create substantial degradation, resulting in massive financial losses. This fungus induces food contamination by causing off flavors, off odors, discoloration, degradation, and functional disorganization. The hazardous metabolites of PR toxins are more common, posing a greater risk to animal and human health. Thus, while preventing fungal growth and toxin production is the most effective method for limiting mold's damaging consequences, inactivation/detoxification of contaminated goods is also critical. Surprisingly, certain strains have developed a unique capacity to resist some chemical treatments and preservatives in recent years. A judicious consideration of LAB strains could halt *P. roqueforti* from growing and in turn help in reducing or eliminating the health-related concerns aligned with mycotoxin exposure in previously contaminated items. *P. roqueforti* was suppressed by practically all LAB strains at a percentage of more than 65.52%, with *Lactobacillus plantarum* C21–41 being the most effective. *L. plantarum* water/salt soluble extract demonstrated antifungal efficacy against *P. roqueforti* CCFM259. The growth of *P. roqueforti* IBT18687 can be firmly hampered by the culture filtrate of *L. plantarum* cultivated on wheat flour hydrolysate. For its probiotic and food-preservation capabilities, a strain of *L. plantarum* TK9 was isolated from the naturally fermented congee of China. Against *P. roqueforti* and other closely related food-rotting organisms, the strain displayed an extensive antifungal spectrum. *B. subtilis* produces a polypeptide that has antifungal action against *P. roqueforti*. Organic extracts from *B. amyloliquefaciens* VJ-1 had antifungal activities that were effective against *P. roqueforti*. *Bacillus* strains produced 3-phenyl lactic acid, which has potent antifungal properties against a variety of molds, including *P. roqueforti*. *B. velezensis* was reported to have strong antagonistic action against *P. roquefortisensulato* (s.l.), the most common fungal contamination in silage. A chemical obtained from *Streptomyces halstedii* K139 inhibits the antifungal function of *P. roqueforti* (Dubey et al., 2018). Yeast species such as *Metschnikowia pulcherrima* and *Wicker hamomycesanomalous* (*Pichia anomala*) can potentially suspend the development of *Penicillium roqueforti* and thus can help in retarding the toxin-producing capacity of the microorganisms (Papp et al., 2021). This yeast's antifungal activity is strongly dependent on initial spore concentration, which is critical for efficient biocontrol.

Mycotoxins can be detoxified via biological conversion into less toxic compounds. Mycotoxins can be detoxified through metabolism or breakdown in the gastrointestinal tract by using microbes or enzyme systems for the treatment of infested commodities. This method of detoxification is irreversible and environmentally beneficial, as it leaves no poisonous residues or undesirable by-products. Some binders can also be used to detoxify roquefortine C present in the rumen. Such binders may include clay minerals and yeast derivatives. These binders can help in reducing roquefortine up to 11%. Some other binders such as bentonite, leonardite, plant extracts, epoxidase, and sepiolite are not found to be efficient for roquefortine detoxification but can detoxify other mycotoxins such as DON, NIV, ZEN, and aflatoxin (Debevere et al., 2020). Detoxification of mycotoxins is usually accomplished by removing or eliminating the contaminated commodities or by physically, chemically, or biologically inactivating the toxins inherent in these commodities (Haque et al., 2020).

8.8 CONCLUSION

Roquefortines can extensively infect varieties of foods and feeds, including cheese; fruits; nuts; silage; and vegetables such as onion, chili, and spices. Numerous studies have shown that roquefortine exposure has negatively impacted human and animal health as well. To prevent roquefortine infestations, a variety of physical, chemical, biological, and genetically modified measures must be used. TLC, LC-PDA-FLD-MS, LC-MS/MS, HPLC, and GC-MS, among other advanced analytical procedures, have been used for the detection of roquefortines, with varying abilities in terms of detection limits. Along with these developments, additional colorimetric, fluorometric, electrochemical, and refractometric methods of defining and detecting roquefortines in food and feed could be used. Nowadays, various advanced strategies like immunoassays, chromatography, and biosensors are used for their proper detection, classification, and estimation. Along with this, controlled atmospheric conditions like high carbon dioxide coupled with low oxygen concentration and several biocontrol measures are considered suitable measures to inhibit or mask the biosynthetic pathways of roquefortine production, ensuring a safe and secure environment for foodstuffs around the world.

REFERENCES

Ali, H., Ries, M. I., Nijland, J. G., Lankhorst, P. P., Hankemeier, T., Bovenberg, R. A., . . . & Driessen, A. J. (2013). A branched biosynthetic pathway is involved in the production of roquefortines and related compounds in *Penicillium chrysogenum*. *PLoS ONE*, *8*(6), e65328.

Bele, A. A., & Khale, A. (2011). An overview of thin-layer chromatography. *International Journal of Pharmaceutical Sciences and Research*, *2*(2), 256.

Betancourt, S. D. P., Carranza, B. V., & Manzano, E. P. (2015). Estimation of mycotoxin multiple contaminations in Mexican hybrid seed maize by HPLC-MS/MS. *Agricultural Sciences*, *6*(9), 1089.

Cakmakci, S., Gurses, M., Hayaloglu, A. A., Cetin, B., Sekerci, P., & Dagdemir, E. (2015). Mycotoxin production capability of *Penicillium roqueforti* in strains isolated from mold-ripened traditional Turkish civil cheese. *Food Additives and Contaminants: Part A*, *32*(2), 245–249.

Chitarra, G. S., Breeuwer, P., Nout, M. J. R., Van Aelst, A. C., Rombouts, F. M., & Abee, T. (2003). An antifungal compound produced by *Bacillus subtilis* YM 10–20 inhibits the germination of *Penicillium roqueforti* conidiospores. *Journal of Applied Microbiology*, *94*(2), 159–166.

Čolović, R., Puvača, N., Cheli, F., Avantaggiato, G., Greco, D., Đuragić, O., . . . & Pinotti, L. (2019). Decontamination of mycotoxin-contaminated feedstuffs and compound feed. *Toxins*, *11*(11), 617.

Coton, E., Coton, M., Hymery, N., Mounier, J., & Jany, J. L. (2020). *Penicillium roqueforti*: An overview of its genetics, physiology, metabolism, and biotechnological applications. *Fungal Biology Reviews*, *34*(2), 59–73.

Debevere, S., Schatzmayr, D., Reisinger, N., Aleschko, M., Haesaert, G., Rychlik, M., . . . & Fievez, V. (2020). Evaluation of the efficacy of mycotoxin modifiers and mycotoxin binders by using an in vitro rumen model as a first screening tool. *Toxins*, *12*(6), 405.

Dell'Orto, V., Baldi, G., & Cheli, F. (2015). Mycotoxins in silage: Checkpoints for effective management and control. *World Mycotoxin Journal*, *8*(5), 603–617.

Demjanová, S., Jevinová, P., Pipová, M., & Regecová, I. (2021). Identification of *Penicillium verrucosum*, *Penicillium commune*, and *Penicillium crustosum* isolated from chicken eggs. *Processes*, *9*(1), 53.

Driehuis, F. (2011). Occurrence of mycotoxins in silage. In *Proceedings of the II International Symposium*.

Driehuis, F., Spanjer, M. C., Scholten, J. M., & Te Giffel, M. C. (2008). Occurrence of mycotoxins in maize, grass, and wheat silage for dairy cattle in the Netherlands. *Food Additives and Contaminants*, *1*(1), 41–50.

Dubey, M. K., Aamir, M., Kaushik, M. S., Khare, S., Meena, M., Singh, S., & Upadhyay, R. S. (2018). PR toxin–biosynthesis, genetic regulation, toxicological potential, prevention and control measures: Overview and challenges. *Frontiers in Pharmacology*, *9*, 288.

Engvall, E., & Perlmann, P. (1971). Enzyme-linked immunosorbent assay (ELISA) quantitative assay of immunoglobulin G. *Immunochemistry*, *8*(9), 871–874.

Erb, M., & Kliebenstein, D. J. (2020). Plant secondary metabolites as defenses, regulators, and primary metabolites: The blurred functional trichotomy. *Plant Physiology*, *184*(1), 39–52.

Erdogan, A., & Sert, S. (2004). Mycotoxin-forming ability of two *Penicillium roqueforti* strains in blue moldy Tulum cheese ripened at various temperatures. *Journal of Food Protection*, *67*(3), 533–535.

Finoli, C., Vecchio, A., Galli, A., & Dragoni, I. (2001). Roquefortines C occurrence in blue cheese. *Journal of Food Protection*, *64*(2), 246–251.

Fontaine, K., Hymery, N., Lacroix, M. Z., Puel, S., Puel, O., Rigalma, K., . . . & Mounier, J. (2015a). Influence of intraspecific variability and abiotic factors on mycotoxin production in *Penicillium roqueforti*. *International Journal of Food Microbiology*, *215*, 187–193.

Fontaine, K., Passeró, E., Vallone, L., Hymery, N., Coton, M., Jany, J. L., . . . & Coton, E. (2015b). Occurrence of roquefortines C, mycophenolic acid, and aflatoxin M1 mycotoxins in blue-veined cheeses. *Food Control*, *47*, 634–640.

Gallo, A., Giuberti, G., Frisvad, J. C., Bertuzzi, T., & Nielsen, K. F. (2015). Review on mycotoxin issues in ruminants: Occurrence in forages, effects of mycotoxin ingestion on health status and animal performance, and practical strategies to counteract their negative effects. *Toxins*, *7*(8), 3057–3111.

García-Estrada, C., & Martín, J. F. (2016). Biosynthetic gene clusters for relevant secondary metabolites produced by *Penicillium roqueforti* in blue cheeses. *Applied Microbiology and Biotechnology*, *100*(19), 8303–8313.

García-Estrada, C., Ullán, R. V., Albillos, S. M., Fernández-Bodega, M. Á., Durek, P., von Döhren, H., & Martín, J. F. (2011). A single cluster of coregulated genes encodes the biosynthesis of the mycotoxins roquefortines C and meleagrin in *Penicillium chrysogenum*. *Chemistry and Biology*, *18*(11), 1499–1512.

Gober, C. M. (2017). *Integration of fermentation and organic synthesis: Studies of roquefortines C and biosynthetic derivatives*.

González-Jartín, J. M., de Castro Alves, L., Alfonso, A., Piñeiro, Y., Vilar, S. Y., Gomez, M. G., ... & Botana, L. M. (2019). Detoxification agents based on magnetic nanostructured particles as a novel strategy for mycotoxin mitigation in food. *Food Chemistry, 294,* 60–66.

Häggblom, P. (1990). Isolation of roquefortines C from feed grain. *Applied and Environmental Microbiology, 56*(9), 2924–2926.

Haque, M. A., Wang, Y., Shen, Z., Li, X., Saleemi, M. K., & He, C. (2020). Mycotoxin contamination and control strategy in human, domestic animal and poultry: A review. *Microbial Pathogenesis, 142,* 104095.

Hymery, N., Vasseur, V., Coton, M., Mounier, J., Jany, J. L., Barbier, G., & Coton, E. (2014). Filamentous fungi and mycotoxins in cheese: A review. *Comprehensive Reviews in Food Science and Food Safety, 13*(4), 437–456.

Ismaiel, A. A., & Papenbrock, J. (2014). The effects of patulin from *Penicillium vulpinum* on seedling growth, root tip ultrastructure, and glutathione content of maize. *European Journal of Plant Pathology, 139*(3), 497–509.

Jouany, J. P. (2007). Methods for preventing, decontaminating, and minimizing the toxicity of mycotoxins in feeds. *Animal Feed Science and Technology, 137*(3–4), 342–362.

Kabak, B., Dobson, A. D., & Var, I. I. L. (2006). Strategies to prevent mycotoxin contamination of food and animal feed: A review. *Critical Reviews in Food Science and Nutrition, 46*(8), 593–619.

Karger, B. L. (1997). HPLC: Early and recent perspectives. *Journal of Chemical Education, 74*(1), 45.

Kirtil, H. E., Metin, B., & Arici, M. (2021). Identification of filamentous fungi in Turkish mold-ripened cheeses and screening of mycotoxin genes of *Penicillium roqueforti* isolates. *Journal of Microbiology, Biotechnology, and Food Sciences, 10*(4), 657–662.

Kokkonen, M., Jestoi, M., & Rizzo, A. (2005). Determination of selected mycotoxins in mold cheeses with liquid chromatography coupled in tandem with mass spectrometry. *Food Additives and Contaminants, 22*(5), 449–456.

Kosalková, K., Domínguez-Santos, R., Coton, M., Coton, E., García-Estrada, C., Liras, P., & Martín, J. F. (2015). A natural short pathway synthesizes roquefortines C but not meleagrin in three different *Penicillium roqueforti* strains. *Applied Microbiology and Biotechnology, 99*(18), 7601–7612.

Kosalková, K., García-Estrada, C., Ullán, R. V., Godio, R. P., Feltrer, R., Teijeira, F., ... & Martín, J. F. (2009). The global regulator LaeA controls penicillin biosynthesis, pigmentation, and sporulation, but not roquefortines C synthesis in *Penicillium chrysogenum*. *Biochimie, 91*(2), 214–225.

Li, S. M. (2009). Evolution of aromatic prenyltransferases in the biosynthesis of indole derivatives. *Phytochemistry, 70*(15–16), 1746–1757.

Li, Y., Tian, G., Dong, G., Bai, S., Han, X., Liang, J., ... & Zhang, H. (2018). Research progress on the raw and modified montmorillonites as adsorbents for mycotoxins: A review. *Applied Clay Science, 163,* 299–311.

Mansfield, M. A., Jones, A. D., & Kuldau, G. A. (2008). Contamination of fresh and ensiled maize by multiple *Penicillium* mycotoxins. *Phytopathology, 98*(3), 330–336.

Maragos, C. M. (2020). Development and characterization of a monoclonal antibody to detect the mycotoxin roquefortines C. *Food Additives and Contaminants: Part A, 37*(10), 1777–1790.

Maragos, C. M. (2021). Roquefortines C in blue-veined and soft-ripened cheeses in the USA. *Food Additives and Contaminants: Part B,* 1–9.

Mezcua, M., Agüera, A., Lliberia, J. L., Cortés, M. A., Bagó, B., & Fernández-Alba, A. R. (2006). Application of ultra-performance liquid chromatography-tandem mass spectrometry to the analysis of priority pesticides in groundwater. *Journal of Chromatography A, 1109*(2), 222–227.

Mohapatra, D., Kumar, S., Kotwaliwale, N., & Singh, K. K. (2017). Critical factors are responsible for fungi growth in stored food grains and non-chemical approaches for their control. *Industrial Crops and Products, 108*, 162–182.

Motloung, L., De Saeger, S., De Boevre, M., Detavernier, C., Audenaert, K., Adebo, O. A., & Njobeh, P. B. (2018). Study on mycotoxin contamination in South African food spices. *World Mycotoxin Journal, 11*(3), 401–409.

Nielsen, K. F., Sumarah, M. W., Frisvad, J. C., & Miller, J. D. (2006). Production of metabolites from the *Penicillium roqueforti* complex. *Journal of Agricultural and Food Chemistry, 54*(10), 3756–3763.

Niu, S., Wang, N., Xie, C. L., Fan, Z., Luo, Z., Chen, H. F., & Yang, X. W. (2018). Roquefortines J, a novel roquefortines alkaloid, from the deep-sea-derived fungus *Penicillium granulatum* MCCC 3A00475. *The Journal of Antibiotics, 71*(7), 658–661.

O'Brien, M., Nielsen, K. F., O'Kiely, P., Forristal, P. D., Fuller, H. T., & Frisvad, J. C. (2006). Mycotoxins and other secondary metabolites produced in vitro by *Penicillium paneum* Frisvad and *Penicillium roqueforti* Thom were isolated from baled grass silage in Ireland. *Journal of Agricultural and Food Chemistry, 54*(24), 9268–9276.

Ogunade, I. M., Martinez-Tuppia, C., Queiroz, O. C. M., Jiang, Y., Drouin, P., Wu, F., . . . & Adesogan, A. T. (2018). Silage review: Mycotoxins in silage: Occurrence, effects, prevention, and mitigation. *Journal of Dairy Science, 101*(5), 4034–4059.

Ohmomo, S., Kitamoto, H. K., & Nakajima, T. (1994). Detection of roquefortiness in *Penicillium roqueforti* isolated from molded maize silage. *Journal of the Science of Food and Agriculture, 64*(2), 211–215.

Overy, D. P., Nielsen, K. F., & Smedsgaard, J. (2005). Roquefortines/oxaline biosynthesis pathway metabolites in *Penicillium* ser. Corymbifera: In planta production and implications for competitive fitness. *Journal of Chemical Ecology, 31*(10), 2373–2390.

Papp, L. A., Horváth, E., Peles, F., Pócsi, I., & Miklós, I. (2021). Insight into yeast–mycotoxin relations. *Agriculture, 11*(12), 1291.

Prakash, B., Kujur, A., Yadav, A., Kumar, A., Singh, P. P., & Dubey, N. K. (2018). Nanoencapsulation: An efficient technology to boost the antimicrobial potential of plant essential oils in the food system. *Food Control, 89*, 1–11.

Prencipe, S., Siciliano, I., Gatti, C., Garibaldi, A., Gullino, M. L., Botta, R., & Spadaro, D. (2018). Several species of *Penicillium* isolated from chestnut flour processing are pathogenic on fresh chestnuts and produce mycotoxins. *Food Microbiology, 76*, 396–404.

Rathnayake, A. U., Saravanakumar, K., Abuine, R., Abeywickrema, S., Kathiresan, K., MubarakAli, D., & Wang, M. H. (2020). Fungal genes encoding enzymes used in cheese production and fermentation industries. In *Fungal Biotechnology and Bioengineering* (pp. 305–329). Springer, Cham.

Reddy, P., Guthridge, K., Vassiliadis, S., Hemsworth, J., Hettiarachchige, I., Spangenberg, G., & Rochfort, S. (2019). Tremorgenic mycotoxins: Structure diversity and biological activity. *Toxins, 11*(5), 302.

Ries, M. I., Ali, H., Lankhorst, P. P., Hankemeier, T., Bovenberg, R. A., Driessen, A. J., & Vreeken, R. J. (2013). Novel key metabolites reveal further branching of the roquefortines/meleagrin biosynthetic pathway. *Journal of Biological Chemistry, 288*(52), 37289–37295.

Rogowska, A., Pomastowski, P., Sagandykova, G., & Buszewski, B. (2019). Zearalenone and its metabolites: Effect on human health, metabolism and neutralization methods. *Toxicon, 162*, 46–56.

Rundberget, T., Skaar, I., & Flåøyen, A. (2004). The presence of *Penicillium* and *Penicillium* mycotoxins in food wastes. *International Journal of Food Microbiology, 90*(2), 181–188.

Sengun, I., Yaman, D. B., & Gonul, S. (2008). Mycotoxins and mold contamination in cheese: A review. *World Mycotoxin Journal, 1*(3), 291–298.

Spadaro, D., Fontana, M., Prencipe, S., Valente, S., Piombo, E., & Gullino, M. L. (2021). Innovative strategies for the management of *Aspergillus* spp. and *Penicillium* spp. on Nuts. In *Postharvest Pathology* (pp. 111–127). Springer, Cham.

Spadaro, D., Prencipe, S., Siciliano, I., Garibaldi, A., & Gullino, M. L. (2017). Mycotoxigenic fungi and mycotoxins in chestnuts and derivatives. *Protezione delle Colture, 2*, 13–18.

Sulyok, M., Krska, R., & Schuhmacher, R. (2010). Application of an LC-MS/MS-based multi-mycotoxin method for the semi-quantitative determination of mycotoxins occurring in different types of food infected by molds. *Food Chemistry, 119*(1), 408–416.

Szulc, J., Kołodziej, A., & Ruman, T. (2021). Silver-109/silver/gold nanoparticle-enhanced target surface-assisted laser desorption/ionisation mass spectrometry—The new methods for an assessment of mycotoxin concentration on building materials. *Toxins, 13*(1), 45.

Tangni, E. K., Pussemier, L., Bastiaanse, H., Haesaert, G., Foucart, G., & Van Hove, F. (2013). Presence of mycophenolic acid, roquefortines C, citrinin, and ochratoxin A in maize and grass silages supplied to dairy cattle in Belgium. *Journal of Animal Science Advances, 3*(12), 598–612.

Taniwaki, M. H., Hocking, A. D., Pitt, J. I., & Fleet, G. H. (2009). Growth and mycotoxin production by food spoilage fungi under high carbon dioxide and low oxygen atmospheres. *International Journal of Food Microbiology, 132*(2–3), 100–108.

Tiwary, A. K., Puschner, B., & Poppenga, R. H. (2009). Using roquefortines C as a biomarker for penitrem A intoxication. *Journal of Veterinary Diagnostic Investigation, 21*(2), 237–239.

Vallone, L., Giardini, A., & Soncini, G. (2014). Secondary metabolites from *Penicillium roqueforti*, a starter for the production of Gorgonzola cheese. *Italian Journal of Food Safety, 3*(3).

Wambacq, E. (2017). *Penicillium roqueforti* sl: *Growth and roquefortines C production in silages* (Doctoral dissertation, Ghent University).

Wambacq, E., Audenaert, K., Höfte, M., De Saeger, S., & Haesaert, G. (2018). *Bacillus velezensis* as antagonist towards *Penicillium roqueforti* sl in silage: in vitro and in vivo evaluation. *Journal of Applied Microbiology, 125*(4), 986–996.

Yang, Y., Li, G., Wu, D., Liu, J., Li, X., Luo, P., … & Wu, Y. (2020). Recent advances on toxicity and determination methods of mycotoxins in foodstuffs. *Trends in Food Science & Technology, 96*, 233–252.

Yousefi, M., Mohammadi, M. A., Khajavi, M. Z., Ehsani, A., & Scholtz, V. (2021). Application of novel non-thermal physical technologies to degrade mycotoxins. *Journal of Fungi*, *7*(5), 395.

FURTHER READINGS

Atungulu, G. G., Mohammadi-Shad, Z., & Wilson, S. (2018). Mycotoxin issues in pet food. *Food and Feed Safety Systems and Analysis*, 25–44.

Habschied, K., Krstanović, V., Zdunić, Z., Babić, J., Mastanjević, K., & Šarić, G. K. (2021). Mycotoxins biocontrol methods for healthier crops and stored products. *Journal of Fungi*, *7*(5), 348.

Leung, M. C., Díaz-Llano, G., & Smith, T. K. (2006). Mycotoxins in pet food: A review on worldwide prevalence and preventative strategies. *Journal of Agricultural and Food Chemistry*, *54*(26), 9623–9635.

Mao, J., Zhou, Y., Lv, G., & Zhou, R. (2022). Simultaneous detoxification of aflatoxin B1, zearalenone, and deoxynivalenol by modified montmorillonites. *Molecules*, *27*(1), 315.

Ricke, S. C., Atungulu, G. G., Rainwater, C., & Park, S. H. (Eds.). (2017). *Food and Feed Safety Systems and Analysis*. Academic Press.

Rustom, I. Y. (1997). Aflatoxin in food and feed: Occurrence, legislation, and inactivation by physical methods. *Food Chemistry*, *59*(1), 57–67.

Spadaro, D., & Garibaldi, A. (2017). Containment of mycotoxins in the food chain by using decontamination and detoxification techniques. In *Practical Tools for Plant and Food Biosecurity* (pp. 163–177). Springer, Cham.

Chapter 9

PR Toxins

*Concerns in Food and Feed with Their
Detection and Management Strategies*

Mousumi Ghosh, Sreemoyee Chakraborty, Sourav
Misra, Shubhangi Srivastava, Sourabh Bondre, Madhu
Kamle, Pradeep Kumar, Dipendra Kumar Mahato

CONTENTS

9.1   Introduction   240
9.2   Occurrence in Food and Feed   241
9.3   Morphological Characteristics   243
9.4   Sources of Toxins   244
9.5   Gene Responsible for Production   245
9.6   Detection Methods of PR Toxins   245
    9.6.1   Conventional Methods   247
    9.6.2   Novel Methods   248
        9.6.2.1   Reversed-Phase High-Performance Liquid
                Chromatography   248
        9.6.2.2   HPLC-NMR and HPLC-MS   248
        9.6.2.3   Molecular Identification Based on Ribosomal
                DNA Sequence Comparison, Random Amplified
                Polymorphic DNA Profiles Using PCR   248
        9.6.2.4   Immunoassays   249
        9.6.2.5   Biosensor-Based Detection   250
9.7   Management and Control Strategies   251
    9.7.1   Methods of Prevention of PR Toxin Formation   251
    9.7.2   Physical Methods of Control   253
    9.7.3   Chemical Methods of Control   253
    9.7.4   Biological Methods of Control   255
9.8   Decontamination through Food/Feed Processing   257
9.9   Conclusion   258

DOI: 10.1201/9781003242208-9

9.1 INTRODUCTION

The word "mycotoxins" comes from two ancient Greek words, *mykes* and *toxon*, meaning "mold" and "poisonous arrow," respectively. They are basically the secondary metabolites of filamentous fungi that can cause dangerous effects in animals and humans even in low concentrations. However, the molds (both genus and species level) considered for spoilage of food generally and also dairy products are divergent in nature, but they are popular for application for various industrial purposes and different food and feed product processing. Application of *Penicillium* spp. in dairy products includes varieties of cheese (blue-veined, mold-ripened), butter, and yogurt. Among various toxigenic species of *Penicillium*, the variant that is part of the *Penicillium roqueforti* complex consists four different species, *Penicillium roqueforti, P. psychrosexualis, P. paneum,* and *P. carneum* (Frisvad et al., 2004; Houbraken and Samson, 2011). It is reported that *P. roqueforti* has been found to be the most dominant of the four. Favorable atmospheric conditions for the growth of these fungi are in foodstuff/silage under low oxygen concentrations, comparatively acidic pH, and low temperature conditions (Wambacq et al., 2018; Driehuis, 2013; Storm et al., 2010). The PR mycotoxins associated with this fungal species can cause acute and chronic toxicity. Several studies reveal that the PR toxin even possesses carcinogenic, mutagenic, teratogenic, and immunotoxic properties. Currently, greater public interest in maintaining a healthy lifestyle and the growing demand for safe food products with superior quality standards emphasize food quality and safety. Some molds that are known to have low toxic potential are commonly used for the preparation of special kind of soft molded cheeses to achieve different sensory characteristics, such as ripened cheeses. However, mold growth on food surfaces is not acceptable and suggests microbial contamination. A few examples of cheeses ripened by mold are Roquefort, gammelost, gorgonzola, and Danish blue, which are prepared by using mainly *P. roqueforti* and *P. camemberti*. The reason behind using *P. roqueforti* for the ripening of this variety of cheeses is that it helps to develop the desired characteristic texture: blue-green spots, flavor, and aroma. In addition to these, it also protects the cheeses against unwanted microbial contamination.

Penicillium roqueforti is a bluish-green–colored spore former frequently found in grain silage and various food matrices stored at low temperatures, such as dairy cheese (shredded), raw meats, wheat, and other cereal products (Hymery et al., 2017; Martín and Coton, 2017). The environment required for the growth of *P. roqueforti* is considerably acidic, with low oxygen and high carbon dioxide levels, which follows microaerophilic and in some special cases low temperature conditions (Hymery et al., 2017). In many reports, it is found that *P. roqueforti* strains are tolerant to weak preservatives, such as 0.5% acetic acid and 9000 ppm. This fungus can also grow well at low salt concentrations

(1% NaCl). The reason behind this fungus being used as seed culture is due to its mechanisms for the maturation of all varieties of blue cheeses by the action of enzymatic reactions (proteolytic and lipolytic) upon hexoses and pentose sugar for ripening and characteristic flavor production. During the production/ripening state of blue cheese, the salt concentration gradient evolves from the core to the surface, which creates a favorable environment to reach equilibrium conditions slowly. Therefore, it promotes the propagation of *P. roqueforti* across other *Penicillium* spp. Studies show that many *Penicillium roqueforti* strains, secluded from various sources of commercially produced cheeses (especially blue cheese), mold contaminated grains and nuts are capable of producing mycotoxins like isofumigaclavin C, patulin, roquefortine, PR toxin, penicillic acid, and botryodiploidin under controlled laboratory testing conditions (Jong and Gantt, 1987).

Very limited or scattered literature available in the context of PR toxin emphasizes the need for comprehensive and up-to-date scientific information, including occurrence, source, morphological characteristics, genetics, diversity, toxicological aspects, detection methodology, and control strategies to maintain food quality and safety. Therefore, this chapter focuses on a thorough review to help researchers and health-conscious consumers gather critical information about PR toxin.

9.2 OCCURRENCE IN FOOD AND FEED

The presence and contamination of PR toxin (Figure 9.1), a bicyclic eremophilane sesquiterpene (7-acetoxy-5,6-epoxy-3,5,6,7,8,8a-hexahydro-carboxaldehyde) mostly found in cereal, corn, forages/grass silage, and dairy cheeses, have been well documented (Table 9.1) (Scott, 1981; O'Brien et al., 2006; Fernández-Bodega et al., 2009; Rasmussen et al., 2010). But the contamination of PR toxin in blue varieties is less common due to its degradation or instability. Most clinical studies have proven that PR toxin has greater toxicokinetic properties, but,

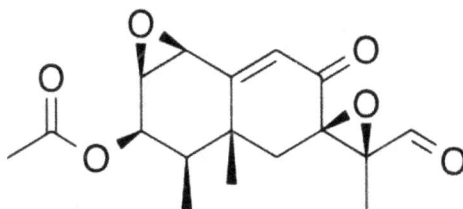

Figure 9.1   Structure of PR toxin.

TABLE 9.1   TOXINS PRODUCED BY *P. ROQUEFORTI* FOUND IN DIFFERENT FOODS

Food Matrix	Secondary Metabolites	Amount	Detection Method
Grass silage	Roquefortine C Myophenolic acid Andrastin A Roquefortines A, B, and D Festuclavine Marcfortine A Agroclavine PR toxin Patulin	Up to 20 mg/kg 0.1–5 mg/kg Not found	HPLC and HPLC-MS
Maize silage	Alternariol monomethyl ether Andrastin A Alternariol Citreoisocoumarin Deoxynivalenol Enniatin B Fumigaclavine A Gliotoxin Marcfortine A and B Mycophenolic acid Nivalenol Roquefortines A and C Zearalenone	1–739 µg/kg	LC-MS/MS
Dairy cheese	Patulin Penicillic acid Citrinin Roquefortine D Macfortine A Eremofortines A, B, C, and D Ergosterol	Not found	Reverse-phased HPLC
	Roquefortine A	Up to 4.7 µg/g	
	Roquefortine C	Up to 6.8 µg/g	
	PR toxin	Not detected (due to unstable nature)	
	PR imine	19–42 µg/kg	
	Roquefortine B Festuclavine	Traces found	
	Mycophenolic acid	Up to 14.3 µg/g	

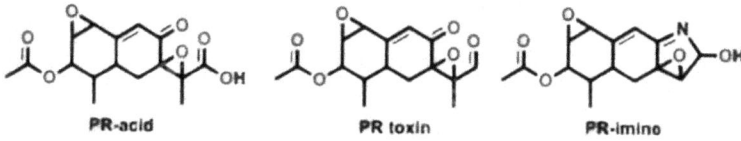

Figure 9.2   Structure of various derivatives of PR toxin.

fortunately, it is also found to be unsteady and is mostly transformed into less poisonous derivatives such as PR imine or expected to be degenerated into PR amide and/or acid based on the growth conditions (Figure 9.2) (Hymery et al., 2014). Furthermore, PR toxin production is not the same in quantity and has been shown to be highly dependent on strain.

9.3 MORPHOLOGICAL CHARACTERISTICS

Penicillium roqueforti is a fungus that grows rapidly. It is distinguished by smooth bright greenish-color colonies with decent to massive population and either chromatic or greyish turquoise colony borders. The underside of the colony is flat with a pale brown or purely black color. A specialized hyphal branch that originated from the subsurface of the fungi that produces conidia called conidiophores bears special distinctive stripes and dense brush-like spores (penicillii) at their extreme end point. Also, the conidiophores are characterized by their branched structure: either three- (terverticillate) or multi-staged (quaterverticillate) and a spore-forming system with rough walls. Depending upon structure, morphological characteristics, molecular level information, and secretion of the type of metabolite (secondary), *P. roqueforti* can be categorized into:

1. *P. carneum*, very commonly found in meat, dairy items (cheese), and bakery products (bread) and the toxins found are mostly patulin, penitrem A, and mycophenolic acid (MPA).
2. *Penicillium paneum*, spread abundantly in bakery products (bread) and forage, brings patulin and botryodiplodin toxins.
3. *P. roqueforti*, perhaps derived mostly in processed foods that are cereal based and also silage. Toxins produced by *P. roqueforti* include PR toxin, marcofortines, and fumigalclavine.

Fernández-Bodega et al. (2009) and Manish et al. (2018) reported that molecular characterization of different members of the *Penicillium* genera can be done based on 300-bp fragments of a specific primer, and it also helps to identify between *P. roqueforti* and *P. carneum* using PCR.

9.4 SOURCES OF TOXINS

Wei et al. (1975) first isolated and partially characterized PR toxins and other similar toxigenic metabolites. Pedrosa and Griessler (2010) reported production of PR toxins occurs via biosynthetic pathways in the presence of precursors like ^{14}C and ^{13}C (Figure 9.3). Eremofortins (Ere) and PR toxin are similar in nature in terms of their structure. They can be identified because in the case of eremofortins (Ere), the (-OH) group (hydroxyl) at the C-12 position is altered by the aldehyde (-CHO) group in PR toxins. This is considered a crucial factor for biological conversion of Ere to PR toxin. Generally, condensation followed by cyclization of farnesyl-diphosphate leads to the formation of PR toxin. These chemical changes are accelerated by aristolochene synthase (encoded by the *ari1* gene) (Cane et al., 1993; Proctor and Hohn, 1993; Manish et al., 2018). It is proven experimentally that among different toxic metabolites secreted by *P. roqueforti* strains, PR toxin can cause significant damage to different organs and has the potential capacity for mutagenic and carcinogenic changes. PR toxin exposure on inter-intestinal cells (Caco-2 and/or THP-1 cells) has been shown to speed up the inflammatory response and necrosis and initiates a certain wide range of biochemical interactions that ultimately cause toxic reactions. Unfortunately, the tolerance level of PR toxin in the case of various raw materials and food products is still uncontrolled and not legally regulated. Hence, the presence of PR toxins raises a question mark when it presents in grains, grass silage, or other food products and feed in terms of safety issues. Since these toxins cannot be degraded by using conventional cooking

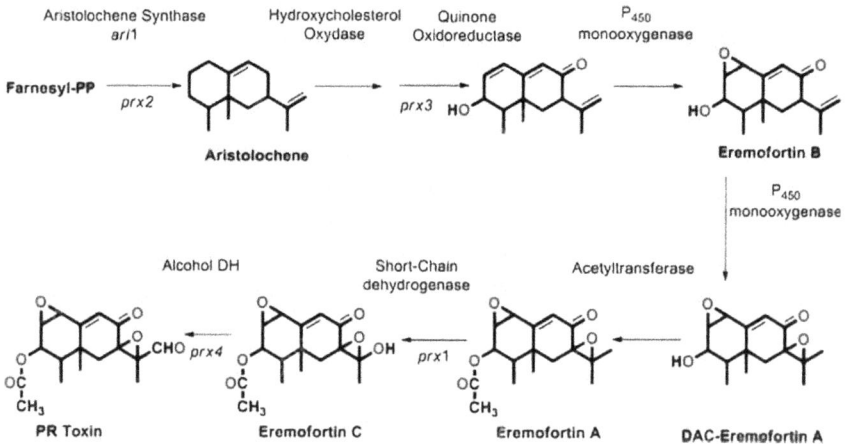

Figure 9.3   PR toxin biosynthesis pathway and its derivatives.

techniques, a potential outbreak could have far-reaching consequences in the health and economic sectors (Storm et al., 2014; Hymery et al., 2017; Kumar et al., 2017; Manish et al., 2018).

9.5 GENE RESPONSIBLE FOR PRODUCTION

In a study done by Geisen et al. (2001), 71 various strains of *P. roqueforti* separated out of different blue cheese seed cultures were identified genetically using the random molecular markers (RAPD) test and displayed a high degree of genetic resemblance. However, considerable structural difference and genetic variations exist among the variants of *P. roqueforti* which help to direct their nomenclature for protected geographical indication (PGI) or protected designation of origin (PDO). The discussion was conducted on the basis of their uses in manufacturing different types of cheese that exhibit a variation of presence in their practicability among the different strains (Ropars et al., 2014; Gillot et al., 2015; Gillot et al., 2017).

The isolates found in grass silage of *P. roqueforti* produce mycophenolic acids as secondary metabolites, the consumption of which may cause health issues for livestock. Environmental conditions favor abundant growth of *P. roqueforti*, such as low oxygen concentration, low pH, and even the presence of organic acids (Pahlow et al., 2003). Therefore, it often acts as a food and feed contaminant for many processed products, including beer, rye bread, other breads, different types of hard cheeses, and olives (Scott et al., 1977; Lafont et al., 1979; Erdogan and Sert, 2004; Moubasher et al., 1979). Incidentally, 90% of the isolates *of P. roqueforti* regularly produce the toxins andrastin A and roquefortine C, but there is a considerable variation in their production of other metabolites like citreoisocoumarin, PR toxin, roquefortine A, and andrastin C (O'Brien et al., 2006). Harmful metabolites such as roquefortine C and mycophenolic acid are routinely isolated from grass silage due to their stability. On the contrary, due to the relatively unstable nature of PR toxin and patulin, their presence in cheese is not proven as harmful for consumers (Figure 9.4).

9.6 DETECTION METHODS OF PR TOXINS

The very first step of detection of PR toxins (PRTs) is detailed sampling coupled with an exhaustive clean-up procedure of the target samples or potentially affected areas (grain silos). This step is often crucial for ensuring reproducibility of results, as faulty sampling techniques cause maximum variation or random errors (~90%). In the case of grain silos, errors are often

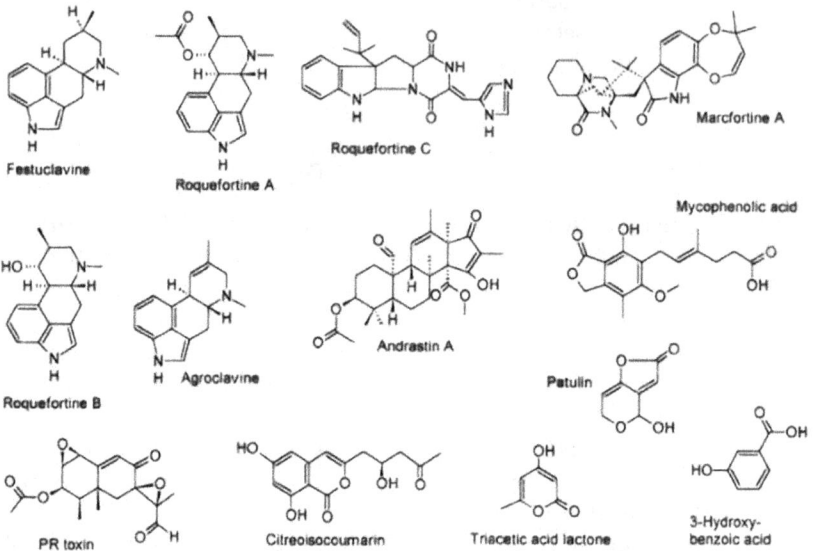

Figure 9.4   Structures of toxic metabolites from *P. roqueforti* and *P. paneum*.

common, as *P. roqueforti* generally produces a very small yet lethal percentage (%) of mycotoxins in small areas, which are called hotspots or pockets, and remain scattered and irregular. In addition to this, the methodology used for detection of *P. roqueforti* is not able to indicate that PR toxin is also present, as it is a part of a large consortium of Roqueforti metabolites. Similarly, the nonappearance of *P. roqueforti* cannot guarantee an absence of PR toxin because it may be due to fungi which have already died while discharging the toxin intact (Möller et al., 1997; Pedrosa and Griessler, 2010; Berge, 2011). To overcome these problems, an exhaustive sampling protocol must be followed to make sure that the collected sample is representative of the whole consignment.

PRT begin as one of the prime mycotoxins produced by *P. roqueforti* and have an LD_{50} (lethal dose or lethality rate of 50%) at a concentration of about 10–15 mg/kg by intraperitoneal administration, which is considered highly toxic per the US EPA's evaluation of premanufacture notices (PMNs). Per the US EPA's final risk assessment report (1997), PR toxins (7–acetoxy-5, 6-epoxy-3,5,6,7,8,8a-hexahydro3', 8, 8a-trimethyl-3-oxaspiro [naphthalene-2 (1H,2'oxirane]-3'-carboxaldehyde) are consistently detected using chromatography by fluorescence under UV light.

9.6.1 Conventional Methods

Most analytical techniques of detection require some pretreatment steps before separation and detection, such as sample selection and preparation, homogenization, extraction, and clean-up processes to isolate interfering components from the sample matrix. Sampling and clean-up are crucial steps taken not only for analysis of PRT but also for any kind of mycotoxin to ensure replicability of results. To do the extraction properly and take a sample, different techniques are in practice, like extraction with organic solvent mixtures, such as chloroform, acetone, ethyl acetate, dichloromethane, acetonitrile, methanol, diluted organic acids, or water (Seymour Shephard, 2008). In addition to this, some modern solvent extraction methods are also employed through instruments like liquid extraction with application of pressure, supercritical fluid extraction (SCFE), extraction assisted with ultrasound, and microwave-assisted extraction.

Next, after the extraction step, an assay for screening is performed to detect mycotoxins. Different assay methods include different chromatographic methods like thin-layer chromatography (TLC), high-performance liquid chromatography (HPLC), LC-MS, GC, GC-MS, immuno-based methods (radio immunoassay, enzyme-linked immunoassay, fluorescence immunoassay), and sensors and various biosensor-based methods (spectrometric, electrochemical, optical, etc.) (Turner et al., 2015; Cigic and Prosen, 2009; Koppen et al., 2010).

A variety of conventional chromatographic techniques have been developed for the detection (Betina, 1985) and semi-quantitative determination of PRT. During the preliminary research on mycotoxin analysis and characterization, thin-layer chromatography was recognized as the most widely accepted method out of all the chromatographic techniques. Before TLC analysis, contaminated samples need to be cleaned up and extracted to obtain authentic results. Two simple methods were employed for extraction, clean-up and identification of PRT from blue cheese and other complex media. In the first method, the sample would be extracted in an alcohol-water mixture, followed by filtration, after which the filtrate was again extracted in chloroform. The final extract was analyzed in TLC for PRT. In the second procedure, the sample was treated in ethyl acetate and partitioned between hexane and acetonitrile. The residue left after evaporation of the acetonitrile layer was dissolved in chloroform for TLC analysis.

The high-performance liquid chromatographic technique has been successfully engineered to measure the concentration of this toxin and other Roqueforti metabolites in different media. It has been successfully used for quantitative detection of PRT at concentrations above 10 ng and qualitative detection at concentrations as low as 2 ng. Even though these simple chromatographic techniques are highly effective in the detection of PRT at very low concentrations, the presence of proteins and amino acids in the test sample matrix may influence the precision of the result (Moreau et al., 1979).

HPLC with ultraviolet detectors helps to detect the light absorbed by the sample at different wavelengths and fluorescence (FLD) due to the occurrence of chromophores in the samples. HPLC-FLD has a precision detection limit of 0.01 to 0.04 µg/kg. On the other hand, a HPLC-PDA (photodiode array) detector works on the principle of separation by hydrophobicity and molecular weight, and it can detect up to 4 ng analyte in 1 g of sample. On different mycotoxins produced by *Aspergillus, Penicillium, Fusarium,* and others the quantified detection limit ranged from 0.031 to 22 µg/kg (Shrivastava and Sharma, 2021).

Gas chromatography is a separation technique based on the analyte's affinity of the stationary phase and mobile phase and uses a spectrometric and/or electron detector. GC-MS is cable of detection of PR toxins and other similar metabolites produced in grain-based products and vegetable oil (Yang et al., 2020).

9.6.2 Novel Methods

9.6.2.1 Reversed-Phase High-Performance Liquid Chromatography

In this method, the mycelium is removed in the first step and the test sample is extracted in chloroform. The filtrate is evaporated and again extracted in the choice of solvent. Reversed-phase HPLC is more specific and sensitive than other chromatographic methods. This technique uses a diverse solvent system for extraction of toxins and is also capable of detecting and quantifying PRT at concentrations as low as 2–3 ng. Also, the composition of the sample matrix, that is, the presence of protein and amino acids, does not affect the detection procedure (Siemens et al., 1992).

9.6.2.2 HPLC-NMR and HPLC-MS

PRT can be detected even in trace amounts using HPLC-MS and HPLC-NMR. However, in the case of HPLC-MS, sometimes in the presence of equal amounts of roqueforti metabolites, mass spectroscopy cannot detect the individual peaks. Also, even though these compounds can easily be differentiated by UV, when there are large differences in concentration, the predominant metabolite can easily mask PRT, which, though lethal, is generally present in very small quantities. In such cases, a slight increase in pH can easily separate the metabolites (Nielsen et al., 2006).

9.6.2.3 Molecular Identification Based on Ribosomal DNA Sequence Comparison, Random Amplified Polymorphic DNA Profiles Using PCR

This method can be successfully used to confirm the presence of the PRT-producing strains of *P. roqueforti*. Similar strains like *Penicillium carneum* and *Penicillium paneum* of the roqueforti group all produce roquefortine C, and both *P. carneum* and *P. paneum* produce patulin, but only *P. roqueforti* produces PRT. Using a combination of ribosomal DNA sequence comparison and the random amplified polymorphic DNA (RAPD) profile method, the isolates can be identified and characterized based on their molecular taxonomy with very little chance of error. The taxonomic information obtained from the DNA sequences can often give sufficient information, though using these sequences as a sole criterion for characterization is not advisable in modern research methodologies. Hence the identification is confirmed by analyzing RAPD and secondary metabolite profile patterns (Boysen et al., 2000).

9.6.2.4 Immunoassays

Immunoassay techniques work based on the principle that targeted specific antigens or haptens are recognized by the antibodies and bind with them to give a visual confirmation of their presence. The enzyme-linked immunosorbent assay (ELISA) method has been applied commercially for mycotoxin detection due to its high sensitivity as well as accuracy for a long time (Kolosova et al., 2006). One drawback of ELISA is that it requires a long time (several hours) for incubation as a preliminary step and complex microplate washing procedures. Hence scientists have been trying to shorten the incubation period while preserving its high-accuracy analytical performance. This led to development of methodology such as fluorescent-linked immunosorbent assays (FLISAs) and chemiluminescence enzyme immunoassays (CLEIAs) based on enhanced optical signals (Li et al., 2014; Lu et al., 2018; Zhang et al., 2018). Although colorimetric immunoassays are cheap and simple methods that do not require any expensive instruments, certain levels of background interference, poor sensitivity, and instability are notable drawbacks of these processes. To overcome these drawbacks, fluorescent nanoparticles have been introduced in their place for rapid detection, higher sensitivity, and enhanced stability. In recent years, one such assay technique has become a popular analytical tool: lateral flow immunoassays (LFIAs) for detection of mycotoxins like PRT, patulin, aflatoxin, fusarium, and others.

LFIA is actually an immunochromatographic assay (ICA) equipped with rapid immuno-biosensing platform. Because of its lower time requirement, low cost, and user-friendly nature, it can be easily commercialized for the purpose of food safety monitoring, disease diagnosis, and point-of-care testing (POCT) (Qian et al., 2020; Gowri et al., 2021). The detection process is integrated on a small chip, and the signal processing is fast and can be done by easily obtained portable instruments. Lateral flow immunoassays use multiple test lines and signal labels for simultaneous detection of multiple mycotoxins. LFIA mostly employs gold nanoparticles (AuNPs) as signal labels because they are inexpensive and easy to synthesize, and when stimulated, they emit visible red to purple light (Li et al., 2021).

9.6.2.5 Biosensor-Based Detection

The development of biosensors for the detection of PRT is a relatively new approach, and it is mostly in the testing phase. The principles for detection are based on analyte sensitivity, specificity, small molecule detection, non-toxicity to the sample matrix with a wide range of detection limits, and cost effectiveness. An amalgamation of all these properties can eventually enable biosensors to integrate into various equipment and testing probes.

Different types of biosensors include electrochemical, acoustical, optical, and mass-sensitive biosensors. These are designed for impromptu real-time detection of toxins and pathogens and other biological/biochemical residue detection in the field and laboratory. A suitable combination and bio-fabrication of sensing elements and transducers can lead to the development of successful detection tools for some major toxins, including aflatoxins, ochratoxin, fumonisins, trichothecenes, PRT, and patulin (Li et al., 2021).

Different sensors used for bio-recognition of mycotoxins like PRT include enzyme sensors, immunosensors, nucleic acid sensors or aptasensors, microbial whole cell sensors, and molecular imprinted sensors. Types of transducers generally coupled with these biosensing elements are electrochemical, amperometric, conductometric, potentiometric, impedimetric, optical, surface plasmon resonance, evanescent wave fluorescence, bioluminescent optical fiber, piezoelectric biosensor (mass-sensitive), and thermometric transducers (Shrivastava and Sharma, 2021).

Electrochemical biosensors are useful sensing tools for PRT because of their high sensitivity and adjustability to be miniaturized; however, background interference, complex electrode modifications, and electrode fouling and degradation are some problems that still need to be solved, as these toxins are generally present in trace amounts along with other metabolites. Electrochemiluminescent and photoelectrochemical sensors are regarded as a combination of the benefits and abilities of electrochemical and optical tools that show low background interference, fast response, and high sensitivity towards a variety of mycotoxins. Piezoelectric biosensors have higher signal stability and the ability to analyze multiple targets at one stretch. Label-free surface plasmon resonance biosensors have not only simplified preparation and detection procedures but also offer direct and real-time detection platforms in the form of tiny chips. Hence, with features like high sensitivity, rapid response time, and new advancements in micro-electromechanical systems (MEMS), sensor devices can be designed to be more proactive and versatile for instantaneous detection and quantification of PRT in the near future.

9.7 MANAGEMENT AND CONTROL STRATEGIES

Several control strategies can be adopted to discard and control the harmful effects of PRT. The key ecological determinants (extrinsic factors) of occurrence of PRTs during before- and after-harvest practices are:

- moisture content
- local climate
- temperature
- improper practices adopted before harvesting
- sloppy harvesting practices
- micronutrient content and bioavailability in the substrate
- improper storage, transportation, marketing, and distribution
- relative humidity
- aeration
- insect/pest infestation
- seed damage
- lack of management and awareness

These practices could facilitate the active growth of molds and ultimately increase the chance of PR toxin contamination (Zain, 2011; Gallo et al., 2015; Wambacq et al., 2016, 2018; Wambacq, 2017). The intrinsic factors responsible for PRT formation include strain occurrence, variation, abundance, spore loads, strain specificity, and microbial interaction (Chang et al., 1991; Boysen et al., 2000; O'Brien et al., 2006; Dell'Orto et al., 2015).

Physical, chemical, and biological control measures can be used to eradicate, control, or detoxify food and feeds contaminated by PR toxin, depending on the conditions. Despite the numerous strategies for fungal degradation and purification of food and feed, the best method is to prevent *P. roqueforti* from growing. The ideal choice of control strategy is to build upon the capability of biological agents (beneficial microbes, enzymes, herbal extracts, essential oils, etc.) to control the growth of mycotoxigenic fungi. Detoxifying enzymes from *P. roqueforti*, such as PR oxidase, PR-amide synthetase, and eremofortin C oxidase, can catalyze PRT breakdown (Chang et al., 2004a, b; García-Estrada and Martín, 2016).

To avoid and control PR toxin contaminations, a variety of traditional techniques or classic hurdle technologies are used alone or in combination. These procedures are categorized in the following, according to the type of treatment.

9.7.1 Methods of Prevention of PR Toxin Formation

To avoid contamination with *P. roqueforti* in a variety of meals and feeds, food safety and quality control processes must be implemented across the food supply chain. However, all actions at each level must be integrated into a single system. Human resource capacity; different types of professionals; and knowledge in the fields of environmental health science, biochemistry, agriculture, horticulture, botany, microbiology, food science, and bioprocess technology, as well as knowledge and skill upgrades through professional courses, are all essential for completing these safety and quality control activities. One of the methods of prevention of formation of mycotoxins that arise during storage is drying crops to bring the moisture content into a range considered safe. To improve food safety management, a hazard analysis and critical control point (HACCP) system coupled with good manufacturing practices/good agricultural practices (GMPs/GAPs) is also critical.

HACCP-based recommended control measures for processed foods and feeds should include measures for prevention, under control, GMPs, and quality control at all phases of production, from farm to fork. Mycotoxins are currently being addressed with HACCP-like methods, and the Food and Agriculture Organization (FAO) has released a comprehensive guide on how to apply the HACCP system to prevent and control mycotoxins (FAO, 2002). HACCP can be used to determine the steps that can be taken to prevent or eliminate mycotoxins, as well as the stages at which monitoring systems can be established.

Mold and mycotoxins can be reduced both before and after harvest using a variety of preventative measures, including appropriate control measures, clean-up, timely harvesting, crop rotation, drying and storage practices, insect infestation management, and the development of fungi-resistant plant cultures, to name a few. Depending on the insect species that attack maize, BT-maize (transgenic corn) has a lower chance of contamination by fumonisin. For generating wheat cultivars or maize hybrids that are resistant to mold, genetic engineering has proven successful in establishing host resistance through the insertion or augmentation of antifungal genes (Bata et al., 2001; Doko et al., 1995).

Controlling mycotoxin contamination is best accomplished through preharvest management. In an ideal world, the risks associated with mycotoxin dangers should be minimized at every stage of the food manufacturing process. When handling mycotoxin-contaminated items, consider the harvesting time, temperature, and moisture during storage and shipping, as well as the humidity of the air in the warehouses (Scott, 1998). Grain and storeroom storage conditions that are ideal in terms of temperature and humidity may considerably reduce the formation of toxigenic mold (Peraica et al., 2002). If the product is to be utilized as human food or animal feed, the dangers associated with specific mycotoxins must be addressed by postharvest measures once mycotoxin contamination occurs. Various physical, chemical, and biological control methods as well as post-harvest processing techniques are examples of such measures that can reduce mycotoxin levels in food or feed.

9.7.2 Physical Methods of Control

Traditional approaches like hurdle technology have achieved the control of PRT contamination to a certain level. These methods include physical sorting of moldy grains, cleaning, washing, drying, winnowing, milling, heat treatment, adsorption, extrusion, irradiation (solar, UV, γ-rays, X-rays, or microwaves), and anaerobiosis. But during most of these physical processes, PR toxin is somewhat stable and survives in final meals and silage. Only treatment methods such as heat treatment, anaerobiosis, and UV irradiation have achieved the highest level of success in eradicating PR toxification (Suttajit, 1989; Scott, 1998; Sengun et al., 2008).

Prior to food processing, all cereals are cleaned, which involves removing dust, broken grains, and other undesired material, as well as physically removing some of the outer layers of grains through abrasion, such as "scouring." Cleaning for mycotoxins has varying degrees of effectiveness, although it is normal to see a drop in concentration of about 40% (Brekke et al., 1975a). Separating corn screenings can considerably reduce the quantities of several toxins in general (Broggli et al., 2002). Another approach for decontaminating grains and other food products such as nuts and legumes based on color is electronic sorting (Pelletier and Reizner, 1992), whereas fluorescence sorting is usually utilized for screening and decontaminating corn, cottonseed, and dried figs. Muller (1983) and Steiner et al. (1988) are two examples of this.

Washing procedures with water or using solutions made by the salt of Na_2CO_3 are occasionally used to lower the amount of certain mycotoxins in grains or corn cultures, although they may only be helpful prior to wet milling due to the high cost of subsequent drying processes. Because most mycotoxins are heat stable, thermal inactivation is not an appropriate strategy for decontamination. Only coffee roasting or microwave treatment can accomplish partial disinfection (Scott, 1998). Different types of radiation have been investigated for the decontamination of some mycotoxins and the control of the growth of some fungi, such as irradiation, X-rays, ultraviolet light, and so on, but they all have drawbacks, because radiation is only effective if it is applied to a very thin layer of material (Peraica et al., 2002).

9.7.3 Chemical Methods of Control

Chemical approaches have long been thought to be the most successful at removing PR poison from contaminated samples. Chemical treatments such as ozonation, use of either acid or base, oxidizing agents, reducing agents, deamination, chlorination, application of ammonia, and formaldehyde treatment have been used to inactivate PRT in affected materials (Jard et al., 2011; Benkerroum, 2016). Inorganic compounds like nitrates and sulfates; weak organic acids; and natural antioxidants such as vitamins A, C, E, selenium, and beta-carotenes have been known to be effective against mycotoxin-producing fungal growth under different conditions (Lind et al., 2005; Ashiq, 2015; Yan et al., 2017).

Formic and propionic acid, ammonia, sodium hydroxide, hydrogen peroxide, ozone, sodium bisulfite, and other chemicals (e.g., chlorine or formaldehyde) have all been researched and found effective enough for PRT decontamination (Peraica et al., 2002).

Peanuts were detoxified using hydrogen peroxide, while oilseeds and corn were detoxified with calcium hydroxide or monomethylamine, respectively, after being contaminated with mycotoxins (Scott, 1998). Another chemical used to detoxify contaminated corn or dried figs is sodium bisulfite (Scott, 1998). Other chemical approaches have garnered the most research interest as a realistic solution to disinfect tainted feed or peanuts, although such procedures are normally not approved for human consumption inside the European Community (EC) (Peraica et al., 2002). Chelkowski et al. (1981) discovered that treating infected grain with ammonia reduced mycotoxin levels to undetectable levels. They discovered that not only does ammoniating grain detoxify mycotoxins like fumonisins, but it also suppresses fungal development (Chelkowski et al., 1981). Although there are certain nutritional changes in the feed, such as a decrease in lysine and sulfur-containing amino acids, there are no negative impacts associated with the ammoniation process in feeding tests (Scott, 1998). In addition, sufficient aeration after ammoniation is essential for animal consumption of the meal.

Various compounds or feed additives have the ability to repair and neutralize mycotoxins or achieve antidote effects against mycotoxin activity, according to Stoev et al. (1999, 2000, 2002, 2004; Stoev, 2008, 2010). Several feed additives bind mycotoxins in the gastrointestinal tract, reducing their bioavailability. Clay and zeolitic minerals contain a diverse family of functionally distinct silicoaluminosilicates that have been found to reduce mycotoxins' bioavailability and toxicities in animals' gastrointestinal tracts. Adding clay or zeolite to the meal, on the other hand, was deemed impractical (Rotter et al., 1989; Ramos et al., 1996; Kubena et al., 1998; Stoev, 2008).

Unlike the nonspecific absorbents discussed previously, cholestyramine appears to be a good absorbent of penicillium mycotoxins in nonruminant animals' gastrointestinal systems. Cholestyramine is an anion exchange resin that has been shown to reduce blood toxin concentrations by 50% when employed at 0.5% in a rat diet containing 1 ppm ochratoxin A (Marquardt and Frolich, 1992).

Another strategy to lessen the toxicity of certain mycotoxins is to gain a better knowledge of their toxicity processes. This strategy entails supplementing the food with specialized antidotes or vitamins to counteract the mycotoxin's specific harmful effects (Stoev, 2008). These specialized food and feed additives include ascorbic acid, water extract of artichoke (*Cynara scolymus* L), roxazyme-G (polyenzyme complement produced by the fungal genus *Trichoderma*), and so on. Antidotes like these, added to feeds as supplements, could be a feasible way to safely use contaminated feed.

Even though these compounds are generally recognized as safe (GRAS), use of some of these fungistatic chemicals is being discouraged due to economic and environmental reasons and food safety issues.

9.7.4 Biological Methods of Control

Unfortunately, most physical and chemical procedures employed in food commodities have limitations, and some of these technologies fail to meet the Food and Agriculture Organization's standards (Beaver, 1991). In addition, certain strains have developed a unique capacity to withstand some chemical treatments and preservatives in recent years. Nonetheless, the possibility for certain fungicides and preservatives to have negative environmental and health effects prompted a search for more natural alternatives. Hence, in recent times, some green alternative strategies have been developed that use the antimicrobial and detoxifying effects of certain natural extracts and microbial strains to effectively control production and initiate removal of PRT.

Selecting plant cultivars that are more resistant to *P. roqueforti* infection is one viable option for lowering plant stress and damage. People are looking for higher-quality, safer, non-toxic, non-preservative, less-processed foods with a longer shelf life, and biopreservatives, or naturally occurring essential oils and plant extracts, have recently gained favor. Valerio et al. highlighted the importance of microorganisms and plant metabolites as natural fungicides, as well as their inhibitory effects against *P. roqueforti*, in their preliminary research on smart and intelligent packaging.

The ability of different lactobacillus (LAB), *Bacillus*, *Streptomyces*, and yeast strains to produce a variety of active metabolites such as alcohol, organic acids, hydrogen peroxide, fatty acids, carbon dioxide, cyclic dipeptides, acetoin, diacetyl, and bacteriocin-like inhibitory substances has been attributed to antimicrobial activity (Höltzel et al., 2000). These organisms' antifungal compounds or secondary metabolites, on the other hand, have a lot of promise for application as natural biological control agents in food preservation to inhibit mold growth and mycotoxin generation. On the other hand, smart packaging is a new frontier, combining environmentally friendly biodegradable biofilm with naturally bioactive substances produced by these organisms as a biopreservative. Furthermore, these bacterial strains' antifungal activity was shown to be comparable to that of calcium propionate, a common chemical preservative ($C_6H_{10}CaO_4$). Yan et al. (2017) investigated *L. plantarum*'s in vitro antagonistic capacity against *P. roqueforti* as an indicator, as well as the impact of its antifungal activities on the shelf life of Chinese steamed bread. According to Valerio et al., many LAB strains reduced *P. roqueforti* growth in various bread goods during storage (2009). The most functional strain was *Lactobacillus plantarum* C21–41. Antifungal compounds derived from *B. subtilis* YM 10–20 can successfully suppress the germination of *P. roqueforti* conidiospores in bread and silage products (Chitarra et al., 2003). Antifungal characteristics have also been discovered in *Streptomyces* and *Saccharomyces* strains, as well as being an effective inhibitor of *P. roqueforti* (Frändberg et al., 2000; Lillbro, 2005).

The culture filtrate of *L. plantarum* cultured on wheat flour hydrolysate showed a strong inhibitory effect on the development of *P. roqueforti* IBT18687, according to Lavermicocca et al. (2000). Lactic acid was created in large amounts during the growth phase, suggesting that this metabolic ability, together with the isolate's low pH, may have contributed to the antifungal phenotype. Zhang et al. investigated the probiotic and food preservation capabilities of an *L. plantarum* TK9 strain isolated from naturally fermented congee in China.

Alboleutin, bacitracin, botrycidin, clorotetain, fengycin, iturins, and rhizocticin are antifungal compounds generated by *Bacillus subtilis* that inhibit or impede the growth of fungi and yeasts. While the bulk of these antifungal metabolites have been examined in relation to fungal mycelial growth, there have been few investigations on their effect on *P. roqueforti* spore survival and germination.

Streptomyces species could also be used to extend the shelf life of food and feed products by preventing spoilage. According to a thorough review of the literature, only a few *Streptomyces* species have been documented to inhibit the growth of resistant *P. roqueforti* strains. Frändberg et al. (2000), for example, investigated the antifungal potential of a compound produced from *Streptomyces halstedii* as an effective inhibitor of *P. roqueforti* and other potentially harmful molds. Several studies previously found that adding live yeast to contaminated feed and food reduced the harmful effects of *P. roqueforti*. The creation of an inhibitor against *P. roqueforti*, which was done with the addition of various carbon sources, considerably improved the inhibitory action of yeast strains.

Extracts from garlic, clove, nutmeg, cinnamon, thyme, mint, anise, oregano, and basil, in addition to microbial strains, have been proven to successfully suppress fungal growth and toxin production. When present in ideal amounts, essential oils from spices, olives, coconut, and palm kernel can function as inhibitors of *P. roqueforti* in processed cereal products (Manish et al., 2018).

9.8 DECONTAMINATION THROUGH FOOD/FEED PROCESSING

Any kind of treatment that may be biological, chemical, or physical done to a food material for the production of the end product is referred to as processing (Scudamore and Banks, 2004). This comprises everything from dried and moist grain milling to various processing like brewing, steaming, extrusion, and baking, as well as feeding grain-based whole meals to livestock for meat and milk production. Chemical and biological activity, as well as temperature, pH, moisture content, pressure, buffering conditions, and the addition of additional chemicals and enzymes, can all affect the stability of mycotoxins.

Even if some penicillium mycotoxins survive processing, their concentration in a finished food item may be substantially lower than in the raw harvested crop. As a result, proper commodity management is required throughout the processing process. Wet and dry milling is a grain-processing method that can be used to remove mycotoxin contamination from food in certain circumstances. This procedure divides the grain into different fractions, and it's crucial to figure out which ones are still dangerous. White flour made from wheat (Osborne et al., 1996) or maize grits (e.g. Broggli et al., 2002) contain much less mycotoxin because mycotoxins are concentrated in the bran during dry milling.

Various processing procedures, such as treating corn with limewater before manufacturing tortillas, can considerably reduce toxin levels (Scott, 1998). During oil refining, any extracted mycotoxins are efficiently removed. Similarly, adding sodium chloride (salt) to unshelled peanuts during pressure boiling will significantly lower toxin levels, while vitamin C in apple juice can gradually reduce patulin levels (Scott, 1998; Peraica et al., 2002). In procedures like baking (Stoloff and Trucksess, 1981) and extrusion (Martinez and Monsalve, 1989), temperature and pH affect toxin heat stability; hence higher temperatures or alkaline operations like leavening agents or tortilla manufacture can reduce aflatoxin content (Price and Jorgensen, 1985; Abbas et al., 1988; De Arrola et al., 1988).

Most mycotoxins, including PR toxin, platulin, deoxynivalenol, zearalenone, and fumonisins, can survive and emerge in maize or wheat beer, but their enzyme breakdown is dependent on the procedure (Scott and Lawrence, 1994; Scott, 1996; Scudamore and Banks, 2004). While malting barley and brewing can help to eliminate some pollutants, they are ineffective against a large number of others (Scott, 1996). High temperatures, acid or alkaline environments, and the presence of enzymes can all break down even the most stable mycotoxins. Patulin and rubratoxin B can be removed by yeast fermentation (Scott, 1998).

Tracing the fate of the mycotoxin at each stage of the process is critical for optimizing control and reducing the amount of various mycotoxins reaching the customer. Implementing excellent farming practices and storage techniques that prevent mycotoxins from accumulating should be the primary goal in eradicating or lowering mycotoxins in the food supply. However, total eradication of mycotoxins is not always attainable. If a particular mycotoxin is regarded to pose a significant risk, laws may be established, as well as some actions targeted at decreasing consumer exposure.

9.9 CONCLUSION

Increasing awareness among consumers and interest in public health and safety have given rise to a demand for superior-quality food products, leading to booming growth of applied biotechnology and industrial microbiology for the development of food processing industries. Since serious toxicological effects on both humans and animals are already reported due to the consumption of PR toxin-contaminated food, recent years have seen the evolution of novel strategies for detection and control of product contamination. The many natural methods of control available can provide a more fruitful and indigenous pathway for inhibiting the toxic strains of *P. roqueforti*. Moreover, new studies exploring the biosynthetic pathway for generation of PR toxin have revealed critical information to regulate the genetic machinery of this fungus to eliminate PR toxin production.

REFERENCES

Abbas HK, Mirocha CJ, Rosiles R, Carvajal M. Effect of tortilla-preparation process on aflatoxin B1 and B2 in corn. *Mycotox Res.* 1988;4:33–36.

Ashiq S. Natural occurrence of mycotoxins in food and feed: Pakistan perspective. *Compr Rev Food Sci Food Saf.* 2015;14:159–175.

Bata A, Rafai P, Kovacs S. Investigation and a new evaluation method of the resistance of maize hybrids grown in Hungary to *Fusarium* moulds. *Phytopathology.* 2001;149:107–111.

Beaver RW. Decontamination of mycotoxin-containing foods and feedstuffs. *Trends Food Sci Technol.* 1991;2:170–173.

Benkerroum N. Biogenic amines in dairy products: Origin, incidence and control means. *Compr Rev Food Sci Food Saf.* 2016;31:D37–D47.

Berge AC. Silage molds affect rumen health. *Bovine Health Quarterly, Feedstuffs,* 2011. Available at: http://fdsmagissues.feedstuffs.com/fds/PastIssues/FDS8315/fds14_8315.pdf

Betina V. Thin-layer chromatography of mycotoxins. *J Chromatogr.* 1985;334:211.

Boysen ME, Jacobsson KG, Schnurer J. Molecular identification of species from the *Penicillium roqueforti* group associated with spoiled animal feed. *Appl Environ Microbiol.* 2000:1523–1526.

Brekke OL, Peplinski AJ, Griffin EL Jr. Cleaning trials for corn containing aflatoxin. *Cereal Chem.* 1975a;52:198–204.

Broggli et al. 2002.

Cane DE, Wu ZH, Proctor RH, Hohn, TM. Overexpression in *Escherichia coli* of soluble aristolochene synthase from *Penicillium roqueforti*. *Arch Biochem Biophys*. 1993;304(2):415–419.

Chang SC, Cheng MK, Wei YH. Production of PR-imine, PR acid, and PR-amide relative to the metabolism of PR toxin by *Penicillium roqueforti*. *Fungal Sci*. 2004a;19:39–46.

Chang SC, Ho CP, Cheng MK. Isolation, purification, and characterization of the PR-amide synthetase from *Penicillium roqueforti*. *Fungal Sci*. 2004b;19:117–123.

Chang SC, Wei YH, Wei DL, Chen YY, Jong SC. Factors affecting the production of eremofortin C and PR toxin in *Penicillium roqueforti*. *Appl Environ Microbiol*. 1991;57:2581–2585.

Chelkowski J, Golinski P, Godlewska B, Radomyska W, Szebiotko K, Wiewiorowska M. Mycotoxins in cereal grains. Part IV. Inactivation of ochratoxin A and other mycotoxins during ammoniation. *Nahrung*. 1981;25:631–637.

Chitarra GS, Breeuwer P, Nout MJ, van Aelst AC, Rombouts FM, Abee T. An antifungal compound produced by *Bacillus subtilis* YM 10–20 inhibits germination of *Penicillium roqueforti* conidiospores. *J Appl Microbiol*. 2003;94(2):159–166.

Cigic IK, Prosen H. An overview of conventional and emerging analytical methods for the determination of mycotoxins. *Int J Mol Sci*. 2009;10(1):62–115.

De Arrola M, Del C, de Porres E, de Cabrera S, de Zepeda M, Rolz C. Aflatoxin fate during alkaline cooking of corn for tortilla preparation. *J Agric Food Chem*. 1988;36:530–533.

Dell'Orto V, Baldi G, Cheli F. Mycotoxins in silage: Checkpoints for effective management and control. *World Mycotoxin J*. 2015;8:603–617.

Doko MB, Rapior S, Visconti A, Schjoth JE. Incidence of levels of fumonisin contamination in maize by genotypes grown in Europe and Africa. *J Agric Food Chem*. 1995;43:429–434.

Driehuis F. Silage and the safety and quality of dairy foods: A review. *Agric Food Sci*. 2013;22:16–34.

Dubey MK, Aamir M, Kaushik MS, Khare S, Meena M, Singh S, Upadhyay RS. PR toxin—Biosynthesis, genetic regulation, toxicological potential, prevention and control measures: Overview and challenges. *Front Pharmacol*. 2018;9:288.

Erdogan A, Sert S. Mycotoxin-forming ability of two *Penicillium roqueforti* strains in blue moldy tulum cheese ripened at various temperatures. *J Food Prot*. 2004 Mar;67(3):533–535.

FAO. *Manual on the Application of the HACCP System in Mycotoxin Prevention and Control*. Joint FAO/WHO Food Standards Programme FAO 2002, Rome, Italy.

Fernández-Bodega MA, Mauriz E, Gómez A, Martín JF. Proteolytic activity, mycotoxins and andrastin A in *Penicillium roqueforti* strains isolated from Cabrales, Valdeón and Bejes–Tresviso local varieties of blue-veined cheeses. *Int J Food Microbiol*. 2009 Nov 30;136(1):18–25.

Frändberg E, Petersson C, Lundgren LN, Schnürer J. Streptomyces halstedii K122 produces the antifungal compounds bafilomycin B1 and C1. *Can J Microbiol*. 2000 Aug;46(8):753–758.

Frisvad JC, Smedsgaard J, Larsen TO, Samson RA. Mycotoxins, drugs and other extrolites produced by species in *Penicillium* subgenus Penicillium. *Stud Mycol*. 2004;49:201–241.

Gallo A, Giuberti G, Frisvad JC, Bertuzzi T, Nielsen KF. Review on mycotoxin issues in ruminants: Occurrence in forages, effects of mycotoxin ingestion on health status and animal performance and practical strategies to counteract their negative effects. *Toxins*. 2015;7:3057–3111.

García-Estrada C, Martín J-F. Biosynthetic gene clusters for relevant secondary metabolites produced by *Penicillium roqueforti* in blue cheeses. *Appl Microbiol Biotechnol*. 2016;100:8303–8313.

Geisen R, Cantor MD, Hansen TK, Holzapfel WH, Jakobsen M. Characterization of *Penicillium roqueforti* strains used as cheese starter cultures by RAPD typing. *Int J Food Microbiol*. 2001 May 10;65(3):183–191.

Gillot G, Jany JL, Coton M, Le Floch G, Debaets S, Ropars J, Lopez-Villavicencio M, Dupont J, Branca A, Giraud T, Coton E. Insights into *Penicillium roqueforti* morphological and genetic diversity. *PLoS ONE*. 2015 Jun 19;10(6):e0129849.

Gillot G, Jany JL, Poirier E, Maillard MB, Debaets S, Thierry A, Coton E, Coton M. Functional diversity within the *Penicillium roqueforti* species. *Int J Food Microbiol*. 2017 Jan 16;241:141–150.

Gowri A, Ashwin Kumar N, Suresh Anand, BS. Recent advances in nanomaterials-based biosensors for point of care (PoC) diagnosis of Covid19—A minireview. *TrAC Trend Anal Chem*. 2021;137:116205.

Höltzel A, Gänzle MG, Nicholson GJ, Hammes WP, Jung G. The first low molecular weight antibiotic from lactic acid bacteria: Reutericyclin, a new tetramic acid. *Angew Chem Int Ed Engl*. 2000 Aug 4; 39(15):2766–2768.

Houbraken J, Samson RA. Phylogeny of penicillium and the segregation of Trichocomaceae into three families. *Stud Mycol*. 2011;70;1–51.

Hymery N, Masson F, Barbier G, Coton E. Cytotoxicity and immunotoxicity of cyclopiazonic acid on human cells. *Toxicol Vitro*. 2014 Aug 1;28(5):940–947.

Hymery N, Puel O, Tadrist S, Canlet C, Le Scouarnec H, Coton E, Coton M. Effect of PR toxin on THP1 and Caco-2 cells: An in vitro study. *World Mycotoxin J*. 2017 Nov 30;10(4):375–386.

Jard G, Liboz T, Mathieu F, Guyonvarch A, Lebrihi A. Review of mycotoxin reduction in food and feed: from prevention in the field to detoxification by adsorption or transformation. *Food Addit Contam*. 2011;A28:1590–1609.

Jong SC, Gantt MJ. *ATCC Catalogue of Fungi/Yeasts*. with an. 1987.

Kolosova AY, Shim WB, Yang ZY, Eremin SA, Chung DH. Direct competitive ELISA based on a monoclonal antibody for detection of aflatoxin B1. Stabilization of ELISA kit components and application to grain samples. *Anal Bioanal Chem*. 2006;384(1):286–294. https://doi.org/10.1007/s00216-005-0103-9.

Koppen R. Determination of mycotoxins in foods: current state of analytical methods and limitations. *Appl Microbiol Biotechnol*. 2010;86(6):1595–1612.

Kubena LF, Harvey RB, Bailey RH, Buckley SA, Rottinghaus GE. Effects of a hydrated sodium calcium aluminosilicate (T-BindTM) on mycotoxicosis in young broiler chickens. *Poult Sci*. 1998;77:1502–1509.

Kumar P, Mahato DK, Kamle M, Mohanta TK, Kang SG. Aflatoxins: A global concern for food safety, human health and their management. *Front Microbiol*. 2017 Jan 17;7:2170.

Lafont PH, Debeaupuis JP, Gaillardin MI, Payen J. Production of mycophenolic acid by *Penicillium roqueforti* strains. *Appl Environ Microbiol*. 1979 Mar;37(3):365–368.

Lavermicocca P, Valerio F, Evidente A, Lazzaroni S, Corsetti A, Gobbetti M. Purification and characterization of novel antifungal compounds from sourdough *Lactobacillus plantarum* strain 21B. *Appl Environ Microbiol.* 2000;66:4084–4090. http://doi.org/10.1128/AEM.66.9.4084-4090.2000

Li R, Wen Y, Wang F, He P. Recent advances in immunoassays and biosensors for mycotoxins detection in feedstuffs and foods. *J Anim Sci and Biotechnol.* 2021;12:108.

Li Y, Liu G, Fu X, He J, Wang Z, Hou J. High-sensitive chemiluminescent ELISA method investigation for the determination of deoxynivalenol in rice. *Food Anal Methods.* 2014;8:656–660.

Lillbro M. *Biocontrol of Penicillium roqueforti on Grain: A Comparison of Mode of Action of Several Yeast Species.* Master thesis, Swedish University of Agricultural Sciences; 2005; Uppsala: 21.

Lind H, Jonsson H, Schnürer J. Antifungal effect of dairy propionibacteria contribution of organic acids. *Int J Food Microbiol.* 2005;98:157–165.

Lu T, Zhan S, Zhou Y, Chen X, Huang X, Leng Y, et al. Fluorescence ELISA based on CAT-regulated fluorescence quenching of CdTe QDs for sensitive detection of FB1. *Anal Methods.* 2018;10(48):5797–5802. https://doi.org/10.1039/C8AY02065E.

Marquardt RR, Frolich AA. A review of recent advances in understanding ochratoxicosis. *J Anim Sci.* 1992;70:3968–3988.

Martín JF, Coton M. Blue cheese: Microbiota and fungal metabolites. In *Fermented Foods in Health and Disease Prevention.* Academic Press, 2017. 275–303.

Martinez and Monsalve, 1989.

Möller T, Kerstrand KA, Massoud T. Toxin-producing species of *Penicillium* and the development of mycotoxins in must and homemade wine. *Nat Toxins.* 1997;5:86–89. http://doi.org/10.1002/(SICI)(1997)5:2<86::AID-NT6>3.0.CO;2-7

Moreau S, Masset A, Biguet J. Resolution of *Penicillium roqueforti* toxin and eremofortins A, B, and C by high-performance liquid chromatography. *ASM J Appl Environ Microbiol.* 1979;37(6).

Moubasher AH, Abdel-Kader MI, El-Kady IA. Toxigenic fungi isolated from Roquefort cheese. *Mycopathologia.* 1979 Jan;66(3):187–190.

Muller HM. A survey of methods of decontaminating mycotoxins. I. Physical methods. *Anim Res Develop.* 1983;18:70–96.

Nielsen KF, Sumarah MW, Frisvad JC, Miller JD. Production of metabolites from the *Penicillium roqueforti* complex. *J Agric Food Chem.* 2006;54(10):3756–3763. http://doi.org/10.1021/jf060114f

O'Brien M, Nielsen KF, O'Kiely P, Forristal PD, Fuller HT, Frisvad JC. Mycotoxins and other secondary metabolites produced in vitro by *Penicillium paneum* Frisvad and *Penicillium roqueforti* Thom isolated from baled grass silage in Ireland. *J Agric Food Chem.* 2006;54:9268–9276.

Osborne BG, Ibe FI, Brown GL, Patagine F, Scudamore KA, Banks JN, Hetmanski MT. The effects of milling and processing on wheat contaminated with ochratoxin A. *Food Addit Contam.* 1996;13:141–153.

Pahlow G, Muck RE, Driehuis F, Elferink SO, Spoelstra SF, Buxton D. *Silage Science and Technology.* Edited by Buxton DR, Munk RE and Harrison JH. 2003:31–94.

Pedrosa K, Griessler K. Toxicity, occurrence and negative effects of PR toxin—The hidden enemy. *Int Dairy Top.* 2010;9:7–9.

Pelletier and Reizner, 1992.

Peraica MC, Domijan A-M, Jurjevic Z, Cvjetkovic B. Prevention of exposure to mycotoxins from food and feed. *Arch Ind Hyg Toxicol.* 2002;53:229–237.

Price RL, Jorgensen KV. Effects of processing on aflatoxin levels and on mutagenic potential of tortillas made from naturally contaminated corn. *J Food Sci.* 1985;50:347–349.

Proctor RH, Hohn TM. Aristolochene synthase. Isolation, characterization, and bacterial expression of a sesquiterpenoid biosynthetic gene (Ari1) from *Penicillium roqueforti. J Biol Chem.* 1993 Feb 25;268(6):4543–4548.

Qian JQ, He QQ, Liu LL, Wang M, Wang BM, Cui LW. Rapid quantification of artemisinin derivatives in antimalarial drugs with dipstick immunoassays. *J Pharm Biomed Anal.* 2020;191:8.

Ramos et al. 1996.

Rasmussen RR, Storm IM, Rasmussen PH, Smedsgaard J, Nielsen KF. Multi-mycotoxin analysis of maize silage by LC-MS/MS. *Anal Bioanal Chem.* 2010 May;397(2):765–776.

Ropars J, López-Villavicencio M, Dupont J, Snirc A, Gillot G, Coton M, Jany JL, Coton E, Giraud T. Induction of sexual reproduction and genetic diversity in the cheese fungus *Penicillium roqueforti. Evol Appl.* 2014 Apr;7(4):433–441.

Rotter RG, Frohlich AA, Marquardt RR. Influence of dietary charcoal on ochratoxin A toxicity in Leghorn chicks. *Can J Vet Res.* 1989;53:449–453.

Scott PM. Toxins of *Penicillium* species used in cheese manufacture. *J Food Prot.* 1981 Sep;44(9):702–710.

Scott. 1996.

Scott PM. Industrial and farm detoxification processes for mycotoxins. *Rev Med Vet.* 1998;149:543–548.

Scott PM, Kennedy BP, Harwig J, Blanchfield B. Study of conditions of production of roquefortine and other metabolites of *Penicillium roqueforti. Appl Environ Microbiol.* 1977 Feb;33(2):249–253.

Scott and Lawrence. 1994.

Scudamore KA, Banks JN. The fate of mycotoxins during cereal processing. In *Meeting the Mycotoxin Menace*, Barug D, van Egmond H, López-García R, van Osenbruggen T and Visconti A (eds), Proceedings of the 2nd World Mycotoxin Forum. Nordwijk, Netherlands. Wageningen Academic Publishers. 2004. 165–181.

Sengun IY, Yaman DB, Gonul S. A. Mycotoxins and mould contamination in cheese: A review. *World Mycotoxin J.* 2008;1:291–298.

Seymour Shephard G. Determination of mycotoxins in human foods. *Chem Soc Rev.* 2008;37(11):2468–2477.

Shrivastava A, Sharma RK. Biosensors for the detection of mycotoxins. *Toxin Rev.* 2021;41(2):618–638. http://doi.org/10.1080/15569543.2021.1894175

Siemens K, Zawistowski J. Determination of *Penicillium roqueforti* toxin by reversed-phase high-performance liquid chromatography. *J Chromatogr.* 1992 Sep 18;609(1–2):205–211. http://doi.org/10.1016/0021-9673(92)80164-p

Steiner WE, Rieker RH, Battaglia R. Aflatoxin contamination in dried figs: Distribution and association with fluorescence. *J Agric Food Chem.* 1988;36:88–91.

Stoev SD. Complex etiology, prophylaxis and hygiene control in mycotoxic nephropathies in farm animals and humans, special issue "Mycotoxins: Mechanisms of toxicological activity—Treatment and prevention", section "Molecular Pathology." *Int J Mol Sci.* 2008;9:578–605.

Stoev SD. Studies on some feed additives and materials giving partial protection against the suppressive effect of ochratoxin A on egg production of laying hens. *Res Vet Sci.* 2010;88:486–491.

Stoev SD, Anguelov G, Ivanov I, Pavlov D. Influence of ochratoxin A and an extract of artichoke on the vaccinal immunity and health in broiler chicks. *Exp Toxicol Pathol.* 2000;52:43–55.

Stoev SD, Anguelov G, Pavlov D, Pirovski L. Some antidotes and paraclinical investigations in experimental intoxication with ochratoxin A and penicillic acid in chicks. *Vet Arhiv.* 1999;69:179–189.

Stoev SD, Djuvinov D, Mirtcheva T, Pavlov D, Mantle P. Studies on some feed additives giving partial protection against ochratoxin A toxicity in chicks. *Toxicol Lett.* 2002;135:33–50.

Stoev SD, Stefanov M, Denev S, Radic B, Domijan A-M, Peraica M. Experimental mycotoxicosis in chickens induced by ochratoxin A and penicillic acid and intervention by natural plant extracts. *Vet Res Commun.* 2004;28:727–746.

Stoloff and Trucksess. 1981.

Storm IM, Kristensen NB, Raun BM, Smedsgaard J, Thrane U. Dynamics in the microbiology of maize silage during whole-season storage. *J Appl Microbiol.* 2010;109:1017–1026.

Storm IM, Rasmussen RR, Rasmussen PH. Occurrence of pre-and post-harvest mycotoxins and other secondary metabolites in Danish maize silage. *Toxins.* 2014 Aug;6(8):2256–2269.

Suttajit M. Prevention and control of mycotoxins. In *Mycotoxin Prevention and Control in Foodgrains,* Semple RL, Frio AS, Hicks PA, Lozare JV (eds). Rome: FAO. 1989.

Turner NW, Bramhmbhatt H, Szabo-Vezse M, Poma A, Coker R, Piletsky SA. Analytical methods for determination of mycotoxins: An update (2009–2014). *Anal Chim Acta.* 2015;901:12–33.

Valerio F, Favilla M, De Bellis P, Sisto A, de Candia S, Lavermicocca P. Antifungal activity of strains of lactic acid bacteria isolated from a semolina ecosystem against *Penicillium roqueforti, Aspergillus niger* and *Endomyces fibuliger* contaminating bakery products. *Syst Appl Microbiol.* 2009 Sep;32(6):438–448.

Wambacq E. *Penicillium roqueforti SL: Growth and Roquefortine C Production in Silages.* Doctoral dissertation, Ghent University, Ghent. 2017.

Wambacq E, Audenaert K, Höfte M, De Saeger S, Haesaert G. *Bacillus velezensis* as antagonist towards *Penicillium roqueforti* sl in silage: In vitro and in vivo evaluation. *J Appl Microbiol.* 2018 Oct;125(4):986–996.

Wambacq E, Vanhoutte I, Audenaert K, De Gelder L, Haesaert G. Occurrence, prevention and remediation of toxigenic fungi and mycotoxins in silage: A review. *J Sci Food Agric.* 2016;96:2284–2302.

Wei RD, Schnoes HK, Hart PA, Strong FM. The structure of PR toxin, a mycotoxin from *Penicillium roqueforti. Tetrahedron.* 1975 Jan 1;31(2):109–114.

Yan B, Zhao J, Fan D, Tian F, Zhang H, Chen W. Antifungal activity of *Lactobacillus plantarum* against *Penicillium roqueforti* in vitro and the preservation effect on Chinese steamed bread. *J Food Process Preserv.* 2017;41:e1296910.

Yang Y, Li G, Wu D, Liu J, Li X, Luo P. Recent advances on toxicity and determination methods of mycotoxins in foodstuffs. *Trend Food Sci Technol.* 2020;96:233–252.

Zain ME. Impact of mycotoxins on humans and animals. *J Saudi Chem Soc.* 2011;15:129–144.

Zhang F, Liu B, Sheng W, Zhang Y, Liu Q, Li S, et al. Fluoroimmunoassays for the detection of zearalenone in maize using CdTe/CdS/ZnS quantum dots. *Food Chem.* 2018;255:421–428.

FURTHER READINGS

Araujo RCZ, Chalfoun SM, Angélico CL, Araujo JBS, Pereira MC. In vitro evaluation of the fungitoxic activity of seasonings on the inhibition of fungi isolated from homemade breads. *Ciênc Agrotechnol Lavras.* 2009;33:545–551. http://doi.org/10.1590/S1413-70542009000200029

Arnold DL, Scott PM, McGuire PF, Hawig J, Nera EA. Acute toxicity studies on roquefortine and P. R. toxin, metabolites of *Penicillium roqueforti* in the mouse 1979. *Cosmet Toxicol.* 1978;16:369–371.

Azzouz MA, Bullerman LB. Comparative antimycotic effects of selected herbs, spices, plant components and commercial antifungal agents. *J Food Prot.* 1982;45:1298–1301. http://doi.org/10.4315/0362-028X-45.14.1298

Bianchini A, Stratton J, Weier S, Cano C, Garcia LM. Use of essential oils and plant extracts to control microbial contamination in pet food products. *J Food Process Technol.* 2014;5:357. http://doi.org/10.4172/2157-7110.1000357

Jelen HH, Mildner S, Czaczyk K. Influence of octanoic acid addition to medium on some volatile compounds and PR-toxin biosynthesis by *Penicillium roqueforti. Lett Appl Microbiol.* 2002;35:37–41. http://doi.org/10.1046/j.1472-765X.2002.01125.x

Milićević DR, Škrinjar M, Baltić T. Real and perceived risks for mycotoxin contamination in foods and feeds: Challenges for food safety control. *Toxins.* 2010;2:572–592.

Pereira MC, Chalfoun SM, Pimenta CJ, Angélico CL, Maciel WP. Spices, fungi mycelial development and ochratoxin A production. *Sci Res Essay.* 2006;1:038–042.

Šimović M, Delaš F, Gradvol V, Kocevski D, Pavlović H. Antifungal effect of eugenol and carvacrol against foodborne pathogens *Aspergillus carbonarius* and *Penicillium roqueforti* in improving safety of fresh cut watermelon. *J Intercult Ethnopharmacol.* 2014;3:91–96. http://doi.org/10.5455/jice.20140503090524

United States. *Environmental Protection Act.* 1997, Attachment I—Final Risk Assessment *Penicillium roqueforti.*

Zargar S. Inhibitory effect of various aqueous medicinal plant extracts on citrinin production and fungal biomass by *Penicillium notatum* and *Aspergillus niger. IAIM.* 2014;1:1–8.

Chapter 10

Occurrence, Production, Determination, Toxicity, and Control Strategies of Cyclopiazonic Acid in Food Products

Sourav Misra, Sitesh Kumar, Pooja Pandey, Shubham Mandliya,
Mousumi Ghosh, Shubhangi Srivastava, Dipendra Kumar Mahato

CONTENTS

10.1	Introduction	266
10.2	Occurrence in Food and Feed	267
10.3	Toxicity of CPA	268
10.4	Gene Responsible for Production	268
	10.4.1 Formation of cAATrp by CpaA, a Hybrid Polyketidenonribosomal Peptide Synthase	269
	10.4.2 Transformation from cAATrp to β-CPA and α-CPA	272
10.5	Detection Methods	272
	10.5.1 Chromatographic Methods	272
	10.5.2 Immunochemical Methods	274
	10.5.3 Capillary Electrophoresis Method	275
10.6	Management and Control Strategies	277
	10.6.1 Grading of Lots	277
	10.6.2 Screening of Samples	278
	10.6.3 Other Physical Separation Techniques	278
10.7	Preventive Measures	278
	10.7.1 Biological Control	279
	10.7.2 Physical Methods	280
	10.7.3 Chemical Methods	280
	10.7.4 Use of Phytochemicals/Essential Oils	281

DOI: 10.1201/9781003242208-10

10.8   Risk Assessment   282
10.9   Summary   282

10.1 INTRODUCTION

Cyclopiazonic acid (CPA), often known as α-CPA (Figure 10.1), is an indole-tetramic acid discovered by Holzapfel (1968). Initially, it was believed to be a poisonous metabolite of *Penicillium griseofulvum* Dierckx as obtained in the isolation process during toxicity characterization of *P. cyclopium* (micro-fungi). Later two non-toxic indole derivatives of CPA produced by *P. cyclopium*, α-cyclopiazonic acid-imine (α-CPA-imine) and bissecodehydro cyclopiazonic acid (β-CPA), were also classified under CPA-type mycotoxins (Ostry et al., 2018). Around 27 different varieties of CPA mycotoxins have been identified, all of which are generated by the widespread genera of molds *Penicillium* and *Aspergillus* (Chang et al., 2009).

These molds colonize and grow on various food, like meat and cheese products, so many agricultural products, food sources, and animal feeds are prone to CPA contamination, but no cases of human poisoning have been reported to date due to the presence of CPA (Chang et al., 2009). A single incident of animal and human toxicity was found in India in Kodo millet that was contaminated with CPA by Rao and Husain (1985). CPA has been detected in the *Aspergillus flavus* and *Aspergillus tamarii* strains. Despite a few incidents of animal

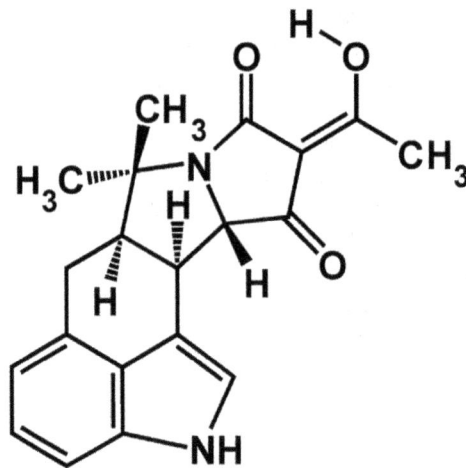

Figure 10.1   Structure of α-cyclopiazonic acid.

mycotoxicosis, the mycotoxin research community has given less attention to CPA as compared to other mycotoxins like aflatoxins, ochratoxins, fumonisins, and trichothecenes in the last two decades.

10.2 OCCURRENCE IN FOOD AND FEED

CPA has been extensively identified in various plant-based food commodities, such as rice, wheat, maize, kodo millet, peanuts, figs, poultry feed, groats, sunflower seeds, and tomato paste and puree (Rao et al., 2021; Ostry et al., 2018). CPA has also been reported in animal-based foods such as milk, salami, and cheese (Table 10.1). CPA quite often co-occurs with aflatoxins, causing turkey X disease and mycotoxicosis in humans, chickens, rats, dogs, and guinea pigs (Rao et al., 2021).

TABLE 10.1   OCCURRENCE OF CYCLOPIAZONIC ACID IN DIFFERENT FOOD PRODUCTS WORLDWIDE

Country	Foodstuffs	CPA Concentration	Reference
Philippines	Corn	27–1510 µg/kg	Hayashi and Yoshizawa (2005)
Thailand	Rice	n.d.	Hayashi and Yoshizawa (2005)
Austria	White mold cheese	Up to 3700 µg/kg	Ansari and Häubl (2016)
Norway	Wheat, peanuts, and rice	25–500 µg/kg	Moldes-Anaya et al. (2009)
Turkey	Dried figs	25–187 ng/g	Heperkan et al. (2012)
Spain	Dry-cured ham	36.1 to 540.1 ng/g	Peromingo et al. (2018)
United States	Maize	5 to 28 µg/kg	Maragos et al. (2017)
United States	Cheese	45 µg/kg	Hossain et al. (2019a)
Italy	White surface cheese	20 to 80 ppb	Zambonin et al. (2001)
Turkey	Dry fig	0.07–398.2 µg/g	Oktay Basegmez and Heperkan (2015)
Brazil	UHT milk	6.4–9.7 µg/L	Oliveira et al. (2006)
Brazil	Peanut	288.0 to 4918.1 µg/kg	Zorzete et al. (2013)
Brazil	Tomato puree	36–117 ng/g	da Motta and Valente Soares (2001)

10.3 TOXICITY OF CPA

Toxicity of CPA has been found in numerous species, like dogs, mice, monkeys, rats, laying hens, guinea pigs, lactating ewes, and chickens (Ostry et al., 2018). The liver, skeletal muscle, kidney, nervous system, and alimentary tract are the major target organs of CPA toxicity, and it has been categorized under potentially serious mycotoxins. CPA causes necrosis and degenerative changes in the spleen, liver, salivary glands, pancreas, skeletal muscles, kidney, and myocardium in rats.

In chickens served diets containing CPA (100 ppm), myocardial inflammation, hepatitis, mucosal necrosis, and proventriculitis, along with more deaths, were observed due to the accumulation of toxins in edible tissue and detected in milk and eggs (Duran et al., 2007).

In dogs, pigs, and rodents (mice and rats), CPA causes diarrhea, degeneration, weight loss, and necrosis of the viscera and muscles as well as convulsions and death. The ability of CPA to impede the sarco (endo)plasmic reticulum Ca^{2+}ATPase (SERCA) and disrupt the normal intracellular calcium flux explains the mechanism of toxicity (Hymery et al., 2014). SERCA causes the translocation of calcium present in the cytosol to the endoplasmic reticulum, which affects the cell fate, enzyme activities, and muscle contraction–relaxation cycle. It has been shown that CPA causes the blockage of calcium access channels and immobilization of four transmembrane helices of ATPase. CPA results in several injuries to the lymphoid organs, particularly the bursa of Fabricius and spleen, but the effect of CPA on the immune system of experimental animals has not been studied yet (Ostry et al., 2018).

10.4 GENE RESPONSIBLE FOR PRODUCTION

The biosynthesis of CPA in *A. flavus* via the mevalonate pathway is shown in Figure 10.2 (Navale et al., 2021). The CPA synthesis commences with mevalonate (primary precursor of dimethylallyl diphosphate; DMAPP), an essential component in terpenoid production. The primary step involves the condensation reaction between acetoacetyl-CoA and acetyl-CoA in which 3-hydroxy-3-methylglutaryl (HMG)-CoA synthase acts as a catalyst, which is then reduced by HMG-CoA reductase. Mevalonic acid is transformed to DMAPP in eukaryotes by phosphomevalonate kinase and mevalonate kinase enzymes (Chang et al., 2009). Mevalonate-5-diphosphate decarboxylase (or diphosphomevalonate decarboxylase) causes decarboxylation of mevalonic acid, after which dehydration yields isopentenyl diphosphate (IPP), a typical intermediate reported in the biosynthetic pathways of steroids, terpenoids, and the prenyl moieties of ergot alkaloids and sirodesmin. After this, IPP isomerase acts on

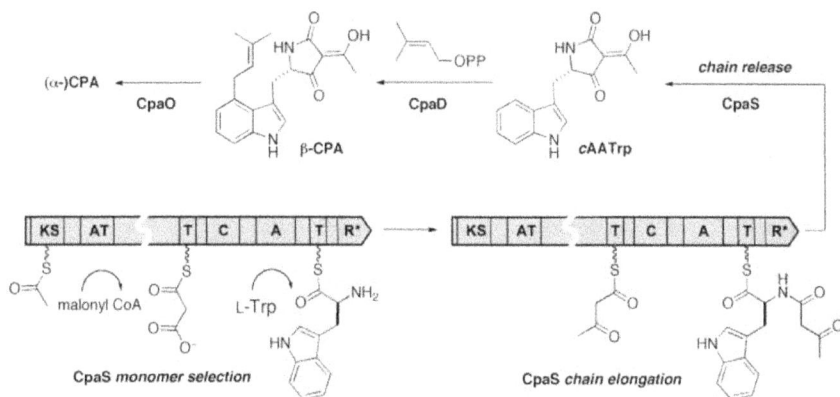

Figure 10.2   Biosynthesis of CPA and conversion of cAATrp to β-CPA and α-CPA.

Source: Adapted from Navale et al. 2021

the IPP, isomerizing it to DMAPP, which may alkylate a range of natural iso-prenoid chemicals, including IPP, to create mycotoxins, plant hormones, pharmaceutical antibiotics, and antibacterial agents (Chang et al., 2009).

In the *A. flavus* NRRL3357 genome, the genes for the production of DMAPP include HMG-CoA reductase, HMG-CoA synthase, and isopentenyl pyrophosphate isomerase. The anticipated enzyme containing a conserved Gly/Ser-rich region possibly participates in binding ATP called GHMP (galacto-, homo-serine, mevalonate, and phosphomevalonate) kinase N terminal domain and GHMP kinase C terminal domain. The diphosphomevalonate decarboxylase gene and mevalonate kinase gene are around 190 kb distant from the mevalonate kinase gene (Chang et al., 2009).

10.4.1 Formation of cAATrp by CpaA, a Hybrid Polyketidenonribosomal Peptide Synthase

In *A. flavus/oryzae*, *cyclo*-acetoacetyl-L-tryptophan (cAATrp) is the primary stable intermediate molecule generated in CPA biosynthesis. The formation of cAATrp necessitates the *pksA-nrps*-encoded hybrid polyketidenonribosomal peptide synthase (PKS-NRPS) present in *A. flavus* NRRL3357 (Figure 10.3 a&b) (Seshime et al., 2009), also known as CPA synthase (CpaA or CpaS), and a PKS-NRPS polypeptide contains four non-functional catalytics and is composed of 3906 amino acids. The transformants produce cAATrp from the *cpaA* (heterologous expression) present in *A. flavus* NRRL3357 and *A. oryzae* M-2-3 due to the effect of α-amylase promoter. Like fungal and bacterial PKSs, the PKS

a

| | KS | AT | (DH) | (MeT) | (ER) (KR) | ACP | C | A | T | R | |

A. flavus CpaA ▨▨▨ ▨▨▨▨▨▨▨▨▨▨▨▨▨▨▨▨▨▨▨▨▨▨▨▨▨▨▨▨▨ (3906 aa)

b

DH domain

```
             *    *    *
LDKS    WLRDHVVGSHIVFPGA
PKSN    WIREHKVQGDILYPGA
FUM1    WLRDHQVLNDVVFPCA
EqiS    WMQGHKLQGQIIFPAT
LNKS    WLDGHALQGQTVFPAA
(CpaA)  WAEGYKEDGRVVLSAA
```

MeT domain SAM binding

```
             *  *  *
LDKS    ARILEIGGGTGGCTQLVVDSL
PKSN    LRILEIGAGTGGTTYHVLERL
FUM1    LRVLEIGAGTGGGAQVILEGL
EqiS    MKILEIGAGTGGTTQATLPSL
LNKS    MDILEIGAGTGGATKYVLATP
(CpaA)  MNVLELDAGTSVVTHQILEVV
```

ER domain NADPH binding

```
             *  *  *
LDKS    GETVLIHAGAGGVGQAAIILAQ
PKSN    GEKVLIHAAAGGVGQAAIMIAQ
FUM1    GQSILIHSACGGIGIAALNLCR
(EqiS)  GRMALVTTAAHLVADNLVHNIP
(LNKS)  AETVISTAKCLGVTDSILVLNP
(CpaA)  SGPILLYEPDELLAAAVEQARE
```

KR domain NADPH binding

```
              *  *  *
LDKS    VSYLVAGGLGGIGRRICEWLVDRGARYLIILSRTAR
PKSN    ATYMIA-GLGGITREIARWLAEKGARYLVFLSRSAA
FUM1    ASYLLVGGLGGLGRAAATWMVESGARYLIFFSRSAG
EqiS    RTYLLIGLSGEVGQSICQWMVSHGARHVVLTSRKPG
LNKS    KTYLLVGLTGDLGRSLGRWMVQHGACHIVLTSRNPQ
(CpaA)  GTYWMIDMATPLGLSILKWMATNGARTFVLAGRNPR
```

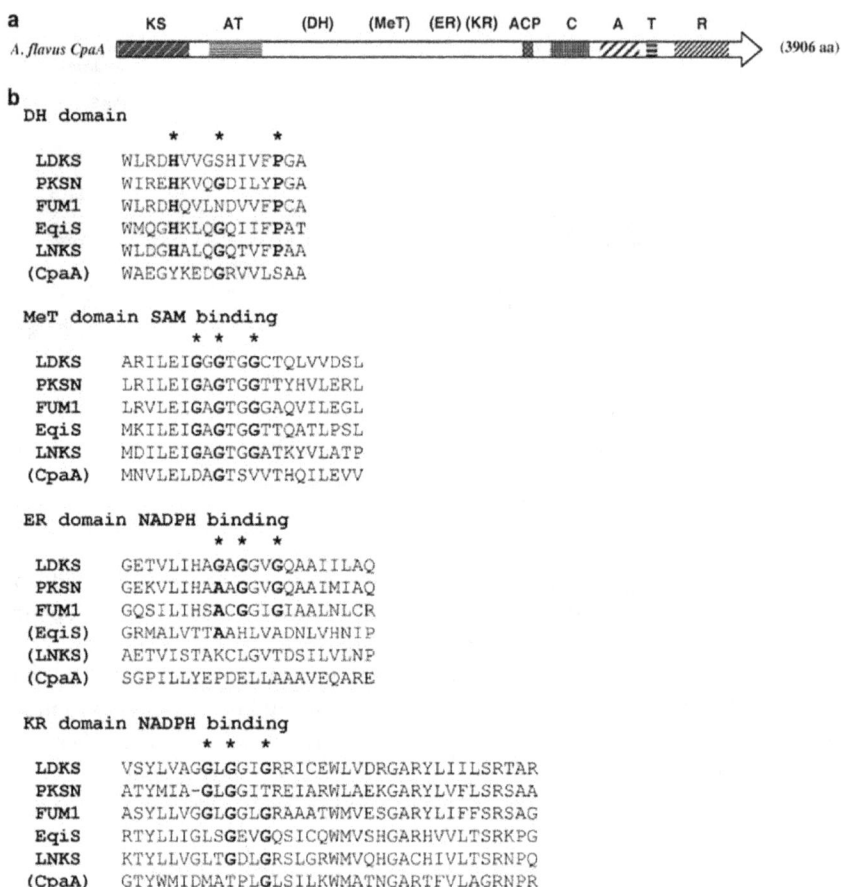

Figure 10.3   Domain architecture of CpaA PKS-NRPS and its non-functional PKS domains. (a) *A. flavus* CpaA architecture; (b) comparison of putative non-functional PKS domains of CpaA with fungal PKSs.

Source: Adapted from Seshime et al. 2009

part of CpaS comprises three functional domains: an acyl transferase (AT), a ketosynthase (KS), and an acyl carrier protein (ACP). The KS domain condenses acetyl-CoA with malonyl-CoA to generate acetoacetyl-CoA, while AT tethers the resultant acetoacetyl moiety to ACP (Figure 10.3 a&b) (Seshime et al., 2009). The PKS component of CpaS contains domains of dehydratase (DH), enoylreductase (ER), methyltransferase (MT), and ketoreductase (KR). The NRPS portion of CpaA has four functional domains: a condensation domain

(C), an adenylation domain (A), a thiolation domain (T) that has phosphopan-
theinylated peptidyl carrier protein (PCP), and a releasing domain (R). The
A-domain triggers the substrate tryptophan (Trp), subsequently transported
to the cofactor (pantetheine) on the adjacent PCP during cAATrp production
(Figure 10.4) (Seshime et al., 2009). Following this, the C domain catalyzes
amide bond formation between the PCP-tethered Trp and the ACP-tethered
acetoacetyl moiety. The resultant substrate is linked to the T domain's PCP
phosphopantetheinyl arm. CpaA's R domain catalyzes a Dieckmann condensation to cyclize acetoacetyl-Trp and liberate it from PCP to create cAATrp
(Figure 10.5 a&b) (Seshime et al., 2009).

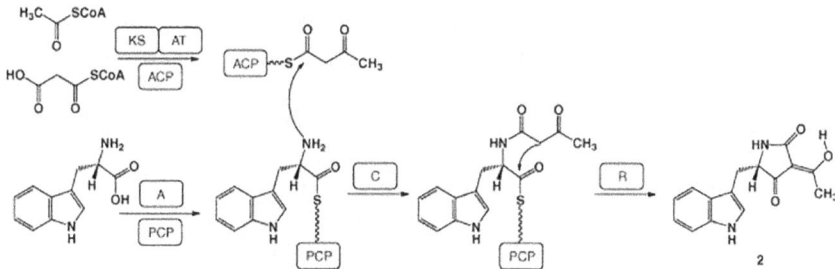

Figure 10.4   Formation of cyclo-acetoacetyl-L-tryptophan catalyzed by CpaA.

Source: Adapted from Seshime et al. 2009

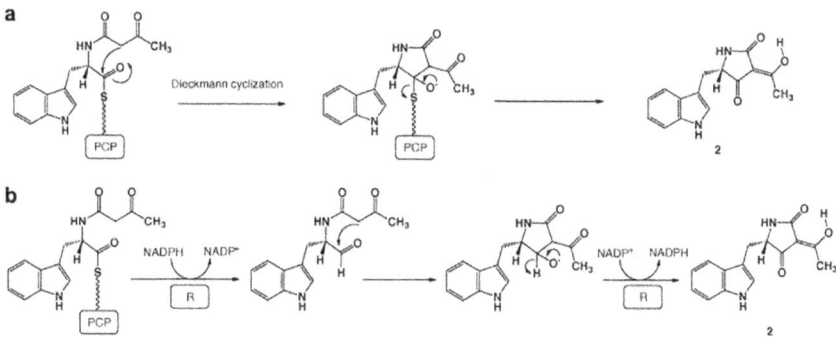

Figure 10.5   Possible roles of CpaA releasing (R) domain in cyclo-acetoacetyl-L-
tryptophan (cAATrp) formation: (a) Dieckmann cyclization/releasing mechanism; (b)
reductive releasing/cyclization/reoxidation mechanism.

Source: Adapted from Seshime et al. 2009

10.4.2 Transformation from cAATrp to β-CPA and α-CPA

The transformation of cAATrp to β-CPA requires two more stages. The alkylation of the Trp indole ring at the C-4 position by DMAPP, a process mediated by cycloacetoacetyltyptophanyl dimethylallyl transferase (DMAT), an enzyme encoded by the dmaT CPA-cluster gene, results in prenylation of cAATrp to β-CPA (Figure 10.2) (Navale et al., 2021). The transferase is also known as CpaD or dimethylallyl cycloacetoacetyl typtophan synthase (DCAT-S). The DMATs from *Aspergillus flavus* and *Aspergillus oryzae* are made up of 437 amino acids with 49 kDa molecular weight, and the functional form is most likely a dimer, as it has been described for several dimethylallyl tryptophan synthases implicated in alkaloid production. Although the *A. flavus* genome contains at least nine dmaT homologs, only the dmaT in the CPA gene cluster is linked to a PKS-NRPS gene. This DMAT seems to be the only transferase capable of catalyzing prenylation of cAATrp since breaking of CPA-cluster dmaT precludes CPA synthesis. This dmaT supports the assembly-line nature of β-CPA biosynthesis, comparable to that seen in prokaryotes for NRPSs (Chang et al., 2009).

The FAD-dependent oxidoreductase CpaO dehydrogenases α-CPA, which results in intramolecular ring closure, which is the last step in the α-CPA biosynthesis. The enzyme involved, CpaO, is a FAD-dependent oxidoreductase with certain monoamine oxidase-like characteristics. The oxidation of α-CPA is likely initiated by CpaO, resulting in a stable cation, and subsequently supports ring closure to yield β-CPA (Chang et al., 2009).

10.5 DETECTION METHODS

Several analytical techniques, such as chromatographic, immunochemical, and capillary electrophoresis methods, have been employed to identify and quantify CPA in foodstuffs and fungal cultures (Table 10.2).

10.5.1 Chromatographic Methods

The creation of the chromatographic technique is based on the ionogenic nature of CPA, which determines the polarity of both the mobile and stationary phases. Most chromatographic techniques include extraction of CPA and then isolation procedures prior to the determinative step. Solvents like dichloromethane and chloroform are often employed for the clean-up and extraction process (Ostry et al., 2018). Combinations of chloroform/methanol and acidified chloroform or mixes of nonchlorinated solvents (like aqueous acetonitrile or methanol) are utilized for the extraction of CPA from commodities. For extracting the toxin in its ionized state, an alkaline aqueous component is

TABLE 10.2   ADVANCED METHODS USED FOR CYCLOPIAZONIC ACID DETERMINATION

Country	Food Products	Analytical Method	Limit	Recovery (%)	Reference
Philippines	Corn	HPLC with UV	LOQ:* 25 ng/g	64.7–75.4	Hayashi and Yoshizawa (2005)
Thailand	Rice	HPLC with UV	LOQ: 25 ng/g	51.4–82.1	Hayashi and Yoshizawa (2005)
Austria	White mold cheese	HPLC–MS/MS	LOD:* 0.2 µg/kg LOQ: 0.5 µg/kg	80–105	Ansari and Häubl (2016)
Norway	Wheat, peanuts, and rice	HPLC–MS/MS	LOD: 5 µg/kg LOQ: 20 µg/kg	Wheat: 70–110 Peanuts: 77–116 Rice: 69–92	Moldes-Anaya et al. (2009)
Turkey	Dried figs	TLC	LOD: 25 µg/kg	74	Heperkan et al. (2012)
Spain	Dry-cured ham	UHPLC-MS/MS	LOD: 3.3 ng/kg LOQ: 10 ng/kg	96.7–99.0	Peromingo et al. (2018)
United States	Maize	Liquid chromatography-fluorescence (LC-FLD)	LOD: 30 µg/kg LOQ: 100 µg/kg	88.6 ± 12.6	Maragos et al. (2017)
United States	Cheese	Imaging surface plasmon resonance (iSPR) assay	LOD: 6 µg/kg	126	Hossain et al. (2019a)
Italy	White surface cheese	HPLC-UV/DAD	LOD: 7 ppb	82–86	Zambonin et al. (2001)

LOD: Limit of detection; LOQ: Limit of quantification

often utilized; in contrast, for neutral extraction, acidified chloroform is used. The type of extraction method (alkaline or acidic) used depends on the downstream clean-up. Recently, methanol:water or acetonitrile:water (7:3, v/v)—along with bicarbonate addition—has gained popularity, most likely due to the absence of halogenated solvents (Hayashi & Yoshizawa, 2005). In this investigation, the authors discovered that diethyl ether was an excellent alternative to chloroform for eliminating interferences from rice and maize samples. Moldes-Anaya et al. (2009) developed an advanced, quick, and effective multiple reaction monitoring (MRM) high-performance liquid chromatography-tandem mass spectrometry (HPLC–MS/MS) method for determining CPA in peanuts, mixed feed, rice, and wheat, which includes extraction of the sample using an alkaline methanol-water mixture, defatting with hexane, and quantification using HPLC-MS/MS. After extraction and isolation process, samples are frequently subjected to a detection step by thin-layer chromatography (TLC) or liquid chromatography (LC). Heperkan et al. (2012) used TLC to evaluate CPA in dried figs and discovered aflatoxin and CPA co-occurrence in 23% of the samples. Ansari and Häubl (2016) used an LC-MS/MS approach to assess CPA in white mold cheese that did not need substantial clean-up processes, and the samples were extracted in acetonitrile containing 1% formic acid. Peromingo et al. (2018) devised an extraction approach as well as the ultra-high performance liquid chromatography-mass spectrometry (UHPLC-MS/MS) method for measuring CPA in dry-cured ham, in which the composition of the mobile phase, gradient-related parameters, flow rates, and solvents utilized for resuspension of dry extracts were investigated. The best peak shape and resolution were obtained in this study by eluting the mobile phase, which comprised acetic acid-ammonium acetate buffer pH 5.75/methanol. Besides these approaches for CPA detection, some approaches have usually depended on detecting the absorbance of toxins in the ultraviolet (UV) region at 279 nm. CPA can be made fluorescent with the help of photolysis. This behavior prompted the invention of an HPLC-fluorescence detection (HPLCFLD) approach for detecting aflatoxin and CPA in fungal cells simultaneously.

10.5.2 Immunochemical Methods

CPA analysis, on the other hand, has been problematic, since it lacks fluorescence and has a broad UV absorption maximum (284 nm). Most chemical procedures need derivatization and substantial sample clean-up, making them time consuming and insensitive. The detection limits varied from 80 to 125 ppb in the case of spectrophotometric and TLC techniques, respectively. HPLC is quite sensitive and accurate and may detect as little as 25 ppb of CPA in a few food commodities; it involves extensive sample preparation and the application of costly instruments (Yu & Chu, 1998).

To address the shortcomings of the biological and chemical approaches, antibodies against CPA were developed, as well as sensitive direct competitive enzyme-linked immunosorbent assays (dc-ELISA), indirect competitive ELISA (idc-ELISA), and immunoaffinity column (IAC). Maragos et al. (2017) lately discovered unique monoclonal antibodies for the detection of CPA in corn with the limits of quantification (LOQ) for ELISA approach of 5 ng/g. This kind of antibody has been used in both IAC and ELISA tests. Unfortunately, because IAC and CPA antibodies are not commercially accessible, most analysts must apply complex isolation and detection protocols. For detection of CPA in maize, mixed feed, and peanuts, dc-ELISA evolved (Yu & Chu, 1998). In this study, two new methods for preparing enzyme markers were used: one involved coupling CPA-bovine serum albumin (CPA-BSA) conjugated to horseradish peroxidase (HRP) using either the periodate (PI) or glutaraldehyde (GA) method. The authors discovered that dc-ELISAs generated with CPA-BSA-HRP using either the PI or GA approach were more successful and utilized to analyze CPA in mixed feed, maize, and peanuts.

Biosensor technology has grown in popularity in agricultural businesses, particularly in the food sector, during the previous decade. Another potent technology is surface plasmon resonance (SPR) used for measuring biomolecular interactions in real time, and it is used in a variety of biosensors. The approach has previously been used to identify a number of mycotoxins, the applications of which have been discussed (Meneely & Elliott, 2014). Imaging SPR (iSPR) is a more contemporary form of classic SPR in which many locations of interest on the sensor surface may be probed at the same time, allowing for more efficient multiplexing (Joshi et al., 2016). iSPR sensors monitor variations in the refractive index of chips with a thin metallic (typically gold) surface. The potential benefits of biosensors include the capacity to reuse biochips, minimize solvent use, and monitor several antigen–antibody interactions in a single test. Hossain et al. (2019a) developed a sensitive and rapid iSPR assay for CPA detection in cheese and maize by combining an indirect competitive immunoassay and signal amplification based on a secondary antibody (Ab2) conjugated with gold nanoparticles and matrix-matched calibration curves.

10.5.3 Capillary Electrophoresis Method

Prasongsidh et al. (1998) devised a capillary electrophoresis technique for measuring CPA in milk. The study did not mention the LOQ for the capillary electrophoresis approach in milk; however, CPA was identified in spiked milk samples with a concentration of 20 ng/ml. Roncada et al. (2003) reported the rapid detection of CPA in goat milk using capillary electrophoresis. This approach is significantly easier since raw goat milk artificially contaminated with CPA is quickly defatted using a micro-centrifuge before injecting into the capillary electrophoresis device. The LOQ was also not stated in this investigation (Roncada et al., 2003).

Several concerns were observed while developing a multi-residue approach for mycotoxin detection, including CPA, that influence the identification of CPA under normal and typical analytical settings (Diaz et al., 2010). These include:

1. *Effect of temperature and ascorbic acid*: Ascorbic acid apparently has a stabilizing impact on CPA, as the unheated standard containing ascorbic acid lost just 5% of its CPA, whereas the unheated standard lacking ascorbic acid lost 13% of its CPA (Diaz et al., 2010). These findings suggest that when exposed to heat, ascorbic acid stabilizes CPA. Prasongsidh et al. (1998) discovered reduced levels of CPA in heated milk considerably and a further decline in CPA upon storing the heated milk at 4°C overnight.

2. *Effect of ascorbic acid and air/oxygen*: It was discovered that there was a significant reduction of CPA when the standard was left out in the open, whereas bubbling air through the standard for a brief period resulted in a lower decrease in CPA concentration (Diaz et al., 2010). After noting that CPA standard concentrations declined dramatically after only a few days of storage at 4°C, the authors speculated that oxygen exposure caused the breakdown of CPA. Because ascorbic acid may operate as an oxygen sink in the solution, this research encouraged the incorporation of ascorbic acid for the stabilization of CPA. The findings of both air exposure trials demonstrated that adding ascorbic acid to CPA resulted in stabilization of CPA. Consequently, ascorbic acid can minimize CPA degradation. CPA instability in solution is a concern because it could result in poor analytical response and repeatability during the analysis. The loss of CPA might be related to the interaction between CPA and dissolved oxygen in the solution, resulting in the formation of a peroxide species following CPA deacetylation; however, this theory has yet to be studied.

3. *Effect of air headspace*: The impact of headspace on CPA is most likely due to CPA reacting with dissolved oxygen in the air (Diaz et al., 2010). Furthermore, when headspace oxygen is depleted as a result of the reaction with CPA, diffusion of extra oxygen into the headspace provides fresh oxygen that reacts with the remaining CPA. The operation is repeated until reaching the equilibrium between the oxygen levels in the headspace vial and the dissolved oxygen (Diaz et al., 2010). CPA loss in vials because of exposure to air might cause issues with high detection limits and poor recoveries in CPA experiments (Diaz et al., 2010). The oxygen present in the air destabilizes the CPA, where CPA was reduced by 13% after 72 hours of storage (Diaz et al., 2010).

4. *Effect of pH level and solvent composition*: CPA was found to degrade per the pH level of the solution, specifically in the presence of formic acid, while ascorbic acid did not affect CPA stabilization in acidic circumstances (Diaz et al., 2010). The authors speculated that the acidic solution's strong destabilizing impact on CPA was due to the weak acid catalysis of a process in the vial containing CPA. Along with exposure to oxygen, this is another combination of circumstances that leads to CPA breakdown. Furthermore, when protons become increasingly available in solution, the rate of catalysis rises, as does the product production rate (Diaz et al., 2010). Due to the increased proton availability, a larger loss of CPA was reported in the vial. Ammonia might be effective in stabilizing CPA from the mild acid catalysis that may occur due to exposure to the optimum mobile phase necessary for analyzing other mycotoxins. This strategy, however, might not be applicable to methods for other mycotoxins since some mycotoxins are not stable at higher pH.
5. *Adsorption to plastic*: When the standard for CPA was prepared in plastic test tubes, there was a considerable CPA loss (35%), indicating that CPA was absorbed by plastic (Diaz et al., 2010). These findings suggest that glass is the preferable material for the preparation of standard solutions of CPA in primarily aqueous solvents.

10.6 MANAGEMENT AND CONTROL STRATEGIES

The bulk of CPA contamination in food items like groundnuts is caused by *A. flavus*, which often co-occurs with aflatoxins. As a result, the controls for preventing CPA contamination are the same as those for avoiding aflatoxin contamination. Management tactics are implemented in stages, beginning with production, harvesting, storage, and customer purchase. Contamination occurs in the field due to drought and heat stress, during product delivery to the point of sale, during improper drying, and during the storage of farmers' stock. The majority of the techniques involve physically separating tainted from uncontaminated items.

10.6.1 Grading of Lots

This entails removing lots with visible *A. flavus* and *A. parasiticus* from the food supply chain, the provision of mycotoxins in collected samples for analysis, and the segregation of lots in each stage of the production line depending on the concentration of CPA detected (Dorner, 2008).

10.6.2 Screening of Samples

Following lot segregation, the initial step in minimizing CPA contamination is screening to isolate particular components with high mycotoxin risk. The significant risk of aflatoxin and CPA is linked to loose shelled kernels being harvested and mishandled. As a result, eliminating high-risk components before shelling and storage may help reduce aflatoxin and CPA contamination in shelled lots during long-term storage. Before the shelling of grains, these separations were done using vibratory perforated screens or belt screens for many years.

10.6.3 Other Physical Separation Techniques

Following shelling, the samples are passed onto a gravity table separator based on density or specific gravity to remove foreign debris and unshelled grains, which reduces the chance of CPA contamination. The grains are sorted according to size by passing through a succession of slotted or round hole screens. The isolation of immature and smaller grains from bigger lots is recommended since they are more vulnerable to aflatoxin and CPA contamination because as the grain size rises, mycotoxin concentrations decrease. Electronic color sorting is the most recently developed grading system for regulating aflatoxin and CPA in commercial shelling operations (Dorner, 2008). It effectively eliminates a more significant proportion of infected, discolored grains, decreasing aflatoxin and CPA concentrations. Blanching preceded by electronic color sorting is one of the most effective procedures for lowering aflatoxin and CPA in shelled peanut lots. It eliminates the testa from kernels, and the electronic sorter effectively separates aflatoxin-contaminated kernels from the blanched lot.

10.7 PREVENTIVE MEASURES

Aflatoxin contamination of wheat before harvest may be mostly prevented with correct and enough irrigation. *A. flavus* and *A. parasiticus* do not colonize developing and mature peanuts until kernel moisture (water activity) starts diminishing due to late-season dry conditions and rising soil warmth. Even if fungal invasion occurs, maintaining high kernel water activity up to harvest supports peanuts' natural defense mechanism (production of phytoalexin) against the development of aflatoxigenic fungus (Dorner, 2008). Unfortunately, many peanut growers lack access to extra irrigation, or the investment is too expensive. The contamination of peanuts may be avoided by promptly drying them until the water activity reaches 0.83, thus preventing aflatoxin development after they are dug and harvested. After then, it is essential to maintain safe storage moisture until the processing of peanuts. Maintaining safe moisture may not be easier during harvest and storage due to environmental conditions. If the means are available, controlling the moisture content of kernels is one of the best strategies to avoid aflatoxin and CPA contamination of peanuts. Therefore, the danger of aflatoxin contamination may be considered, allowing for earlier-than-usual harvesting to reduce aflatoxin contamination (Dorner, 2008).

10.7.1 Biological Control

An advanced biological control method has been evolved to avoid most aflatoxin and CPA contamination of peanuts. This control is done by introducing a competitive exclusion that can be attained by introducing a non-toxigenic competitive strain of *A. flavus* into the soil of growing peanuts. Biological control has been advantageous for preventing both pre- and post-harvest contamination with aflatoxin and CPA. This technique has been successfully commercialized, and the biopesticide Afla-Guard was registered for aflatoxin decontamination of peanuts by the US Environmental Protection Agency in 2004 (Dorner, 2008).

CPA decontamination in hulled barley has been achieved by covering it with non-toxic *A. flavus* conidia (NRRL 21882). Not only does the strain lack aflatoxin production, but it also lacks CPA and other precursors of aflatoxin biosynthesis (Dorner, 2008). A loss of the complete aflatoxin gene cluster was discovered during a genetic study of the strain (Chang et al., 2009). Afla-Guard is best applied 60–80 days after planting or immediately after the peanut crop's canopy has closed. The coated conidia germinate and proliferate after being applied and absorbing moisture, creating copious sporulation that is spread into the soil to compete with naturally occurring toxigenic strains. According to research, this biological management technique may reduce aflatoxin contamination by around 80–90% (Dorner, 2008). Delgado et al. (2019) applied a novel biological control technique to avoid CPA contamination in dry-fermented sausage produced by *Penicillium griseofulvum* and found that *P. chrysogenum* produces the PgAFP protein, significantly decreasing CPA formation on dry-fermented sausages. Because the metabolites generated by *P. chrysogenum* are generally regarded as safe (GRAS), using this strain with good manufacturing practices (GMPs) in Hazard Analysis Critical Control Point (HACCP) protocols to reduce consumer exposure to CPA from dry-fermented sausages may be studied. Biological control is a better option, but biological agents must meet specific criteria, including being non-toxic and segregated from dry-cured beef products.

10.7.2 Physical Methods

The effectiveness of physical methods for controlling CPA in foods has not been explored yet by researchers. The bulk of CPA contamination in food items is caused by *A. flavus*, which often co-occurs with aflatoxins. Thus, to prevent CPA and aflatoxin decontamination, physical methods can also be proposed. Conventional thermal methods such as roasting, cooking, baking, extrusion, and frying coupled with pre-and post-harvest management strategies have been employed for aflatoxin decontamination. Processing time and temperature, moisture content, and properties of the food matrix are the critical influencing factors in traditional thermal processing (Sipos et al., 2021). Other novel processing technologies such as microwave heating, cold plasma, pulsed light, pulsed electric field, and gamma and electron beam irradiation have been successfully applied to prevent mycotoxins in food materials.

The application of microwave irradiation (with a frequency from 300 MHz to 300 GHz and wavelengths from 1 to 1000 mm) on stored cereals, pulses, and oilseeds is an advantageous process due to the volumetric nature of electromagnetic radiation, rapid drying, and high drying efficiency (Misra et al., 2022). It was observed that longer exposure to microwave radiation at a lower power level or vice versa was effective for aflatoxin as well as CPA degradation in foodstuffs, and an internal temperature of >150°C resulted in a 90% reduction of aflatoxin (Pankaj et al., 2018).

Ionizing radiations such as ultraviolet, gamma rays, and electron beams have shown a pronounced effect on the destabilization of aflatoxins in peanuts, almonds, corn, dried figs, and dates (Sipos et al., 2021). Both the number of fungi and concentrations of mycotoxins were reduced due to the exposure to gamma radiation with a dose of 1 to 10 kGy in naturally contaminated corn kernels (Serra et al., 2018). Aflatoxin degradation by gamma rays might be attributed to free radical reaction due to radiolysis of water, which can readily attack AFB_1 but the efficacy of degradation could be improved when combined with other physical treatments (Pankaj et al., 2018). A similar effect by electron beam radiation on foods was observed, but the variations were in the methods of application and depth of penetration (Assuncao et al., 2015). Aflatoxins are more susceptible to degradation when exposed to ultraviolet radiation at 362 nm, but their application is limited due to the narrower wavelength, lower depth of penetration, and residual toxicity retention by the degraded product (Pankaj et al., 2018).

10.7.3 Chemical Methods

Currently, various chemical methods for the reduction of aflatoxin and CPA are based on chemical agents such as acids, enzymes, and ozone treatment. Acidification and ozonation are the two primary chemical treatment methods to reduce the aflatoxin and CPA in food products. Several organic and inorganic acids depending on the type of food products were used as a preventive measure to stop the efficacy of CPA and aflatoxins. Acidification of foods contaminated with lactic, citric, hydrochloric, and tartaric acid was more effective than acids such as ascorbic, acetic, formic, and succinic against the aflatoxin present in foods (Rushing & Selim, 2019).

Lee et al. (2015) concluded that tartaric acid and succinic acid provided the highest and lowest degradation of aflatoxin in soybean with a treatment time of 18 h, respectively. Rushing and Selim (2016) found that only hydrochloric acid (85%) and citric acid (73.4%) provided a good reduction in aflatoxin content with a treatment time of 24 h.

Ozonation is one of the most common chemically used methods to reduce the aflatoxin content in foods. For a shorter treatment time, ozonation in the range of 6–90 mg/L has effectively reduced aflatoxins (Rushing & Selim, 2019). Inan et al. (2007) investigated that in red pepper, ozonation (66 mg/L) reduced aflatoxin by 93%. Zorlugenç et al. (2008) observed the ozonation effect (13.8 mg/L) on dried figs and found 95.21% degradation in aflatoxin content. El-Desouky et al. (2012) studied the ozonation effect (40 ppm) on wheat and found 86.75% in only 20 min treatment time. Several researchers tried the ozonation effect on peanuts and found degradation from 65.9 to 89.4% based on different ozonation levels (Chen et al., 2014; Diao et al., 2013; de Alencar et al., 2012). These chemical methods were found to be more effective when combined with heat treatment (Rushing & Selim, 2019).

10.7.4 Use of Phytochemicals/Essential Oils

Foods are contaminated with different fungal toxins during their growth. These lead to major health concerns and economic loss. Thus, various strategies such as physical, chemical, or biological methods have been used to preserve the foods from toxins. However, both physical and chemical methods are expensive (Mahato et al., 2019), whereas biological methods interact with nutrients (Udomkun et al., 2017). Thus, there is a need for sustainable and novel alternative methods for the preservation of food against toxins with marginal or no mycotoxin residue (Haque et al., 2020). Essential oils and their bioactives from spices and herbs are novel and green alternatives to reduce mycotoxins and preserve foods with the lowest cost and have a generally recognized as safe status (Prakash et al., 2020). Essential oils and bioactives have many advantages over traditional methods due to their pharmacological properties and inexpensiveness, thereby increasing the shelf life as well as preserving foods (Kumar et al., 2022). However, their high volatility and poor solubility often act as a barrier to their usefulness (Ju et al., 2019). Another major challenge with essential oil is their adverse effect on the organoleptic property that needs to be preserved. Thus, micro- or nano-encapsulation is required to preserve their properties without affecting their characteristics.

The major fungi related to CPA are *Penicillium* and *Aspergillus,* which are mainly found in rice, wheat, peanut, maize, and so on. Various researchers used anise-, thyme-, and peppermint-based essential oils/nanoemulsions to preserve the rice from *Aspergillus* fungi that produce CPA (Das et al., 2021; Wan et al., 2019; Hossain et al., 2019b). Pepperina- and rose-scented geranium–based nanogels/nanoparticles were used to preserve maize from the *Aspergillus* fungi (Kujur et al., 2020; López-Meneses et al., 2018).

10.8 RISK ASSESSMENT

In published trials, the benchmark dose (BMD), "no observed adverse effect level" (NOAEL), and "lowest observed adverse effect level" (LOAEL) for CPA for any outcome could not be found. There are very little data available on CPA toxicity, but they are important for risk assessment. As a result, neither the European Food Safety Agency (EFSA) nor the Joint FAO/WHO Expert Committee on Food Additives (JECFA) have conducted any risk assessments. Nonetheless, following a pig study, Burdock and Flamm (2000) calculated an acceptable daily intake (ADI) of 10 g/kg body weight (BW) based on the NOEL value of 1.0 mg/kg BW/day. Overall, it was determined that the statistics from pertinent sub-chronic inquiries on CPA in experimental animals are insufficient for TDI calculation.

10.9 SUMMARY

The molds producing CPA *Penicillium* and *Aspergillus* colonize and grow on various food, like meat and cheese products, many agricultural products, food sources, and animal feeds. CPA toxicity has been found in several animal species, but no cases of human poisoning have been reported to date due to the presence of CPA. Management practices should be implemented in various stages, beginning with production, harvesting, storage, and customer purchase, and suitable detection methods should be standardized for the assessment of the toxicity and severity of CPA and health risk assessment for animals and humans.

REFERENCES

Ansari, P., & Häubl, G. (2016). Determination of cyclopiazonic acid in white mold cheese by liquid chromatography–tandem mass spectrometry (HPLC–MS/MS) using a novel internal standard. *Food Chemistry, 211,* 978–982.

Assuncao, E., Reis, T. A., Baquiao, A. C., & Correa, B. (2015). Effects of gamma and electron beam radiation on Brazil nuts artificially inoculated with *Aspergillus flavus*. *Journal of Food Protection, 78*(7), 1397–1401.

Burdock, G. A., & Flamm, W. G. (2000). Safety assessment of the mycotoxin cyclopiazonic acid. *International Journal of Toxicology, 19*(3), 195–218.

Chang, P. K., Ehrlich, K. C., & Fujii, I. (2009). Cyclopiazonic acid biosynthesis of *Aspergillus flavus* and *Aspergillus oryzae*. *Toxins, 1*(2), 74–99.

Chen, R., Ma, F., Li, P. W., Zhang, W., Ding, X. X., Zhang, Q. I., . . . & Xu, B. C. (2014). Effect of ozone on aflatoxins detoxification and nutritional quality of peanuts. *Food Chemistry, 146*, 284–288.

da Motta, S., & Valente Soares, L. M. (2001). Survey of Brazilian tomato products for alternariol, alternariol monomethyl ether, tenuazonic acid and cyclopiazonic acid. *Food Additives & Contaminants, 18*(7), 630–634.

Das, S., Singh, V. K., Dwivedy, A. K., Chaudhari, A. K., & Dubey, N. K. (2021). Nanostructured *Pimpinella anisum* essential oil as novel green food preservative against fungal infestation, aflatoxin B1 contamination and deterioration of nutritional qualities. *Food Chemistry, 344*, 128574.

de Alencar, E. R., Faroni, L. R. D. A., Soares, N. D. F. F., da Silva, W. A., & da Silva Carvalho, M. C. (2012). Efficacy of ozone as a fungicidal and detoxifying agent of aflatoxins in peanuts. *Journal of the Science of Food and Agriculture, 92*(4), 899–905.

Delgado, J., Peromingo, B., Rodríguez, A., & Rodríguez, M. (2019). Biocontrol of *Penicillium griseofulvum* to reduce cyclopiazonic acid contamination in dry-fermented sausages. *International Journal of Food Microbiology, 293*, 1–6.

Diao, E., Hou, H., Chen, B., Shan, C., & Dong, H. (2013). Ozonolysis efficiency and safety evaluation of aflatoxin B1 in peanuts. *Food and Chemical Toxicology, 55*, 519–525.

Diaz, G., Thompson, W., & Martos, P. (2010). Stability of cyclopiazonic acid in solution. *World Mycotoxin Journal, 3*(1), 25–33.

Dorner, J. W. (2008). Management and prevention of mycotoxins in peanuts. *Food Additives and Contaminants, 25*(2), 203–208.

Duran, R. M., Cary, J. W., & Calvo, A. M. (2007). Production of cyclopiazonic acid, aflatrem, and aflatoxin by *Aspergillus flavus* is regulated by veA, a gene necessary for sclerotial formation. *Applied Microbiology and Biotechnology, 73*(5), 1158–1168.

El-Desouky, T. A., Sharoba, A. M. A., El-Desouky, A. I., El-Mansy, H. A., & Naguib, K. (2012). Effect of ozone gas on degradation of aflatoxin B1 and *Aspergillus flavus* fungal. *Journal of Environmental and Analytical Toxicology, 2*(1), 128.

Haque, M. A., Wang, Y., Shen, Z., Li, X., Saleemi, M. K., & He, C. (2020). Mycotoxin contamination and control strategy in human, domestic animal and poultry: A review. *Microbial Pathogenesis, 142*, 104095.

Hayashi, Y., & Yoshizawa, T. (2005). Analysis of cyclopiazonic acid in corn and rice by a newly developed method. *Food Chemistry, 93*(2), 215–221.

Heperkan, D., Somuncuoglu, S., Karbancioglu-Güler, F., & Mecik, N. (2012). Natural contamination of cyclopiazonic acid in dried figs and co-occurrence of aflatoxin. *Food Control, 23*(1), 82–86.

Holzapfel, C. W. (1968). The isolation and structure of cyclopiazonic acid, a toxic metabolite of *Penicillium cyclopium* Westling. *Tetrahedron, 24*(5), 2101–2119.

Hossain, F., Follett, P., Salmieri, S., Vu, K. D., Fraschini, C., & Lacroix, M. (2019a). Antifungal activities of combined treatments of irradiation and essential oils (EOs) encapsulated chitosan nanocomposite films in in vitro and in situ conditions. *International Journal of Food Microbiology*, *295*, 33–40.

Hossain, Z., Busman, M., & Maragos, C. M. (2019b). Immunoassay utilizing imaging surface plasmon resonance for the detection of cyclopiazonic acid (CPA) in maize and cheese. *Analytical and Bioanalytical Chemistry*, *411*(16), 3543–3552.

Hymery, N., Masson, F., Barbier, G., & Coton, E. (2014). Cytotoxicity and immunotoxicity of cyclopiazonic acid on human cells. *Toxicology in Vitro*, *28*(5), 940–947.

Inan, F., Pala, M., & Doymaz, I. (2007). Use of ozone in detoxification of aflatoxin B1 in red pepper. *Journal of Stored Products Research*, *43*(4), 425–429.

Joshi, S., Segarra-Fas, A., Peters, J., Zuilhof, H., van Beek, T. A., & Nielen, M. W. (2016). Multiplex surface plasmon resonance biosensing and its transferability towards imaging nanoplasmonics for detection of mycotoxins in barley. *Analyst*, *141*(4), 1307–1318.

Ju, J., Chen, X., Xie, Y., Yu, H., Guo, Y., Cheng, Y., . . . & Yao, W. (2019). Application of essential oil as a sustained release preparation in food packaging. *Trends in Food Science & Technology*, *92*, 22–32.

Kujur, A., Kumar, A., Yadav, A., & Prakash, B. (2020). Antifungal and aflatoxin B1 inhibitory efficacy of nanoencapsulated *Pelargonium graveolens* L. essential oil and its mode of action. *LWT*, *130*, 109619.

Kumar, P., Mahato, D. K., Gupta, A., Pandhi, S., Mishra, S., Barua, S., . . . & Kamle, M. (2022). Use of essential oils and phytochemicals against the mycotoxins producing fungi for shelf-life enhancement and food preservation. *International Journal of Food Science & Technology*, *57*, 2171–2184.

Lee, J., Her, J. Y., & Lee, K. G. (2015). Reduction of aflatoxins (B1, B2, G1, and G2) in soybean-based model systems. *Food Chemistry*, *189*, 45–51.

López-Meneses, A. K., Plascencia-Jatomea, M., Lizardi-Mendoza, J., Fernández-Quiroz, D., Rodríguez-Félix, F., Mouriño-Pérez, R. R., & Cortez-Rocha, M. O. (2018). *Schinus molle* L. essential oil-loaded chitosan nanoparticles: Preparation, characterization, antifungal and anti-aflatoxigenic properties. *LWT*, *96*, 597–603.

Mahato, D. K., Lee, K. E., Kamle, M., Devi, S., Dewangan, K. N., Kumar, P., & Kang, S. G. (2019). Aflatoxins in food and feed: an overview on prevalence, detection and control strategies. *Frontiers in Microbiology*, 2266.

Maragos, C. M., Sieve, K. K., & Bobell, J. O. H. N. (2017). Detection of cyclopiazonic acid (CPA) in maize by immunoassay. *Mycotoxin Research*, *33*(2), 157–165.

Meneely, J. P., & Elliott, C. T. (2014). Rapid surface plasmon resonance immunoassays for the determination of mycotoxins in cereals and cereal-based food products. *World Mycotoxin Journal*, *7*(4), 491–505.

Misra, S., Rayaguru, K., Dash, S. K., Mohanty, S., & Panigrahi, C. (2022). Efficacy of microwave irradiation in enhancing the shelf life of groundnut (*Arachis hypogaea* L.). *Journal of Stored Products Research*, *97*, 101957.

Moldes-Anaya, A. S., Asp, T. N., Eriksen, G. S., Skaar, I., & Rundberget, T. (2009). Determination of cyclopiazonic acid in food and feeds by liquid chromatography–tandem mass spectrometry. *Journal of Chromatography A*, *1216*(18), 3812–3818.

Navale, V., Vamkudoth, K. R., Ajmera, S., & Dhuri, V. (2021). *Aspergillus* derived mycotoxins in food and the environment: Prevalence, detection, and toxicity. *Toxicology Reports*, *8*, 1008–1030.

Oktay Basegmez, H. I., & Heperkan, D. (2015). Aflatoxin, cyclopiazonic acid and β-nitropropionic acid production by *Aspergillus* section Flavi from dried figs grown in Turkey. *Quality Assurance and Safety of Crops & Foods, 7*(4), 477–485.

Oliveira, C. A., Rosmaninho, J., & Rosim, R. (2006). Aflatoxin M1 and cyclopiazonic acid in fluid milk traded in São Paulo, Brazil. *Food Additives and Contaminants, 23*(2), 196–201.

Ostry, V., Toman, J., Grosse, Y., & Malir, F. (2018). Cyclopiazonic acid: 50th anniversary of its discovery. *World Mycotoxin Journal, 11*(1), 135–148.

Pankaj, S. K., Shi, H., & Keener, K. M. (2018). A review of novel physical and chemical decontamination technologies for aflatoxin in food. *Trends in Food Science & Technology, 71*, 73–83.

Peromingo, B., Rodriguez, M., Nunez, F., Silva, A., & Rodríguez, A. (2018). Sensitive determination of cyclopiazonic acid in dry-cured ham using a QuEChERS method and UHPLC–MS/MS. *Food Chemistry, 263*, 275–282.

Prakash, B., Kumar, A., Singh, P. P., & Songachan, L. S. (2020). Antimicrobial and antioxidant properties of phytochemicals: Current status and future perspective. In *Functional and Preservative Properties of Phytochemicals* (pp. 1–45). Academic Press.

Prasongsidh, B. C., Kailasapathy, K., Skurray, G. R., & Bryden, W. L. (1998). Analysis of cyclopiazonic acid in milk by capillary electrophoresis. *Food Chemistry, 61*(4), 515–519.

Rao, B. L., & Husain, A. (1985). Presence of cyclopiazonic acid in kodo millet (*Paspalum scrobiculatum*) causing 'kodua poisoning' in man and its production by associated fungi. *Mycopathologia, 89*(3), 177–180.

Rao, V. K., Navale, V., Ajmera, S., & Dhuri, V. (2021). *Aspergillus* derived mycotoxins in food and the environment: Prevalence, detection, and toxicity. *Toxicology Reports, 8*, 1008–1030.

Roncada, P., Cretich, M., Chiari, M., & Greppi, G. F. (2003). Fast identification of cyclopiazonic acid in milk by capillary electrophoresis. *Italian Journal of Animal Science, 2*(sup1), 551–553.

Rushing, B. R., & Selim, M. I. (2016). Effect of dietary acids on the formation of aflatoxin B2a as a means to detoxify aflatoxin B1. *Food Additives & Contaminants: Part A, 33*(9), 1456–1467.

Rushing, B. R., & Selim, M. I. (2019). Aflatoxin B1: A review on metabolism, toxicity, occurrence in food, occupational exposure, and detoxification methods. *Food and Chemical Toxicology, 124*, 81–100.

Serra, M. S., Pulles, M. B., Mayanquer, F. T., Vallejo, M. C., Rosero, M. I., Ortega, J. M., & Naranjo, L. N. (2018). Evaluation of the use of gamma radiation for reduction of aflatoxin B1 in corn (*Zea mays*) used in the production of feed for broiler chickens. *Journal of Agricultural Chemistry and Environment, 7*(1), 21–33.

Seshime, Y., Juvvadi, P. R., Tokuoka, M., Koyama, Y., Kitamoto, K., Ebizuka, Y., & Fujii, I. (2009). Functional expression of the *Aspergillus flavus* PKS–NRPS hybrid CpaA involved in the biosynthesis of cyclopiazonic acid. *Bioorganic & Medicinal Chemistry Letters, 19*(12), 3288–3292.

Sipos, P., Peles, F., Brassó, D. L., Béri, B., Pusztahelyi, T., Pócsi, I., & Győri, Z. (2021). Physical and chemical methods for reduction in aflatoxin content of feed and food. *Toxins, 13*(3), 204.

Udomkun, P., Wiredu, A. N., Nagle, M., Müller, J., Vanlauwe, B., & Bandyopadhyay, R. (2017). Innovative technologies to manage aflatoxins in foods and feeds and the profitability of application–A review. *Food Control*, 76, 127–138.

Wan, J., Zhong, S., Schwarz, P., Chen, B., & Rao, J. (2019). Enhancement of antifungal and mycotoxin inhibitory activities of food-grade thyme oil nanoemulsions with natural emulsifiers. *Food Control*, 106, 106709.

Yu, W., & Chu, F. S. (1998). Improved direct competitive enzyme-linked immunosorbent assay for cyclopiazonic acid in corn, peanuts, and mixed feed. *Journal of Agricultural and Food Chemistry*, 46(3), 1012–1017.

Zambonin, C. G., Monaci, L., & Aresta, A. (2001). Determination of cyclopiazonic acid in cheese samples using solid-phase microextraction and high performance liquid chromatography. *Food Chemistry*, 75(2), 249–254.

Zorlugenç, B., Zorlugenç, F. K., Öztekin, S., & Evliya, I. B. (2008). The influence of gaseous ozone and ozonated water on microbial flora and degradation of aflatoxin B1 in dried figs. *Food and Chemical Toxicology*, 46(12), 3593–3597.

Zorzete, P., Baquião, A. C., Atayde, D. D., Reis, T. A., Gonçalez, E., & Corrêa, B. (2013). Mycobiota, aflatoxins and cyclopiazonic acid in stored peanut cultivars. *Food Research International*, 52(1), 380–386.

Chapter 11

Tremorgenic Mycotoxin Concerns in Food and Feed

Detection and Management Strategies

Himani, Mitali Madhumita, Mohit Singla, Pramod K Prabhakar

CONTENTS

11.1 Introduction   287
11.2 Occurrence of Tremorgenic Mycotoxins in Food and Feed   290
    11.2.1 Mechanism of Tremorgen Production   292
11.3 Sources of Toxins   293
    11.3.1 *Penicillium*-Synthesized Tremorgens   293
    11.3.2 *Aspergillus*-Synthesized Tremorgens   294
    11.3.3 *Claviceps*-Synthesized Tremorgens   294
    11.3.4 *Epichloë*-Synthesized Tremorgens   295
11.4 Gene Responsible for Production of Tremorgenic Mycotoxins   296
11.5 Detection Methods for Tremorgenic Mycotoxins   298
    11.5.1 High-Performance Liquid Chromatography-Tandem
            Mass Spectrometry   299
    11.5.2 Liquid Chromatography/Mass Spectrometry   300
    11.5.3 Thin-Layer Chromatography   300
    11.5.4 Gas Chromatography/Mass Spectrometry   301
    11.5.5 Issues in the Detection of Tremorgenic Mycotoxins   301
    11.5.6 Recent Developments   301
11.6 Management and Control Strategies   302
11.7 Conclusion   303

11.1 INTRODUCTION

Fungi are microorganisms that produce diversified secondary metabolites, which are non-essential to the producing microorganism. These secondary

DOI: 10.1201/9781003242208-11

metabolites may vary in their structure as well as pharmacological properties (Evans and Gupta, 2018; Gerald and Gloer, 2016). Tremorgenic mycotoxins are one of such secondary metabolites produced by filamentous fungi
belonging to the genera *Penicillium, Aspergillus, Claviceps,* and *Neotyphodium.*
They are known as indole diterpenes and are produced from geranylgeranyl
diphosphate (GGPP) and a tryptophan-derived indole moiety biosyntheti-
cally (Maragos, 2020; Gao et al., 2011; Saikia et al., 2008; Byrne et al., 2002).
Tremorgens induce intermittent or sustained muscle tremors in humans and
several animal groups (Reddy et al., 2019; Kozák et al., 2019). In addition, they
also cause convulsions, ataxia, and confusion and in extreme cases can lead to
death (Awuchi et al., 2021). Although tremorgens are produced by unassociated
fungal species, they all consist of a modified indole moiety that characterizes
the structural and biological features (Reddy et al., 2019; Gao et al., 2011).

Tremorgens were first isolated in 1964 (Gordon et al., 1993). These indole diterpenes can be broadly classified into two groups, paxilline-like compounds and
those that do not have a paxilline-like core. The paxilline-like classes include
more than 70 of these indole diterpenes identified to date (Saikia et al., 2008; Lane
et al., 2000). These classes include the six structural groups: paxilline, penitrems,
lolitrems, janthitrems, aflatrem, and paspaline (Saikia et al., 2008; Steyn and
Vleggaar, 1985). Further, these classes may include others like terpendoles produced by *Neotyphodium lolli* isolated from perennial ryegrass (Saikia et al., 2008;
Gatenby et al., 1999) and *Albophoma yamanashiensis* (syn. Chaunopycnis alba)
(Saikia et al., 2008; Huang et al., 1995), shearinines produced from *Eupenicillum
shearii* (Saikia et al., 2008; Belofsky et al., 1995), and sulpinines from *Aspergillus
sulphureus* (Saikia et al., 2008; Laakso et al., 1992). These compounds have structures with different patterns of hydroxylation, acetylation, chlorination, and
epoxidations and also differ in ring stereochemistry (Saikia et al., 2008).

The non-paxilline indole diterpenes involve emindoles produced by
Emericella spp. (Saikia et al., 2008; Nozawa et al., 1988), petromindoles produced by *Petromyces muricatus* (Saikia et al., 2008; Xiong et al., 2003), nodulisporic acid produced by *Nodulisporium* sp. (Saikia et al., 2008; Hensens et al.,
1999), aflavinines produced by *Aspergillus flavus* from its sclerotia, nominines
from the sclerotia from *A. nomius* (Saikia et al., 2008; Gloer, 1995), radarins
produced from the sclerotia from *A. sulphureius* (Saikia et al., 2008; Laakso et al.,
1992), and thiersinines produced from *Penicillium thiersii* (Saikia et al., 2008; Li
et al., 2002). Around 20 of the mycotoxins with a tryptophan-derived indole
moiety are identified for exhibiting tremorgenic effects in animals and humans,
which include penitrems, paspaline, paspalicine, paspalinine, lolitrems, paxilline, aflatrem, janthitrems, and paspalitrem A and B (Evans and Gupta, 2018;
Bennett and Klich, 2003). Some other fungal metabolites such as roquefortine
C, paspaline, paspalacine, and cyclopiazonic, however, have a similar chemical
structure to these mycotoxins but are not known to be tremorgenic in nature
(Evans and Gupta, 2018; Tiwary et al., 2009; Gordon et al., 1993). Further, tremorgenic mycotoxins can also be categorized into four subgroups on the basis of
nitrogen atoms per molecule (Figure 11.1) (Sabater-Vilar et al., 2003).

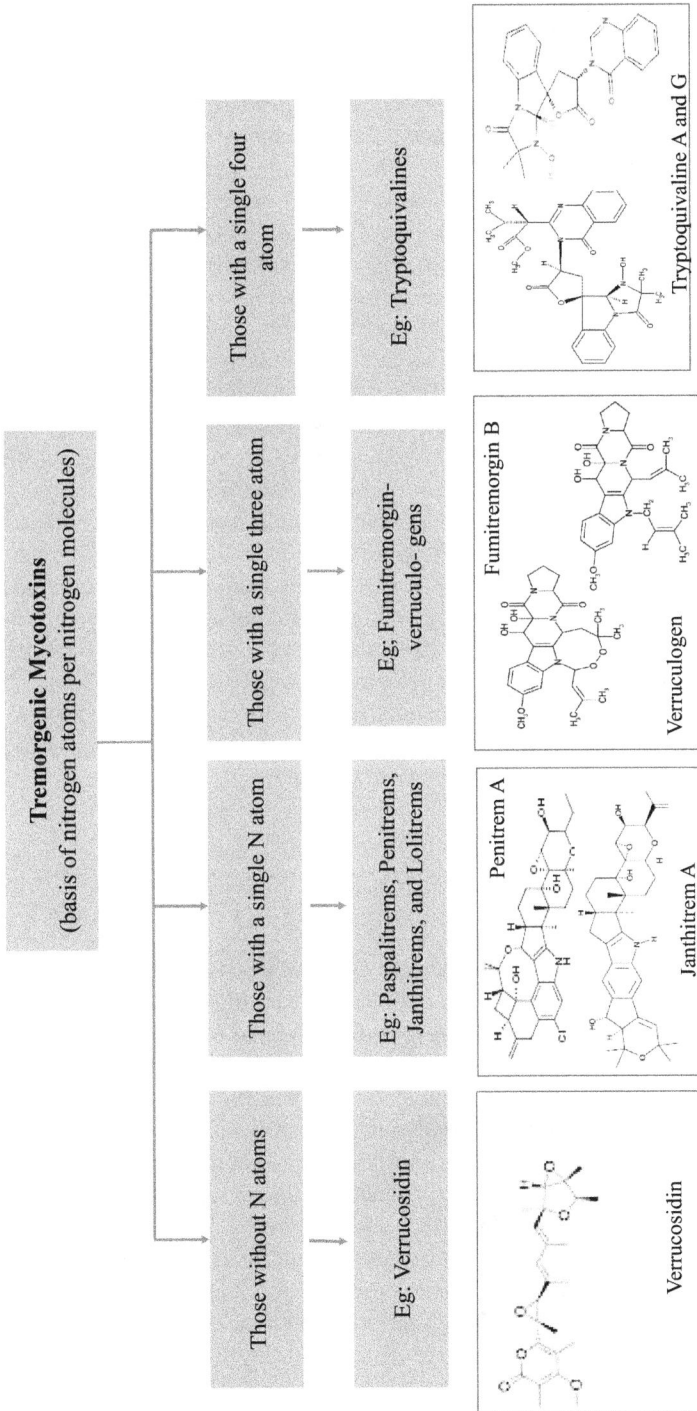

Figure 11.1   Classification of tremorgenic mycotoxins.

Tremorgenic mycotoxins are a concern to livestock/agriculture and can lead to neurological conditions commonly known as "staggers syndrome" (Awuchi et al., 2021; Uhlig et al., 2009). It can pose numerous neurological conditions in humans like muscle tremors, mental confusion, and even death; thus the effect may vary from mild symptoms to life threatening (Awuchi et al., 2021; Evans and Gupta, 2018; Barker et al., 2013). The tremorgens are known to cause various animal diseases like ryegrass staggers, corn staggers, paspalum staggers, and bermudagrass tremors (Evans and Gupta, 2018; Cole, 1993).

Tremorgenic mycotoxins exist in very small concentrations in nature, which can cause difficulty in their identification in mold-contaminated animal feed or other synthetic growth media. This difficulty mainly arises due to complex matrix formation due to numerous other secondary metabolites, lipids, and lipoproteins (Selala et al., 1991). However, several secondary fungal metabolites such as secopenitrem D are identified using newly emerged Nuclear Magnetic Resonance (NMR), High Performance Liquid Chromatography Mass Spectrometry (HPLC-MS), and Liquid Chromatography with tandem mass spectrometry (LC-MS-MS) spectroscopic methods; however, their pathogenicity with respect to tremorgenic symptoms has still not been determined. This difficulty in identification of their nature is influenced by the diversified fungal matrix of fungi, the propensity for myco- and phytotoxins to be present in complex mixtures, and the difficulties inherent in relating xenobiotic structure to function and concentration to biological relevance (Evans and Gupta, 2018). This chapter discusses tremorgenic mycotoxins, their occurrence in food and feed, their primary sources, genes responsible for their production, detection methods, and management and control strategies.

11.2 OCCURRENCE OF TREMORGENIC MYCOTOXINS IN FOOD AND FEED

The occurrence of tremorgenic toxins in food and food is rarely due to the ingredients but occurs mostly due to food spoilage (Atungulu et al., 2018). Tremorgen-producing fungi can contaminate a wide variety of food and feedstuffs. The fungi can contaminate various agricultural products, including grain, faragas, corn, silage, and so on (Gao et al., 2011; Cole, 1993). The products contaminated by tremorgens may include milk and milk or cereal products, stored cereals, and nuts and feedstock including legumes and grasses, and tremorgens can even be found in food and beverage byproducts, garbage, and compost (Evans and Gupta, 2018; Tola and Kebede, 2016; Moldes-Anaya et al., 2012). Further, some fungal strains are isolated from a variety of traditional fermented meats; however, tremorgenic mycotoxins are less investigated (Sabater-Vilar et al., 2003). Table 11.1 identifies types of tremorgenic mycotoxins, the fungi responsible for producing them, and the food source (Evans and Gupta, 2018).

TABLE 11.1   TYPES OF TREMORGENIC MYCOTOXINS IN FOOD AND FEED

Type of Tremorgenic Mycotoxin	Fungi Responsible	Food/Feed Source
Alfatrem	*Aspergillus flavus*	Corn
Janthitrems A, B, and C	*Penicillium janthinellum*	Perennial ryegrass
Lolitriol	*Neotyphodium lolii*	Perennial ryegrass
Lolitrems A, B, C, and D	*Neotyphodium lolii*	Perennial ryegrass
Paspalitrems A, B, and C	*Claviceps paspali*	Dallisgrass
	Claviceps cinerea	Bahiagrass
Paspalinine	*Claviceps paspali, Claviceps cinerea*	Dallisgrass, bahiagrass
Paxilline	*Neotyphodium lolii*	Perennial ryegrass
Penitrem A	*Penicillium crustosum*	Meat, cereals, nuts, cheeses, eggs, fruits, processed/refrigerated foods, refuse, compost
	P. crustosum	
	P. commune	
	Penicillium spp.	
Roquefortines	*Penicillium roqueforti*	Grains, processed food, nuts, fruits, meat, eggs, refuse, compost
Territrems A and B	*Aspergillus terreus*	Cereals and cereal products
Tryptoquivaline tremorgens	*Penicillium* spp.	Soil, seeds, cereal crops
	Aspergillus clavatus	Sprouting cereal grains, malting byproducts
Verruculogen	*Penicillium* spp.	Soil, seeds, cereal crops
	Aspergillus spp.	

Food products may become contaminated with tremorgens with the fungal strains used in their production. One such example is the contamination of blue cheese with *Penicillium roqueforti* employed as a starter culture for producing blue-veined cheese and ultimately resulting in the synthesis of tremorgen roquefortine C and PR toxin (Dobson, 2017; Abbas and Dobson, 2011; Naudé et al., 2002). However, this tremorgens does not have a severe effect on health, and moreover, it can be avoided by using *P. roqueforti* stains that do not produce such secondary metabolites (Abbas and Dobson, 2011). Further, roquefortine tremorgens have been known to result in human intoxication due to consumption of moldy beer contaminated with *P. crustosum* (Naudé et al., 2002).

Likewise, these tremorgens may get into animal feed due to mold contamination. It has been mentioned by several studies that this mycotoxin may poison dogs on the consumption of walnuts and other discarded food products infected with *Penicillium* spp. (Evans and Gupta, 2018; Munday et al., 2011). Pet animals are normally exposed to tremorgens when they consume moldy garbage (Boysen et al., 2002). Studies have specified that infection in dogs is frequently reported due to ingestion of moldy feed or food service waste contaminated with *P. crustosum*, which may lead to tremors, ataxia, and vomiting (Eriksen et al., 2010; Schell, 2000). Roquefortine C has also been reported to poison cattle fed on barley grain feed at a Swedish farm. The barley grain was found to be highly infested by *P. roqueforti* with a dense mycelium growth (Abramson, 1999).

While tremorgens may enter into the animal or human system through the direct consumption of fungus-infected food, they may also enter through foodstuffs grown on infected land, especially grass usually consumed by animals while grazing (Gordon et al., 1993). One clinical study has also identified a case of human poisoning due to consumption of beer contaminated with a fungus of the genus *Penicillium* and later detected with tremorgenic syndrome (Lewis et al., 2005). Although there are numerous cases of tremors and convulsions in humans due to mold-contaminated food, dogs are relatively more prone to intoxication due to indiscriminate appetite and uncontrolled roaming (Atungulu et al., 2018; Naudé et al., 2002). Among all the tremorgens, the penitrem A type and roquefortines are frequently noted as harming companion animals (Atungulu et al., 2018; Eriksen et al., 2010; Puschner, 2002).

Toxin production differs on the basis of seasonal growing conditions and well as the species of the fungi producing the mycotoxins (Lucas, 2017). Moreover, a mold infestation can aggressively increase during humid climates, exposure to the outside environment, over storage, and with unhygienic conditions (Rafiuddin et al., 1999). Several studies have highlighted that these mycotoxins are difficult to detect in food, as they usually occur in low concentrations (Selala et al., 1991).

11.2.1 Mechanism of Tremorgen Production

The mechanism of producing tremorgens is still unclear. However, several action mechanisms have been put forward for different types of tremorgenic mycotoxins (Boysen et al., 2002; Gordon et al., 1993). It was noticed that tremorgenic mycotoxins, even those with chemical similarity, may exert neurotoxicity with several mechanisms; however, impaired gamma-aminobutyric acid (GABA) and glycine-mediated inhibitory pathways and excessive excitatory neurotransmitter release seem to have a significant role in the onset of clinical symptoms. A few studies have specified that the effect of tremorgens is temporary and fully reversible; however, limited studies are available determining the effect of individual compounds on animal models (Reddy et al., 2019; Gordon et al., 1993).

11.3 SOURCES OF TOXINS

Tremorgens are secondary metabolites, which are produced by two classes of ascomycete fungi: Eurotiomycetes, including *Penicillium* and *Aspergillus*, and Sordariomycetes including *Claviceps* and *Epichloë* (Tola and Kebede, 2016; Nicholson et al., 2015). These indole-diterpene compounds are found predominantly in natural environments. The key source of tremorgenic mycotoxins can be mold-contaminated food products. Furthermore, it has been noticed that moldy products due to contamination by *Penicillium* species are the most common source of tremorgens (Seyedmousavi et al., 2018; Barker et al., 2013). The key sources of tremorgenic mycotoxins are discussed in the following.

11.3.1 *Penicillium*-Synthesized Tremorgens

Although several classes of fungi are known to produce tremorgenic symptoms and affect the central nervous system (CNS), fungi belonging to *Penicillium* species are most frequently encountered to produce this mycotoxin (Schell, 2000). Dogs can be exposed to tremorgenic mycotoxins by eating walnuts and other discarded foodstuffs infected with *Penicillium* species (Evans and Gupta, 2018; Tiwary et al., 2009), and several cases of field poisoning have been found in sheep, cattle horses, and so on (Munday et al., 2011; Young et al., 2003).

Among the *Penicillium* species, *P. crustosum* is a commonly occurring foodborne fungi that produces spoilage in foods and thus produces the tremorgen penitrem A (Reddy et al., 2019; Evans and Gupta, 2018). Penitrem A, a potent neurotoxin, causes prolonged tremors and, at high doses, results in convulsions and, in severe cases, death of lab and farm animals. *Penicillium crustosum* has been shown to cause spoilage of food, including cereals, nuts, fruits, cheese, meat, eggs, and several types of processed and refrigerated food (Talcott, 2013; Lewis et al., 2005). On ingestion of moldy food due to *P. crustosum* or inhalation of moldy hay dust, humans have experienced tremors followed by headache, vomiting, weakness, diplopia, and bloody diarrhea (Reddy et al., 2019; Lewis et al., 2005). Moreover, *Penicillium* species can also produce roquefortines. It has been found that *P. crustosum* and most of the *Penicillium* species can produce roquefortines together with penitrem A after growth and sporulation (Evans and Gupta, 2018). Although penitrem A is known to have more severe tremorgenic effect than roquefortine C, it is more slowly metabolized by humans and animals than penitrem A (Tor et al., 2006).

Roquefortine is known to be a weak neurotoxin. Per studies on ruminants, it may result in weak muscles and lack of coordination during the consumption of contaminated silage. Roquefortine, when injected intraperitoneally into mice, can induce convulsive seizures (Abbas and Dobson, 2011). In fact, roquefortine C, which is widely produced by various *Penicillium* species, mainly by *P. roqueforti*, is known to cause anorexia and paralysis. Moreover, it is recognized as a reliable biomarker of the tremorgen penitrem A, specifically where those mycotoxins were present in low concentration or slightly less than their detection level (Tiwary et al., 2009). In addition, other tremorgens, janthitrem A, B, and C, can also be produced, and they are known to cause stagger outbreaks in sheep grazing ryegrass (Evans and Gupta, 2018; Burrows and Tyrl, 2001).

11.3.2 *Aspergillus*-Synthesized Tremorgens

Verruculogen, another tremorgen that causes tremor syndrome in mice, rats, and farm animals, is synthesized by fungi belonging to *Aspergillus* and *Penicillium* species (Reddy et al., 2019; Evans and Gupta, 2018). Verruculogen is known to block the potassium channel, modulating GABA and thus causing tremors (Kosalec and Pepeljnjak, 2005). Likewise, species of *Aspergillus* and *Penicillium* also produce tryptoquivaline and related quinazoline ring–containing indole alkaloids (Gao et al., 2011). *A. clavatus* is involved in the production of tryptoquivaline (Hocking, 2006).

The tremorgens synthesized by *A. terreus* are called territrem A and B, earlier designated C1 and C2, respectively (Evans and Gupta, 2018; Hocking, 2006). The territrems produced are known to elicit tremors in rats and mice and affect the central nervous system (Hocking, 2006). It has been noticed that *A. flavus* (a soil fungus), in addition to the production of aflatoxin, can also synthesize the tremorgens aflatrem and β-alfatrem. It can also produce other mycotoxins containing indole diterpenes like paspalinine, which may occur in contaminated corn (Burrows and Tyrl, 2001). Alfatrem produces fast and sharp tremors but does not result in the same level of sustained tremors as those produced by lolitrem B (Reddy et al., 2019).

11.3.3 *Claviceps*-Synthesized Tremorgens

Claviceps purpurea is usually linked with the occurrence of classic ergotism in humans and animals and occasionally with tremors or convulsions (i.e., convulsive or nervous ergotism) in livestock (Evans and Gupta, 2018; Burrows and Tyrl, 2001). Other *Claviceps* species produce toxicosis, usually identified by the occurrence of tremorgenic stagger syndrome in sheep, horses, and cattle. It has been reported that these animals can be exposed to grass stagger syndrome after multiple days ingesting mature dallisgrass (*P. dilatatum*) or bahiagrass

(*P. notatum*) poisoned with the sclerotia of *C. paspali* (Evans and Gupta, 2018; Burrows and Tyrl, 2001). Dallisgrass and dahiagrass staggers are mostly encountered in the southeastern United States; South and Central America; some European countries; and South Africa, New Zealand, and Australia.

Conventionally, it was thought that ergot alkaloids are largely produced by *C. purpurea* and in smaller amounts by *C. paspali*, inducing the tremors observed in the combination of exposure to both species of *Claviceps* (Evans and Gupta, 2018; Cheeke, 1998). However, it was later understood that with regard to the stagger syndrome induced by *C. paspali* and less commonly by *C. cinerea*, the large concentrations of indole diterpene tremorgens (e.g. paspalinine and paspalitrems A, B, and C) found in the sclerotia of these species are responsible for the neurotoxicity, while for other grass staggers, *Claviceps*-related tremors are usually associated with exercise-exacerbated nervousness, tremors, "wild" facial expressions, aggressive attitude, convulsions, ataxia, and in severe cases death (Burrows and Tyrl, 2001)

Furthermore, cattle are usually exposed to periodic tremors on ingestion of Bermudagrass (*Cynadon dactylon*) (Evans and Gupta, 2018; Burrows and Tyrl, 2001). Several *Claviceps* species produce ergot-type alkaloids like ergine, ergonovine, and ergonovinine and have been found in numerous cases of Bermudagrass staggers (Evans and Gupta, 2018; Cheeke, 1998). Studies have mentioned that during clinical cases of this disease, a higher concentration of paspalitrem-type indole alkaloids was detected (Uhlig et al., 2009).

11.3.4 *Epichloë*-Synthesized Tremorgens

The asexual form of endophyte fungus *Epichloë festucae* variety lolli (earlier designated as *Acremonium lolii* and *Neotyphodium lolii*) forms a symbiotic relation together with perennial ryegrass (*L. perenne*). Perennial ryegrass contaminated by this endophyte contains the lolitrems produced by *E. festucae* var. lolli (called LpTG-1) (Reddy et al., 2019; Evans and Gupta, 2018; White and Torres, 2009; Tor-Agbidye et al., 2001). Among the lolitrems synthesized by the symbiotic relationship of perennial grass with LpTG-1e, lolitrem B is most abundantly synthesized and has multiple analogues and precursors that induce tremors in animals. There is a lack of clarity about the structure–activity relationship; however, it has been suggested by the literature that analogues of lolitrem and biosynthetic intermediates like paxilline and terpendole C are recognized to induce tremors in grazing animals.

The synthesis of lolitrems induces grass stagger syndrome in animals like deer, horse, cattle, and especially sheep, which is commonly called perennial ryegrass staggers and is different from annual ryegrass toxicosis (Grancher et al., 2004; Burrows and Tyrl, 2001). *E. festucae* var. lolli is present in perennial ryegrass in the seed (caryopsis) and the outer and lower leaf sheaths. Staggers syndrome does not appear immediately but after a couple of days of exposure to the infected ryegrass (Repussard et al., 2014; Cheeke, 1998). *E. festucae* var. lolli also produces ergovaline and other ergot alkaloids causing fescue toxicosis (Vassiliadis et al., 2019; Cheeke, 1998), but the classical neurological signs of perennial ryegrass staggers are similar to those linked to *Penicillium* and *Aspergillus* species and the adverse effects of neurotoxic tremorgens lolitrems A, C, and D; lolitrem precursors (e.g., paxilline and lolitriol); and especially lolitrem B (Repussard et al., 2014; Burrows and Tyrl, 2001).

11.4 GENE RESPONSIBLE FOR PRODUCTION OF TREMORGENIC MYCOTOXINS

The type of tremorgenic mycotoxin and the quantity produced depend on the genetic ability of the different fungi responsible for production of the tremorgens. It has been shown by studies that genetic diversity may lead to differences in genetic production (Eltariki et al., 2018). Factors like climate, environment, storage, and other chemical and biological characteristics of the plant or seed may result in genetic diversity. Further, genetic diversity may also be due to differences in breeding method in fungi or coexistence with fungi or other microorganisms. Thus, different gene clusters with a definite set of genes can result in the production of different tremorgenic mycotoxins (Eltariki et al., 2018; Saikia et al., 2008). Moreover, the number of genes may alter the toxin amount produced and thus the toxicity (Eltariki et al., 2018). To understand the biosynthesis of the indole-diterpene tremorgens by filamentous fungi, several attempts have been made to determine the gene cluster behind tremorgenic mycotoxins through gene cloning and characterization of genes (Saikia et al., 2008).

So far, studies have identified the gene clusters for the production of tremorgenic mycotoxins involving paxilline, lolitrems, aflatrem, papaline, and so on. Gene cloning and characterization of gene clusters from *Penicillium paxilli*, which is responsible for the production of paxilline, have enabled understanding of the genetics of these classes of mycotoxins (Liu et al., 2013; Saikia et al., 2007; Young et al., 2001). Analysis of the pax gene cluster initially identified 17 genes behind the biosynthesis of paxilline. Some have been identified to have similarity to monooxygenases and prenyltransferases, including GGP synthase (paxG); two cytochrome P450 monooxygenases, paxP and paxQ; a prenyltransferase (paxC); FAD-dependent monooxegenase (paxM); and a dimethylallyltryptophan (DMAT) synthase (paxD) (Young et al., 2001). Further, extended sequencing identified other genes like paxN, which is a FAD-dependent monooxyegenase; two putative transcription factors containing Zn (II)2 Cys6 DNA-binding motifs, paxR and paxS; a metabolite transporter (paxT); dehydrogenase (paxH); and an oxidoreductase (paxO). In addition, five genes of unknown functionality were also identified: paxU, paxV, paxW, paxX, and paxY. (Saikia et al., 2007; Young et al., 2001; Pan and Coleman, 1990). The pax gene cluster analysis further helped in the isolation of pax gene orthologues from *Neotyphodium lolli* and *Aspergillus flavus* that produce the more complex diterpenes lolitrems and aflatrem, respectively (Saikia et al., 2007; Young et al., 2005). Further, gene disruption and chemical complementary studies has shown a cluster of seven genes is essential for the biosynthesis of paxilline: paxG, paxM, paxC, paxP, paxQ, paxA, and paxB (Liu et al., 2013; Saikia et al., 2007). Additionally, four genes among these genes, paxG, paxB, paxM, and paxC, are essential for the biosynthesis of the tremorgen paspaline (Saikia et al., 2007). Further steps involved in biosynthesis of paxilline involve two cytochromes, paxP and paxQ, which utilize paspaline and 13-desoxypaxilline as their substrates, respectively (Liu et al., 2013; Saikia et al., 2007). These compounds, in combination with paspaline B, PC-M6, and β-paxitriol, have been known to act as a metabolic grid for the biosynthesis of paxilline (Saikia et al., 2007). Thus, it was identified that paxilline formation can happen either through 13-desoxypaxilline or β-paxitriol, which suggests a bifurcation towards the final step of the biosynthesis of paxilline (Saikia et al., 2007).

Studies have further identified that, based on molecular analysis and genetic analysis, three symbiotic genes are required for the lolitrem biosynthesis from Neotyphodium lolli and *Epichloë festucae* (Figure 11.2) (Young et al., 2005). These identified genes are ltmG, ltmM, and ltmK. Further, it is a well-known fact that lolitrem is similar to paxilline, and the isolation of paxilline gene clusters from *P. paxilli* helped researchers understand the biosynthesis of lolitrems and other tremorgens. On comparison of the lolitrem gene cluster with the pax gene cluster, it was found that ltm G and ltmM are the functional orthologues of paxG and paxM, while, for the third gene in the lolitrem cluster, it was identified that there is no orthologue of ltmK in the paxilline gene cluster, and it was thus identified to be unique in the biosynthesis of lolitrem (Ludlow et al., 2019; Saikia et al., 2008; Young et al., 2005). Among these three genes, the gene ltmG is recognized to be involved in encoding a GGPP synthase that acts as a catalyst in the biosynthesis of lolitrem (Young et al., 2005). Further, a similar pathway of biosynthesis is followed by janthitrems (Ludlow et al., 2019), and it has been also recognized that in the case of *Epichloë* endophytes, terpendoles are the precursor for the synthesis of epoxy-janthitrems, which is knows as a precursor in the case of biosynthesis of lolitrem B as well, including the genes ltmG, ltmC, ltmM, and ltmB, which are present in paspaline synthesis (Ludlow et al., 2019; Schardl et al., 2012). The preliminary steps involved in epoxy-janthitrems synthesis require LtmP and LtmQ for demethylation and hydroxylation of paspaline, which produce β-paxitriol. Further, JtmD and JtmQ are needed for the formation of distinct A/B rings in the janthitremane indole diterpenes by the process of diprenylation and oxidative cyclization. LtmF and LtmK help in prenylation and cyclization of the epoxy-janthitrems Ii, III, and IV, similar to biosynthesis of lolitrem. Further, P450 monooxygenase (Jtm01) facilitates the epoxidation process, and O-acyltransferase,

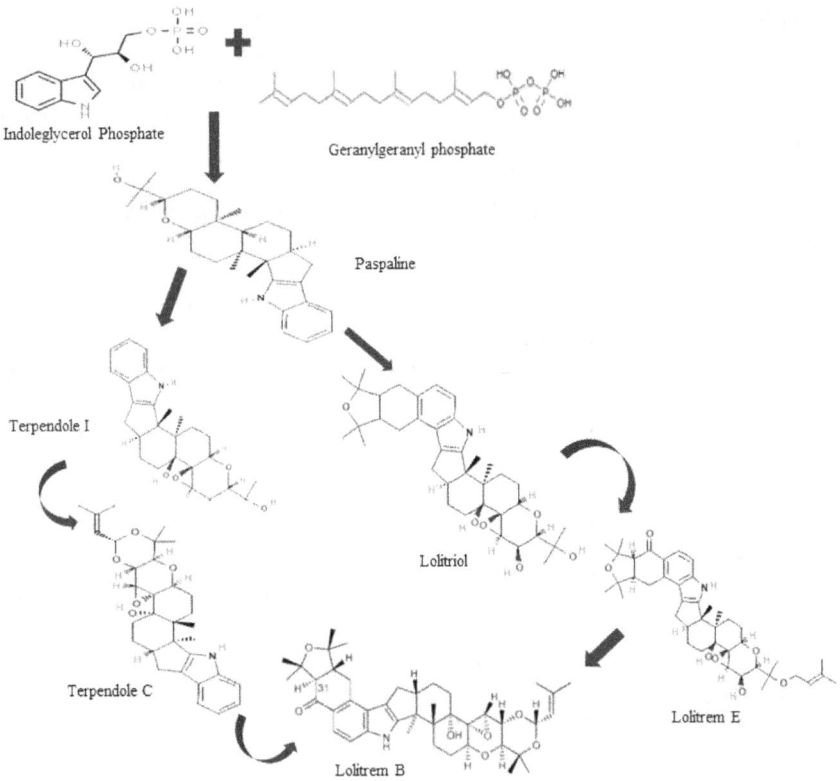

Figure 11.2   Biosynthesis of lolitrems.

which is membrane bound (Jtm02), may be involved in the acetylation of epoxy-janthitrems and janthitrems (Ludlow et al., 2019; Andersen et al., 1997). It is well known in the current literature that a gene cluster with numerous genes can be responsible for the production of particular tremorgenic mycotoxins, and bio-synthesis can follow similar pathways. However, there are still literature gaps regarding the identification of unknown genes in a well-identified gene cluster behind the production of a specific tremorgen.

11.5 DETECTION METHODS FOR TREMORGENIC MYCOTOXINS

Detection methods for tremorgenic mycotoxins in suspected food or feed, gas-trointestinal tract contents, vomitus, urine, and bile are the basis for diagnosis of infection by them (Evans and Gupta, 2018). Although moldy food is a source of tremorgenic mycotoxins, detection of molds without detecting tremorgens does not confirm the occurrence of tremorgenic mycotoxicosis. There have been numerous cases of tremorgen poisoning and their frequent occurrence in foodstuffs, especially penitrems A, demanding a reliable quantitation method of detection in food (Kalinina et al., 2018).

Tremorgens can be detected using high-performance liquid chromatography-tandem mass spectrometry (HPLC-MS/MS), liquid chromatography/mass spectrometry (LC/MS), thin-layer chromatography (TLC), and single gas chromatography/mass spectrometry (GC/MS) (Evans and Gupta, 2018; Tiwary et al., 2009; Tor et al., 2006). Along with analysis of other tremorgenic mycotoxins, analysis of penitrems A and roquefortines are frequently available to clinicians (Evans and Gupta, 2018; Tiwary et al., 2009). Among all the methods of detection, TLC is the most frequently used, and the preferred agent for detection is van Urk's reagent (4-dimethyl-amino-benzaldehyde in 50% ethanolic sulfuric acid) (Selala et al., 1991). The toxins can be separated by elution through the second, third, or fourth volume to remove undesirable components, resulting in another fraction of eluent. The crude mixture still contains a high number of other metabolites. The desired secondary metabolite in medium to high purity can be isolated through the precipitation reaction, preparative TLC, and so on, and later purification can be done through liquid chromatography (LC), TLC, and recrystallization. These detection procedures are often complicated, tedious, and labor intensive, requiring a large solvent volume (Selala et al., 1991). A few of the methods are discussed in the following.

11.5.1 High-Performance Liquid Chromatography-Tandem Mass Spectrometry

This method involves a mass spectrometer equipped with an LC system. Positive mode electrospray ionization is employed for the detection. The HPLC-MS/MS detection method has the following steps (Kalinina et al., 2018):

1. Initially, the temperature of the source is adjusted to 500°C, while the pressure of the curtain gas is adjusted to 35 psi, and the pressure of ion source gases 1 and 2 are adjusted to 35 and 45 psi, respectively. The voltage for ion spray is 4500 V, and the collision is set to high frequency.

2. The dwell time for multiple reaction monitoring transitions is usually kept at 40 ms.

3. The separation of the tremorgen is performed using a Reprosil Gold RP-18 HPLC column (150 mm × 2.0) and guard column (5 × 2 mm) with the same material. The separation is executed with a binary gradient made of MeCN and H_2O, both containing 0.1% of formic acid. The temperature of the column oven is set at 40°C, with a flow rate of 350 µL/min.

4. The analysis of extraction is carried out with the Analyst software (version 1.6.2). The analysis generates the calibration curves with the help of the plot of peak area versus the tremorgen concentration.
5. The measurement of the calibration solution is duplicated, and the average of the peak areas is taken. The tremorgen concentration is calculated by taking the slope and intercept of the curves. The sample weight is taken, and the result is based on three independent measurements.

11.5.2 Liquid Chromatography/Mass Spectrometry

Liquid chromatography is usually combined with the UV absorption, amperometric detection, and fluorescence detection (FLD) stage with either pre-column or post-column derivatization. LC combined with MS is usually used for the detection of tremorgens (Singh and Mehta, 2020) and is known as a powerful technique (Núñez and Lucci, 2020).

LC/MS effectively combines two selection techniques that isolate tremorgens in a highly complex structure for measurement. In liquid chromatography, differentiation is based on physicochemical characteristics, while in mass spectrometry, differentiation is done by mass, mainly mass and charge ratio. The MC not only functions as an LC detector but also helps in the identification of the species indicated by the respective chromatographic peak using the mass spectrum. While LC utilizes high pressure to separate a liquid phase, thereby producing a high gas load, MS is conducted under vacuum and requires a limited gas load (Sargent, 2013; Pitt, 2009).

The LC-MS/MS method is well suited for veterinary diagnostics and public health lab setups where rapid antemortem diagnosis is required for detecting tremorgenic exposure (Tor et al., 2006).

11.5.3 Thin-Layer Chromatography

This is the most common method of detection of tremorgenic mycotoxins in food and feed. A variety of absorbents and formats can be used for conducting thin-layer chromatography. For the detection of mycotoxins like tremorgens, a silica gel TLC plate is the most frequently used absorbent, as it effectively separates the toxin of interest from the matrix. The plates can either be pre-coated or self-coated.

Different formats of thin-layer plates can be used. Most often a square plate of dimensions 20 × 20 cm is used, although 10 × 10–cm and 7 × 7–cm self-cut plates can yield good results as well. The use of smaller-sized plates thus saves the amount of sample required, especially in two-dimensional separation. In two-dimensional separation, the sample extract is placed at the corner of a TLC plate, and two developments are conducted using two different solvents on the two sides in parallel. These two solvents should have compatibility and must be independent; that is, the retention patterns in both cases should be different, which otherwise would lead to agglomeration along the bisector of the TLC plate (Braselton and Johnson, 2003; Selala et al., 1991).

11.5.4 Gas Chromatography/Mass Spectrometry

Like LC/MS, this method also combines two techniques of selection. The method combines gas chromatography that results in the separation of the components, and individual characterization is done by mass spectroscopy. The combination of techniques can evaluate several tremorgens both qualitatively and quantitatively (Adebo et al., 2020; Krone et al., 2010).

11.5.5 Issues in the Detection of Tremorgenic Mycotoxins

Since tremorgenic mycotoxins occur in small concentrations in nature, this causes problems in their identification and detection. The problems that may arise in their analysis may include clean-up, separation and isolation, identification, and detection of small concentrations in complex organic material. Thus, the analytical methods employed for the detection of different types of tremorgens should be extremely sensitive and specific (Bauer et al., 2017; Selala et al., 1991).

11.5.6 Recent Developments

For detecting an indole-diterpene alkaloid paxilline (PAX) and similar groups of tremorgens, techniques like enzyme immunoassay (EIA) have been developed. This technique has used been used in analyzing ergot sclerotia collected from barley and rye fields (Bauer et al., 2017). The method uses ethyl acetate as an extraction solvent. The sample (1 g) is dissolved in ethyl acetate (10 mL) and allowed to stand in an incubator overnight in the dark. The extract thus obtained is then agitated for 10 min and kept aside for settlement of solid particles. One mL of the portion is taken and centrifuged for 20 min at 10,000 rpm. The supernatant of the extract is made up to 2 mL, the solvent is dried with a rotary evaporator at 40°C, and the residue thus received is again dissolved in 2 mL of methanol. For EIA analysis of PAX, the extract is diluted with phosphate buffered saline (PBS) to make a methanol/PBS buffer (20%). The solution was further diluted in a ratio of 1:4 using 20% methanol/PBS (pH 7.3). Using this procedure, the EIA analysis would have a minimum sample extract dilution factor of 200. The value of absorbance is marked on a standard curve, which is used to determine the mycotoxin amount in the sample. It has been noted that the long-term analysis of the standard curves of EIA enables highly sensitive and easily reproducible detection of tremorgens (Bauer et al., 2017; Maragos, 2015).

Hence, it is evident that common separation methods can easily be used for the determination of frequent tremorgenic mycotoxins with modifications per the type of tremorgen the test is meant to determine. This modification not only provides highly sensitive and effective results but also quick analysis.

11.6 MANAGEMENT AND CONTROL STRATEGIES

Mycotoxins, including tremorgens that occur in food and feed, are usually thermally and chemically stable compounds (Atungulu et al., 2018). Thus, most of the management and control strategies are aimed at effectively reducing exposure of plants, grains, or processed food itself to mycotoxin-producing fungi. Thus, avoiding fungal or mold growth and controlling the tremorgenic concentration can help in management practices. Furthermore, tremorgen-associated risk can also be prevented by using contaminated grains or plants for less sensitive uses (Kharayat and Singh, 2018; Tola and Kebede, 2016; Garcia and Heredia, 2006). In most grain-based food products, contamination can be effectively reduced by simple techniques like sieving, washing, and pearling (Atungulu et al., 2018).

In general, management strategies for tremorgenic mycotoxins can be applied during all the stages, pre-harvest, harvest, and post-harvest, which includes detoxification approaches (Kharayat and Singh, 2018; Neme and Mohammed, 2017; Tola and Kebede, 2016). Food grains or plants can be contaminated with tremorgens due to improper pre-harvest and post-harvest management of the grains (Awuchi et al., 2021, Neme and Mohammed, 2017). Pre-harvest management methods often rely on good agricultural practices (GAPs) involving optimized irrigation and employing disease-free seeds as well as application of fungicides if required (Awuchi et al., 2021). However, each pre-harvest method can effectively reduce contamination with fungi and thus reduce produced tremorgens, but an integrated approach used by a combination of several preventive measures can provide a better result (Rose et al., 2018). Some of the integrated preventive measures that are applied pre-harvesting for reducing fungus contamination may include management of fertilization, crop rotation, protection against insect attack and overwintering, use of resistant crops, optimum humidity level, and removal of debris from the previous harvest (Awuchi et al., 2021; Rose et al., 2018). Pre-harvest management with GAP can include the gene modification process for depressing the synthesis of tremorgens; analysis of the soil to determine the fertilizer requirements; optimum seedbed treatment; crop rotation techniques; and use of insecticides, herbicides, or fungicides for weed management and for protection from infection by fungi and insect attack (Awuchi et al., 2021; Kharayat and Singh, 2018).

Post-harvest management techniques play a crucial role in controlling tremorgenic production by good management practices in the food chain during the time of harvest, cleaning and drying, and further storage and processing (Neme and Mohammed, 2017; Magan and Aldred, 2007). The key good management practices (GMPs) that are important during the storage of grains are sanitation, screening, proper aeration, and monitoring (Neme and Mohammed, 2017). A good post-harvest process can be initiated by harvesting grains at low moisture or drying the grain to reduce the moisture level for restraining fungal growth or drying the grains to maintain such a low moisture level (Neme and Mohammed, 2017; Bullerman and Bianchini, 2014). Further, maintaining GMPs in grain elevators and implementing hazard analysis critical control points (HACCPs) helps in eliminating chances of contamination during the food supply chain (Neme and Mohammed, 2017; Channaiah and Maier, 2014). Additionally, techniques like modified atmosphere storage and the use of aliphatic acid-based preservatives can be used for effective post-harvest management of food grains. Modified atmosphere storage usually involves modification of oxygen and carbon dioxide levels, and O_2-free nitrogen can also be used. Further, aliphatic acid-based preservatives are used in moist grain, which is normally used for animal feed. Preservatives based on sorbic acid and propionic salts are available on a commercial level for this purpose (Magan and Aldre, 2007).

The control methods that have been used in animals to effectively reduce the effect of tremorgen toxicity are controlling seizures, decontaminating the gastrointestinal tract, and stabilizing the animal. In the case of animals, treatment with diazepam, methocarbamol, and barbiturates such as pentobarbital depending on the severity of tremors can be used to control tremorgen-induced tremors. Once the tremors are under control, the GI tract is decontaminated, which stabilizes the animal. The GI tract can be decontaminated by performing gastric lavage. Further toxin absorption can be limited by using activated charcoal and a cathartic such as sorbitol (70%) or magnesium sulfate. In addition, support therapy can be provided like intravenous fluids and corticosteroids for shock management and thermoregulation (Evans and Gupta, 2018; Barker et al., 2013; Schell, 2000).

11.7 CONCLUSION

Tremorgenic mycotoxins are one of the most complex compounds produced by filamentous fungi derived from tryptophan and containing an indole moiety. They are found in low concentrations in food and are responsible for causing tremors in animals as well as humans. Among these tremorgens, penitrems and roquefortines are the most frequently encountered and are produced by the *Penicillium* species. The use of fungal strains can easily contaminate food and feed species with tremorgens. Further, mishandling of grain during pre- and post-harvesting stages can make food and feed more prone to infection by fungal tremorgens. In comparison to humans, animals, especially stray dogs, are more exposed to contamination due to their indiscriminate food habits. The neurotoxicity of tremorgens can follow multiple pathways; however, pathways mediated by impaired GABA and glycine and enhanced release of neurotransmitters such as glutamate and aspartate are known to have a major effect in the emergence of clinical symptoms. The particular gene cluster within the fungus determines the type of tremorgen and the concentration produced by it, following similar or different pathways. Tremorgenic mycotoxins are detected by techniques like HPLC-MS/MS, LC/MS, TLC, and GC/MS depending on the type of tremorgen to be detected. In addition, these techniques can be combined or modified for better detection of the tremorgen. Tremorgen contamination is usually avoided by adopting management and control strategies at pre-harvesting and post-harvesting with the help of good agriculture and manufacturing practices. The tremors induced by tremorgens are controlled by a combination of methods depending on the severity of the effect and involve the process of seizure control, decontamination of the GI tract, and stabilizing the animal. Secondary metabolites occur in low concentrations and follow several unidentified pathways of production, which may create hindrance in their identification and thus appropriate health treatment. Henceforth, several studies are required to identify their exact production mechanism, composition of the gene cluster, more sensitive detection methods, and control strategies.

REFERENCES

Abbas, A., & Dobson, A.D.W. (2011). Yeasts and Molds | *Penicillium camemberti.* In *Encyclopedia of Dairy Sciences*, Second ed., pp. 776–779. Elsevier, Academic Press, Cambridge, Massachusetts, United States.

Abramson, D. (1999). Mycotoxins, toxicology. In *Encyclopedia of Food Microbiology*, pp. 1539–1547. Elsevier, Academic Press, Cambridge, Massachusetts, United States.

Adebo, O.A., Oyeyinka, S.A., Adebiyi, J.A., Feng, X., Wilkin, J.D., Kewuyemi, Y.O., Abrahams, A.M., & Tugizimana, F. (2020). Application of gas chromatography–mass spectrometry (GC-MS)-based metabolomics for the study of fermented cereal and legume foods: A review. *International General of Food Science+ Technology*, 56(4): 1514–1534.

Andersen, J.F., Walding, J.K., Evans, P.H., Bowers, W.S., & Feyereisen, R. (1997). Substrate specificity for the epoxidation of terpenoids and active site topology of house fly cytochrome P450 6A1. *Chemical Research in Toxicology*, 10: 156–164.

Atungulu, G.G., Mohammadi-Shad, Z., & Wilson, S. (2018). Mycotoxin issues in pet food. In *Food and Feed Safety Systems and Analysis*, pp. 25–44. Academic Press, Elsevier, Cambridge, Massachusetts, United States.

Awuchi, C.G., Ondari, N.E., Ogbanna, C.U., Upadhyay, K.U., Baran, K., Okpala, R.O.C., Małgorzata Korzeniowska, M., & Guine, F.P.R. (2021). Mycotoxins affecting animals, foods, humans, and plants: Types, occurrence, toxicities, action mechanisms, prevention, and detoxification strategies: A revisit. *Foods*, 10(6): 1279.

Barker, A.K., Stahl, C., & Ensley, S.M., et al. (2013). Tremorgenic mycotoxins. *Compendium: Continuing Education for Veterinarians*: E1–E6.

Bauer, J., Gross, M., Cramer, B., Wegner, S., Hausmann, H., Hamscher, G., & Usleber, E. (2017). Detection of the tremorgenic mycotoxin paxilline and its desoxy analog in ergot of rye and barley: a new class of mycotoxins added to an old problem. *Analytical and Bioanalytical Chemistry*, 409(21): 5101–5112.

Belofsky, G.N., Gloer, J.B., Wicklow, D.T., & Dowd, P.F. (1995). Antiinsectan alkaloids: shearinines A-C and a new paxilline derivative from the ascostromata of *Eupenicillium shearii*. *Tetrahedron*, 51: 3959–3968.

Bennett, J.W., & Klich, M. (2003). Mycotoxins. *Clinical Microbiology Reviews*, 16(3): 497–516.

Boysen, S.R., Rozanski, E.A., Chan, D.L., Grobe, T.L., Fallon, M.J., & Rush, J.E. (2002). Tremorgenic mycotoxicosis in four dogs from a single household. *Journal of the American Veterinary Medical Association*, 221: 1441–1444.

Braselton, W.E., & Johnson, M. (2003). Thin layer chromatography convulsant screen extended by gas chromatography-mass spectrometry. *Journal of Veterinary Diagnostic Investigation*, 15(1): 42–45.

Bullerman, L.B., & Bianchini, A. (2014). Good food-processing techniques: stability of mycotoxins in processed maize-based foods. In *Mycotoxin Reduction in Grain Chains*, pp. 89–100. Wiley, Hoboken, New Jersey.

Burrows, G.E., & Tyrl, R.J. (2001). *Toxic Plants of North America*. Iowa State University Press, Ames, IA, pp. 1–1342.

Byrne, K.M., Smith, S.K., & Ondeyka, J.G. (2002). Biosynthesis of nodulisporic acid A: precursor studies. *Journal of the American Chemical Society*, 124: 7055–7060.

Channaiah, L., & Maier, D.E. (2014). Best stored maize management practices for the prevention of mycotoxin contamination. In *Mycotoxin Reduction in Grain Chains*, pp. 78–88. Wiley, Hoboken, New Jersey.

Cheeke, P.R. (1998). *Natural Toxicants in Feeds*. Second ed. Interstate Publishers, Inc, Danville, IL, pp. 1–479.

Cole, R.J. (1993) Fungal tremorgens. *Prikladnaia Biokhimiia Mikrobiologiia*, 29(1): 44–50.

Dobson, A.D.W. (2017). Mycotoxins in cheese. *Cheese*, 4: 595–601.

Eltariki, F.E.M., Tiwari, K., Ariffin, I.A., & Alhoot, M.A. (2018). Genetic diversity of fungi producing mycotoxins in stored crops. *Journal of Pure and Applied Microbiology*, 12(4): 1815–1823.

Eriksen, G.S., Hultin-Jäderlund, K., Moldes-Anaya, A., Schönheit, J., Bernhoft, A., Jæger, G., Rundberget, T., & Skaar, I. (2010). Poisoning of dogs with tremorgenic *Penicillium* toxins. *Medical Mycology*, 48(1): 188–196.

Evans, T. J., & Gupta, R. C. (2018). Tremorgenic mycotoxins. In *Veterinary Toxicology* (pp. 1033–1041). Academic Press, Elsevier, Academic Press, Cambridge, Massachusetts, United States.

Garcia, S., & Heredia, N. (2006). Mycotoxins in Mexico: epidemiology, management, and control strategies. *Mycopathologia*, 162(3): 255–264.

Gatenby, W.A., Munday-Finch, S.C., Wilkins, A.L., Miles, C.O., & Terpendole, M. (1999). A novel indole-diterpenoid isolated from *Lolium perenne* infected with the endophytic fungus *Neotyphodium lolii. Journal of Agricultural and Food Chemistry*, 47: 1092–1097.

Gerald, F.B., G.F., & Gloer, J.B. (2016). Biologically active secondary metabolites from the fungi. *Microbiol Spectr*, 4(6): 1087–1119.

Gloer, J.B. (1995). Antiinsectan natural products from fungal sclerotia. *Accounts of Chemical Research*, 28: 343–350.

Gao, X., Chooi, Y.H., Ames, B.D., Wang, P., Walsh, C.T., & Tang, Y. (2011). Fungal indole alkaloid biosynthesis: genetic and biochemical investigation of tryptoquialanine pathway in penicillium aethiopicum. *Journal of the American Chemical Society*, 133(8): 2729–2741.

Gordon, K.E., Masotti, R.E. & Waddell, W.R. (1993). Tremorgenic encephalopathy: a role of mycotoxins in the production of CNS disease in humans. *Canadian Journal of Neurological Sciences*, 20(03): 237–239.

Grancher, D., Jaussaud, P., Durix, A., Berthod, A., Fenet, B., Moulard, Y., Bonnaire, Y., & Bony, S. (2004). Countercurrent chromatographic isolation of lolitrem B from endophyte-infected ryegrass (*Lolium perenne* L.) seed. *Journal of Chromatography A*, 1059: 73–81.

Hensens, O.D., Ondeyka, J.G., Dombrowski, A.W., Ostlind, D.A., & Zink, D.L. (1999). Isolation and structure of nodulisporic acid A1 and A2, novel insecticides from *Nodulisporium* sp. *Tetrahedron Letters*, 40: 5455–5458.

Hocking, A.D. (2006). Aspergillus and related teleomorphs. In *Food Spoilage Microorganisms*, Ch.17, pp. 451–487 .Woodhead Publishing Ltd. Cambridge, UK.

Huang, X.H., Tomoda, H., Nishida, H., Masuma, R., & Omura, S. (1995). Terpendoles, novel ACAT inhibitors produced by *Albophoma yamanashiensis*. I. Production, isolation and biological properties. *Journal of Antibiotics*, 48: 1.

Kalinina, S.A., Jagels, A., Hickert, S., Marques, L.M.M., Cramer, B., & Humpf, H. (2018). Detection of the cytotoxic penitrems A–F in cheese from the European single market by HPLC-MS/MS. *Journal of Agricultural and Food Chemistry*, 66(5): 1264–1269.

Kharayat, B.S., & Singh, Y. (2018). Mycotoxins in foods: Mycotoxicoses, detection, and management. In *Microbial Contamination and Food Degradation*, pp. 395–421. Academic Press, Elsevier, Cambridge, Massachusetts, United States.

Kosalec, I., & Pepeljnjak, S. (2005). Mycotoxigenicity of clinical and environmental *Aspergillus fumigatus* and *A. flavus* isolates. *Acta Pharm*, 55(4): 365–375.

Kozák, L., Szilágyi, Z., Toth, L., Pocsi, I., & Molnár, I. (2019). Tremorgenic and neurotoxic paspaline-derived indolediterpenes: biosynthetic diversity, threats and applications. *Applied Microbiology Biotechnology*, 103(4): 1599–1616.

Krone, N., Hughes, B.A., Lavery, G.G., Stewart, P.M., Arlt, W., & Shackleton, C.H.L. (2010). Gas chromatography/mass spectrometry (GC/MS) remains a pre-eminent discovery tool in clinical steroid investigations even in the era of fast liquid chromatography tandem mass spectrometry (LC/MS/MS). *The Journal of Steroid Biochemistry and Molecular Biology*, 121(3–5): 496–504.

Laakso, J.A., Gloer, J.B., Wicklow, D.T., & Dowd, P.F. (1992). Sulpinines A–C and secopenitrem B: New anti-insectan metabolites from the sclerotia of *Aspergillus sulphureus*. *Journal of Organic Chemistry*, 57: 2066–2071.

Lane, G.A., Christensen, M.J., & Miles, C.O. (2000). Coevolution of fungal endophytes with grasses: the significance of secondary metabolites. In Bacon CW, White JFJ (Eds.), *Microbial Endophytes Marcel Dekker*, pp. 341–388. CRC Press, Taylor & Francis Group. 5 Howick Place, London, England.

Lewis, P.R., Donoghue, M.B., Hocking, A.D., Cook, L., & Granger, L.V. (2005). Tremor syndrome associated with a fungal toxin: sequelae of food contamination. *The Medical Journal of Australia*, 182(11): 582–584.

Li, C., Gloer, J.B., Wicklow, D.T., & Dowd, P.F. (2002). Thiersinines A and B: novel anti-insectan indole diterpenoids from a new fungicolous *Penicillium* species (NRRL 28147). *Organic Letters*, 4: 3095–3098.

Liu, C., Noike, M., Minami, A., Oikawa, H., & Dairi, T. (2013). Functional analysis of a prenyltransferase gene (paxD) in the paxilline biosynthetic gene cluster. *Applied Microbiology and Biotechnology*, 98(1): 199–206.

Lucas, J.A. (2017). Fungi, food crops, and biosecurity: Advances and challenges. *Advances in Food Security and Sustainability*, 2: 1–40.

Ludlow, E.J., Vassiliadis, S., Ekanayake, P.N., Hettiarachchige, I.K., Reddy, P., Sawbridge, T.I., Rochfort, S.J., Spangenberg, G.C., & Guthridge, K.M. (2019). Analysis of the indole diterpene gene cluster for biosynthesis of the epoxy-janthitrems in epichloë endophyte. *Microorganisms*, 7(11): 560.

Magan, N., & Aldred, D. (2007). Post-harvest control strategies: minimizing mycotoxins in the food chain. *International Journal of Food Microbiology*, 119 (1–2): 131–139.

Maragos, C.M. (2015). Development and evaluation of monoclonal antibodies for paxilline. *Toxins*, 7(10): 3903–3915.

Maragos, C.M. (2020). Development and characterisation of a monoclonal antibody to detect the mycotoxin roquefortine C. In *Food Additives & Contaminants: Part A*, pp. 1–14.CRC Press, Taylor & Francis Group. 5 Howick Place, London, England.

Miles, C.O., di Menna, M.E., Jacobs, S.W., Garthwaite, I. Lane, G.A., Prestidge, R.A., Marshall, S.L., Wilkinson, H.H., Schardl, C.L., Ball, O.J., & Latch, G.C. (1998). Endophytic fungi in indigenous Australasian grasses associated with toxicity to livestock. *Applied and Environmental Microbiology*, 64(2): 601–606.

Moldes-Anaya, A., Rundberget, T., Fæste, C.K., Eriksen, G.S., & Bernhoft, A. (2012). Neurotoxicity of *Penicillium crustosum* secondary metabolites: Tremorgenic activity of orally administered penitrem A and thomitrem A and E in mice. *Toxicon*, 60: 1428–1435.

Moyano, S.A., Lanuza, F.A., Torres, A.B., Cisternas, E.A., & Fuentes, M.V. (2009). Implementation of a method to determine lolitrem B in ryegrass (*Lolium perenne* L) by liquid chromatography (HPLC). *Chilean Journal of Agricultural Research*, 69: 455–459. Avda. Vicente Méndez 515, Casilla 426, Chillán, Chile.

Munday, J.S., Thompson, D., Finch, S.C., Babu, J.V., Wilkins, A.L., Menna, M.D., & Miles, C.O. (2011). Presumptive tremorgenic mycotoxicosis in a dog in New Zealand, after eating mouldy walnuts. *New Zealand Veterinary Journal*, 56(3): 145–147.

Naudé, T.W., O'Brien, O.M., Rundberget, T., McGregor, A.D.G., Rouxb, C., and Flåøyen, A. (2002). Tremorgenic neuromycotoxicosis in 2 dogs ascribed to the ingestion of penitrem A and possibly roquefortine in rice contaminated with *Penicillium crustosum*. *Journal of the South African Veterinary Association*, 73(4): 211–215.

Neme, K., & Mohammed, A. (2017). Mycotoxin occurrence in grains and the role of post-harvest management as a mitigation strategies. A review. *Food Control*, 78: 412–425.

Nicholson, M., Eaton, C.J., Stärkel, C., Tapper, A.B., Cox, M.P., & Scott, B. (2015). Molecular cloning and functional analysis of gene clusters for the biosynthesis of indole-diterpenes in *Penicillium crustosum* and *P. janthinellum*. *Toxins*, 7: 2701–2722.

Nozawa, K., Nakajima, S., Kawai, K., & Udagawa, S. (1988). Isolation and structures of indoloditerpenes, possible biosynthetic intermediates to the tremorgenic mycotoxin, paxilline, from *Emericella striata*. *Journal of the Chemical Society, Perkin Transactions*, 1: 2607–2610.

Núñez, O., & Lucci, P. (2020). Application of Liquid Chromatography in Food Analysis. *Foods*, 9(9): 1277.

Pan, T., & Coleman, J.E. (1990). GAL4 transcription factor is not a 'zinc finger' but forms a Zn (II)2Cys6 binuclear cluster. *Proceedings of the National Academy of Sciences of the United States of America*, 87: 2077–2081.

Pitt, J.J. (2009). Principles and applications of liquid chromatography-mass spectrometry in clinical biochemistry. *Clinical Biochemist Reviews*, 30(1): 19–34.

Puschner, B. (2002). Mycotoxins. *Veterinary Clinics of North America: Small Animal Practice*, 32(2): 409–419.

Rafiuddin, M., Girisham, S., & Reddy, S.M. (1999). Natural incidence of tremorgenic mycotoxins in bakery products. *Indian Phytopath*, 52(3): 259–262.

Reddy, P., Guthridge, K., Vassiliadis, S., Hemsworth, J., Hettiarachchige, I., Spangenberg, G., & Rochfort, G. (2019). Tremorgenic mycotoxins: Structure diversity and biological activity. *Toxins (Basel)*, 11(5): 302.

Repussard, C., Tardieu, D., Alberich, M., & Guerre, P. (2014). A new method for the determination. *Animal Feed Science and Technology*, 193: 141–147.

Rose, L.J., Okath, S., Flett, B.C., Rensburg, B.J., & Viljoen, A. (2018). Preharvest management strategies and their impact on mycotoxigenic fungi and associated mycotoxins. *Mycotoxins—Impact and Management Strategies*, pp. 41–57.

Sabater-Vilar, M., Nijmeijer, S., & Fink-Gremmels, J. (2003). Genotoxicity assessment of five tremorgenic mycotoxins (fumitremorgen B, paxilline, penitrem A, verruculogen, and verrucosidin) produced by molds isolated from fermented meats. *Journal of Food Protection*, 66(11): 2123–2129.

Saikia, S., Nicholson, J.M., Young, C., Parker, J.E., & Scott, B. (2008). The genetic basis for indole-diterpene chemical diversity in filamentous fungi. *Mycological Research*, 112: 184–199.

Saikia, S., Parker, E.J., Koulman, A., & Scott, B. (2007). Defining paxilline biosynthesis in penicillium paxilli: Functional characterization of two cytochrome P450 monooxygenases. *The Journal of Biological Chemistry*, 282(23): 16829–16837.

Sargent, M. (2013). Guide to achieving reliable quantitative LC-MS measurements. *RSC Analytical Methods Committee*, 1: 1–31.

Schardl, C.L., Young, C.A., Faulkner, J.R., Florea, S., & Pan, J. (2012). Chemotypic diversity of *Epichloë*, fungal symbionts of grasses. *Fungal Ecology*, 5: 331–344.

Schell, M.M. (2000). Tremorgenic mycotoxin intoxication. *Veterinary Medicine*, 95(4): 283.

Selala, M.I., Musuku, A., & Schepens, P.J.C. (1991). Isolation and determination of paspalitrem-type tremorgenic mycotoxins using liquid chromatography with diode-array detection. *Analytica Chimicu Acta*, 244: 1–8.

Seyedmousavi, S., Bosco, S.M.G., Hoog, S., Ebel, F., Elad, D., Gomes, R.R., Jacobsen, I.D., Jensen, H.E., Martel, A., Mignon, B., Pasmans, F., Pieckova, E., Rodrigues, A.M., Singh, K., Vicente, V.A., Wibbelt, G., Wiederhold, N.P., & Guillot, J. (2018). Fungal infections in animals: A patchwork of different situations. *Medical Mycology*, 56: S165–S187.

Singh, J., & Mehta, A. (2020). Rapid and sensitive detection of mycotoxins by advanced and emerging analytical methods: A review. *Food Science & Nutrition*. Wiley Periodicals, Inc.

Steyn, P.S., & Vleggaar, R. (1985). Tremorgenic mycotoxins. *Fortschr Chem Org Naturst*, 48: 1–80.

Talcott, P.A. (2013). Mycotoxins. In *Small Animal Toxicology*, Third ed., pp. 677–682.

Tiwary, A.K., Puschner, B., & Poppenga, R.H. (2009). Using roquefortine C as a biomarker for penitrem A intoxication. *Journal of Veterinary Diagnostic Investigation*, 21: 237–237. Elsevier, , Cambridge, Massachusetts, United States.

Tola, M., & Kebede, B. (2016). Occurrence, importance and control of mycotoxins: A review. *Cogent Food & Agriculture*, 2: 1191103.

Tor, E.R., Pushner, B., Filigenzi, M.S., Tiwary, A.K., & Poppenga, R.H. (2006). LC–MS/MS screen for penitrem A and roquefortine C in serum and urine samples. *Analytical Chemistry*, 78(13): 4624–4629.

Tor-Agbidye, J., Blythe, L.L., & Craig, A.M. (2001). Correlation of endophyte toxins (ergovaline and lolitrem B) with clinical disease: fescue foot and perennial ryegrass staggers. *Veterinary Human Toxicology*, 43: 140–146.

Uhlig, S., Botha, C.J., Vrålstad, T., Rolén, E., & Miles, C.O (2009). Indole-diterpenes and ergot alkaloids in *Cynodon dactylon* (Bermuda grass) infected with *Claviceps cynodontis* from an outbreak of tremors in cattle. *Journal of Agriculture and Food Chemistry*, 57(23): 11112–11119.

Vassiliadis, S., Elkins, A.C., Reddy, P., Guthridge, K.M., Spangenberg, G.C., & Rochfort, S.J. (2019). A simple LC–MS method for the quantitation of alkaloids in endophyte-infected perennial ryegrass. *Toxins*, 11(11): 649.

White, J.F., & Torres, M.S. (2009). Is plant endophyte-mediated defensive mutualism the result of oxidative stress protection? *Physiologia Plantarum*, 138: 440–446.

Xiong, Q., Zhu, X., Wilson, W.K., Ganesan, A., & Matsuda, S.P. (2003). Enzymatic synthesis of an indole diterpene by an oxidosqualene cyclase: Mechanistic, biosynthetic, and phylogenetic implications. *Journal of the American Chemical Society*, 125: 9002–9003.

Young, C.A., Bryant, M.K., Christensen, M.J., Tapper, B.A., Bryan, G.T., & Scott, B. (2005). Molecular cloning and genetic analysis of a symbiosis-expressed gene cluster for lolitrem biosynthesis from a mutualistic endophyte of perennial ryegrass. *Molecular Genetics and Genomics*, 274: 13–29.

Young, C.A., McMillan, L., Telfer, E., & Scott, B. (2001). Molecular cloning and genetic analysis of an indole–diterpene gene cluster from *Penicillium paxilli*. *Molecular Microbiology*, 39: 754–764.

Young, K.L., Villar, D., Carson, T.L., Imerman, P.M., Moore, R.A., & Bottoff, M.R. (2003). Tremorgenic mycotoxin intoxication with penitrem A and roquefortine in two dogs. *Journal of the American Veterinary Medical Association*, 222: 52–53.

FURTHER READINGS

Bressolle, F., Bromet-Petit, M., & Audran, M. (1996). Validation of liquid chromatographic and gas chromatographic methods. *Journal of Chromatography B: Biomedical Sciences and Applications*, 686(1): 3–10.

Moyano, S.A., Lanuza, F.A., Torres, A.B., Cisternas, E.A., & Fuentes, M.V. (2008). Implementation of a method to determine lolitrem-B in ryegrass (*Lolium perenne* L.) by liquid chromatography (HPLC). *Chilean Journal of Agricultural Research*, 69(3): 455–459.

World Health Organization. (2004). *Concise International Chemical Assessment— Chloroform*. Document 58, World Health Organization, Geneva, pp. 19–24.

Chapter 12

Zearalenone in Food and Feed

*Occurrence, Biosynthesis, Detection,
and Management Strategies*

Shikha Pandhi, Ashok Kumar Yadav, Vidhi Tyagi, Saloni,
Akansha Gupta, Surabhi Pandey, Dipendra Kumar Mahato,
Pradeep Kumar, Arun Kumar Pandey, Arvind Kumar

CONTENTS

12.1 Introduction   311
12.2 Major Sources of Zearalenone   313
12.3 Occurrence of Zearalenone in Food and Feed   313
12.4 Biosynthesis of Zearalenone   318
12.5 Zearalenone Chemical and Physical Properties   320
    12.5.1 Storage and Preservation   320
    12.5.2 Zearalenone Production   320
12.6 Effects of Processing on Zearalenone   322
    12.6.1 Effects of Thermal Processing   322
    12.6.2 Effect of Drying and Washing   322
    12.6.3 Effect of Milling   323
    12.6.4 Effect of Extrusion   323
    12.6.5 Effect of Cooking   324
12.7 Detection Methods of Zearalenone in Food and Feed   325
12.8 Management and Control Strategies   326
12.9 Conclusion   328

12.1 INTRODUCTION

Mycotoxins are harmful chemicals formed by molds belonging to the genera *Aspergillus*, *Penicillium*, *Alternaria*, *Claviceps*, and *Fusarium*. Mycotoxins are widely found in a wide variety of plant-based agricultural commodities,

DOI: 10.1201/9781003242208-12

including grains, nuts, fruits, vegetables, and forage. If animals consume mycotoxin-contaminated feedstuffs, these toxins may be discovered in animal-derived products (e.g., meat, eggs, and milk) (Shi et al., 2018). Among a large number of mycotoxins produced by fungi, aflaoxins (Mahato et al., 2019), fumonisins (Kamle et al., 2019), zearalenone (Mahato et al., 2021), and ochratoxins (Kumar et al., 2020) are known to be of higher concern, as they affect crops and livestock to a larger extent. Because of its significant pathogenicity and ubiquitous distribution in animal feed components and agricultural goods, zearalenone (ZEN) has emerged as a global public health problem. Zearalenone, a mycotoxin, is a secondary metabolite produced by *Fusarium* species (primarily *Fusarium roseum, Fusarium culmorum*, and *Fusarium graminearum*) that causes global contamination of cultivated and/or stored crops such as wheat, sorghum, oat, rye, and maize under humid and warm conditions (Reinholds et al., 2020). (Hao et al., 2022). Previous studies have shown that ZEN has immunosuppressive (Niazi et al., 2019), genotoxic (Virk et al., 2020), hepatotoxic (Faisal et al., 2020), and neurotoxic (Makowska et al., 2017) effects, and it has been classed as a category III carcinogen by the International Agency for Research on Cancer (Radi et al., 2020). ZEN (Figure 12.1) is a [6-(10-hydroxy-6-oxo-trans-1-undecenyl) b-resorcylic-acid-lactone] that is soluble in alkaline solutions, ether, benzene, acetonitrile, methyl chloride, chloroform, acetone, and alcohols but not water (Gromadzka et al., 2008). It is a white crystalline powder with the chemical formula $C_{18}H_{22}O_5$ and a molecular weight of 318.364 g/mol (De Rycke et al., 2021). It is thermostable, which implies that it will not degrade whether milled, extruded, stored, or heated. This mycotoxin accumulates in grains mostly before harvesting, although it can also build up after harvesting under poor storage conditions, making removal and/or breakdown from food difficult (Ropejko and Twarużek, 2021). Among the several cereals impacted by ZEA, maize and related products are the most commonly infected by this mycotoxin. As the use of cereals and cereal by-products by people and animals increases, appropriate analytical approaches for the regular management of this pollutant are required (Llorent-Martinez et al., 2019).

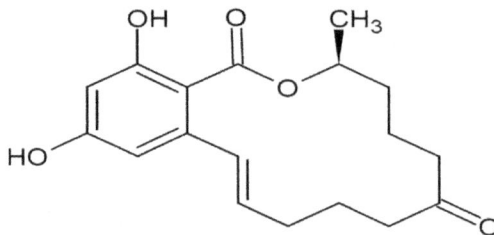

Figure 12.1   Structure of zearalenone.

ZEN is a xenoestrogen, which are exogenous chemical compounds with a structure similar to naturally occurring estrogens. By attaching to cellular estrogen receptors, they bioaccumulate and impede hormone secretion, synthesis, metabolism, transport, and action (all of which are crucial for homeostasis, growth, and reproduction). ZEN has strong estrogenic activity and can bind to estrogen receptors in mammalian cells because of its comparable structure to estradiol, the main hormone produced in the ovaries (Rogowska et al., 2019; De Rycke et al., 2021). Because of its severe pathogenicity and widespread distribution in feed components and agricultural products, ZEN exposure in both animals and humans is a major public health concern. Because ZEN has major harmful consequences, it must be kept at a safe level. As a result, an outdoor on-site assessment of ZEN in feed components and agricultural products requires a portable, rapid, sensitive, and specific analytical approach that does not require a substantial investment or energy inputs (Hao et al., 2022). This chapter provides readers with information on the many sources of ZEN, its presence in food and feed, and its biosynthesis mechanism, with an emphasis on the effect of processing on ZEN concentration, as well as detection and management strategies.

12.2 MAJOR SOURCES OF ZEARALENONE

Whole grains like corn, barley, sorghum, wheat, and rye are high in ZEN mycotoxin. Wheat, rye, and oats are the most common causes of ZEN contamination in European nations, whereas corn and wheat are the most common sources of ZEN contamination in the United States and Canada. High humidity and low temperatures are ideal for ZEN production. ZEN, a mycotoxin, has been detected in wheat, rice, sorghum, maize, barley, rye (Gil-Serna et al., 2014, Mally et al., 2016), malt, sesame seed, flour, soybeans, beer (Zinedine, 2014), and corn oil. This toxin can also be found in grain-derived by-products such as human food grains, baked goods such as cookies and bread (Fink-Gremmels and Malekinejad, 2007), breakfast cereals (Mally et al., 2016), and pasta. Temperature, humidity, time in vegetation, and fungal strain contamination all have an impact on the level of ZEN in crops (Jiménez and Mateo, 1997).

12.3 OCCURRENCE OF ZEARALENONE IN FOOD AND FEED

Many essential crops, including maize, wheat, sorghum, barley, oats, sesame seed, hay, and corn silage, have been shown to contain zearalenone and its derivatives. The largest levels of zearalenone produced by *Fusarium* were found below 25°C, with a high amplitude of daily temperature and 16% humidity (moisture) (Gromadzka et al., 2008). The presence of mycotoxin ZEN in food has been studied in detail because of its highly toxic nature. ZEN has been reported to be found in cereal crops and other food items in many parts of the world such as Asia, Africa, and Europe (Pleadin et al., 2012; Queiroz et al., 2012). Research has shown that among the cereals with the highest contamination of ZEN are maize and wheat, while oats, wheat, and barley have also been reported to be contaminated with this toxin often (Zinedine et al., 2007). The levels of ZEN in food are regulated in 16 countries because of its high levels of contamination. Bearing in mind the health risks of the consumer, certain regulations for the amount of ZEN in various food items have been provided by the European Union (EU). The EU has defined the maximum authorized levels for ZEN in food items in unprocessed whole cereals should be 100–200 µg/kg; in processed cereals, it should be around 75 µg/kg; in cereal snacks, the value should be around 50 µg/kg; and the value should be around 60 µg/kg in wheat and its flour. Moreover, because of the high amounts of ZEN in cereal-derived food items, children and vegan populations were suspected to be at a greater risk of exposure to its ill effects. In some cases, cows and buffalo consume foods spoiled with ZEN; therefore, this mycotoxin can be found in the milk (Prelusky et al., 1990; Coffey and Cummins, 2009), and hence ZEN becomes accessible in the human food chain.

Compiled research data on the presence of ZEN in food items are shown in Table 12.1. From this data, it can be concluded that the most adulterated food items are grain, maize, corn, beans, and animal feed mixtures, while the least adulterated food items are peas and infant feed based on cow's milk, beer, and barley. This shows that grains and other feedstuffs are highly exposed to the presence of ZEN.

TABLE 12.1   OCCURRENCE OF ZEARALENONE IN FOOD AND FEED

Continent	Country	Food Grains and Products	ZEN Range	References
Europe	Bulgaria	Maize	Max. 148 µg/kg	Manova and
		Wheat	Max. 120 µg/kg	Mladenova,
		Barley	Max. 36.6 µg/kg	2009
		Maize	Max. 1700 µg/kg	Vrabcheva
				et al., 1996
				Manova and
				Mladenova,
				2009
				Domijan et al.,
				2005

Continent	Country	Food Grains and Products	ZEN Range	References
	Croatia	Maize Wheat Barley Oats	About 27.7–1182 µg/kg About 13–50 µg/kg About 35–80 µg/kg Max. 18 µg/kg	Klari_c et al., 2009
	France	Maize	About 3–165 µg/kg	Scudamore and Patel, 2009
	Germany	Maize Maize by-products Maize silage Maize germ oil Wheat	Max. 860 µg/kg Max. 1362 µg/kg Max. 1790 µg/kg About 3–1730 µg/kg About 0.001–8040 µg/kg	Schollenberger et al., 2006 Muller et al., 1997
	Scotland	Stored barley (3 months–1 year)	About 2100–26,500 µg/kg	Gross and Robb, 1975
	Yugoslavia	Maize Dairy cattle feed	About 43–10,000 µg/kg About 140–960 µg/kg	Balzer et al., 1977 Skrinjar et al., 1995
	Italy	Corn Maize	About 4–150 µg/kg Max. 453 µg/kg	Visconti and Pascale, 1998 Pietri et al., 2004
	Hungary	Moldy stored corn	About 10–11,800 µg/kg	Fazekas et al., 1996
	Slovakia	Poultry feed mixture	About 3–86 µg/kg	Labuda et al., 2005
	Switzerland	Wheat Wheat	About 10–121 µg/kg About 10–18 µg/kg	Bucheli et al., 1996 Noser et al., 1996
	UK	Corn Corn feed	About 4–180 µg/kg About 2–180 µg/kg	Scudamore et al., 1998

(Continued)

TABLE 12.1   OCCURRENCE OF ZEARALENONE IN FOOD AND FEED   (CONTINUED)

Continent	Country	Food Grains and Products	ZEN Range	References
	Yugoslavia	Dairy cattle feed barley	About 140–960 μg/kg	Skrinjar et al., 1995
		Oats	About 21–30 μg/kg	Hietaniemi and
		Corn	About 30–860 μg/kg	Kumpulainen,
			About 430–10 μg/kg	1991
				Hietaniemi and
				Kumpulainen,
				1991
				Balzer et al., 1977
Africa	Egypt	Cereals	About 5–45 μg/kg	Abd Alla, 1997
		Maize	About 9800–38,400 μg/kg	El-Maghraby et al., 1995
	Morocco	Maize	About 135–165 μg/kg	Zinedine et al., 2006
	Nigeria	Moldy acha	About 200–600 μg/kg	Gbodi et al., 1986a
		Maize	Max. 17,500 μg/kg	Gbodi et al., 1986b
		Beer	About 245–1320 mg/l	Okoye, 1987
	Portugal	Cereal grains	Max 930 μg/kg	Marques et al., 2008
	South Africa	Cereal/animal feed	About 50–8000 μg/kg	Dutton and Kinsey, 1996
South America	Argentina	Grains and food	About 200–7500 μg/kg	Lopez and Tapia, 1980
		Poultry feeds	About 327–5850 μg/kg	Dalcero et al., 1998
		Maize	About 100–1560 μg/kg	Roige et al., 2009
	Brazil	Wheat	About 40–210 μg/kg	Furlong et al., 1995
		Maize	About 653–9830 μg/kg	Sabino et al., 1989
		Corn	About 368–719 μg/kg	Vargas et al., 2001
	Uruguay	Dried fruits and vegetables	About 100–200 μg/kg	Pineiro et al., 1996a
		Barley and malt	About 100–200 μg/kg	Pineiro et al., 1996b

Continent	Country	Food Grains and Products	ZEN Range	References
North America	Canada	Corn Barley Wheat and barley Feed corn	About 5–647 µg/kg About 4–21 µg/kg Max. 300 µg/kg Approx. <100–14,100	Scott, 1997 Scott, 1997 Stratton et al., 1993 Funnell, 1979
	United States	Maize Wheat Sorghum	About 100–21,600 µg/kg About 360–11,050 µg/kg About 46–1480 µg/kg	Park et al., 1996 Shotwell et al., 1977 Bagneris et al., 1986
Asia	China	Maize Maize bran Wheat bran Wheat middings Wheat Wheat Barley Rice bran	Max. 624.3 µg/kg Max. 1268 µg/kg Max. 439 µg/kg Max. 1195.5 µg/kg Max. 1400 µg/kg About 10.1 to 3048.8 µg/kg About 393 µg/kg Max. 879 µg/kg	Ma et al., 2018 Ma et al., 2018 Ma et al., 2018 Ma et al., 2018 Li et al., 2002 Ji et al., 2014 Ma et al., 2018 Ma et al., 2018
	India	Straw Mixed concentrate samples Maize Wheat and rice	Avg. 422 µg/kg Max. 843 µg/kg 100–400 µg/kg Max. 6000 µg/kg	Phillips et al., 1996 Phillips et al., 1996 Janardhana et al., 1999 JECFA, 2000
	Iran	Maize Maize snacks Maize	About 100–212 µg/kg Max. 1471 µg/kg Max. 889 µg/kg	Hadiani et al., 2003 Reza Oveisi et al., 2005 Reza Oveisi et al., 2005
	Thailand	Maize	Max 923 µg/kg	Yamashita et al., 1995
	Japan	Barley Barley Wheat	About 10–658 µg/kg About 11,000–15,000 µg/kg About 53–510 µg/kg	Sugiura et al., 1993 Yoshizawa, 1997 Yoshizawa, 1997

TABLE 12.1   OCCURRENCE OF ZEARALENONE IN FOOD AND FEED   (CONTINUED)

Continent	Country	Food Grains and Products	ZEN Range	References
	Indonesia	Maize-based products	Max. 589 µg/kg	Nuryono et al., 2005
	Korea	Maize	About 4–386 µg/kg	Kim et al., 1993
		Maize	About 21–47 µg/kg	Park et al., 2005
		Barley	About 4–1416 µg/kg	Kim et al., 1993
		Barley foods	About 34–120 µg/kg	Park et al., 2002
	Philippines	Maize	About 59–505 µg/kg	Yamashita et al., 1995
	Qatar	Rice	About 1.8–1.4 µg/kg	Abdulkadar et al., 2004
		Wheat	About 2.1–21 µg/kg	
		Cornflakes	About 3.8–6.81 µg/kg	

12.4 BIOSYNTHESIS OF ZEARALENONE

Zearalenone is produced via the acetate-polymalonate process, which results in a nonaketide precursor that is subsequently subjected to various cyclization and modifications. *F. graminearum* had a 50 kb gene cluster, and 11 genes were discovered at the molecular level (Nahle et al., 2021). However, within this cluster, zearalenone production is restricted to four genes: two PKS 4 and PKS 13 polyketide synthase genes, a gene that is comparable to isoamyl alcohol oxidase (ZEB1), and a regulatory protein gene (ZEB2). For particular detection and analysis of zearalenone-generating species, molecular approaches were devised. When the weather is warm and the air humidity is at or above 20%, more is produced (Agriopoulou et al., 2020). Due to its toxic effect and broad distribution in animal feed, farm animals' exposure to zearalenone is of global health concern. *Fusarium* species infect grains and contribute to the accumulation of ZEN before harvesting (D'Mello et al., 1999), resulting in a large level of ZEN found in animal's natural feed samples and also due to inappropriate storage (Zhang et al., 2018; Olsen et al., 1981).

Production of ZEN takes place by hydroxylation, catalyzed in vivo, in the presence of 3-hydroxy-steroid-dehydrogenases enzyme (HSDs) and combining of ZEN with glucuronic acid, which is catalyzed by uridine diphosphate glucuronyl transferases enzyme (Zinedine et al., 2007). As ZEN and its conjugate compounds are very rare to date, there is a need to develop biosynthesis pathways to develop ZEN without a great need for a lab or equipment. In 2014, *Rhizopus* and *Aspergillus* were known for being capable of ZEN conjugate formation (Brodehl et al., 2014). ZEN was biosynthesized by *F. graminearum* and could be used for its conjugate formation, like ZEN-14-S, ZEN-14-G, and ZEN-16-G by specified strains of *Rhizopus* and *Aspergillus* under controlled predefined conditions (Figure 12.2) (Borzekowski et al., 2018). For ZEN-14-S, ZEN-14-G, and ZEN-16-G production, a simple and cost-effective biosynthesis process has been developed with the help of fungal strains with varying ZEN formation and metabolization capabilities. Strains that could consistently produce ZEN at a lower cost were sought. On humid rice flour, *F. graminearum* was identified as a fungal strain that can produce ZEN in the mg/g range. This biosynthesis process was combined with another stage of association in a straightforward manner in conjugate compound formation (Brodehl et al., 2014).

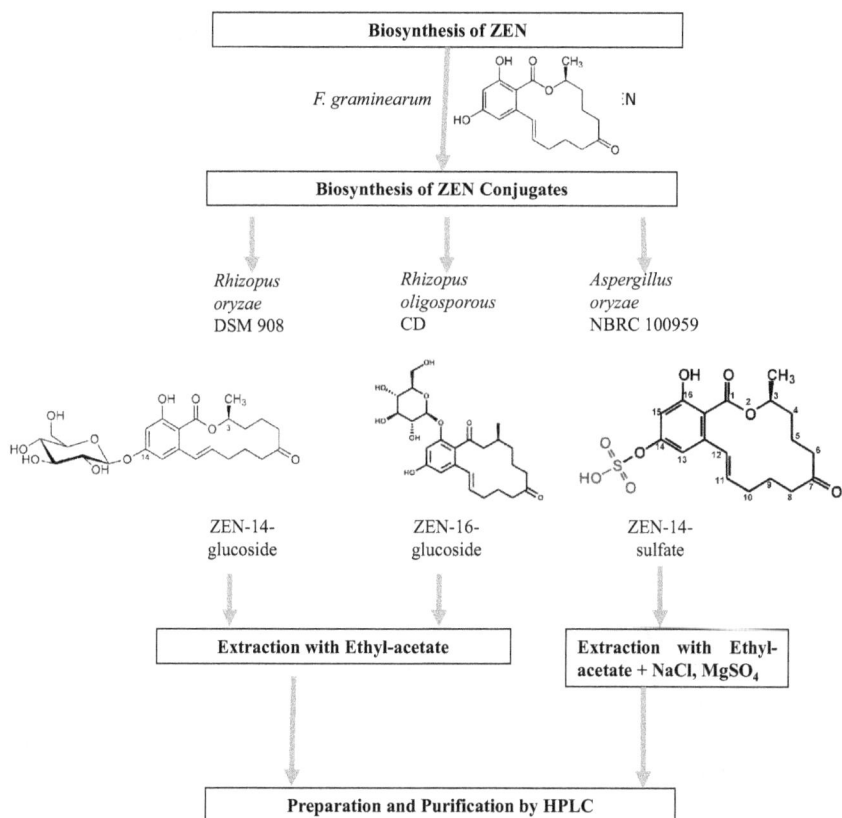

Figure 12.2 Flow chart of biosynthesis, isolation, and purification of zearalenone-14-glucoside, ZEN-16-glucoside, and ZEN-14-sulfate (Borzekowski et al., 2018).

12.5 ZEARALENONE CHEMICAL AND PHYSICAL PROPERTIES

Zearalenone is a crystalline solid that is white. It is a lactone of resorcyclic acid. When stimulated by long-wavelength ultraviolet (UV) light of 360 nm wavelength, it emits a blue-green fluorescent light, and when excited with short wavelength UV light (365 nm), it emits a more intense green fluorescent light of 260 nm wavelength. ZEN is marginally dissolved in hexane and mildly dissolved in acetone, acetonitrile, benzene, ethanol, methanol, and methylene chloride. It is also dissolved in an alkali aqueous solution (Ferrer et al., 2009).

12.5.1 Storage and Preservation

Zearalenone can be developed in maize when it is in the field, as well as during the time of storage, especially if stored with higher moisture content and if not properly dried before storage (Mahato et al., 2021). In Hungary, 88% of stored corn samples were found to be moldy and tested positive for zearalenone, and the converse is observed, such as contamination of mycotoxin being found to be lower in non-moldy and dried corn and other cereals (Agriopoulou et al., 2020)

12.5.2 Zearalenone Production

Zearalenone, which is also known as F2 toxin, is a sex hormone–promoting chemical that causes vulvovaginitis and estrogenic reactions in pigs. *Fusarium* species create the compound zearalenone. In late autumn and winter, zearalenone has been discovered spontaneously in maize with a high moisture content, predominantly from *Fusarium graminearum* and *Fusarium culmorum* development (Bennet and Klich, 2003; Mahato et al., 2021).

Although the chemical is not particularly hazardous, physiological reactions in swine can be induced at concentrations of 1–5 ppm. In sows' milk, zearalenone can be passed on to new piglets, causing estrogenism in young pigs. Moldy hay, high-moisture content maize, pre-infected corn, and pelleted feed rations have all been discovered to contain zearalenone (Mostrom, 2016). Although there is no evidence that zearalenone is involved in human toxicoses, it is classed as an endocrine disrupter and is regarded as potentially dangerous to humans. Humid condition, fluctuation in temperature, and then slightly lower temperature promote the production of zearalenone. During the autumn harvest, these conditions are common in temperate climates (Bullerman, 2003).

A trial was done on a pig in which it was observed that after 30 minutes feeding zearalenone to the animal, ZEN conjugate compounds were observed in the blood plasma, sex organs like ovaries, adipose cells, and also in the intestinal regions (Kuiper-Goodman et al., 1987). The half-life of zearalenone in test animals after oral dose administration has been found to be 86 hours (Biehl et al., 1993). There are two types of metabolism for zearalenone production, phase I and phase II, with the first biotransformation step catalyzed with 3-α and 3-β hydroxysteroid dehydrogenase enzymes. Zearalenol is produced when the reduction at c6 takes place of the keto group during phase-I metabolism. The C11–C12 double bonds are further reduced, resulting in α and β zearalanol. The difference in zearalenone susceptibility in different species may be due to different hepatic biotransformation. Pig hepatic microsomes produce the greatest amount of α-zearalenol, while chicken microsomes produce more β-zearalenol (Malekinejad et al., 2005). Humans can convert zearalenone to α-zearalenol, which is more estrogenic and dangerous.

Glucuronidation is the process by which there is an easy conjugation of all absorbed zearalenone and α-zearalenol in pigs. Liver is the main organ responsible for glucuronidation. Along with the liver, the intestinal mucosa is also involved in the process. In sow, the conversion takes place in intestinal mucosal cells in the duodenum and jejunum, where zearalenone is converted into α- and β-zearalenol (Olsen et al., 1987). Gut bacteria can also help in the breakdown of zearalenone. Zearalenone can be degraded in the rumen, resulting in production of α-zearalenol and β-zearalenol in lower amounts (Kiessling et al., 1984). Zearalenone absorption and toxic effects in animals are influenced by the GI tract, liver metabolism, and receptor sites for cytosolic estrogen.

In almost all animals, zearalenone undergoes enterohepatic circulation and biliary elimination. Animals excrete zearalenone mostly through their feces; however, rabbits excrete it primarily through their urine. The majority of zearalenone in a dose is eliminated within 72 hours. Within 72 hours after 10 mg/kg body weight of radiolabeled zearalenone administered orally to laying hens, approximately 94% of the radiolabeled zearalenone was removed through the excreta (Dailey et al., 1980), although there was no significant deposition of radiolabeled activity in lipophilic metabolites observed in egg yolk at a concentration of approximately 2 mg/kg after 72 hours of dosing.

The maximum accumulation of radioactive zearalenone was observed in the liver after 30 minutes of intubation in 7-week-old broiler chickens, and it showed traces of radioactive compounds after 48 hours of intubation (Mostrom, 2016). After 48 hours of administration, zearalenone was found in muscles at a 4 ppb concentration, which proves that residual levels of zearalenone in edible tissue are of negligible amounts.

12.6 EFFECTS OF PROCESSING ON ZEARALENONE

Fusarium graminearum and *Fusarium culmorum* are common molds that are responsible for zearalenone mycotoxin contamination in major cereals (Ferrigo et al., 2016). The activity of this mycotoxin in animals and humans is related to its estrogenic activity, which ultimately disturbs endocrine function. Various research efforts have tried to reduce the ZEN level through various chemical, physical, and biological methods. Some chemical methods also show effectiveness against zearalenone content in contaminated food items (Awuchi et al., 2021). Zearalenone comes under the category of xenoestrogen, which has a structure similar to estrogen, which causes hormonal imbalance in the human body and several reproduction-related diseases (Mahato et al., 2021)

The effect of any physical processing methods can be determined by the dispersion of ZEN in food samples and with the help of chemical attributes like heat stability and solubility. For example, flour obtained after milling of wet corn was found to be ZEN free, while the by-products of the same milling process like bran and germ were found to have a concentrated portion of ZEN, which, if used as feed for animals, is ultimately toxic for them (Magallanes et al., 2019).

12.6.1 Effects of Thermal Processing

Zearalenone is frequently found in cereals and grains and their by-products. Heat treatment methods are used in the food processing industry that have had promising results in reduction of mycotoxin. Several techniques such as cooking, roasting, and extrusion at varied temperatures ranging from 100°C to 200°C for 10 to 30 minutes decrease the level of mycotoxin level in terms of zearalenone in infected wheat, maize, and oat. The reduction in ZEN can easily be verified by enzyme-linked immunosorbent assay (ELISA) (Pleadin et al., 2019; Cazzaniga et al., 2001).

There was a minor reduction in ZEN due to cooking, that is, an 11% reduction in ZEN level was observed, but a 46% reduction in ZEN concentration was observed in roasting, and an 80% reduction was seen in extrusion cooking (Pleadin et al., 2019; Cazzaniga et al., 2001).

12.6.2 Effect of Drying and Washing

The effects of drying and washing on the concentration of zearalenone in naturally contaminated wheat samples were studied (Yener and Koksel, 2013). Wheat grain was washed for 1 to 2 minutes with high-pressure water, chlorinated water, sodium carbonate, and sodium hydroxide solutions. Then washed wheat samples were dried using three different methods: low-temperature infrared drying, oven drying, and microwave drying. During the experiment, a

21.1% reduction in ZEN level was observed when the wheat samples were pressure-washed with water and dried in the oven. ZEN concentration was reduced to 89.0% when infrared and microwave drying were employed after pressurized water washing. When the wheat samples were washed with chlorinated water, sodium carbonate, and sodium hydroxide solutions, the reductions in ZEN levels were 31.6% to 83.6%. Thus, pressure washing and microwave and infrared drying are effective ways to reduce mycotoxin levels (Yener and Koksel, 2013).

12.6.3 Effect of Milling

The effect of milling on zearalenone in wheat products was analyzed. Three break flours, three middling flours, and two outer layer fractions (bran and short) were milled from two different grain samples, norin-61 and chikugoizumi. Patent flour was created from break and middling grains for human consumption, while low-grade flour for animal feed was made from break and outer layer milling of wheat. Solvent extraction, multifunctional cartridges, and HPLC-fluorescence detection were used to determine the amount of ZEN in these different flour samples. In the patent flour, a more than 50% reduction in ZEN was observed (Zheng et al., 2014).

12.6.4 Effect of Extrusion

The extrusion technique is widely used for cereal snacks and animal feed. European Union legislation regulates the permissible levels of mycotoxins in cereals and related products, making public health concerns of utmost importance. However, very few research has been done so far on the removal or reduction of *Fusarium* mycotoxins during processing. Using pilot-scale equipment for extrusion, naturally contaminated maize flour and grits were investigated. The *fumonisins* are lost to variable degrees during extrusion cooking, while ZEN remains rather stable. When extrusion processing of maize is done with less moisture content, it reduces ZEN concentrations. Mycotoxin levels can also be affected by the addition of sugar and sodium chloride. Moisture content is a more important parameter than temperature during extrusion processing on *fumonisin* reduction, although dried materials require more energy in extruder operations. The reduction level in *Fusarium* toxins is dependent on initial concentration and the type of process employed (Kamle et al., 2019).

Extrusion cooking is one of the best alternatives to reduce the ZEN level. In extrusion cooking, raw materials are exposed to high temperature, high pressures, and shear stress, lowering the ZEN levels and estrogenic activity. ZEN levels were reduced when foods were fermented with bacteria and yeast. Fermentation can convert ZEN to its conjugate compound α-zearalenol.

Further, more research is needed to identify effective methods for the removal/ detoxification of ZEN (Ryu et al., 2002).

A twin-screw extruder was used to analyze the stability effect of zearalenone in maize grits 4.4 micrograms in size. Variables like two types of screw extruder (mixing and non-mixing), temperature (120, 140, and 160°C), and moisture content (18, 22, and 26%) were evaluated to study the stability of ZEN (Ryu et al., 2000). High-performance liquid chromatography was used to test both extruded and extruded samples of maize. Extrusion heating of maize grits resulted in considerable ZEN reductions in both kinds of extrusions, but mixing screws gave slightly better results, with a 66 to 83% reduction in ZEN, while non-mixing screws reduced ZEN from 65 to 77%. A higher reduction in ZEN level was also found between 120 and 140°C than 160°C. Moisture content did not show any effect on the reduction of mycotoxin either with mixing or non-mixing types of extrusion (Ryu et al., 2000).

12.6.5 Effect of Cooking

Several mycotoxins were present in edible oils due to refining flaws. These can be easily removed or reduced by refining processes. Many local oil production units do not refine oil properly, causing mycotoxins to be present in oils (Bhat and Reddy, 2017). The effect of cooking on the stability of mycotoxins is less investigated. The effects of deep-frying, microwave, and oven cooking on zearalenone in maize oil were studied. High-resolution liquid chromatography with fluorescence detectors was used in the analysis. All three methods of cooking show a major reduction in ZEN concentration with time-temperature combinations of frying at 130–190°C for 2.5 to 5 minutes, oven cooking at 110–230°C for 2.5 to 5 minutes, and microwave exposure at 150°C for 10 minutes. Microwave cooking follows first-order kinetics, where the maximum zearalenone concentration degraded up to 38% in 10 minutes. The degradation in mycotoxin levels is dependent on the initial concentration and cooking method. These findings could be useful in determining the chemical safety of edible oils and dishes cooked with them (Sadeghi et al., 2020).

The effects of heat, moisture, and pH were studied on the stability of zearalenone. The effect of pH, salt, and temperature on the stability of other mycotoxins was promising in the aspect of degradation. But treatment with a sodium bicarbonate solution for 12 days at 110°C could not lower the ZEN content (Lauren and Smith, 2001). Because ZEN has so many negative effects on animals, plants, and humans, there is a need for the development of effective techniques for identifying, analyzing, and degrading these mycotoxins that ensure the quality and safety of food and food items. Therefore, still more suitable and appropriate procedures are required in processing cereals that could reduce the maximum zearalenone.

12.7 DETECTION METHODS OF ZEARALENONE IN FOOD AND FEED

Determination of mycotoxin levels in food samples is typically conducted through the following steps: sampling, preparation, extraction, cleaning, and detection, which are carried out using a variety of instrumental and non-instrumental approaches (Assefa and Geremew, 2018). Among the instrumental methods used to detect ZEN in food and feed are high-performance liquid chromatography (HPLC), liquid chromatography-mass spectroscopy (LC-MS)/MS, and gas chromatography-mass spectroscopy (GC-MS). In mycotoxin testing, GC is a typical detection tool. Capillary gas chromatography with electron-capture detection, despite its excellent sensitivity, has a problem with mycotoxin confirmation. To address this issue, gas chromatography-mass spectrometry is employed to determine the existence of mycotoxins. Recent research has demonstrated the effectiveness of the GC-MS technique for ZEA analysis in cereals, commodities, and feeds during routine operations (Gromadzka et al., 2008). Combining high-speed counter-current chromatography with a macroporous resin column yielded an improved method for purifying ZEN from *Fusarium graminearum* rice culture (Wang et al., 2019). Although these instrumental techniques offer high sensitivity and specificity, the main limitations include time-consuming sample preparation, long analysis periods, high prices, and the inability to do an on-site rapid assessment. To overcome these constraints, immunoassay techniques for easy, rapid, and in-field monitoring of large-scale sample screening and detection at low cost and high sensitivity were developed. The most common immunoassay techniques include ELISA, rapid immunochromatographic assays (ICAs), lateral flow immunoassay (LFA), fluorescence polarization immunoassays, immunochip, immunosensor, multiplex dipstick immunoassay, and suspension array. Despite their great sensitivity and specificity, these techniques require expert personnel to operate (Mahato et al., 2021). A double-label immunochromatographic assay (DL-ICA) based on Eu^{3+}nanoparticles (EuNP) intended to measure the amounts of zearalenone in cereals showed high sensitivity, reproducibility, and selectivity, with low detection limits (0.21–0.25 g/kg) (Wang et al., 2018). A 3-D printed device for detection based on a smartphone integrated with a solid phase latex microsphere immunochromatography platform (SIAP) was devised for quantitative and sensitive detection of ZEN at low detection limits of 0.08 and 0.18 g/kg, respectively, in cereal foods and feed for on-site monitoring (Li et al., 2019).

Aside from these, a variety of electrochemical-based detection methodologies, including voltammetric, potentiometric, conductivity, and amperometric approaches, are utilized. The changes in voltage, current, and impedance produced by a chemical or biological molecule's oxidation/reduction process are studied in electrochemical systems with three electrodes. Electrochemical biosensors are frequently employed by researchers for ZEN detection due to their sensitivity, selectivity, low cost, and simplicity. The bulk of electrochemical ZEN sensors are electrochemical immunosensors, which are based on high-affinity interactions between antigen and antibodies. Electrodes, immobilization layer, primary antibody, secondary antibody, labeled enzymes, and reaction substrate are the fundamental components of electrochemical immunosensors (Caglayan and Üstünda, 2020).

Another effective detection approach is the use of biosensors. A biosensor is a device that has an active biological sensing element linked to a transducer that translates observable physical or chemical changes into a quantifiable signal. Among biosensors, immunosensors are the most commonly utilized analytical devices for mycotoxin detection. To identify the presence of mycotoxin contamination, several immunosensors such as optical immunosensors, label-based immunosensors, and piezoelectric immunosensors have been utilized. Aptamers are also used to detect mycotoxins because of benefits such as long shelf life, reversible denaturation, ease of production, and chemical modification (Cerchia and de Franciscis, 2020). Further efforts are being made to improve biosensor qualities such as sensitivity, selectivity, rapid response, and low cost; as a result, nanomaterials (nanoparticles, nanorods, nanotubes, and nanowires) are being included in biosensors. The use of nanoparticles has the advantage of either expanding the surface area of the sensor appropriate for biomolecule immobilization or strengthening the signal provided by immunocomplex formation (Majer-Baranyi et al., 2021).

12.8 MANAGEMENT AND CONTROL STRATEGIES

Fusarium growth and ZEN accumulation in crops have been major issues for food and feed quality and safety, resulting in economic losses. As a result, strategies for reducing contamination must be developed. Prevention is perhaps the most effective technique; up to 90% of mycotoxins discovered in feeds are produced during plant development. Such operations include plant cultivation, harvesting, storage, transportation, marketing, and processing (Rogowska et al., 2019). These programs make use of good agricultural practices (GAPs) and good manufacturing procedures (GMPs). Weed control; soil analysis; pest and fungal eradication with herbicides, fungicides, and insecticides; seedbed treatment; seed decontamination; crop rotation; tillage and ploughing; fertilizers for nutrient enrichment; and genetically modified plants for mycotoxin suppression are all examples of pre-harvest strategies to control mycotoxin contamination (Mahato et al., 2021). Despite precautionary efforts, crop contamination with mycotoxins cannot be completely prevented. As a result, many methods for inactivating and removing these contaminants have been developed. Decontamination is a term that is commonly used to describe these treatments. However, it is only permitted for use in feed and cannot be utilized to manufacture raw materials or food products. Detoxification treatments are critical for improving food safety, reducing economic losses, and recovering diseased goods (Zinedine et al., 2007). Physical and chemical purification techniques are the most often employed for feed and food. The most common physical operations include washing, sorting, hulling, grinding, heat treatment, irradiation with UV or gamma radiation, and adsorption (Zhang et al., 2016). Extrusion cooking appears to be a successful approach as well. The addition of inert sorbents such as kaolinite, clays, activated carbon, zeolites, bentonite, or aluminosilicates are among the latest approaches (Kowalska et al., 2016; Makowska et al., 2017; Wang et al., 2018). Chemical approaches rely on the use of chemical compounds such as sodium hypochlorite, sulfur dioxide, hydrogen peroxide, or ammonia to react with mycotoxins and degrade them. The grains are then washed with sodium carbonate or water to remove *Fusarium* toxins. According to Polak et al. (2009), adding 2% sodium carbonate to ZEN-contaminated feed decreases the amount of this toxin (Rogowska et al., 2019). Although these technologies are useful in a variety of contexts, they have several drawbacks, the most obvious of which are high costs and the loss of important nutrients from food (Zhang et al., 2016). As a result, there is a growing interest in the use of microbiological technologies, which appear to be safer, more ecologically friendly, and less expensive than traditional techniques (Zhang et al., 2016). Mycotoxin biodegradation appears to be a viable technology that does not impair the flavor or nutritional content of food and, most importantly, poses no harm to human or animal health (Zhang et al., 2016). Mycotoxin adsorption onto the walls of microbial cells or mycotoxin breakdown mediated by microbial secretases are the primary methods of biological detoxification. The use of non-pathogenic microorganisms for ZEN decontamination is a current research topic, and it has the potential to be a unique strategy for accomplishing mycotoxin detoxification in practical situations in the future. The use of probiotics as mycotoxin detoxifying agents is becoming increasingly common. As of now, the synthesis of ZEN-degrading enzyme genes in probiotic host cells has been the primary focus of their application in ZEN toxin decontamination (eventually meant to assure food safety for consumption by animals and humans). Further, technological advances in genomics, marker creation, and genetic engineering have the potential to enhance food safety against mycotoxin contamination. Due to the polygenic and complicated resistance to mycotoxin contamination, marker identification is also required to accelerate the transfer of resistance characteristics into agronomically viable genetic backgrounds (Assefa and Geremew, 2018). It is quite unusual to detect many kinds of mycotoxins in a single sample while examining the safety of food and feed. As a result, research on the co-degradation of several mycotoxins is becoming increasingly crucial (Wang et al., 2019).

12.9 CONCLUSION

The widespread contamination of various foods and feed with mycotoxins endangers human and animal health as well as global commercial trade. Zearalenone is a mycotoxin produced by *Fusarium* species that is extremely prevalent in food and feed. It causes a wide range of alterations and abnormalities in the reproductive system, resulting in huge economic losses. ZEN is harmful to human and animal health due to its mutagenic, teratogenic, carcinogenic, nephrotoxic, immunotoxic, and genotoxic effects. Because this toxin is very resistant to diverse processing conditions, efficient detection and control measures are required to limit its negative effects and ensure food safety. This chapter contains detailed information about the origins, occurrence, and biosynthetic mechanism of ZEN, which will aid in the development of enhanced detection and control measures. Various detection and control strategies have been discussed. Further improvement in these technologies, as well as the implementation of innovative detection and decontamination procedures, is required to carry out the aim of achieving global food and feed safety and security. To eliminate the possibility of zearalenone contamination from "farm to fork," systematic testing of raw materials and finished goods, as well as the effective execution of preventive tools such as good agricultural practices, good manufacturing practices, and hazard analysis and critical control points (HACCPs), is required. More emphasis should be focused on epidemiological research and the development of dependable, low-cost, simple-to-use, portable diagnostic tools. Microbial control agents and immobilized enzymes for mycotoxin decontamination and degradation can be marketed based on their availability for food enterprises. Furthermore, employing genetic engineering to produce crops with greater fungal resistance is a viable future option. Furthermore, future research should focus on multi-mycotoxin detection to deal with the co-occurrence of two or more mycotoxins at the same time in the near future.

REFERENCES

Agriopoulou, S., Stamatelopoulou, E. and Varzakas, T. (2020). Advances in occurrence, importance, and mycotoxin control strategies: Prevention and detoxification in foods. *Foods*, 9(137): 1–48. http://doi.org/10.3390/foods9020137.

Assefa, T. and Geremew, T. (2018). Major mycotoxins occurrence, prevention and control approaches. *Biotechnology and Molecular Biology Reviews*, 12(1): 1–11.

Awuchi, C. G., Ondari, E. N., Ogbonna, C. U., Upadhyay, A. K., Baran, K., Okpala, C., Korzeniowska, M. and Guine, R. (2021). Mycotoxins affecting animals, foods, humans, and plants: Types, occurrence, toxicities, action mechanisms, prevention, and detoxification strategies: A revisit. *Foods* (Basel, Switzerland), 10(6): 1279–1327.

Bennet, J. W. and Klich, M. (2003). Mycotoxins. *Clinical Microbiology Reviews*, 16(3): 497–516.

Bhat, R. and Reddy, K. R. N. (2017). Challenges and issues concerning mycotoxins contamination in oil seeds and their edible oils: Updates from last decade. *Food Chemistry*, 215: 425–437.

Biehl, M. L., Prelusky, D. B., Koritz, G. D., Hartin, K. E., Buck, W. B. and Trenholm, H. L. (1993). Biliary excretion and enterohepatic cycling of zearalenone in immature pigs. *Toxicology and Applied Pharmacology*, *121*(1): 152–159.

Borzekowski, A., Drewitz, T., Keller, J., Pfeifer, D., Kunte, H. J., Koch, M., Rohn, S. and Maul, R. (2018). Biosynthesis and characterization of zearalenone-14-sulfate, zearalenone-14-glucoside and zearalenone-16-glucoside using common fungal strains. *Toxins, 10*: 104–118.

Brodehl, A., Moeller, A., Kunte, H.-J., Koch, M. and Maul, R. (2014). Biotransformation of the mycotoxin zearalenone by fungi of the genera *Rhizopus* and *Aspergillus*. *FEMS Microbiology Letters, 359*: 124–130.

Bullerman, L. B. (2003). Mycotoxins classifications. In *Book Encyclopedia of Food Sciences and Nutrition*. Academic Press, pp. 4080–4089. http://doi.org/10.1016/B0-12-227055-X/00821-X.

Caglayan, M. O. and Üstündağ, Z. (2020). Detection of zearalenone in an aptamer assay using attenuated internal reflection ellipsometry and its cereal sample applications. *Food and Chemical Toxicology, 136*: 111081.

Cazzaniga, D., Basilico, J. C., Gonzalez, R. J., Torres, R. L. and de Greef, D. M. (2001). Mycotoxin inactivation by extrusion cooking of corn flour. *Letters in Applied Microbiology, 33*(2): 144–147.

Cerchia, L. and de Franciscis, V. (2020). Targeting cancer cells with nucleic acid aptamers. *Trends Biotechnology, 28*: 517–525.

Coffey, R., Cummins, E. and Ward, S. (2009). Exposure assessment of mycotoxins in dairy milk. *Food Control, 20*: 239–249.

Dailey, R. E., Reese, R. E. and Brouwer, A. (1980). Metabolism of [14C] Zearalenone in laying hens. *Journal of Agricultural and Food Chemistry, 28*: 286–291.

De Rycke, E., Foubert, A., Dubruel, P., Bol'hakov, O. I., De Saeger, S. and Beloglazova, N. (2021). Recent advances in electrochemical monitoring of zearalenone in diverse matrices. *Food Chemistry, 353*: 129342.

D'Mello, J. P. F., Placinta, C. M. and Macdonald, A. M. C. (1999). *Fusarium* mycotoxins: A review of global implications for animal health, welfare and productivity. *Animal Feed Science and Technology, 80*: 183–205.

Faisal, Z., Garai, E., Csepregi, R., Bakos, K., Fliszár-Nyúl, E., Szente, L., . . . and Poór, M. (2020). Protective effects of beta-cyclodextrins vs. zearalenone-induced toxicity in HeLa cells and Tg (vtg1: mCherry) zebrafish embryos. *Chemosphere, 240*: 124948.

Ferrer, E., Juan-Garcia, A., Font, G. and Ruiz, M. J. (2009). Reactive oxygen species induced by beauvericin, patulin and zearalenone in CHO-K1 cells. *Toxicol in Vitro, 23*(8): 1504–1509.

Ferrigo, D., Raiola, A. and Causin, R. (2016). *Fusarium* toxins in cereals: Occurrence, legislation, factors promoting the appearance and their management. *Molecules, 21*(5): 627–661.

Fink-Gremmels, J. and Malekinejad, H. (2007). Clinical effects and biochemical mechanisms associated with exposure to the mycoestrogen zearalenone. *Animal Feed Science and Technology, 137*: 326–341.

Gil-Serna, Y., Vázquez, C., González-Jaén, M. T. and Patiño, B. (2014). Discrimination of the main ochratoxin A-producing species. In *Encyclopedia of Food Microbiology*, 2nd edn, vol. 1, eds. Batt C. A., Tortorello M. A. London: Elsevier, pp. 887–892.

Gromadzka, K., Waskiewicz, A., Chelkowski, J. and Golinski, P. (2008). Zearalenone and its metabolites: Occurrence, detection, toxicity and guidelines. *World Mycotoxin Journal, 1*(2): 209–220.

Hao, W., Ge, Y., Qu, M., Wen, Y., Liang, H., Li, M., . . . and Xu, L. (2022). A simple rapid portable immunoassay of trace zearalenone in feed ingredients and agricultural food. *Journal of Food Composition and Analysis, 107*: 104292.

Jiménez, M. and Mateo, R. (1997). Determination of mycotoxins produced by *Fusarium* isolates from banana fruits by capillary gas chromatography and high-performance liquid chromatography. *Journal of Chromatography A, 778*(1–2): 363–372.

Kamle, M., Mahato, D. K., Devi, S., Lee, K. E., Kang, S. G. and Kumar, P. (2019). Fumonisins: Impact on agriculture, food, and human health and their management strategies. *Toxins, 11*(6): 328–350.

Kiessling, K. H., Pettersson, H., Sandholm, K. and Olsen, M. (1984). Metabolism of aflatoxin, ochratoxin, zearalenone and three trichothecenes by intact rumen fluid, rumen protozoa and rumen bacteria. *Applied and Environmental Microbiology, 47*(5): 1070–1073.

Kowalska, K.; Habrowska-Górczy´nska, D. E. Piastowska-Ciesielska, A. W. (2016). Zearalenone as an endocrine disruptor in humans. *Environmental Toxicology and Pharmacology, 48*: 141–149.

Kuiper-Goodman, T., Scott, P. M. and Watanabe, H. (1987). Risk assessment of the mycotoxin zearalenone. *Regul Toxicol Pharmacol, 7*(3): 253–306.

Kumar, P., Mahato, D. K., Sharma, B., Borah, R., Haque, S., Mahmud, M. C., . . . & Bui, S. (2020). Ochratoxins in food and feed: Occurrence and its impact on human health and management strategies. *Toxicon, 187*: 151–162.

Lauren, D. R. and Smith, W. A. (2001). Stability of the fusarium mycotoxins nivalenol, deoxynivalenol and zearalenone in ground maize under typical cooking environments. *Food Additives & Contaminants, 8*(11): 1011–1016.

Li, R., Meng, C., Wen, Y., Fu, W. and He, P. (2019). Fluorometric lateral flow immunoassay for simultaneous determination of three mycotoxins (aflatoxin B1, zearalenone and deoxynivalenol) using quantum dot microbeads. *Microchimica Acta, 186*(12): 1–9.

Llorent-Martinez, E. J., Fernández-Poyatos, M. P. and Ruiz-Medina, A. (2019). Automated fluorimetric sensor for the determination of zearalenone mycotoxin in maize and cereals feedstuff. *Talanta, 191*: 89–93.

Magallanes, A., Manthey, F. and Simsek, S. (2019). Wet milling of deoxynivalenol contaminated wheat: Effect on physicochemical properties of starch. *Cereal Chemistry, 97*(2): 293–303.

Mahato, D. K., Lee, K. E., Kamle, M., Devi, S., Dewangan, K. N., Kumar, P., & Kang, S. G. (2019). Aflatoxins in food and feed: an overview on prevalence, detection and control strategies. *Frontiers in microbiology, 10*: 2266.

Mahato, D. K., Devi, S., Pandhi, S., Sharma, B., Maurya, K. K., Mishra, S., Dhawan, K., Selvakumar, R., Kamle, M. and Mishra, A. K. (2021). Occurrence, impact on agriculture, human health, and management strategies of zearalenone in food and feed: A review. *Toxins, 13*(2): 92–115.

Majer-Baranyi, K., Adányi, N. and Székács, A. (2021). Biosensors for deoxynivalenol and zearalenone determination in feed quality control. *Toxins, 13*(7): 499.

Makowska, K., Obremski, K., Zielonka, L. and Gonkowski, S. (2017). The influence of low doses of zearalenone and T-2 toxin on calcitonin gene related peptide-like immunoreactive (CGRP-LI) neurons in the ENS of the porcine descending colon. *Toxins, 9*(3): 98.

Malekinejad, H., Maas-Bakker, R. F. M. and Fink-Gremmels, J. (2005). Bioactivation of zearalenone by porcine hepatic biotransformation. *Veterinary Research, 36*(5–6): 799–810.

Mally, A., Solfrizzo, M. and Degen, G. H. (2016). Biomonitoring of the mycotoxin zearalenone: current state-of-the art and application to human exposure assessment. *Archives of Toxicology, 90*(6): 1281–1292.

Mostrom, M. (2016). Mycotoxins: Classification. In *Book Encyclopedia of Food and Health. Reference Module in Food Science*. Academic Press, pp. 29–34.

Nahle, S., Khoury, A. and Atoui, A. (2021). Current status on the molecular biology of zearalenone: Its biosynthesis and molecular detection of zearalenone producing *Fusarium* species. *European Journal of Plant Pathology, 159*: 247–258.

Niazi, S., Khan, I. M., Yu, Y., Pasha, I., Shoaib, M., Mohsin, A., . . . & Wang, Z. (2019). A "turnon" aptasensor for simultaneous and time-resolved fluorometric determination of zearalenone, trichothecenes a and aflatoxin B1 using WS2 as a quencher. *Microchimica Acta, 186*(8): 1–10.

Olsen, C. R., Simpson, H. J., Peng, T. H., Bopp, R. F. and Trier, R. M. (1981). Sediment mixing and accumulation rate effects on radionuclide depth profiles in Hudson estuary sediments. *Journal of Geophysical Research, 86*(c11): 11020–11028.

Olsen, K. H., Baldridge, W. S. and Callender, J. F. (1987). Rio Grande rift: An overview. *Tectonophysics, 143*(1–3): 119–139.

Pleadin, J., Babic, J., Vulic, A., Kudumija, N., Aladic, K., Kis, M. and Subaric, D. (2019). The effect of thermal processing on the reduction of deoxynivalenol and zearalenone cereal content. *Croatian Journal of Food Science and Technology, 11*(1): 44–51.

Pleadin, J., Zadravec, M., Perši, N., Vulić, A., Jaki, V. and Mitak, M. (2012). Mould and mycotoxin contamination of pig feed in northwest Croatia. *Mycotoxin Research, 28*(3): 157–162.

Polak, M., Gajęcki, M., Kulik, T., Łuczyński, M. K. and Obremski, K. (2009). The evaluation of the efficacy of sodium. *Polish Journal of Veterinary Sciences, 12*(1): 103–111.

Prelusky, D. B., Scott, P. M., Trenholm, H. L. and Lawrence, G. A. (1990). Minimal transmission of zearalenone to milk of dairy cows. *Journal of Environmental Science and Health. Part. B, Pesticides, Food Contaminants, and Agricultural Wastes, 25*(1): 87–103.

Queiroz, V. A. V., de Oliveira Alves, G. L., da Conceição, R. R. P., Guimarães, L. J. M., Mendes, S. M., de Aquino Ribeiro, P. E. and da Costa, R. V. (2012). Occurrence of fumonisins and zearalenone in maize stored in family farm in Minas Gerais, Brazil. *Food Control, 28*(1): 83–86.

Radi, A. E., Eissa, A. and Wahdan, T. (2020). Molecularly imprinted impedimetric sensor for determination of mycotoxin zearalenone. *Electroanalysis, 32*(8): 1788–1794.

Reinholds, I., Bogdanova, E., Pugajeva, I., Alksne, L., Stalberga, D., Valcina, O. and Bartkevics, V. (2020). Determination of fungi and multi-class mycotoxins in *Camelia sinensis* and herbal teas and dietary exposure assessment. *Toxins, 12*(9): 555.

Rogowska, A., Pomastowski, P., Sagandykova, G. and Buszewski, B. (2019). Zearalenone and its metabolites: Effect on human health, metabolism and neutralisation methods. *Toxicon, 162*: 46–56.

Ropejko, K. and Twarużek, M. (2021). Zearalenone and its metabolites—General overview, occurrence, and toxicity. *Toxins, 13*(1): 35.

Ryu, D., Hanna, H. A. and Bullerman, L. B. (2000). Stability of zearalenone during extrusion of corn grits. *Journal of Food Protection, 62*(12): 1482–1484.

Ryu, D., Jackson, L. S. and Bullerman, L. B. (2002). Effects of processing on zearalenone. In *Mycotoxins and Food Safety. Advances in Experimental Medicine and Biology*, vol. 504, eds. DeVries J. W., Trucksess M. W., Jackson L. S. Boston, MA: Springer.

Sadeghi, E., Oskoei, L. B., Nejatian, M. and Mehr, S. S. (2020). Effect of microwave, deep frying and oven cooking on destruction of zearalenone in spiked maize oil. *World Mycotoxin Journal, 13*(4): 515–522.

Shi, H., Li, S., Bai, Y., Prates, L. L., Lei, Y. and Yu, P. (2018). Mycotoxin contamination of food and feed in China: Occurrence, detection techniques, toxicological effects and advances in mitigation technologies. *Food Control, 91*: 202–215.

Virk, P., Al-Mukhaizeem, N. A. R., Morebah, S. H. B., Fouad, D. and Elobeid, M. (2020). Protective effect of resveratrol against toxicity induced by the mycotoxin, zearalenone in a rat model. *Food and Chemical Toxicology*, *146*: 111840.

Wang, M., Yin, L., Hu, H., Selvaraj, J. N., Zhou, Y., Zhang, G. (2018). Expression, functional analysis and mutation of a novel neutral zearalenone-degrading enzyme. *International Journal of Biological Macromolecules*, *118*: 1284–1292.

Wang, N., Wu, W., Pan, J. and Long, M. (2019). Detoxification strategies for zearalenone using microorganisms: A review. *Microorganisms*, *7*(7): 208.

Yener, S. and Koksel, H. (2013). Effects of washing and drying applications on deoxynivalenol and zearalenone levels in wheat. *World Mycotoxin Journal*, *6*(3): 335–341.

Zhang, H., Dong, M., Yang, Q., Apaliya, M. T., Li, J., & Zhang, X. (2016). Biodegradation of zearalenone by Saccharomyces cerevisiae: Possible involvement of ZEN responsive proteins of the yeast. *Journal of Proteomics*, *143*: 416–423.

Zhang, G. L., Feng, Y. L., Song, J. L. and Zhou, X. S. (2018). Zearalenone: A mycotoxin with different toxic effect in domestic and laboratory animals' granulosa cells. *Frontiers in Genetics*, *9*: 667.

Zheng, Y., Md. Hossen, S., Sago, Y., Yoshida, M., Nakagawa, H., Nagashima, H., Okadome, H., Nakajima, T. and Kushiro, M. (2014). Effect of milling on the content of deoxynivalenol, nivalenol, and zearalenone in Japanese wheat. *Food Control*, *40*: 193–197.

Zinedine, A. and Ruiz, M.-J. (2014). Zearalenone. Mycotoxins implic. *Food Safety*: 52–66.

Zinedine, A., Soriano, J. M., Carlos Molto, J. and Manes, J. (2007). Review on the toxicity, occurrence, metabolism, detoxification, regulations and intake of zearalenone: An oestrogenic mycotoxin. *Food and Chemical Toxicology*, *45*(1): 1–18.

FURTHER READINGS

Katzenellenbogen, B. S. and Korach, K. S. (1997). A new actor in the estrogen receptor drama enter ER-β. *Endocrinology*, *138*(3): 861–862.

Kushiro, M. (2008). Effects of milling and cooking processes on the deoxynivalenol content in wheat. *International Journal of Molecular Sciences*, *9*(11): 2127–2145.

Lee, U. S., Lee, M. Y., Park, W. Y. and Ueno, Y. (1992). Decontamination of *Fusarium* mycotoxins, nivalenol, deoxynivalenol, and zearalenone, in barley by the polishing process. *Mycotoxin Research*, *8*(1): 31–36.

Mirocha, C., Robison, T. S., Pawlosky, R. J. and Allen, N. K. (1982). Distribution and residue determination of [3H] zearalenone in broilers. *Toxicology and Applied Pharmacology*, *66*: 77–87.

Mokoena, P., Chelule, P. K. and Gqaleni, N. (2005). Reduction of fumonisin B 1 and zearalenone by lactic acid bacteria in fermented maize meal. *Journal of Food Protection*, *68*(10): 2095–2099.

Scudamore, K. A., Baillie, H., Patel, S. and Edwards, S. G. (2007). Occurrence and fate of *Fusarium* mycotoxins during commercial processing of oats in the UK. *Food Additives & Contaminants*, *24*(12): 1374–1385.

Scudamore, K. A., Guy, R. C. E., Kelleher, B. and MacDonald, S. J. (2008). Fate of *Fusarium* mycotoxins in maize flour and grits during extrusion cooking. *Food Additives and Contaminants Part A Chemistry Analysis Control Exposure & Risk Assessment*, *25*(11): 1374–1384.

Chapter 13

Mycotoxins in Stored Foods

Preeti Sharma, Kanishka Chawla, Kamakshi Kalia,
Vanshika Saini, Aastha Bhardwaj, Vasudha Bansal,
Nitya Sharma

CONTENTS

13.1 Introduction   334
13.2 Mycotoxins in Foods and Their Health Implications   335
    13.2.1 Aflatoxins   336
    13.2.2 Trichothecene   338
    13.2.3 Ochratoxin   339
    13.2.4 Patulin   340
    13.2.5 Fumonisins   341
    13.2.6 Zearalenone   342
13.3 Prevalence of Mycotoxins in Stored Foods   344
    13.3.1 Grains and Oilseeds   344
    13.3.2 Pulses and Legumes   345
    13.3.3 Fruits and Vegetables   345
    13.3.4 Processed Foods (Flours, Juices, Spices, etc.)   346
13.4 Factors Influencing Mycotoxin Development in Foods during
    Storage   346
13.5 Control Measures   350
    13.5.1 Using Ozone   350
    13.5.2 Other Control Measures   351
        13.5.2.1 Physical Detoxification   351
        13.5.2.2 Biological Detoxification 352
        13.5.2.3 Chemical Detoxification   352
13.6 Control Strategies to Prevent Mycotoxin Contamination
    during Storage   354

DOI: 10.1201/9781003242208-13

13.1 INTRODUCTION

Mycotoxins are biological toxins produced by the fungus family as secondary metabolites (Murphy et al., 2006). The World Health Organization defines mycotoxins as toxic compounds naturally produced by certain types of molds, especially fungi. About 25% of world crops has been estimated to be infected by molds and fungal growth (Pandya and Arade, 2016). Mycotoxins are considered toxic since they are produced by fungi, but some fungi products that may be toxic to bacteria are helpful in medicinal uses, such as penicillin. Mycotoxins grow out of fungi commonly found in crops, food, and storage across globe and are a major growing concern in food and feeds. The World Health Organization (WHO) and Food and Agricultural Organization (FAO) of the UN have jointly formed an international expert committee, the Joint FAO/WHO Expert Committee on Food Additives (JECFA) to assess the health risks posed by natural toxins, such as mycotoxins, which in turn has facilitated forming the international standards and code of practice Code Alimentarious Commission, which details limits of exposure to a variety of mycotoxins in various types of cereals, fruits, nuts, and other food items. These toxins, even in low concentrations, pose serious concern over a short time. Much research is being carried out to contain this global risk in the food and storage industry.

A Malaysian case study by Afsah-Hejri et al. (2013) stated that foodborne mycotoxin contamination is a global issue that poses a major threat to human and animal health, as well as huge economic losses in both developing and wealthy countries. Selective mycotoxins have the potential to be chemical warfare agents. (Ráduly et al., 2020). Pitt and Miller explored the history of mycotoxins and found that mycotoxins have existed since humans started cultivating crops and storing them for future needs. Coppock & Dziwenka (2019) investigated the effects of mycotoxins in humans and animals using biomarkers in toxicology. They discovered that mycotoxins could contaminate a variety of foods and beverages, that some mycotoxins penetrate the placenta and are present in the fetus at birth, and that some are excreted in milk.

Mycotoxins are created during the exponential growth phase and have no apparent impact on the growth and metabolism of generating organisms (James et al., 2007). When humans or animals consume them, absorb them through their skin, or inhale biological poisons, it can result in serious sickness or even death in living organisms. Molds, for example, may grow in a wide variety of temperatures, although their growth and mycotoxin production rates are influenced by temperature and water availability (water activity). As a result, mycotoxins might be created in favorable conditions during crop storage or cultivation. Mycotoxin-producing fungi can be found in a variety of foods, including cereal grains, tree nuts, apples, beer, peanuts, and other fruits, as well as coffee beans, oilseeds, cocoa beans, wine, and spices. The majority of mycotoxins are chemically stable and can withstand cooking. With the exception of fusarium plant

TABLE 13.1   CLASSIFICATION OF MYCOTOXINS

Mycotoxins	Fungi
Aflatoxins B1, B2, G,1 and G2	*Aspergillus flavus* *Aspergillus parasiticus*
Sterigmatocystin	*Aspergillus versicolor* *Penicillium nidulans*
Fumonisin	*Fusarium moniliforme*
Ochratoxin	*Aspergillus ochraceus* *Penicillium viridicatum*
Zearalenone	*Fusarium tricinctum* *Fusarium roseum* *Graminearum*
Trichothecenes Nivalenol DON T-2 Diacetoxycirpenol	*Fusarium* spp. *Trichothecium* spp.

pathogens, naturally occurring fungal flora are dominated by genera such as *Penicillium*, *Aspergillus*, and *Fusarium*, which can harbor both dangerous and non-harmful bacteria (Murphy et al., 2006). There are hundreds of different types of mycotoxins, 14 of which are carcinogenic, with aflatoxins being the most dangerous, followed by patulins, fumonisins, ochratoxins, nivalenol/deoxynivalenol, and zearalenone (Table 13.1). Consuming diseased food or animal products, such as milk from animals fed contaminated feed, can expose you to these mycotoxins (Zain, 2011). When these fungal metabolites are eaten in large quantities, they produce toxic effects that range from short-term to long-term tetragenic and mutagenic effects, with symptoms ranging from skin irritation to immunological deficiency, congenital abnormalities, neurotoxicity, and death (Zain, 2011). The effectiveness of mycotoxins is divided into five groups: 1, 2A, 2B, 4, and 5. Group 1 contains aflatoxins, which are human carcinogens. Group 2B refers to substances that are likely to cause cancer, such as ochratoxins.

13.2 MYCOTOXINS IN FOODS AND THEIR HEALTH IMPLICATIONS

Several hundred mycotoxins have been known to be found in foods, and about a dozen have been identified to cause severe illness symptoms, which may appear immediately after consumption of infected food or over a long-term period depending upon their type, exposure, and duration. These mycotoxins have been categorized into select groups of major concern.

13.2.1 Aflatoxins

These are the most widely studied mycotoxins and are derived from *Aspergillus flavus* toxin (A-fla-toxin). These are derivatives of difuranocoumarin in which a group of bifuran is attached to the nucleus of coumarin with pantanon/lacton ring. B1, B2, G1, and G2 (shown in Figure 13.1) are the most common aflatoxins (James et al., 2007). Aflatoxin B1 has a chemical formula of $C17H1206$, while aflatoxin G1 has a molecular formula of $C17H1207$. Aflatoxins B2 and G2 are dihydro derivatives of the original molecules, $C17H406$ and $C17H1407$. The melting point of B1, B2, G1, and G2 are 268–269°C, 286–289°C, 244–246°C and 237–240°C, respectively (Wogan, 1966).

Aflatoxins are the most toxic mycotoxins and are usually found in soil, soiled crops, grains, legumes, oilseeds, and spices and nuts (Table 13.2). Crops of cereals like corn, sorghum, maize, wheat, rice, ragi, and jowar; oilseeds like sunflower, soybean, and peanuts; spices like peppers, turmeric, ginger, and coriander; and nuts like almonds, pistachio, walnut, and coconut are frequently infected by species of *Aspergillus*. Among this category, *A. bombycis*, *A. ochraceoroseus*, *A. nomius*, and *A. pseudotamari* are also known to be infrequent aflatoxin-producing species (Benkerroum, 2020).

Consumption of infected feed has resulted in detection of *Aspergillus* in milk produced by cattle. Aflatoxin in milk is known as the key product to have caused aflatoxin outbreaks in history. Consumption of aflatoxin-infected milk by humans may lead to serious illness or poisoning and has resulted in life-threatening situations and long-term effects to vital organs like the liver, heart, and kidney. Further, infected cattle feed and livestock have also been found to affect the growth and life of poultry and cause serious consequences to the poultry industry (Adilah and Redzwan, 2017). India has reported several outbreaks of aflatoxicosis in chickens (Ditta et al., 2018). Aflatoxins have been found to alter and damage DNA, which may cause cancer in humans and livestock, and aflatoxins have been attributed as one of the main factors responsible for liver cancer. Environmental factors, exposure amount, and exposure duration, as well as age, health, and dietary nutritional status, can all influence toxicity. Maleki et al. (2015) did a study on aflatoxins secreted in breast milk of lactating women. The results revealed that aflatoxins can have various detrimental effects on infants such as growth impairment, underweight, and infections.

Out of the four types of aflatoxins B1, B2, G1, and G2, B1 is considered the most harmful and carcinogenic (Awuchi et al., 2020). Aflatoxin M1 (AFM1) is a significant metabolite of aflatoxin B1 that can be found in the milk of animals

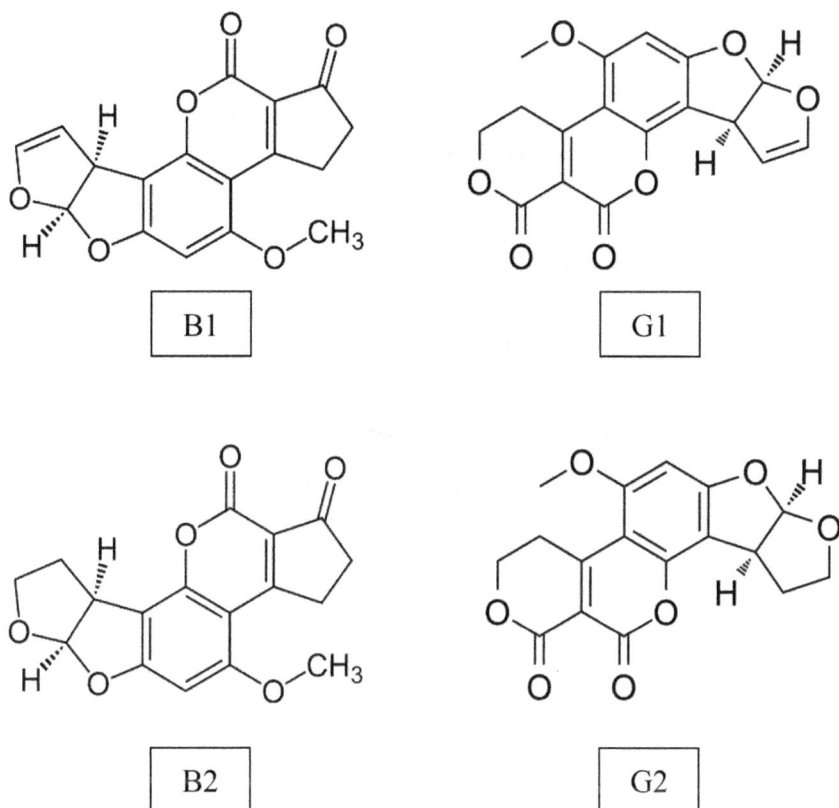

Figure 13.1   Structures of aflatoxin B1, B2, G1, and G2 (Attia and Harisa, 2016).

TABLE 13.2   AFLATOXINS AND FOOD SOURCES

Aflatoxin	Food Source	References
B1 (0.15 ngml⁻¹)*, B2 (0.10 ngml⁻¹)*	Groundnuts	Hajian and Ensafi, 2009
B1, B2, G1, G2, M1 (2.1–3.2 ngml⁻¹)*	Peanuts	Li et al., 2009
B1, B2, G1, G2 (unknown)	Chili	Oriordan and Wilkinson, 2009
B1 (15 to 500 µg/kg)*	Red chili powder	Tripathi & Mishra, 2009
B1 (0.6 ngml⁻¹)*	Corn	Piermarini et al., 2009
B1 (0.1–308 µg/kg)*	Rice	Tan et al., 2009
*(LOD)		

that have eaten aflatoxin B1-contaminated feed. AFM1 has been discovered to have harmful and carcinogenic effects in experimental animals, and as a result, it is designated as a class 2B human carcinogen (Min et al., 2021). In the case of AFM1, even pasteurization, storage, and processing do not alter its stability, and it poses a serious health risk to humans, particularly children, who are heavy milk consumers (Marchese et al., 2018).

13.2.2 Trichothecenes

Trichothecenes (TCNs) are the mycotoxins produced by *Fusarium*, *Myrothecium*, and *Stachybotrys* species of fungi. They have been categorized into four groups (Table 13.3) (Figure 13.2).

TABLE 13.3   TYPES OF TRICHOTHECENES

Type A	T-2, HT-2, diacetoxyscirpenol (DAS), harzianum A, neosolaniol (NEO), and trichodermen
Type B	Deoxynivalenol (DON), nivelenol (NIV), trichothecin, and fusarenon (X)
Type C	Crotocin
Type D	Satratoxin G and H, roridin A, and verrucarin A

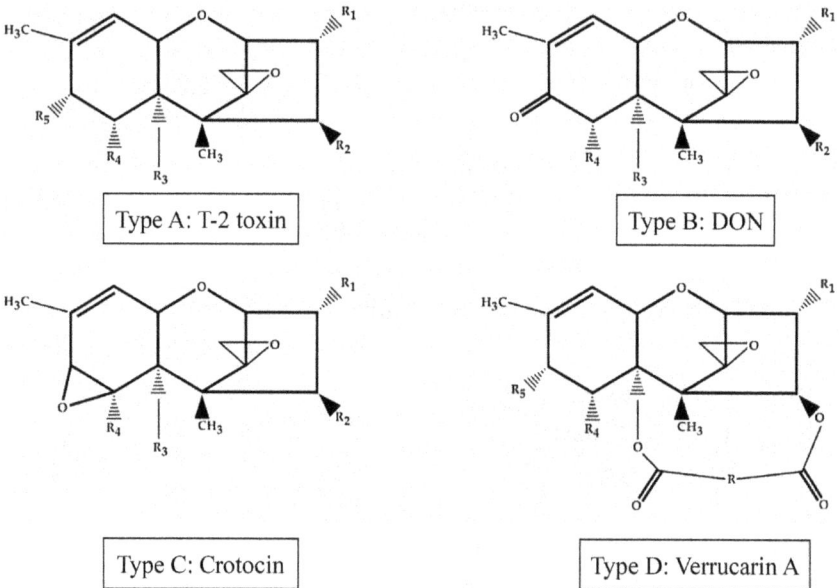

Type A: T-2 toxin

Type B: DON

Type C: Crotocin

Type D: Verrucarin A

Figure 13.2   Structures of trichothecenes (Types A, B, C, and D) (Wu et al., 2017).

The genus *Fusarium* is known to infect plants, including wheat, corn, barley, oats and forage grown in temperate climates, and causes widespread contamination. Mahato et al. (2022) provided an overview of TCN types and food sources, the associated biochemical pathways and genes involved for its production in food and feed, and the effect of processing conditions and environment on the production of TCNs. The authors also discussed detection techniques and management strategies for TCNs. The toxicological activity of tricothecenes is due to the presence of an epoxide at the C12, C13 sites. Trichothecene consumption leads to hemorrhage in the intestinal mucosa, vomiting, and diarrhea and is a known cause of dermatitis. The principal mycotoxins in this group are T-2 toxin (Type A) and DON (Type B), both of which induce toxicity in humans and animals when consumed orally (Nathanail et al., 2015).

DON (deoxynivalenol), 3a,7a, 15-trihydroxy-12, 13-epoxytochothec 15-trihydroxy-12, 13-epoxy to clothes-9-en-s-one, along with nivalenol and T-2 toxin, forms the largest group of mycotoxins naturally produced by *Fusarium* spp. (mainly *F. gramine arum* and *F. culmorum*). It is also referred to as vomitoxin because of its strong emetic effects after feeding, as it is transported to the brain, where dopaminergic receptors are found (Sobrova, et al., 2010). These mycotoxins are the most dangerous to animal and human health due to their high toxicity and high occurrence of fungi species that synthesize them. The mentioned DON prevalence worldwide was discovered through Canady et al. (2001). DON is a significant pre-harvest problem in international locations where grains are dried to prevent mold growth at <13% moisture content. In areas where the moisture content of stored grains is low, DON may be produced. DON elaboration and concurrent contamination largely depend on weather, humidity, and temperature. Hence, year to year and region to region, DON levels in wheat, barley, and corn can vary extensively.

13.2.3 Ochratoxin

Ochratoxin (OT) is another category of mycotoxin produced by *Aspergillus* and *Penicillium* species like *Aspergillus alliaceus, Aspergillus auricomus, Aspergillus carbonarius, Aspergillus glaucus, Aspergillus melleus, Aspergillus niger*, as well as *Penicillium verrucosum*. It is further classified into three types; A, B, and C, which have a similar structure and are equally harmful (Zahra et al., 2019) (Figure 13.3). Among the three ochratoxin varieties, ochratoxin A (OTA) has been found to hold the maximum toxicity for both humans and animals. OTA is genotoxic (a chemical or agent that can cause DNA or chromosomal damage) and is a product of toxigenic fungi. Schrenk et al. (2020) updated their position on ochratoxin A, stating that it is genotoxic both in vitro and in vivo, although the causes of genotoxicity are unknown. Tumor formation could be aided by both direct and indirect genotoxic and non-genotoxic mechanisms of action.

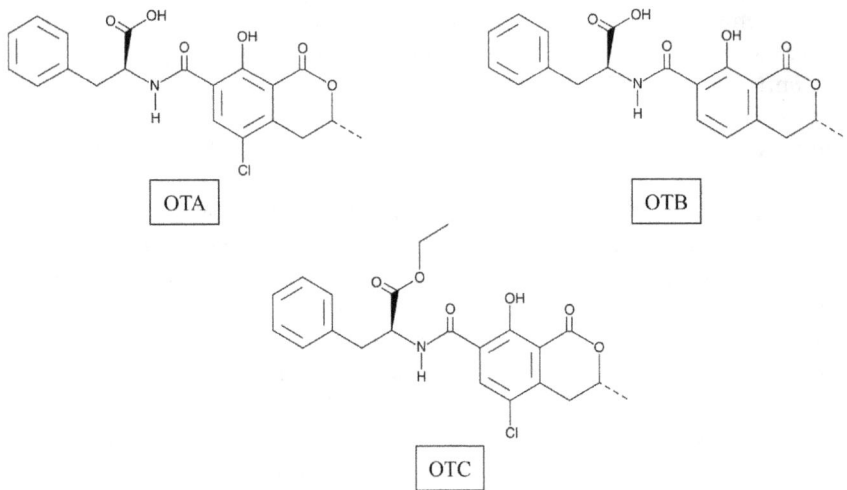

Figure 13.3   Structure of OTA, OTB, and OTC (Kőszegi and Poór, 2016).

Along with genotoxicity, it is also characterized by other detrimental effects like hepatotoxicity (destructive to liver cells), immunotoxicity (having an adverse effect on the functioning of both local and systemic immune systems), carcinogenicity, and nephropathy (Luo et al., 2018). Tunisian nephropathy and human Balkan endemic nephropathy (a chronic tubulointerstitial disease marked by a high rate of urothelial atypia that can lead to renal pelvis and urethral tumors) have both been linked to OTA. OTA has well-known effects on the fetus because of its ability to cross the placenta. It is teratogenic (causing birth defects) and leads to deformation of the brain and central nervous system (Omotayo et al., 2019).

OTAs have been detected in a variety of foods grown in both cool and warm temperature zones, like coffee, wine, and maize. OTA has the highest affinity to grow at the time of product storage, at temperatures ranging between 25 and 30°C, at a water activity level of 0.98 (Zahra et al., 2019), and mainly in foods relating to grapes. In countries like China, the highest OTA concentration was found to be in maize, from the cereals family. OTA not only contaminates cereals and grapes but also chocolate/cocoa plantations (Luo et al., 2018).

13.2.4 Patulin

Patulin (PAT) is a polyketide lactone (4-hydrox-y-4H-furo [3,2-c] pyran-2(6H)-one). Often found in rotting apples and their byproducts like apple juice, pickles and jellies, marmalade, and fruits and grains, patulin has been found to be obtained from molds like *Aspergillus, Penicillium*, and *Byssochlamys*. Fruits and vegetables are commonly used substrates for PAT occurrence. It can also be present in pears, grapes, oranges, and their derivatives, in addition to apples and apple-based goods (Saleh and Goktepe, 2019; Vidal et al., 2019). The main cause of fungal contamination is the usage of overripe fruit, which results in brown rot in peaches, bananas, apricots, and pineapples. PAT has been discovered not only in rotten fruits and vegetables but also in visually beautiful fruits. The genus *Penicillium*, namely *Penicillium expansum*, is a major PAT producer that poses a serious health and economic danger. A prevalent post-harvest illness known as blue mold rot is caused by the mycotoxin (Mahato et al., 2021). *P. expansum*, a wound parasite fungal mycotoxin, enters the damaged surface of the fruits, which could be caused by insect and bird infestation, harsh climatic conditions, or injuries sustained while harvesting and transportation (Li et al., 2020). In addition to fruits and fruit-based commodities, PAT can be isolated from cereals and cereal products (Babaali et al., 2017). Furthermore, two more *Penicillium antarcticum* strains have been recovered from patulin potato dextrose and malt extract agar, both of which have been found in seafood such as shellfish (Wright, 2015).

As reviewed by Mahato et al. (2021), patulin was isolated in 1943 and was utilized as an antibacterial agent against both gram-positive and gram-negative bacteria. The International Agency for Research on Cancer, however, classified it as Group 3 after discovering its harmful effects (IARC, 2018). Other recognized harmful effects on health include immunotoxicity, hepatotoxicity and gastrointestinal and neurological disorders, despite the lack of proof that PAT is carcinogenic (Pal et al., 2017).

13.2.5 Fumonisins

These are secondary metabolites that are obtained by *Fusarium verticillioides, Fusarium proliferatum*, and related species. They are a common contaminant of corn, corn products, and cereals like rice, wheat, barley rye, oat, and millet (Kamle et al., 2019). Fumonisin is obtained from *fusarium* species of infected maize in the pre-harvest period. Fumonisin production has also been documented during the post-harvest period; nevertheless, under unfavorable storage conditions, fumonisin production has been observed (Chulze, 2010). There are over 15 fumonisin homologues, with fumonisins A, B, C, and P being the most frequent (Braun and Wink, 2018). The most common forms of fumonisin B are FB1, FB2, and FB3, with FB1 being the most dangerous form that can coexist with other fumonisin forms like FB2 and FB3 (Damiani et al., 2019). The three categories of food pollutants (FB1, FB2, and FB3) are the most common. The diester FB1 is built up of propane-1,2,3-tricarboxylic acid (TCA) and

2-amino-12,16-dimethyl-3,5,10,14,15-pentahydroxyleicosane, with hydroxyl (OH-) groups at the C-14 and C-15 positions connecting with TCA's carboxyl groups (-COOH). FB2 and FB3 are, on the other hand, the C-5 and C-10 dehydroxy counterparts of FB1 (Figure 13.4) (Shephard, 1998). FB1 is formed mainly from fungi found on crops before harvest, unlike ochratoxins and aflatoxins, so the formation of the fungi is often difficult to halt or prevent completely. It mainly affects and destroys corn and corn products. FB1 also affects rice and other cereals, such as oats, rye, barley, and wheat, but to a lesser extent. There have been studies that conclude that a hot and humid environment results in the growth of FB1. Fumonisins also has showed high stability at temperatures from 28.97 to 32.14°C, humidity from 27.29 to 32.14%, and pH from 5.5 in maize plants. Thus, FB1 is more dangerous to plants and animals in tropical regions. The effect of FB1 on humans depends primarily on the dietary habits in the region where they reside. Wheat products, which have lower levels of FB1, are less harmful to humans compared to corn and maize. The defense mechanisms and reactions of plants can lead to change in the chemical structure of mycotoxins, resulting in bound mycotoxins, extractable or non-extractable, or mycotoxin metabolites.

Fumonisins have been related to a variety of health problems, including esophageal cancer, and have also been found to be toxic to the liver and nephron (Chu and Li, 1994). Furthermore, fumonisin B1, the most toxic compound of the family, has been linked to hepatocarcinoma, immune system activation and repression, neural-tube abnormalities, nephrotoxicity, and a variety of other diseases. Animal models have been shown to generate hepatocarcinoma and to have symbiotic association with aflatoxin B1 (AFB1) at two stages, cancer initiation and progression (Kamle et al., 2019). The International Agency for Research on Cancer has categorized FB1 as a group 2B probable human carcinogen.

13.2.6 Zearalenone

ZEN is another form of commonly found mycotoxin generated by *Fusarium* species, mainly *F. graminearum* and *F. culmorum*, known to cause reproductive and infertility issues in domestic animals. These *Fusarium* species infect both plants and animal diets and are typically found on plants growing in temperate areas (Rogowska et al., 2019). Various cereal crops like maize, oats, sorghum, barley, rice, and wheat, which make up a significant portion of animal feed, are particularly vulnerable to contamination with ZEN mycotoxin. Under ideal humidity and temperature circumstances, mycotoxin contamination of these crops poses a major threat to the health of living beings. Furthermore, foods like milk, meat, and its products, when produced from infected plants and animals, offer the greatest risk of

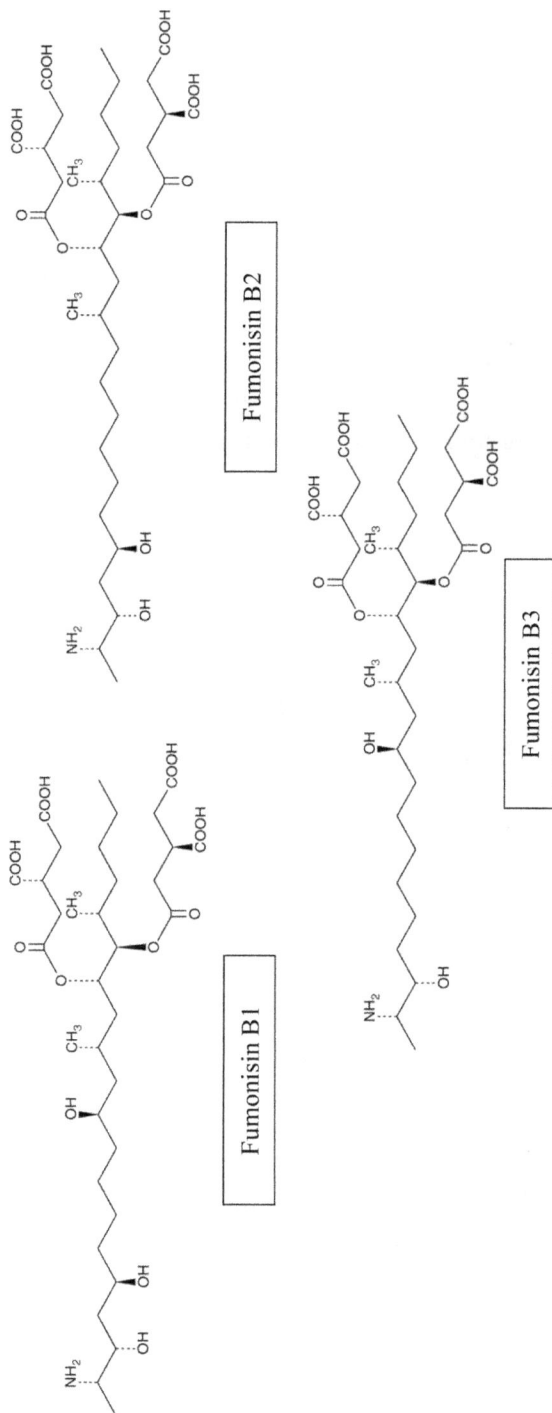

Figure 13.4 Structure of fumonisins B1, B2, and B3 (Ren et al., 2011).

contamination from mycotoxins. If left neglected, ZEN can occur at the vegetal stage as well as during prolonged storage. Other cereal- and milk-based products also have documented presence of ZEN (Rogowska et al., 2019). According to Chang, animal feed composed of protein from grains and vegetables can be a major source of fungal growth, thereby threatening its safety. ZEN contamination in feeds can occur under both unsuitable pre-harvest and storage conditions. The accessibility of substrates (such as magnesium, cobalt, and zinc), raw material moisture content above 15%, and a relative humidity of 70% or more are ideal circumstances for mycotoxin formation. Other parameters that influence fungal growth include pH, optimal temperature (20–30°C), and oxygen availability (Rogowska et al., 2019).

As reviewed by Mahato et al. (2021), ZENs are xenoestrogens with a molecular structure similar to natural estrogens, allowing them to attach to estrogenic receptor sites, resulting in increased estrogenicity. In animals such as cows, pigs, and rats, exposure to this mycotoxin is associated with lower levels of progesterone and serum testosterone, resulting in infertility and lower pregnancy rates. Even at low concentrations, ZEN has shown immunotoxic effects. By affecting multiple metabolic activities such as cell growth and death, ZEN toxicity causes several alterations in the target cells. It has been linked to reproductive syndromes and hyperactive estrogenic illnesses in in farm animals and humans, respectively, on several occasions. Many studies have shown that ZEN metabolites can cause estrogenic effects in farm animals, with pigs being the most vulnerable to its toxicity. For instance, Zhang et al. (2018) carried out *in vitro* and *in vivo* studies on animals and found dominant estrogenic activity in mice, swine, *Equus asinus*, and cattle and reviewed the detrimental effects of ZEA from early to final oogenesis stages of mammalian folliculogenesis. The International Agency for Research on Cancer has classified ZEN as a Group 3 carcinogen due to its unclassifiable carcinogenicity in humans and insufficient data.

13.3 PREVALENCE OF MYCOTOXINS IN STORED FOODS

13.3.1 Grains and Oilseeds

Because of their negative effects on humans, poultry, and animals, aflatoxins, ochratoxins, and fusarial toxins (fumonisins) are among the most important mycotoxins. Toxins are created in both the field and storage of wheat grains.

Because of their negative effects on humans, poultry, and animals, aflatoxins, ochratoxins, and fusarial toxins (fumonisins) are among the most important mycotoxins. Toxins are created in both the field and the storage of wheat grains. Numerous conventional and modern chromatography-based methods have been utilized for detection and quantification of mycotoxins in several foods. For instance, Kim et al. (2017) applied an instrumental LC-MS/MS method to validate and determine the prevalence of eight mycotoxins, including aflatoxins (B1, B2, G1, and G2), fumonisins (B1, B2), zearalenone, and ochratoxcin A in 134 grains, nuts, and oilseed samples. The limit of detection (LOD) and limit of quantitation (LOQ) for the eight mycotoxins ranged from 0.14 to 8.25 µg/kg and 1.08 to 7.21 µg/kg, respectively.

13.3.2 Pulses and Legumes

Legumes continue to constitute a good source of protein and carbohydrates globally for all strata of life. The demand and supply of legumes and pulses are increasing. Like other food items prone to mycotoxins, pulses and legumes are no different. Though the demand is ever increasing, little data are available on mycotoxins in pulses and legumes. The Food Industry Sanitation Auditor (FISA) found that, as with grains, aflatoxins, zearalenone, trichothecenes, or phomopsin A (PHOA) can also contaminate pulses, thereby causing detrimental effects. Some of the main fungal species that have been isolated from chickpea seeds and chickpea-based products are *Aspergillus, Fusarium, Penicillium, Alternaria*, and *Rhizopus*, which also change under storage conditions. However, fungal species like *Alternaria, Cladosporium, Botrytis*, and *Fusarium* are carried within seeds from the field and are an important source of mycotoxins.

13.3.3 Fruits and Vegetables

Natural soil is a reservoir of microscopic fungi and acts as a medium from which fungi enter plants and form soil humus by destroying compounds. Some species of microscopic fungi produce toxic compounds in fruits and vegetables. Out of the mycotoxins, some of them formed in fruits and vegetables include aflatoxins, patulin, ochratoxin, alternariol, and citrine. The majority of them are formed in the pre-harvest stage, and select ones form in the transportation/storage and decaying process. Studies have shown that patulinin found in apples and their products is produced 100% by the strains *Penicillium expansum*. Often, the fungus also attacks fruits damaged by the phytopathogenic *Botrytis cinerea*. The optimal temperature at which patulin is synthesized is 21–30°C. The optimal humidity is 90%. But the most common fungus producer,

P. expansum, survives and grows at much lower temperatures and usually affects fruit still on the tree (Vitkova et al., 2021).

13.3.4 Processed Foods (Flours, Juices, Spices, etc.)

Mycotoxin presence in processed foods is a growing concern for human health, as even low doses may lead to severe health issues and toxicity. Intense quality measures need to be adopted by the food processing industry, which include almost all types of foods, be it from cereals, juices, spices, and so on. Regulatory controls setting the limit of mycotoxins available have been created by most countries and the food processing industry. Good agricultural practices, effective management of plant diseases, and measures during storage are the some of the prerequisites of today's food processing industry; however, mycotoxin production and carriage cannot be eliminated completely from the wide variety of processed foods. In their work, Karlovsky et al. (2017) found that mycotoxin levels can be further eliminated by decontaminating and physically removing them by transforming them into less harmful compounds using chemical or enzymatic methods. Removal of mycotoxins using physical methods like manual and automated sorting can substantially reduce the amount of mycotoxins in processed foods. This has been considered the most coherent method for the elimination of mycotoxins during post-harvest processing and has the greatest scope for further development. In a study on mycotoxins in maize food processing, Schaarschmidt & Fauhl-Hassek (2021) discovered that the effect of processing is generally dependent on numerous parameters, which is especially evident in the case of cleaning. The level of decrease during dry milling is also dependent on particle size, at least for *Fusarium* toxins. When applying good practices to flour, adherence to EU maximum levels may not always be guaranteed. Commercial wet milling reduces mycotoxin levels in starch with a high degree of effectiveness. Table 13.4 summarizes a few recent studies on various mycotoxins, their prevalence in foods, and potential health applications.

13.4 FACTORS INFLUENCING MYCOTOXIN DEVELOPMENT IN FOODS DURING STORAGE

Numerous factors are responsible for the growth and production of mycotoxins in stored foods and are discussed in detail in the following. Ziska (2018) stated that tropical circumstances such as high temperatures and dampness,

TABLE 13.4   SUMMARIZING MYCOTOXIN SPECIES, WITH THEIR ASSOCIATED HEALTH IMPLICATIONS AND PERMISSIBLE LIMITS

Mycotoxin	Fungal Species	Detection	Found In	Health Implications	Permissible Limits	References
AFB1, AFB2, AFG1, AFG2	A. parasiticus, A. flavus, etc.	LC-MS/MS	Nuts, cereals, legumes, fruits, vegetables	• Liver cancer • Targets DNA • Hepatocellular carcinoma etc.	4–30 µg/kg human consumption; AFB1 and total AFs not exceeding 2 and 4 µg/kg, respectively.(EU)	Mahato et al. (2019).
Trichothecenes (nivalenol/ deoxynivalenol, crotocin, etc.)	F. crookwellense, F. culmorum, F. graminearum, F. poae), Trichoderma, Myrothecium, etc.	LC-MS/MS	Barley; wheat; maize; rice; oats; vegetables; and animal foods like eggs, milk, liver, and kidney	• Inhibition of DNA, RNA • Lipid peroxidation • Neurotransmitter changes • Alimentary toxic aleukia (ATA) in humans, etc.	1 ppm limit on deoxynivalenol (USFDA)	Chen et al. (2020)
OTA, OTB, OTC	Species of Aspergillus and Penicillium like A. ochraceus, A. niger, A. carbonarius, P. verrucosum	TLC, HPLC, MS, and immuno-chemical methods	Fruits and vegetables, seeds, nuts, cereals, legumes	• Immunotoxic • Teratogenic • Neurotoxic • Hepatotoxic • Nephrotoxic • Urothelial tumors	Max. 10 and 5 µg/kg for instant and roasted coffee, respectively; 2 µg/kg for grape juice and wine; 3 µg/kg for processed cereal food products; 5 µg/kg for unprocessed cereal g rains (EU)	Meulenberg (2012)

(Continued)

TABLE 13.4   SUMMARIZING MYCOTOXIN SPECIES, WITH THEIR ASSOCIATED HEALTH IMPLICATIONS AND PERMISSIBLE LIMITS (CONTINUED)

Mycotoxin	Fungal Species	Detection	Found In	Health Implications	Permissible Limits	References
Patulin	*Penicillium expansum, P. patulum, P. crustosum, A. clavatus*	HPLC-DAD GC-MS LC-MS Biosensors	Apples and apple products, fruits and vegetables, legumes, cereals, nuts and seeds	• Mutagenicity • Teratogenicity • Immunotoxicity • Neurotoxicity • Carcinogenesis • Acute chronic effects on cell cultures	10 to 50 µg/kg depending on the type of food (EU)	Mahato et al. (2021)
Fumonisins (B1, B2, B3, etc.)	*Fusarium* species like *F. verticillioides, F. oxysporum; Aspergillus awamori, A. niger,* etc.	LC-MS/MS	Corn and its food products, sorghum, beer, rice, beans, soybeans, etc.	• Inhibition of sphingolipids synthesis associated with atherosclerosis in monkeys • Esophageal and liver cancer in humans • Liver cancer in rodents, etc.	800–1000 µg/kg for maize and its products (EU), 2 to 4 mg/kg for corn and its products, and 3 mg/kg in the case of popcorn (USFDA) 2 µg/kg for fumonisins B1, B2, and B3, in combination or alone in a day (JECFA)	Kim et al. (2017)
Zearalenone (F2 toxin)	*Fusarium* species such as *F. crookwellense, F. cerealis, F. equiseti, F. culmorum,* etc.	LC-MS/MS	Rice, rye, oats, barley, wheat, maize, grain products, etc.	• Recognized for binding transcription factors to pregnane X receptors, which are involved in expression of enzymes in biosynthetic pathways • Can be used to treat uterine fibroids, pituitary adenomas, and other conditions	0.25 µg/kg per day (EFSA and JECFA)	Kim et al. (2017)

monsoons, excessive rain during harvest, and flash floods can encourage fungus growth and mycotoxin production.

1. **Temperature and humidity:** Temperature and humidity are some of the key factors influencing production of fungi-producing mycotoxins. It has been found that fungi make mycotoxins at optimum temperatures in the range of 24 to 28°C; however, growth has been observed up to temperature of 33°C as well. At temperatures below 8°C, the poisons are not created. After incubation at a low temperature of 12–14°C, *Fusarium* species produce more toxin (Perincherry et al., 2019). Field-produced fungi, storage-produced fungi, and advanced decay fungi are the three categories of mycotoxin-producing fungi. Species of the genera *Alternaria* and *Fusarium*, which require grain moisture of 22 to 25%, are among the field fungi. Storage fungi, which are mostly *Aspergillus* and *Penicillium* species, require grain moisture of 13 to 18%. Advanced decay fungi, which include *Cladosporium*, *Fusarium*, and *Trichoderma* species, thrive in environments with more than 18% moisture content (Fleurat-Lessard, 2017).

2. **Atmosphere:** The growth of mycotoxins is directly affected by the carbon dioxide and oxygen levels and has been found be inversely related; that is, raised carbon dioxide and diminished oxygen level in the controlled environment have an inhibitory effect on the growth of mycotoxins (Elkenany and Awad, 2020).

3. **Nature of Food:** The type of crop/food product also affects the growth of mycotoxins at pre- and post-harvest levels. Milk of the animals for consumption and its byproducts can be the base of toxin infection.

4. **Animal Species:** Monogastric farm animals who are fed on cereals in their food, like chicken and pigs, are more vulnerable to mycotoxins, particularly aflatoxins, in comparison to other ruminants who are less vulnerable because rumenal flora have the capacity to transform some of the mycotoxins into slightly carcinogenic metabolites or biologically inactive compounds (Elkenany and Awad, 2020).

5. **pH**: pH has a direct effect on the growth and production of mycotoxins, as acidic pH contributes to faster growth of mycotoxins.

6. **Light:** Ultraviolet and fluorescent light are known to have the capacity to detoxify mycotoxins. When mycotoxins produced by *Aspergillus parasiticus*, *Fusarium verticillioides*, *Scopulariopsis fusca*, and *Verticillium lecanii* were exposed to fluorescent light and short and long UV and kept at room temperature for 3 weeks under different relative humidity (50–80%), they were completely eliminated (Elkenany and Awad, 2020)

7. **Other factors:** In subtropical and tropical nations with insufficient infrastructure, such as processing facilities, transportation, storage, and skilled human resources, mycotoxin contamination is more common (Elkenany and Awad, 2020).

13.5 CONTROL MEASURES

13.5.1 Using Ozone

Ozone is the safest method for decontamination. The various reasons for using ozone are:

- Food industries have been using ozone for sanitation and surface decontamination.
- O_3 gas has a short half-life and decomposes fast to create oxygen, leaving little residue on food.
- Ozone is a broad-range decontaminator, killing microbes such as mycotoxin-producing fungi without leaving any residue.
- Its high oxidizing property makes it a safe antibacterial agent in the food industry. It works by breaking down fungal cells. It attacks the polyunsaturated fatty acids of the cell wall, oxidizing the sulfhydryl and amino acid groups of enzymes.
- It changes the molecular structures of mycotoxin molecules by reacting with their functional groups, resulting in the formation of end products with lower molecular weight, a smaller number of double bonds, and less toxicity.
- Ozone leads to complete inactivation of *Fusarium*, followed by *Aspergillus* and *Penicillium*. Ozone fumigation lowers spore germination levels and toxin production.
- Its penetrative capacity is limited, and decomposition is very quick.

Ozone decontamination is successful depending upon the various conditions:

- the time for which it is exposed to contaminant,
- the concentration of ozone applied,
- the temperature at which decontamination is conducted,
- the moisture content of the product, and
- the relative humidity of the environment in which decontamination occurs.

Ozone treatment is known to be a safe and green decontamination technique. Ozone has advantages in relation to other chemical treatments:

1. Ozone precursors are abundant,
2. Application of ozone is in both gaseous and aqueous form,

3. No residue left after exposure,
4. It can be produced/generated at the same site at which the decontamination procedure occurs, and
5. It releases no hazardous by-products.

Hazard analysis critical control points (HACCPs) is a systematic approach to food safety in production processes that protects against biological, chemical, and physical risks. It is the foundation of good quality management principles.

1. Good agricultural practice,
2. Good manufacturing practice,
3. Good hygienic practice, and
4. Good storage practice.

This system protects consumers against mycotoxin-contaminated food. The various contamination-preventive approaches are followed during harvest.

1. At pre-harvest stages:
   • prevention techniques at the farm level, and
   • Selecting fungal and pest-resistant seeds, selection of seeds with high tolerance, resistance and integrity, correct application of fungicides and pesticides, rodent control, and soil management approaches.
2. At post-harvest stages:
   • Procedures for safe harvesting, processing, drying, and storing, and
   • Dehydration of crops and storage in a low-moisture environment (Afsah-Hejri et al., 2020).

13.5.2 Other Control Measures

13.5.2.1 Physical Detoxification

The physical methods that are currently used for mycotoxin decontamination are thermal processes (application of a particular temperature for a particular period of time), irradiation (exposure to radiation), and adsorption techniques.

• Extrusion cooking involves a combination of high pressure and high temperature for a short time, which helps in reduction of AFs and DON levels in corn flour. But this can only be used for temperature-stable foods (whose chemical composition does not change with a change in temperature) and can denature fat and protein molecules in food.
• Adsorbents like active carbon and mycotoxin-selective clay lower food product quality. These are most efficient when used for liquids like oils or milk.

- Even though irradiation can be helpful in reduction of mycotoxin levels in food, it can lead to potential molecular reactions (Afsah-Hejri et al., 2020)

13.5.2.2 Biological Detoxification

These methods include fermentation (breaking down of sugar molecules into simpler compounds by the action of enzymes to produce chemical energy) and microbial metabolization (process by which the microbe obtains energy and nutrients to live and reproduce). Microorganisms that change their chemical structures can degrade mycotoxins into less harmful compounds using *Flavobacterium aurantiacum* (belonging to genus *Flavobacterium*, which are gram-negative, rod-shaped bacteria), *Nocardia corynebacterioides* (belonging to genus *Nocardia*, weakly staining gram-positive bacteria), *Mycobacterium fluoranthenivorans* (belonging to phylum Actinobacteria and genus *Mycobacterium*), *Lactobacillus rhamnosus* (belonging to phylum Bacillota, also found in the human intestine), *Saccharomyces cerevisiae* (also known as brewer's yeast), and *Enterococcus faecium* (a gram-positive bacteria belonging to phylum Bacillota). Saprophytic yeasts or atoxigenic strains of *Aspergillus* prevent growth of AF. But these can release their metabolites in food and also absorb some nutrients (Afsah-Hejri et al., 2020)

13.5.2.3 Chemical Detoxification

This involves the use of chemical compounds or ozone for decontamination of mycotoxins. Ammoniation, neutral electrolyzed oxidizing water, citric acid, sodium hydrosulfite, antioxidants, alkaline solutions, and salts are chemical treatments used for mycotoxin removal. Nanomaterials (materials with at least one external dimension that measures 100 nm or less) and metallic nanoparticles (sub-micron–scale entities made of pure metals like Zn, Fe, Pt, etc.) acting as antifungal agents inhibit mycotoxin production. Some chemical treatments cause changes in the molecular structure of the mycotoxin or even make the food product inedible by leaving residue on its surface (Afsah-Hejri et al., 2020).

Mycotoxins being the cause of food contamination has led to development of certain permissible limits, and developing sensitive and reliable methods for their detection has become a top priority. Contaminated samples have to undergo many extractions and purificatory and cleansing procedures before other detection, spotting, analysis, and quantification procedures. Sample preparation strategies include reducing analysis time, using tiny solvent quantities, and using extraction procedures on a large scale. Some of the processes used to remove pollutants include liquid–liquid extraction (LLE), solid–liquid extraction (SLE), accelerated solvent extraction (ASE), supercritical fluid extraction (SFE), and microwave-assisted extraction (MAE). Thin-layer chromatography (TLC), high-performance liquid chromatography (HPLC) in combination with various detectors (e.g., fluorescence, diode array, UV), liquid chromatography coupled with mass spectrometry (LC–MS), liquid chromatography–tandem mass spectrometry (LC–MS/MS), and gas chromatography–tandem mass spectrometry (GC–MS/MS) are the most widely used for mycotoxin analysis. When quick examination of mycotoxins is necessary, immunoassay-based technologies such as enzyme-linked immunosorbent assay (ELISA) and lateral-flow devices (LFDs) are useful. Biosensors are also useful for detecting mycotoxins, with proteomic and genomic approaches and molecular procedures (Agriopoulou et al., 2020).

1. In cereals, HSI can be used for risk management of *Fusarium* pathogens and DON. The various advantages of this technique are: it is inexpensive, original characteristics of the sample are retained, it takes less time to perform, and it analyses each pixel space to obtain data regarding the microbial contamination. HSI is quite efficient and can be used as a replacement for other techniques used during analysis and screening of crops like cereals.

2. Electronic nose analysis. As the name suggests, it works on the principle of detecting volatile compounds that are a by-product of contaminated or spoilt food via solid-state sensor. The sensors that are used in this analysis are highly responsive to the unique fingerprint for each food and help in creation of a specific taste and aroma. The characteristic odor is used as the data to analyze product metabolites. This technology is used in various fruit cultures.

3. Proteomic (large-scale study of protein) methods involve the following procedures:
   - Proteins are extracted from the food (mold peptide extraction).
   - Later, the food is analyzed by matrix-assisted laser desorption or ionization time-of-flight mass spectrometry (MALDI-TOF MS). These techniques are very quick to detect fungal isolates, and they are very precise in the identification. They are used as an alternative to chromatography methods.

4. Aggregation-induced emission (AIE). Certain organic luminophores exhibit this effect (fluorescent dyes). AIE takes advantage of the fact that some organic luminophores have a stronger fluorescence in aggregated form than in solution. Restriction in the intramolecular rotations detected in the combined state may result in the intensity of fluorescence produced by these dyes. It has high efficacy in recognizing OTA. AIE has been quite successful in detection and analysis of ochratoxins in wine and coffee

5. The high sensitivity, accuracy, and reliability of the LC/MS-MS technique makes it an efficient procedure for analysis of a collection of toxins produced by fungi. It is used in experiments used to analyze the different colonies of mycotoxins in the same matrix.

6. Nuclear magnetic resonance (NMR) technologies aid in the understanding of the *Fusarium* mycotoxin fusarin C's rearrangement process. It is based on the metabolomics principle (multiple endogenous metabolites are detected and statistically interpreted simultaneously in a biological system).

7. Grains (cereals and legumes) play an important role in achieving food and nutrition security. Grains are highly susceptible to fungal contamination and production of metabolic by-products known as mycotoxins. Good farming practices are responsible for developing strategies and procedures that need to be adopted for prevention of contamination in food—preharvest or post-harvest. The hazard analysis and critical control point complements good agricultural practices (GAPs). In the year 2001, the WHOs Five Keys to Safer Foods (WHO-FKSF) communicated and formulated reasons and preventive measures to eliminate common food handling errors seen globally.

13.6 CONTROL STRATEGIES TO PREVENT MYCOTOXIN CONTAMINATION DURING STORAGE

Molds that produce mycotoxins can develop on a variety of foods and crops, penetrating deeply into the food at all stages of production, including pre-harvest, post-harvest, and storage. It has been found that mold usually does not grow in properly dried and stored foods. Prevention of mycotoxins during the pre-harvest stage is considered the most effective strategy to prevent mycotoxin formation in agricultural produce. Effective drying and storage is considered one of the important methods of preventing mycotoxin formation in crops and food items. The following approaches can be used to prevent mycotoxin contamination in agricultural goods (Figure 13.5).

Inspection of whole grains (particularly corn, sorghum, wheat, and rice), dried figs, and nuts is recommended by the World Health Organization, as they are frequently contaminated with aflatoxins, as evidenced by mold development, discoloration, and shriveling. Grain damage should be prevented before, during, and after drying and storage since damaged grain is more prone to mold invasion and consequently mycotoxin contamination. Purchase grains and nuts as soon as they are available; ensure that foods are properly stored, which includes keeping them free of insects, dry, and not too warm. By not storing goods for long periods of time before using them, and by

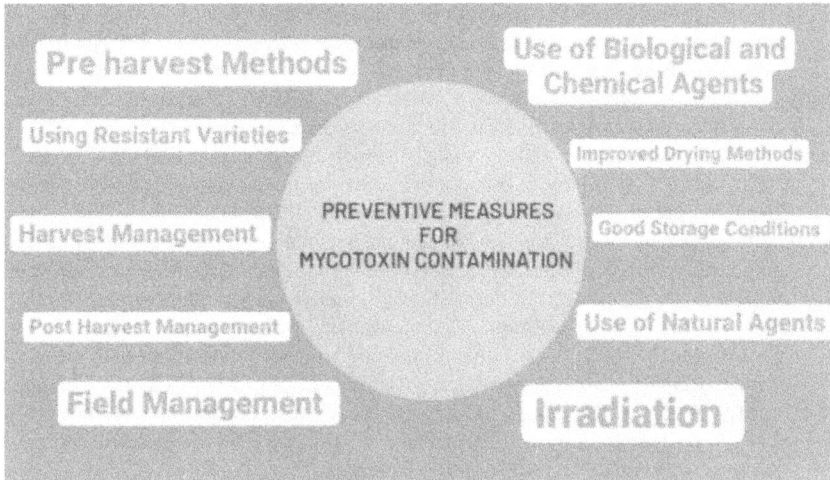

Figure 13.5   Methods for prevention of mycotoxin contamination.

maintaining a wide range of foods. Neme and Mohammed (2017) studied the role of post-harvest management in the prevention of mycotoxins. Mechanical damage, insect infestation, harvesting time, drying method, storage structure and conditions, handling, and processing are the key post-harvest factors that produce grain mycotoxin contamination. Post-harvest mitigation strategies have been found to be an important and cost-effective method of controlling the cause. Rapid and proper drying, insect control after harvest, proper transportation and packaging, ideal storage conditions, use of natural and chemical agents, and irradiation are the main post-harvest grain treatments used as mycotoxin mitigation measures. Mycotoxin levels have been shown to be reduced by sorting, cleaning, grinding, fermenting, baking, roasting, flaking, nixtamalization, and extrusion cooking. In general, a system approach to good manufacturing practice and implementation based on HACCPs is critical to mitigating risks.

REFERENCES

Adilah, Z. N., & Redzwan, S. M. (2017). Effect of dietary macronutrients on aflatoxicosis: A mini-review. *Journal of the Science of Food and Agriculture, 97*(8), 2277–2281.

Afsah-Hejri, L., Hajeb, P., & Ehsani, R. J. (2020). Application of ozone for degradation of mycotoxins in food: A review. *Comprehensive Reviews in Food Science and Food Safety, 19*(4), 1777–1808.

Afsah-Hejri, L., Jinap, S., Hajeb, P., Radu, S., & Shakibazadeh, S. H. (2013). A review on mycotoxins in food and feed: Malaysia case study. *Comprehensive Reviews in Food Science and Food Safety, 12*(6), 629–651.

Agriopoulou, S., Stamatelopoulou, E., & Varzakas, T. (2020). Advances in analysis and detection of major mycotoxins in foods. *Foods, 9*(4), 518.

Attia, S. M., & Harisa, G. I. (2016). Risks of environmental genotoxicants. In M. L. Larramendy & S. Soloneski (Eds.), *Environmental Health Risk—Hazardous Factors to Living Species.* IntechOpen. https://doi.org/10.5772/62454

Awuchi, C. G., Amagwula, I. O., Priya, P., Kumar, R., Yezdani, U., & Khan, M. G. (2020). Aflatoxins in foods and feeds: A review on health implications, detection, and control. *Bulletin of Environment, Pharmacology and Life Sciences, 9*, 149–155.

Awuchi, C. G., Ondari, E. N., Ogbonna, C. U., Upadhyay, A. K., Baran, K., Okpala, C. O. R., … & Guiné, R. P. (2021). Mycotoxins affecting animals, foods, humans, and plants: Types, occurrence, toxicities, action mechanisms, prevention, and detoxification strategies: A revisit. *Foods, 10*(6), 1279.

Babaali, E., Abbasi, A., & Sarlak, Z. (2017). Risks of patulin and its removal procedures: A review. *International Journal of Nutrition Sciences, 2*(1), 10–15.

Benkerroum, N. (2020). Chronic and acute toxicities of aflatoxins: Mechanisms of action. *International Journal of Environmental Research and Public Health, 17*(2), 423.

Braun, M. S., & Wink, M. (2018). Exposure, occurrence, and chemistry of fumonisins and their cryptic derivatives. *Comprehensive Reviews in Food Science and Food Safety, 17*(3), 769–791.

Canady, R. A., Coker, R. D., Egan, S. K., Krska, R., Olsen, M., Resnik, S., & Schlatter, J. (2001). T-2 and HT-2 toxins. *Safety Evaluation of Certain Mycotoxins in Food. WHO Food Additives Series, 47*, 557–597.

Chen, P., Xiang, B., Shi, H., Yu, P., Song, Y., & Li, S. (2020). Recent advances on type A trichothecenes in food and feed: Analysis, prevalence, toxicity, and decontamination techniques. *Food Control, 118*, 107371.

Chu, F. S., & Li, G. Y. (1994). Simultaneous occurrence of fumonisin B1 and other mycotoxins in moldy corn collected from the People's Republic of China in regions with high incidences of esophageal cancer. *Applied and Environmental Microbiology, 60*(3), 847–852.

Chulze, S. N. (2010). Strategies to reduce mycotoxin levels in maize during storage: A review. *Food Additives and Contaminants, 27*(5), 651–657.

Coppock, R. W., & Dziwenka, M. M. (2019). Biomarkers of petroleum products toxicity. In *Biomarkers in Toxicology* (pp. 561–568). Academic Press.

Damiani, T., Righetti, L., Suman, M., Galaverna, G., & Dall'Asta, C. (2019). Analytical issue related to fumonisins: A matter of sample comminution? *Food Control, 95*, 1–5.

Ditta, Y. A., Mahad, S., & Bacha, U. (2018). Aflatoxins: Their toxic effect on poultry and recent advances in their treatment. In *Mycotoxins-Impact and Management Strategies.*

EFSA Panel on Contaminants in the Food Chain (CONTAM), Schrenk, D., Bodin, L., Chipman, J. K., del Mazo, J., Grasl-Kraupp, B., … & Bignami, M. (2020). Risk assessment of ochratoxin A in food. *EFSA Journal, 18*(5), e06113.

Elkenany, R., & Awad, A. (2020). Types of mycotoxins and different approaches used for their detection in foodstuffs. *Mansoura Veterinary Medical Journal, 22*(1), 25–32.

Fleurat-Lessard, F. (2017). Integrated management of the risks of stored grain spoilage by seedborne fungi and contamination by storage mould mycotoxins—An update. *Journal of Stored Products Research, 71*, 22–40.

Hajian, R., & Ensafi, A. A. (2009). Determination of aflatoxins B1 and B2 by adsorptive cathodic stripping voltammetry in groundnut. *Food Chemistry, 115*(3), 1034–1037.

IARC. (2018). International agency for research on cancer. *Agents Classified by the IARC Monographs*, pp. 1–104. https://monographs.iarc.fr/agents-classified-by-the-iarc/.

James, B., Adda, C., Cardwell, K., Annang, D., Hell, K., Korie, S., . . . & Houenou, G. (2007). Public information campaign on aflatoxin contamination of maize grains in market stores in Benin, Ghana and Togo. *Food Additives and Contaminants, 24*(11), 1283–1291.

Kamle, M., Mahato, D. K., Devi, S., Lee, K. E., Kang, S. G., & Kumar, P. (2019). Fumonisins: Impact on agriculture, food, and human health and their management strategies. *Toxins, 11*(6), 328.

Karlovsky, P., Suman, M., Berthiller, F., et al. (2017). Impact of food processing and detoxification treatments on mycotoxins contamination. *Mycotoxin Research, 32*, 179–205.

Kim, J. K., Kim, Y. S., Lee, C. H., Seo, M. Y., Jang, M. K., Ku, E. J., . . . & Yoon, M. H. (2017). A study on the safety of mycotoxins in grains and commonly consumed foods. *Journal of Food Hygiene and Safety, 32*(6), 470–476.

Kőszegi, T., & Poór, M. (2016). Ochratoxin A: Molecular interactions, mechanisms of toxicity and prevention at the molecular level. *Toxins, 8*(4), 111.

Li, B., Chen, Y., Zhang, Z., Qin, G., Chen, T., & Tian, S. (2020). Molecular basis and regulation of pathogenicity and patulin biosynthesis in *Penicillium expansum*. *Comprehensive Reviews in Food Science and Food Safety, 19*(6), 3416–3438.

Li, P., Zhang, Q., Zhang, W., Zhang, J., Chen, X., Jiang, J., & Zhang, D. (2009). Development of a class-specific monoclonal antibody-based ELISA for aflatoxins in peanut. *Food Chemistry, 115*(1), 313–317.

Luo, Y., Liu, X., & Li, J. (2018). Updating techniques on controlling mycotoxins: A review. *Food Control, 89*, 123–132.

Mahato, D. K., Kamle, M., Sharma, B., Pandhi, S., Devi, S., Dhawan, K., . . . & Kumar, P. (2021). Patulin in food: A mycotoxin concern for human health and its management strategies. *Toxicon, 198*, 12–23.

Mahato, D. K., Lee, K. E., Kamle, M., Devi, S., Dewangan, K. N., Kumar, P., & Kang, S. G. (2019). Aflatoxins in food and feed: an overview on prevalence, detection and control strategies. *Frontiers in Microbiology*, 2266.

Mahato, D. K., Pandhi, S., Kamle, M., Gupta, A., Sharma, B., Panda, B. K., . . . & Kumar, P. (2022). Trichothecenes in food and feed: Occurrence, impact on human health and their detection and management strategies. *Toxicon*.

Maleki, F., Abdi, S., Davodian, E., Haghani, K., & Bakhtiyari, S. (2015). Exposure of infants to aflatoxin M1 from mother's breast milk in Ilam, Western Iran. *Osong Public Health and Research Perspectives, 6*(5), 283–287.

Marchese, S., Polo, A., Ariano, A., Velotto, S., Costantini, S., & Severino, L. (2018). Aflatoxin B1 and M1: Biological properties and their involvement in cancer development. *Toxins, 10*(6), 214. http://doi.org/10.3390/toxins10060214

Meulenberg, E. P. (2012). Immunochemical methods for ochratoxin A detection: A review. *Toxins, 4*(4), 244–266.

Min, L., Fink-Gremmels, J., Li, D., Tong, X., Tang, J., Nan, X., . . . & Wang, G. (2021). An overview of aflatoxin B1 biotransformation and aflatoxin M1 secretion in lactating dairy cows. *Animal Nutrition, 7*(1), 42–48.

Murphy, P. A., Hendrich, S., Landgren, C., & Bryant, C. M. (2006). Food mycotoxins: An update. *Journal of Food Science, 71*(5), R51–R65.

Nathanail, A. V., Syvähuoko, J., Malachová, A., Jestoi, M., Varga, E., Michlmayr, H., . . . & Peltonen, K. (2015). Simultaneous determination of major type A and B trichothecenes, zearalenone and certain modified metabolites in Finnish cereal grains with a novel liquid chromatography-tandem mass spectrometric method. *Analytical and Bioanalytical Chemistry, 407*(16), 4745–4755.

Neme, K., & Mohammed, A. (2017). Mycotoxin occurrence in grains and the role of post-harvest management as a mitigation strategies. A review. *Food Control, 78,* 412–425.

Omotayo, O. P., Omotayo, A. O., Mwanza, M., & Babalola, O. O. (2019). Prevalence of mycotoxins and their consequences on human health. *Toxicological Research, 35*(1), 1–7.

Oriordan, M. J., & Wilkinson, M. G. (2009). Comparison of analytical methods for aflatoxin determination in commercial chilli spice preparations and subsequent development of an improved method. *Food Control, 20*(8), 700–705.

Pal, S., Singh, N., & Ansari, K. M. (2017). Toxicological effects of patulin mycotoxin on the mammalian system: An overview. *Toxicology Research, 6*(6), 764–771.

Pandya, J. P., & Arade, P. C. (2016). Mycotoxin: A devil of human, animal and crop health. *Advances in Life Sciences, 5,* 3937–3941.

Perincherry, L., Lalak-Kańczugowska, J., & Stępień, Ł. (2019). *Fusarium*-produced mycotoxins in plant-pathogen interactions. *Toxins, 11*(11), 664.

Piermarini, S., Volpe, G., Micheli, L., Moscone, D., & Palleschi, G. (2009). An ELIME-array for detection of aflatoxin B1 in corn samples. *Food Control, 20*(4), 371–375.

Ráduly, Z., Szabó, L., Madar, A., Pócsi, I., & Csernoch, L. (2020). Toxicological and medical aspects of *Aspergillus*-derived mycotoxins entering the feed and food chain. *Frontiers in Microbiology,* 2908.

Ren, Y., Zhang, Y., Han, S., Han, Z., & Wu, Y. (2011). Simultaneous determination of fumonisins B1, B2 and B3 contaminants in maize by ultra high-performance liquid chromatography tandem mass spectrometry. *Analytica Chimica Acta, 692*(1–2), 138–145.

Rogowska, A., Pomastowski, P., Sagandykova, G. and Buszewski, B. (2019). Zearalenone and its metabolites: Effect on human health, metabolism and neutralisation methods. *Toxicon, 162:* 46–56.

Saleh, I., & Goktepe, I. (2019). The characteristics, occurrence, and toxicological effects of patulin. *Food and Chemical Toxicology, 129,* 301–311.

Schaarschmidt, S., & Fauhl-Hassek, C. (2021). The fate of mycotoxins during the primary food processing of maize. *Food Control, 121,* 107651.

Shephard, G. S. (1998). Chromatographic determination of the fumonisin mycotoxins. *Journal of Chromatography A, 815*(1), 31–39.

Sobrova, P., Adam, V., Vasatkova, A., Beklova, M., Zeman, L., & Kizek, R. (2010). Deoxynivalenol and its toxicity. *Interdisciplinary Toxicology, 3*(3), 94.

Tan, Y., Chu, X., Shen, G. L., & Yu, R. Q. (2009). A signal-amplified electrochemical immunosensor for aflatoxin B1 determination in rice. *Analytical Biochemistry, 387*(1), 82–86.

Tripathi, S., & Mishra, H. N. (2009). Studies on the efficacy of physical, chemical and biological aflatoxin B1 detoxification approaches in red chilli powder. *International Journal of Food Safety, Nutrition and Public Health, 2*(1), 69–77.

Vidal, A., Ouhibi, S., Ghali, R., Hedhili, A., De Saeger, S., & De Boevre, M. (2019). The mycotoxin patulin: An updated short review on occurrence, toxicity and analytical challenges. *Food and Chemical Toxicology, 129,* 249–256.

Vitkova, T. G., Enikova, R. K., & Stoynovska, M. R. (2021). Medical evaluation of the potential biological and chemical dangers in high-risk foods. *Journal of IMAB–Annual Proceeding Scientific Papers, 27*(3), 3924–3929.

Wogan, G. N. (1966). Chemical nature and biological effects of the aflatoxins. *Bacteriological Reviews*, *30*(2), 460–470.

Wright, S. A. (2015). Patulin in food. *Current Opinion in Food Science*, *5*, 105–109.

Wu, Q., Wang, X., Nepovimova, E., Wang, Y., Yang, H., Li, L., . . . & Kuca, K. (2017). Antioxidant agents against trichothecenes: New hints for oxidative stress treatment. *Oncotarget*, *8*(66), 110708.

Zahra, N., Saeed, M. K., Sheikh, A., Kalim, I., Ahmad, S. R., & Jamil, N. (2019). A review of mycotoxin types, occurrence, toxicity, detection methods and control. *Biological Sciences-PJSIR*, *62*(3), 206–218.

Zain, M. E. (2011). Impact of mycotoxins on humans and animals. *Journal of Saudi Chemical Society*, *15*(2), 129–144.

Zhang, G. L., Feng, Y. L., Song, J. L., & Zhou, X. S. (2018). Zearalenone: A mycotoxin with different toxic effect in domestic and laboratory animals' granulosa cells. *Frontiers in Genetics*, 667.

Ziska, L. H. (2018). *Agriculture, Climate Change and Food Security in the 21st Century: Our Daily Bread*. Cambridge Scholars Publishing.

FURTHER READINGS

Bayman, P., Baker, J. L., Doster, M. A., Michailides, T. J., & Mahoney, N. E. (2002). Ochratoxin production by the *Aspergillus ochraceus* group and *Aspergillus alliaceus*. *Applied and Environmental Microbiology*, *68*(5), 2326–2329.

Fernández-Cruz, M. L., Mansilla, M. L., & Tadeo, J. L. (2010). Mycotoxins in fruits and their processed products: Analysis, occurrence and health implications. *Journal of Advanced Research*, *1*(2), 113–122.

Gizachew, D., Szonyi, B., Tegegne, A., Hanson, J., & Grace, D. (2016). Aflatoxin contamination of milk and dairy feeds in the Greater Addis Ababa milk shed, Ethiopia. *Food Control*, *59*, 773–779. https://doi.org/10.1016/j.foodcont.2015.06.060

Ismaiel, A. A., & Papenbrock, J. (2015). Mycotoxins: producing fungi and mechanisms of phytotoxicity. *Agriculture*, *5*(3), 492–537.

Levasseur-Garcia, C., Bailly, S., Kleiber, D., & Bailly, J. D. (2015). Assessing risk of fumonisin contamination in maize using near-infrared spectroscopy. *Journal of Chemistry*. https://doi.org/10.1155/2015/485864

Liew, W. P. P., & Mohd-Redzwan, S. (2018). Mycotoxin: Its impact on gut health and microbiota. *Frontiers in Cellular and Infection Microbiology*, *8*, 60.

Pitt, J. I., & Miller, J. D. (2017). A concise history of mycotoxin research. *Journal of Agricultural and Food Chemistry*, *65*(33), 7021–7033.

Ramirez, M. L., Cendoya, E., Nichea, M. J., Zachetti, V. G. L., & Chulze, S. N. (2018). Impact of toxigenic fungi and mycotoxins in chickpea: A review. *Current Opinion in Food Science*, *23*, 32–37.

Sharma, V., & Patial, V. (2021). Food mycotoxins: Dietary interventions implicated in the prevention of mycotoxicosis. *ACS Food Science & Technology*, *1*(10), 1717–1739.

Chapter 14

Climate Change's Impact on Mycotoxin Production in Food and Feed

Monika Mathur, Raveena Kargwal, Raman Selvakumar,
Dipendra Kumar Mahato, Madhu Kamle, Pradeep Kumar

CONTENTS

14.1	Introduction	363
14.2	Impact of Climate Change	366
14.3	Impact of Climate Change on Mycotoxin Production	372
14.4	Impact of Climate Change on Agriculture and Associated Food Safety Issues	372
	14.4.1 Soil Quality and Degradation	374
14.5	Effects of Climate Change on Crop Pests and Their Populations	375
14.6	Fungal Growth and Mycotoxin Production	376
	14.6.1 Temperature, Water Activity, and Relative Humidity	376
	14.6.2 pH	379
	14.6.3 Substrate	379
14.7	Effects of Climate Change on Mycotoxins	380
14.8	Safety and Microbial Spoilage at Various Stages of the Food Chain	382
	14.8.1 Mycotoxin Control and Prevention Strategies	382
	14.8.2 Proper Field Practices	383
	14.8.3 Field Preparation and Management prior to Planting	383
	14.8.4 Field and Crop Management after Planting	384
14.9	Impact of Climate Change on Animal Production and Associated Food Safety Issues	384
	14.9.1 Livestock Production and Husbandry Practices	385
14.10	Early Detection of Fungal Species	385
	14.10.1 Chemical Control	385
	14.10.2 Biological Control	386

DOI: 10.1201/9781003242208-14

14.10.3   Proper Practices during Harvest   386
14.10.4   Proper Practices during Drying   387
14.10.5   Proper Storage Practices   387
14.11 Methods for Detoxification and Decontamination of Mycotoxins   388
14.11.1   Physical Decontamination   390
14.11.2   Chemical Decontamination   392
14.11.3   Biological Decontamination   392
14.11.4   Effect of Processing on Mycotoxins   393
14.12 Effects of Increasing Antimicrobial Resistance Are Expected
       to Change the Transmission Pathways of Pests   394
14.12.1   Vectors   395
14.12.2   Correlation between Climatic Factors and Water and
          Food-Borne Diseases   395
14.12.3   Salmonellosis   396
14.13 Climatic-Driven Emerging Risks   396
14.14 Impact of Climate Change on Food Spoilage   397
14.15 Impact of Global Warming on the Spoilage Risk of Non-
       Refrigerated Processed Food Products   397
14.16 Impact of Increased Precipitation Events and Humidity on
       Bulk Dried Food and Cereal Grains   398
14.16.1   Toxin Detection Methods: Classical Mycotoxin
          Quantifying/Detecting Technology   399
14.16.2   Analysis of Thin-Layer Compounds   400
14.16.3   Gas Chromatography   400
14.16.4   High-Performance Liquid Chromatography   400
14.16.5   Solid Phase Extraction or MycoSep Columns   401
14.16.6   Liquid Chromatography/Mass Spectrometry   401
14.16.7   Enzyme-Linked Immunosorbent Assay   402
14.16.8   Emerging Technologies for Mycotoxin Analyses   402
          14.16.8.1 Lateral Flow Devices   402
          14.16.8.2 Fluorescence Polarization Immunoassay   403
14.16.9   Infrared Spectroscopy   403
14.16.10 Capillary Electrophoresis   404
14.16.11 Fiber-Optic Immunosensors   404
14.16.12 Biosensors   405
14.16.13 Competitive Surface Plasmon Resonance-Based
          Immunoassays   405
14.16.14 Competitive Electrochemical ELISA Based on
          Disposable Screen-Printed Carbon Electrodes   405
14.16.15 Molecularly Imprinted Polymers   406
14.17 Management Strategies   406
14.17.1   Rapid Quantification Methods   408
14.18 Conclusion   410

14.1 INTRODUCTION

The National Aeronautics and Space Administration (NASA) defines climate change as "a broad spectrum of global phenomena mostly produced by the combustion of fossil fuels, which contributes heat-trapping gases to the earth's atmosphere." Along with the increased temperature trends associated with global warming, these phenomena include changes in sea level, Greenland's ice mass loss, mountain glaciers worldwide, the Arctic and Antarctica, fluctuations in flower/plant blossoming, and severe weather situations. In part, this is due to population growth and the emission of greenhouse gases (Sulaiman & Abdul-Rahim, 2018; Rehman et al., 2021; Mikhaylov et al., 2020).

Since human activity is mostly responsible for climate change, it is a global phenomenon that is characterized by alterations in the planet's normal climate (temperature, precipitation, and wind). Due to the disruption of the planet's climate, the planet's environment, as well as humanity's destiny and the global financial system's stability, are jeopardized. Carbon dioxide (CO_2) levels in the atmosphere have increased by 40% since pre-industrial times, to 391 parts per million (ppm) in 2011. This rise is the result of increased emissions from the combustion of fossil fuels and alterations in land use. Anthropogenic and natural forces combined to produce a net increase in energy flux in 2011, resulting in the largest annual CO_2 input to the climate system (Cavicchioli et al., 2019; Ahmed, 2020; Fu et al., 2020, resulting in a 1- to 2-degree increase in the earth's temperature over the last century (Chakraborty et al., 1998; Shindell & Smith, 2019; Basu et al., 2020; Nadeau et al., 2022; Jones et al., 2021; Böhringer et al., 2021). Additionally, this does not end here; as a result of human irresponsibility, it continues to rise. Heat waves are expected to grow more frequent and intense as the global temperature rises. This is because the frequency and duration of these occurrences will rise (IPCC, 2013), leading to increased desertification (Medina et al., 2017a).

Global warming–induced changes in the global water cycle will be very variable between locations and seasons. The contrast between rainy and dry zones will diminish in consistency. In the majority of places, increasing atmospheric moisture results in a drop in mean precipitation. The El Niño–Southern Oscillation's variability will likewise increase (IPCC, 2013). The IPCC's Working Group II on Climate Change identified a high probability that climate change will have an effect on numerous areas of food security. The assessment's breadth demonstrated that the hazards presented by climate change are systemic and global in scope. Global food security is in jeopardy because of climate change, and the world's population is expected to increaseto 9.2 billion people by 2050 (Godfray et al., 2010; Lal, 2013).

A survey conducted by the National Health Officials Association revealed that many city and county health officials believe that climate change is not happening and that they do not have the expertise to address the issue

(Krueger et al., 2015). The IPCC's Fourth Report on Climate Change was widely recognized worldwide due to its numerous scientific findings. According to the public, global warming cannot be avoided despite the efforts made by governments to reduce greenhouse gas emissions (Sheehan et al., 2017).

In 1984, the WMO and UNEP declared carbon dioxide the main cause of global warming. This was followed by the formation of the IPCC. The Club of Rome Report on global warming was published in 1972. In 1985, the World Meteorological Organization and the United Nations Environment Program concluded that carbon dioxide was the primary cause of global warming. The IPCC report stated that global warming is already causing significant impacts on the environment. It is predicted that by 2100, the average global temperature would rise by up to 6.4°C if human activity continues at its current rate. In addition, the sea level will rise by 59 cm. Natural disasters are becoming more frequent and more intense as a result of human-caused global warming. Climate change is also affecting various other natural phenomena globally. Some of these include the rise of sea level, changes in glaciers, and changes in animal habitats. Climate change mitigation is a broad term that refers to efforts to reduce greenhouse gas emissions and minimize their effects on the environment. For agriculture, the IPCC has shifted its focus to adaptation and mitigation. As a result of the vulnerability of agriculture to climate change, we may expect the agriculture sector to be affected by climate change throughout the next few decades. That is why the entire industry needs to prepare for its inevitable impacts. Climate change is threatening the stability of East Asian agriculture. It is therefore very urgent that the region's agricultural policies be adapted to address this issue. This report will help local government officials and farming households develop long-term agricultural development programs and plans.

Global warming is projected to accelerate the creation of new illnesses and diseases by altering the environmental conditions of the organisms that transmit these diseases. This could result in the emergence of novel strains.

Agriculture is one of the areas which is mainly affected by extremes of temperature conditions such as those caused by climate change and natural disasters. Due to this, climate change is expected to have a significant impact on field production, crop and animal quality (Rosenzweig et al., 2001). As a result, there are growing concerns about food security and the ability to feed the world's expanding population. Plants and pathogens that produce pests and disease are both impacted by the quality and quantity of agricultural goods. It also has a significant effect on host-microorganism interactions. Climate change and the rise of insect populations have the potential to diminish farmland productivity (Porter et al., 2014; Rosenzweig, 2011; Donatelli et al., 2017). Pathogens have significant and widespread detrimental effects on food quality and safety (Oerke, 2006; Savary et al., 2011).

Food safety and security are becoming increasingly affected by the growth of plant diseases, which are closely linked to the spread of toxic fungal varieties.

Fungi-produced secondary metabolites produce a wide range of harmful chemicals that harm the quality of cereal crops and cause financial losses for farmers. When mycotoxins negatively impact human and animal health, as well as the commercial trade of staple foods like grains, it results in enormous economic losses for the world's food supply chain (Mathur et al., 2020).

Climate change is jeopardizing the world food supply by impairing the growth of several critical crops, including wheat, rice, and maize. These crops account for approximately 30% of the food calories required by the developing countries (Hellin et al., 2012). Contamination with mycotoxins has developed into a difficult concern for plant pathologists (Esker et al., 2012; Savary et al., 2012). It has an effect on the management of *Fusarium* head blight in particular. In terms of disease burden produced by *Fusarium* head blight, climate change–related factors have been recognized as the primary contributors to the illness's severity (Bottalico & Perrone, 2002; Osborne & Stein, 2007). Understanding the precise genetics and epidemiology of a disease outbreak can aid in its management. Several studies have been conducted on this topic (Chakraborty et al., 2006; Miedaner et al., 2008). Crop losses caused by mycotoxigenic fungi are seen as a significant hindrance to achieving food security (McMullen et al., 1997; Mcmullen et al., 2012). *Fusarium* head blight is a resurgent disease that has wreaked havoc on wheat and barley crops worldwide (Nganje et al., 2004; Cowger & Sutton, 2005). Mycotoxins harm 25% of the biosphere's food yields (Eskola et al., 2020), and 30%–100% of food and animal feed samples are polluted with mycotoxins (Sanzani et al., 2016; Rodrigues & Naehrer, 2012). *Fusarium, Aspergillus,* and *Penicillium* are the most prominent mycotoxigenic fungi, producing ochratoxin (*Aspergillus* and *Penicillium*), aflatoxin (*Aspergillus*), trichothecenes, and fumonisins (*Fusarium*) (Bryden, 2007; Mahato et al., 2022; Mahato et al., 2019; Kumar et al., 2017; Kumar et al., 2020; Chhaya et al., 2021). Contamination of food with mycotoxin poses a danger to food safety because it can result in chronic or acute disorders (Bryden, 2007). Mycotoxin-tainted food may also increase the risk of food insecurity among those who eat it. Additionally, mycotoxins' economic influence on livestock production is frequently related to decreased animal output and the eradication of contaminated feed (Magnoli et al., 2019; Binder et al., 2007).

Nuts, for example, are just one of many agricultural products (Kluczkovski, 2019), many products like fruits, and vegetables that are both fresh and dried (Sanzani et al., 2016; Gonçalves et al., 2019), cereals such as rice, wheat, and maize (Varzakas, 2016), liquids such as grape juice, wine (Welke, 2019), and beer (Pascari et al., 2018), dairy products and milk (Viegas et al., 2020), all these products are the most affected products from mycotoxin contamination. Coffee and cocoa (Bessaire et al., 2019b; Huertas-Perez et al., 2017), spices and herbs (Gambacorta et al., 2019), and feed (Kebede et al., 2020), storage and feed chain (Misiou & Koutsoumanis, 2021) can be infected with mycotoxins at different stages of the food processing, production, harvesting. Consumption of

TABLE 14.1   COMMODITIES FOUND TO BE CONTAMINATED WITH MYCOTOXINS (PATERSON & LIMA, 2010)

Mycotoxin	Commodity
Aflatoxins	Various foods, groundnuts, milk, eggs, figs, cheese, maize, wheat, cottonseed, copra, nuts
Citrinin	Cereal grains like wheat, barley, corn, rice
Cyclopiazonic acid	Kodo millet, corn, groundnut, milk products like cheese
Ochratoxin A	Fungus-contaminated groundnut, milk cheese, swine tissue, coffee, raisins, grapes, dried fruits, wine, cocoa, cereal grain like wheat, barley, oats, maize, dry beans
Patulin	Wheat straw residue, spoiled apples, apple juice, moldy feed
Penicillic acid	Cereal grains like wheat, barley, corn, oat, dried beans, moldy tobacco, and stored corn
Penitrem	Walnuts, hamburger buns, beer, moldy cream cheese
Sterigmatocystin	Cottonseed, hard cheese, peas, green coffee, moldy wheat, grains
Trichothecenes	Barley, oats, corn, wheat, mixed feeds, commercial cattle feed
Zearalenone	Water systems, corn, rotten hay, commercial feed

contaminated food by young children, the elderly, and patients with compromised immunity can have a detrimental effect on their health (Bryden, 2012). Additionally, a toxic response is typically triggered by a brief exposure to a substance, whereas chronic toxicity refers to a long-term exposure that can result in cancer and further detrimental effects (Bennett & Klich, 2003). Monitoring of the environment or biological systems can be utilized to ascertain human exposure to mycotoxins. Human exposure to mycotoxin compounds is assessed by the presence or absence of their metabolites and residual levels (Hsieh, 1988); toxins can have a negative result on both animal and human health (Bryden, 2012). Mycotoxins can cause a wide range of health issues in people, including cancer, hormone imbalances, and a diminished ability to fight against infections (Marin et al., 2013). Table 14.1 lists commodities contaminated with mycotoxins.

14.2 IMPACT OF CLIMATE CHANGE

Due to the presence of layers of atmosphere surrounding the planet, growth in greenhouse gases eventually transforms the world into a gas chamber, where greenhouse gases are unable to move or escape into space. Due to rising temperatures, the Earth's climate is becoming more and more unstable as time goes on (Figure 14.1(a). This has a three-fold effect on the earth system (Figure 14.1(b).

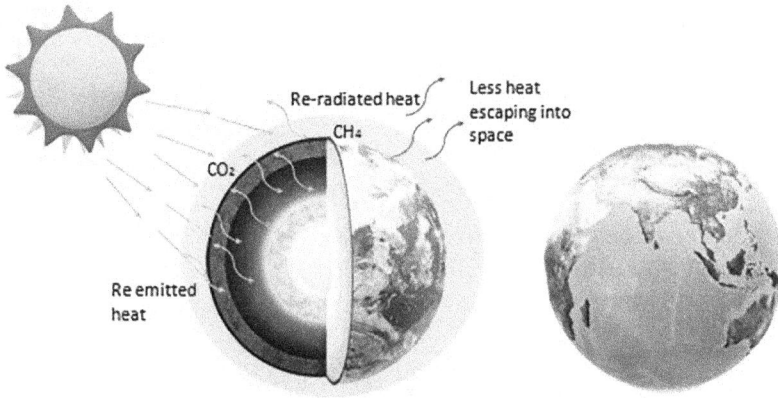

Figure 14.1   (a) Effect of greenhouse gases in solar system.

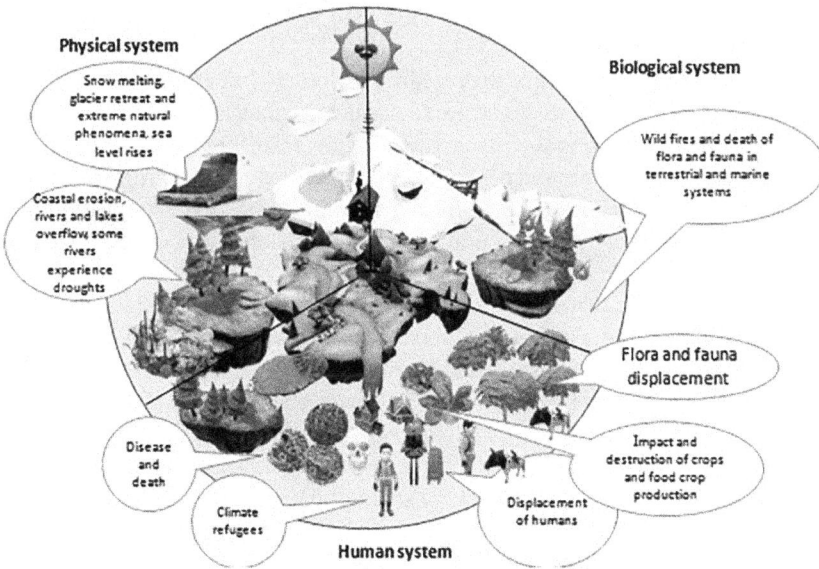

Figure 14.1   (b) Impact of climate change on physical, biological and human systems.

Physical System

1. Snow melting
2. Permafrost warming and thaw
3. Glacier retreat

4. Extreme natural phenomenon
5. Coastal erosion
6. Sea-level rise
7. Rivers and lakes overflow

Biological System

1. Wildfires
2. Death of flora and fauna in the terrestrial and marine ecosystem
3. Flora displacement
4. Fauna displacement

Human System

1. Climate refugees
2. Disease and deaths
3. Destruction and loss of economic means of substance
4. Impact and distribution of crops and food production

There has already been significant damage done to Earth's natural systems as a result of climate change. Several of these include glacier retreat and sea level rise. Additionally, the seas absorb carbon dioxide, resulting in ocean acidification. Ecosystems and humans have already been impacted by climate change. In many places, it exacerbates food insecurity and water scarcity. Additionally, it exacerbates land degradation and drought. Additionally, climate change is projected to boost migration.

Climate change's future consequences are contingent upon governments' efforts to avoid and mitigate greenhouse gas emissions. The focus on climate change in the shortrange obscures some of the longstanding consequences for society. For example, actions implemented over the next many decades will have a profound effect on ecosystems and human cultures. Strict mitigation policies can contribute to limiting global warming to less than 2°C by 2100. Without these regulations, rising energy demand and fossil fuel consumption might result in global warming of more than 4°C.

Among the mycotoxins with a diverse series of poisonous biotic activities (Agriopoulou et al., 2020), aflatoxins, the most-studied mycotoxins, have been shown to be hazardous, mutagenic, teratogenic, and immunosuppressant (Agriopoulou et al., 2016; Mahato et al., 2019; Kumar et al., 2017), while aflatoxin AFB1 has been categorized as a type 1 carcinogen (carcinogenic to humans) by the International Agency for Research on Cancer (Loprieno, 1975). Nearly 400 distinct mycotoxins have been found (Ji et al., 2016). The current food supply chain is teeming with poisons that can result in food illness (Mahato et al., 2019; Kumar et al., 2017). Aflatoxins (AFs), ochratoxin A (OTA), patulin, fumonisins, citrinin, ergot alkaloids, trichothecenes such as deoxyni-valenol (DON) and T-2 toxin (T-2), and zearalenone (ZEN) are only a few of these (Mahato et al., 2021; Mahato et al., 2022; Anfossi et al., 2016; Kumar et al., 2020; Mahato et al., 2019; Kumar et al., 2017; Mahato et al., 2021; Kamle et al., 2019; Kamle et al., 2022). Environment change has the potential to have a important effect on the food chain (Misiou & Koutsoumanis, 2021; Kumar et al., 2017). It has the potential to alter food availability and other essential elements such as food access, use, and stability (Wheeler & Von, 2013). Mycotoxins are formed as a result of ecological conditions such as humidity, temperature, and nutrition (Geisen et al., 2017). For example, the presence of aflatoxin M1 in milk prod-ucts can be attributed to animals eating feed contaminated with aflatoxin B1 (Danesh et al., 2018).

Temperature, water activity (a_w), relative humidity (RH), pH, mycologi-cal strain, and substrate all have a significant effect on mycotoxin formation. Water availability and temperature are two factors that influence mycotoxi-genic mold growth and development (Magan & Aldred, 2007). These factor interactions result in the infection of parameters such as germination, growth, sporulation, and mycotoxins (Sanchis & Magan, 2004). These variables all have an effect on the many parameters of these organisms. A warmer environment can result in an increase in the frequency of heatwaves and an increase in the growth of certain plant diseases. There is a greater chance of climate change, such as increased precipitation or drought, as methane, carbon dioxide, nitrous oxide, and chlorofluorocarbon concentrations rise in the atmosphere.

This phenomenon may have a higher impact on the formation and manu-facturing of mycotoxins, as the majority of plant diseases thrive in optimal growth and production circumstances. There are numerous food safety con-cerns linked to climate change, including mycotoxins. Fungi and plants pro-duce them, depending on the environment in which they thrive. Fumonisin and DON crops will no longer be able to sustain themselves as a result of the consequences of climate change. The ideal temperatures for mycotoxin genera-tion and growth in vitro of several significant plant pathogenic fungi are listed in Table 14.2. When temperatures and humidity levels are ideal, fungal infec-tions can invade at various stages, this can usually take place in different parts of the plant depending on the stage when the invasion begins (Perdoncini et al., 2019; Joubrane et al., 2020).

Mycotoxins are fungi-produced chemicals that are present in a diversity of food items. They can take on a variety of forms, including seeds, fruits, and spices. Mycotoxins can also develop as a result of improper food preservation and handling. Exposure to warm, damp, and humid circumstances might result in mold growth. Mycotoxins are a class of compounds that can endure a varied range of temperatures and can be safely preserved in food. Over a hundred distinct mycotoxins are known to be toxic to people and livestock.

Aflatoxin, ochratoxin A, and patulin are only a few of these. Mold infection can have a detrimental effect on crop development and quality. It can potentially enter the food chain via milk-contaminated animal feed. Certain mycotoxins are extremely hazardous and can result in serious illness within seconds of consuming contaminated food. Others have the potential to have lasting health consequences, such as cancer induction. Approximately a dozen mycotoxins with some carcinogenic effects have been identified thus far, which can invade different crops and contaminate them. Their prevalence in food has raised widespread concern.

Aflatoxins are very dangerous mycotoxins produced by fungi found in soil, decaying plants, and cereals. They have been shown to cause severe allergic responses in humans. Additionally, *Aspergillus* spp. can grow on certain seeds and oils. Aflatoxins have been detected in the milk of animals fed tainted grain. They can induce acute poisoning and are lethal in high doses. Aflatoxins have been linked to cancer in both animals and humans. Additionally, they have the potential to damage DNA (Mahato et al., 2019; Kumar et al., 2017). Khodae et al. (2021) conducted a survey and concluded that the aflatoxin B1 danger in wheat, maize, and rice is significant, since it was consistently much higher than the European Union limit in the majority of investigations. Worldwide environmental change is altering the fungus population and mycotoxin patterns in various places and crops. As a result, it is critical to establish practical control and management measures to ensure crop safety.

Mycotoxins are fungal pathogens that can contaminate food (Misiou & Koutsoumanis, 2021). Fungi from the *Aspergillus* and *Penicillium* genera are responsible for the production of these substances. Ochratoxin A is a mycotoxin that is present in a variety of food products (Kumar et al., 2020). Cereals, coffee beans, and fruit juice are just a few of these. Ochratoxin A is a toxin that can be created as a result of agricultural storage (Kumar et al., 2020). One of the most common side effects in animals has been renal impairment.

Even though it has been linked to cancer and renal damage, the link between toxins and humans is not yet clear. Patulin is a mycotoxin produced by various molds. It can be found in decaying apples and other food products. Apples and juice are major sources of patulin. This bacterium can cause acute illness in animals. There have been many cases of vomiting, nausea, and stomach issues in humans. Portions of patulin are believed to be carcinogenic.

Fusarium fungus is commonly found in soil and produces various toxins, such as deoxynivalenol and nivalenol. It can also be used to make poisons. Certain kinds of cereal are known to contain *Fusarium* poisons such as ZEN and DON. They can cause severe allergic reactions and can also suppress the immune system.

Fusarium head blight (FHB) is a highly destructive disease that can severely harm tiny cereal grains. It has been recorded in more than 40 countries.

Climate change may result in the spread of FHB disease throughout Canada's prairie provinces. This lethal sickness is already wreaking havoc in Canada. Deoxynivalenol is a significant mycotoxin generated by the *Candida albiculae* fungus. It is a substance that is frequently found in wheat grain. FHB is a disease whose pathogenesis is heavily influenced by temperature. A little variation in temperature is believed to have an effect on the severity and occurrence of the condition. In North America, environmental change has been connected to population movement. Each province in Canada has seen a growth in the number of producers of 3-ADON. 3-ADON is most prevalent in western Canada. According to studies, the variety of FHB pathogens has changed toward producers of 3-ADON. The number of producers of 3-ADON has expanded dramatically in western Canada. This indicates that they have a selective disadvantage when compared to the indigenous 15-ADON species. High temperatures and dry conditions favor *Aspergillus flavus* growth and development, hence limiting the fungus's spread. This reduces crops' capacity to develop and produce aflatoxin. Fungi that produce aflatoxin are indigenous to warm, arid, and tropical climates. The quantity of fungi that can produce aflatoxin could be drastically reduced as a result of climate change, which could then spread to new areas.

The Food and Agriculture Organization estimates that mycotoxins infect around 25% of agricultural products globally each year (Eskola et al., 2020). Through ingestion and inhalation, mycotoxins impair the quality of food and animal products, which has a negative impact on the worldwide economy.

TABLE 14.2   OPTIMAL TEMPERATURE (°C) FOR MYCOTOXIN PRODUCTION AND IN VITRO GROWTH OF IMPORTANT PLANT PATHOGENS (LIMA ET AL., 2021)

Fungus Species	Type of Mycotoxin	Optimum Conditions for Mycotoxin Production		Optimum Conditions for In Vitro Growth	
		°C	a_w	°C	a_w
Alternaria alternata	Alternaria toxins	25	0.95	23	0.95–0.90
Aspergillus flavus	Aflatoxins	33	0.99	35	0.99
Fusarium verticilliodes	Fumonisins	15–30	0.95	30	0.95–0.98
Fusarium graminearum	Deoxynivalenol	30	0.99	20–22	0.98–0.99
Fusarium culmorum	Deoxynivalenol	26	0.90	20–25	0.90–0.98
Penicillium verrucosum	Ochratoxin A	25	0.95	26	<0.95
Claviceps sp.	Ergot alkaloids	23–26	0.91	30	0.90–0.9

14.3 IMPACT OF CLIMATE CHANGE ON MYCOTOXIN PRODUCTION

One of the most significant threats to global food supply has been identified as climate change. However, its impact on food safety has received little attention (Al et al., 2008). Agriculture accounts for around 40% of the terrestrial surface area, according to the World Bank. The agricultural sector is expected to be the most vulnerable to the effects of climate change due to its vast surface area (Miraglia et al., 2009). Climate change is anticipated to have a substantial influence on agricultural and food safety. Numerous factors, including rising temperatures, droughts, and CO_2-enriched atmosphere, will have a direct effect on food production and consumption (Al et al., 2008).

Changes in climate are expected to affect soil microbial communities and the evolution of disease-carrying vectors. The characteristics and nutrients of the soil are also impacted. There may be food safety concerns about mycotoxins and pesticide residues in the food supply (WHO, 2019). A wide variety of fungi produce mycotoxins. If they aren't handled properly, they can cause serious illness and even death. Exposure to mycotoxin has a direct correlation to the severity of the symptoms. For example, chronic sickness and death might result from exposure to excessive levels. Mycotoxins are the primary constituents of the fungi that make up the *Aspergillus* family of organisms. They can be harmful both acutely and chronically, and they pose the greatest risk of human exposure to aflatoxins. Mycotoxins produced by *Fusarium*, *Penicillium*, and *Aspergillus* are the most prevalent (Council for Agricultural Science and Technology, 2003). They are the most dangerous and can result in a variety of ailments, including cancer and genotoxicity. Climate change is anticipated to increase the frequency of heatwaves and droughts in agricultural regions of wheat and maize. This increases the risk of aflatoxin contamination and threatens to reduce average global wheat yields by 6% as a result (Zhao et al., 2017). In recent years, aflatoxin contamination has become a major food safety issue. Increases are possible, especially in developing countries.

14.4 IMPACT OF CLIMATE CHANGE ON AGRICULTURE AND ASSOCIATED FOOD SAFETY ISSUES

According to the World Bank, farming accounts for nearly half of the world's land surface. Agriculture and livestock productivity are expected to be severely affected by climate change. Variations in the weather can have a significant impact on this industry (Miraglia et al., 2009). Environmental factors like air pollution and extreme heat can have an impact on the soil's microbiome (Figure 14.2).

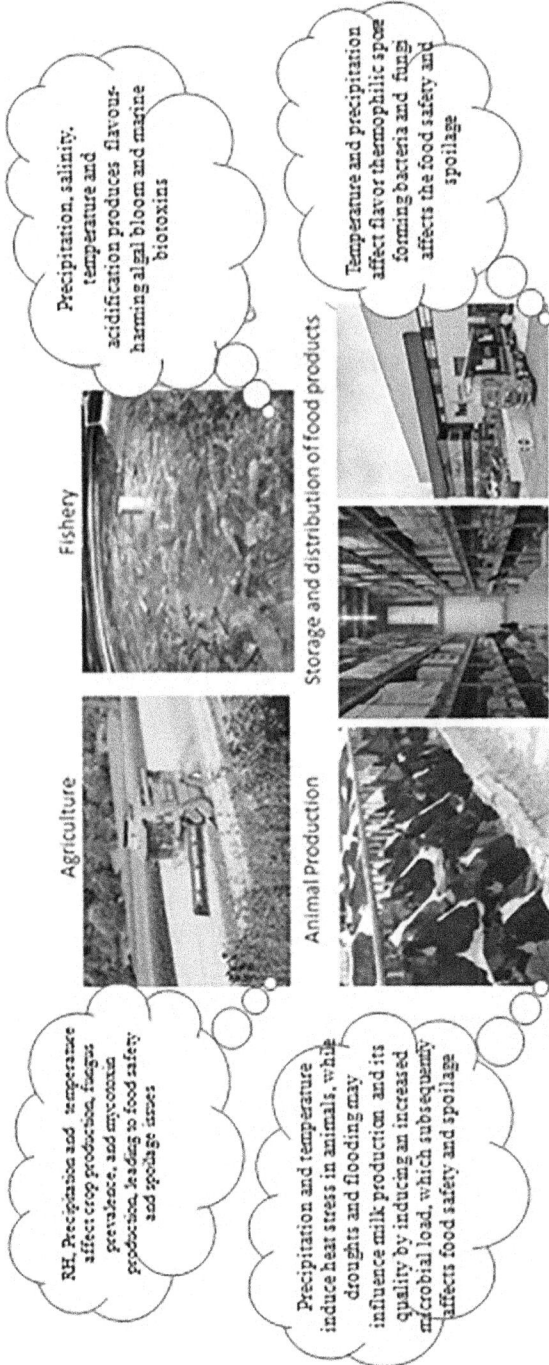

Figure 14.2 Factors that are expected to change due to projected climate change, along with their potential impacts on food safety and spoilage.

Secondary metabolites of various fungi are used by plants to produce toxic fungi and mycotoxins. People and animals alike can be poisoned by mycotoxins. Mycotoxin concentration and dose affect the severity of symptoms. Mycotoxins produced by *Fusarium*, *Penicillium*, and *Aspergillus* are the most dangerous, as they can cause cancer and deteriorate the immune system. Additionally, they can be discovered in food that has been contaminated with animal feed. Drought and heatwaves are expected to spread aflatoxin contamination as a result of climate change. This could have an effect on the world's food supply. Climate change will exacerbate the problem of food insecurity, as a 1°C rise in temperature would reduce global wheat production by 6% on average.

Aflatoxin contamination is a food safety concern that is projected to become more prevalent as mycotoxins levels rise. It is possible that it is caused by a combination of decreased yields and increased aflatoxin contamination.

14.4.1 Soil Quality and Degradation

As a result of climate change, soil quality has been shown to be significantly impacted. In contrast, the impact on soil microbial populations and plant growth is still unknown. Concerns have been expressed by soil scientists about the complexity of interactions between various components, such as chemicals and minerals, in the soil (Sparks, 2001). Soil quality and climate change are intertwined, and it is important to know how these two factors affect each other.

It's expected that climate change will have a significant impact on soil trace element transport and bioavailability to plants. It's possible that more frequent and severe rainstorms could cause soil erosion and leaching (Smith et al., 2005). Erosion and leaching have the potential to obliterate soil minerals (Miraglia et al., 2009). Significant elements affecting plant growth and development are the interactions between soil factors and plant communities. The microbial population, mineral distribution, and bioavailability are all expected to suffer as a result of climate change. Increased phosphorus and nitrogen levels in plants may result in a loss of microbial biodiversity (Dai et al., 2018). A deficiency of nutrients in the soil might weaken a plant's resistance to some pests and illnesses (FAO, 2008). This could potentially result in the evolution of fungi capable of infecting and producing mycotoxins (Godfray et al., 2010). Plants' ability to thrive in dry environments can be hampered by the prevalence of fungus and bacteria in soil microbial populations (Bahram et al., 2018; Maestre et al., 2015; Jing et al., 2015). Climate change will also have an effect on the conditions of agricultural land.

Climate change is widely regarded to have a detrimental influence on agricultural output and the biogeographical scenario of cultivated plants. Water scarcity and the frequency of extreme weather events are also known to influence how plants adapt to their biogeographical environment (Giorgi & Lionello, 2008). Within the next few years, the areas suitable for agricultural production in Northern Europe are likely to expand (Maracchi et al., 2005; Messerli et al., 2000; Olesen & Bindi, 2002). This is due to the anticipated increase of arable land.

14.5 EFFECTS OF CLIMATE CHANGE ON CROP PESTS AND THEIR POPULATIONS

The natural habitats and habits of insect populations may be altered as a result of climate change (Rosenzweig et al., 2001). Crop yield and food safety could be affected by changes in pest and disease dynamics, and the impact of climate change on pathogens is unknown (Petzoldt & Seaman, 2006).

Increasing fungal resistance, as well as decreased stress resistance, are both predicted outcomes as a result of climate change. These effects have the potential to have an effect on the growth and development of a variety of plant species and crops. Insect attacks can have a detrimental effect on the development and life cycle of plants. Additionally, they can promote fungus growth by causing damage to the fruit and kernel (Tirado et al., 2010; Gonçalves et al., 2020). Early stages of pest resistance development can promote fungus growth and mycotoxin generation. The maritime environment is already feeling the effects of climate change (Bar-On et al., 2018; Cavicchioli et al., 2019). This is mostly related to marine biomass decline (Behrenfeld, 2014). Additionally, the use of chemicals and natural pesticides for crop protection is inadvisable during periods of harsh weather (Bailey, 2004; Muriel et al., 2000). Additionally, these solutions deteriorate rapidly and may not provide adequate protection against pests. As food demand and farming skills improve, pesticide use is likely to increase. This could result in an increase in residual pesticide levels in food and crops. Certain insecticides have been shown to have a damaging result on the growth and reproduction of pests and other animals (Rosenzweig et al., 2005). Additionally, climate change is anticipated to jeopardize the safety of food produced by farmers (Tirado et al., 2010). Environment change has resulted in a large decline in marine biomass, which covers around 70% of the Earth's surface (Bar-On et al., 2018; Cavicchioli et al., 2019). This environmental issue has ramifications for food safety and the ecosystem (Behrenfeld, 2014). Climate change is projected to have a profound effect on the maritime ecosystem. These consequences are almost certain to have an effect on the safety and health of seafood consumers.

The risk of waterborne infections such as SARS and *Vibrio* spp. is one of the microbiological risks offered by water pathogens. Toxicity to humans and the environment is increased when a large number of microscopic algae bloom (Anderson, 1995; Erdner et al., 2008). Certain dangerous algal species release poisons that are toxic to humans. These toxins have the potential to build up in the food chain and eventually reach the general population. Diarrheic and alkaloid shellfish poisoning are also potential causes of food poisoning, as are azapsir acid and diarrheic shellfish poisoning. These algal species are most of time utilized as food additives and can result in a variety of ailments, including paralytic and diarrheic shellfish poisoning. It They have also been linked to illnesses in humans who handle shellfish that are amnestic, azapsir acid shellfish poisoning, neurotoxic shellfish poisoning (NSP), and paralytic shellfish poisoning, among other toxins. While food preparation techniques can assist in avoiding the consumption of these poisons, they cannot be used to detect their existence (Fleming et al., 2006).

14.6 FUNGAL GROWTH AND MYCOTOXIN PRODUCTION

Numerous mycotoxigenic fungi are known to produce a variety of mycotoxins in a variety of crops. Additionally, under some conditions, they can manufacture their own mycotoxins (Richard et al., 2003). Numerous elements, including temperature, soil conditions, and environmental factors, might influence the growth and generation of mycotoxins. Due to this mechanism, food contamination can also occur at many points of the food chain (Richard et al., 2003). While fungi can contaminate food, their presence does not always imply that the pollutants are linked to mycotoxins (Table 14.3) (Kochiieru et al., 2020; Perdoncini et al., 2019).

The initial and secondary phases of fungal development are distinct. The main phase entails the synthesis of organic chemicals that are used to synthesize primary metabolites. Mycophenolic acids are secondary metabolites produced by fungus to decrease their concentration. These metabolites are formed as a result of excess primary metabolites that might be accumulated to aid in the growth of certain organisms (Perdoncini et al., 2019). Due to the fact that mycotoxins are created by a diverse range of organisms, their environment can change considerably. Several of these criteria are referred regarded as the key determinants of mycotoxin contamination.

14.6.1 Temperature, Water Activity, and Relative Humidity

When it comes to determining the level of fungal colonization and activity, climate factors play an important role (Smith et al., 2005). The environment in which they operate has a significant impact on their level of success (Magan & Olsen, 2004).

TABLE 14.3   MAJOR MYCOTOXINS, THEIR PRODUCING FUNGI, AND AFFECTED FOOD TYPES (DAOU ET AL., 2021)

Mycotoxin	Producing Fungi	Affected Foodstuff
Aflatoxin B1, B2, G1, and G2	*Aspergillus parasiticus* *Aspergillus flavus* *Aspergillus nomius*	Wheat, rice, maize, nuts, peanuts, spices, cottonseed and oilseeds
Aflatoxin M1	Metabolite of aflatoxin B1	Milk, dairy products, etc.
Ochratoxin A	*Aspergillus niger* *Aspergillus* *Aspergillus ochraceus carbonarius* *Penicillium nordicum* *Penicillium verrucosum* *Penicillium cyclopium*	Coffee beans, cocoa beans, dried fruits, fruits and fruit juice, oats, barley, wheat, and wine
Patulin	*Penicillium expansum* *Byssochlamys nivea* *Aspergillus clavatus*	Cheese, fruit and fruit juices, and wheat
Trichothecenes	*Fusarium* *Fusarium langsethiae sporotrichiodes* *Fusarium culmorum* *Fusarium graminearum* *Fusarium cerealis*	Wheat, maize, oats, barley, animal feed, and grains
Zearalenone	*Fusarium graminearum* *Fusarium equiseti* *Fusarium verticilliodes* *Fusarium cerealis* *Fusarium culmorum* *Fusarium incarnatum*	Wheat, maize, rye, barley, and animal feed
Fumonisin B1, B2, B3	*Fusarium proliferatum* *Fusarium verticillioides*	Maize, wheat, rice, barley, sorghum, and oats

In 2003, Doohan et al. stated that mycotoxigenic fungal growth and survival are influenced by temperature and humidity. Toxin buildup and plant health are both affected by these factors. Plant growth, strength, and health are all affected by temperature and humidity, and mycotoxigenic fungi's ability to compete is also influenced (Richard et al., 2003).

The characteristics of mycotoxigenic fungi, such as their survival and distribution, influence the accumulation of toxins and plant growth. Each fungus has its own optimal temperature and water activities range. These ranges can be used to induce fungal activity. Due to differences in the environment's

conditions and growth requirements, mycotoxin production is more prevalent in regions where the climate is subtropical or tropical (Leslie et al., 2008;
Mannaa & Kim, 2017a). A favorable environment can induce fungal invasion.
This can happen at different stages depending on the conditions and the stage
of the invasion (Joubrane et al., 2020). To grow and produce their mycotoxins,
most fungi in the wild are hydrophilic, requiring a humidity level of at least
90%. At relative humidity levels of 80 to 90% for *Aspergillus* spp. and *Penicilliums*
spp. and lower, mycotoxins begin to germinate, grow, and be produced by these
fungi after harvest (Mannaa & Kim, 2017a). Storage in a humid environment
has an impact on food's moisture content. Fungal growth and mycotoxin production can be facilitated by this (Table 14.2).

Most fungal species require a temperature range of 5 to 35°C to grow well
(Dix & Webster, 1995). Some fungal species can tolerate low temperatures.
These include psychrophiles and thermophiles (Magan & Olsen, 2004). The
ideal temperature ranges for fungi are known to promote the development
of their organisms. However, when the temperature drops below these conditions, the reaction rate may stop or even reverse (Kamil et al., 2011) Even if the
conditions that favor fungal growth don't lead to the production of mycotoxin,
they can still encourage the development of fungi. Water activity of more than
0.78 meters per second is considered favorable for fungal development in temperatures between 25 and 30°C (Thanushree et al., 2019). For *Aspergillus*, this
can mean that the conditions are right for the organism to promote germination. This can also be compared to the conditions that support fungal growth
(Mannaa & Kim, 2017a). Climate variables are critical in affecting the level of
fungal colonization and activity (Smith et al., 2016). Their environment has a
significant impact on their level of success (Magan & Olsen, 2004).

Temperature and humidity are the two parameters that influence the development and survival of mycotoxigenic fungus (Doohan et al., 2003), These
factors all contribute to the buildup of toxins and the health of the plant. The
competitiveness of mycotoxigenic fungi is influenced by temperature and
humidity as well as plant growth, strength, and health (Richard et al., 2003).

The survival and dispersion of mycotoxigenic fungi have an effect on the
buildup of toxins and plant development. Each fungus has a temperature and
water activity range that is best for it. These ranges may be used to stimulate
fungal growth. Mycotoxin generation is more widespread in locations with
subtropical or tropical climates because of variances in environmental circumstances and growth needs (Leslie et al., 2008; Mannaa & Kim, 2017a). A
conducive setting may facilitate fungal invasion. This can occur at various levels depending on the circumstances and stage of the invasion (Joubrane et al.,
2020). In the field, the majority of fungi are hydrophilic, which means they
require at least 90% relative humidity to thrive and generate their mycotoxins. There are no more mesophilic or xerophylic fungi after harvest, and their

place is taken over by those that germinate, thrive, and produce mycotoxins at relative humidity levels as low as 80%, such as *Aspergillus* and *Penicillium* species (Mannaa & Kim, 2017a). The humidity level in the storage environment has an effect on the moisture content of the food. This can make the food more susceptible to fungus development and mycotoxin generation (Table 14.2).

14.6.2 pH

The pH value is present and has an effect on the development and generation of fungi and mycotoxins at various stages of their lifetime. By secreting acids or alkalis, fungi can regulate the ambient pH. *Penicillium* and *Aspergillus*, for example, can release acidifying enzymes (Vylkova, 2017). Fungi's ability to regulate the pH within the host enables them to grow. It can, however, influence their metabolic processes and temperature connections (Wang et al., 2017). Genes that play a role in the production of OTA have been shown to be affected by pH (Brzonkalik et al., 2012).

Influence of pH and carbon to nitrogen ratio on mycotoxin production by *Alternaria alternata* in submerged cultivation (Brzonkalik et al., 2012). On the other hand, while it is unknown which mycotoxins will benefit from acidic circumstances, it is widely thought that these conditions will boost mycotoxin growth and production. Aflatoxin is synthesized in a low-pH environment. For example, a pH value of 4.0 is required to create a high synthesis rate (Reverberi et al., 2010; Perdoncini et al., 2019). Fumonisin B1, a form of resin, is not stable in an acidic environment and must be manufactured at a pH of 4.0 to 5.0. A pH value of 4.0 is required for alkaloid synthesis. This is because the lower the pH, the higher the rate of synthesis. Generation of mycotoxin is determined by the toxicity of a particular fungus species. Although it can be restricted to specific strains within a species, the precise amount of mycotoxin produced is determined by a variety of circumstances.

Different fungal strains can produce a variety of mycotoxins, depending on their toxicity. Certain mycotoxins are even species specific. Different species' features can also influence the growth and generation of mycotoxins. For example, various strains of *Aspergillus flavus* can generate toxins at different temperatures (15–44°C) (Mannaa & Kim, 2017a).

14.6.3 Substrate

Although fungi can grow on a variety of different substrates, the precise reason they predominate on specific food products remains unknown. They thrive mostly on carbohydrate-containing foods (Kokkonen et al., 2005). While fungal growth may be useful for mycotoxin formation, it is not always practical to rely solely on fungal growth substrates.

Mycotoxins are mostly formed through the reaction of mycobacteria with a substrate. The following elements have all been determined to contribute to their production: pH, temperature, and composition (Özcelik & Özcelik, 1990). The combination of numerous elements inside a substrate inhibits certain organisms' fungal growth and mycotoxin generation. Osmotic pressure has been shown to have a significant effect on the growth and synthesis of mycotoxin by fungi. Additionally, it has an effect on the formation and synthesis of secondary metabolites (Duran et al., 2010). Osmotic stress in fungi has been demonstrated to alter their physiology in order to promote their survival and adaption. While sugars are made up of carbon molecules, they can be hydrolyzed. As a result, they have the potential to boost the economy and generate renewable energy (Hamad et al., 2014). Simple sugars are easily degraded. This results in an increase in fungal growth, as they take less digestion to digest. Mycotoxins can also be synthesized in response to simple carbohydrates. For instance, Liu et al. (2016) showed that, in cell culture, increasing the concentration of soluble carbohydrates such as glucose can lead to an increase in AFB1 synthesis. Uppala et al. demonstrated that increasing the sugar content of the medium increased AFB1 synthesis by *A. flavus* (Uppala et al., 2013).

14.7 EFFECTS OF CLIMATE CHANGE ON MYCOTOXINS

Temperatures are expected to rise 1.5 to 4.5°C by the end of the 21st century as a result of climate change (Van der et al., 2016). As a result of climate change, drought, heat waves, and flooding are expected to become more frequent and severe. In addition, it is expected to raise the concentration of greenhouse gases in the atmosphere (Medina et al., 2015). Climate change and global warming have been shown to have a significant impact on food security. Apart from impairing crop quality and raising concerns about food safety, these variables can also impact the profitability of various food items (Table 14.4) along the food chain (Miraglia et al., 2008). Climate change is projected to have a range of consequences on different places in a variety of ways. According to the European Food Safety Authority, the consequences of rising temperatures on specific places will be beneficial or deleterious (Battilani et al., 2012). Climate change is projected to have a major impact on mycotoxin production and consumption in Europe (Medina et al., 2017a).

According to Table 14.4, different environmental elements such as climate change and humidity might influence mycotoxin growth and production. For example, some mycotoxins are more frequent in tropical and warm climates, whereas others are less prevalent in cold climates. Additionally, certain mycotoxins can be formed at low temperatures, while others can be created at elevated temperatures. *Aspergillus flavus* colonization and aflatoxin production

TABLE 14.4   AGRICULTURE, MYCOTOXIN PRODUCTION AND STORAGE, AND THE CONSEQUENCES OF CLIMATE CHANGE ON THE ATMOSPHERE (DAOU ET AL., 2021)

Impact of Climate Change on Different Elements	Specific Effect
Weather impact Global temperature rise	Increased precipitation; extreme weather (protracted periods of warm or cold); droughts; atmospheric gas accumulation (CO_2); flooding
Impact on agriculture	Increased pest and insect population, dissemination, and attacks; lower yields; lower crop quality; early crop maturation; lowered plant vigor; crop pathology shifts
Impact on mycotoxins	Mycotoxin generation varies by geography and the predominance of mycotoxin-producing environmental factors such as temperature and humidity
Impact on storage	Increased risk of fungal invasion and mycotoxin generation; hotspot formation; intragranular CO_2 increase; grain respiration; insect and pest attack

were induced by a series of hot and dry events in 2003 and 2004 in Italy (Giorni et al., 2008; Valencia-Quintana et al., 2020). Since the ideal conditions for a particular mycotoxin's production vary, each mycotoxin will be affected differently.

Climate change may also have an impact on the spread of pests and illnesses around the world. Additionally, it has the potential to reduce plant resilience and modify host reaction to CO_2 (Skendžić et al., 2021; Medina et al., 2015; Paterson & Lima, 2010). Mycotoxins' response to climate change will require more investigation because fungi have been shown to be resilient (Medina et al., 2017b). Climate change can also have an effect on agricultural harvesting and drying. In some places, abrupt severe rains may prompt farmers to keep immature kernels without drying, increasing the risk of mycotoxin contamination. In places where warm weather is forecast, earlier harvest and faster crop growth cycles result in decreased drying requirements (Moses et al., 2015; Medina et al., 2017b; Paterson & Lima, 2010). Grain deterioration can be sped up by climate change, which affects food storage. The presence of fungi capable of causing fungal invasions in developing countries is expected to lead to deterioration in the storage conditions of many crops. Additionally, these organisms (such as *Aspergillus flavus*) may thrive in warm climes and create mycotoxins such as aflatoxins, and in places with high humidity and moist atmospheres, the danger of fungal invasion in stored grains increases (Moses et al., 2015).

Due to the degradation of grain quality caused by environmental changes and the growing number of storage hotspots, fungal growth and the formation of mycotoxins will occur during storage. This problem raises the likelihood of fungal infection and food contamination (Moses et al., 2015). It is critical to store grains and crops in climate-controlled facilities to ensure a safe and dry environment. For the sake of grain quality, these facilities should have adequate aeration systems in place.

14.8 SAFETY AND MICROBIAL SPOILAGE AT VARIOUS STAGES OF THE FOOD CHAIN

Environmental residues from a variety of industries, mycotoxin and marine biotoxin contamination, and zoonotic infections have all been identified as climate-driven emerging risks to human life (Misiou & Koutsoumanis, 2021).

14.8.1 Mycotoxin Control and Prevention Strategies

Due to the nature of mycotoxins, the majority of food produced worldwide is at danger of contamination. This is why it is critical to regulate mycotoxins throughout the food chain. Mycotoxins can be created in a variety of ways and exhibit a wide range of properties. Mycotoxins can be applied to a variety of crops and are easily distributed.

To minimize the presence of mycotoxins in food, a number of simple steps can be taken. An integrated food safety system that incorporates proper quality procedures at each stage of manufacturing can help limit the frequency of mycotoxins in the final product (Figure 14.3).

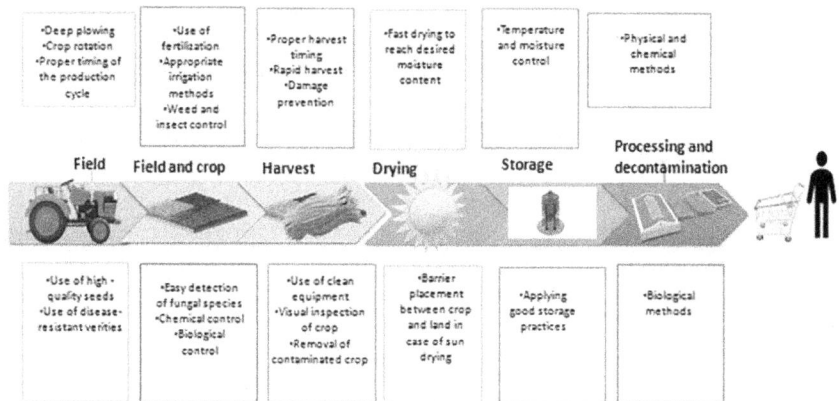

Figure 14.3   Proper practices to minimize mycotoxin contamination along the food chain.

14.8.2 Proper Field Practices

The majority of fungus are considered phytopathogens, and it is critical to manage contamination during the pre-harvest phases to avoid the introduction of mycotoxins into food. The most prevalent fungi in the field are *Fusarium* spp., *Cladosporium* spp., and *Alternaria* spp.

Although the majority of mycotoxins are found in nature, they can also be detected in trace amounts under certain situations. This means that under some conditions, the extent of contamination is greater.

On the other hand, *Aspergillus* spp. and *Penicillium* spp. can be discovered in the field; the level of contamination varies according to the environment and the type of fungus to which they are exposed (Joubrane et al., 2011). At the same time, while it is impossible to completely eliminate mycotoxin production during pre-harvest, it is critical to design specific programs aimed at reducing contamination during this phase.

Proper harvesting practices can aid in reducing the pre-harvest inoculum concentration. Given the critical role of toxigenic fungi in crop quality, correct approaches and practices should be used to decrease their prevalence (Mannaa & Kim, 2017b).

Numerous variables in the field, such as drought stress, pest invasion, heat, poor soil health, and delayed harvesting, can all contribute to the production of mycotoxins.

14.8.3 Field Preparation and Management prior to Planting

It is critical to prepare the farm for fungal attack and its toxins before planting. Tilling, deep ploughing, and crop rotation are all examples of this. Preparing entails soil preparation for tillage, crop rotation, and the application of high-quality seeds.

Deep ploughing is critical for removing residue from the land during a good crop rotation. It allows the soil to heal and protects fresh crops from fungal contamination. Ploughing buries waste underground, rendering it uninhabitable by fungi (Golob, 2007). Additionally, it can assist in minimizing water loss caused by soil compaction (Munkvold, 2014). Additionally, crop rotation can help reduce the danger of mycotoxin contamination by preventing fungal species from growing (Mannaa & Kim, 2017b; Mahuku et al., 2019).

Mycotoxin contamination has been reported to increase in plots where the same crops have been grown for multiple years. This is because repeated exposure to the same plant can result in the growth of molds (Golob, 2007; Rose et al., 2018). Effective seeds are critical for plant growth because they contribute to the formation of healthy plants that are resistant to fungal attack. By introducing resistant cultivars, resistance to disease or toxin contamination can be developed. These variations, however, may not offer protection against all known fungi. Seeds that are partially resistant to cold can be used in colder climates. Seeds that are not resistant to disease are more expensive and may be unavailable in marketplaces. This may have an effect on the quality of seeds available to farmers in developing countries.

14.8.4 Field and Crop Management after Planting

Following planting, the correct techniques and equipment are employed to encourage healthy crop growth and development. This procedure entails the use of fertilizers and a variety of additional chemicals in order to manage pests and diseases. Climate change may also have an impact on the distribution and survival of some plant species. This may result in an increase in their generation number, changed interspecific interactions, and an increased risk of pest invasion (Skendžić et al., 2021; Shrestha, 2019). Fertilizers aid in the maintenance of the plant's resistance to fungal infection. Additionally, they increase the plant's vigor and give necessary nutrients for growth. Because fertilizer treatments are typically scheduled and the amount applied is proportional to the plant's stress, they should be applied with caution to avoid generating insect and fungal infestations. Additionally, excessive fertilizer use might be dangerous to humans. Because fertilizers contain heavy metals (lead, cadmium, chromium, and arsenic) in addition to necessary nutrients, they may accumulate in the plant or contaminate subsurface water. They can also cause a variety of respiratory and health problems, including coughing, chest tightness, difficulty breathing, skin rashes, and dermatitis, as a result of repeated exposure (Nganchamung et al., 2017). Irrigation can assist in minimizing the formation of fungus spores and plant cracks. Additionally, it can assist in minimizing plant stress and preventing fungal spread (Golob, 2007).

Weed control is also critical for crop protection against fungal invasion and disease. Weeds contaminate crops by competing for nutrients and water (Reboud et al., 2016). While weed control is critical, it is as critical to avoid insects and diseases caused by insects. They have the potential to cause physical damage to some grains. It is critical to remove all plant debris from a facility in order to prevent rodents from gaining access to the facility's food source. This will assist in keeping the region pest free. Additionally, it is critical to use the proper amount of insecticides to manage pests.

14.9 IMPACT OF CLIMATE CHANGE ON ANIMAL PRODUCTION AND ASSOCIATED FOOD SAFETY ISSUES

The influence of climate change on animal productivity has long been recognized in the marine and agricultural industries. However, the same issue has not been acknowledged in the same way with reference to animal production. Global demand for cattle is anticipated to quadruple by 2050 as a result of climate change (Rojas-Downing et al., 2017). This is a critical issue for the livestock business. It is well established that the various meteorological elements affecting livestock output have a substantial impact on a farm's profitability. Climate change's effects on farm animals vary according to the type of animal and the amount of accessible water. Drought and flooding can also have an effect on milk output (Van der et al., 2012).

14.9.1 Livestock Production and Husbandry Practices

Climate change is projected to increase the frequency with which antimicrobial treatments are administered to animals with infectious diseases. This could result in a greater amount of residual pharmaceuticals being used in the food of origin (Tirado et al., 2010). A change in husbandry practices may promote the spread of infectious animal diseases.

14.10 EARLY DETECTION OF FUNGAL SPECIES

Fungi may suggest an elevated risk of contamination since mycotoxin formation requires specific conditions. Early detection of fungi in crops is critical to preventing their behavior from deviating. Early diagnosis of filamentous fungus is critical for crop disease control. Several of these techniques are mycological, proteomic, and genomic in nature (Rodríguez et al., 2015). Mycological cultivation procedures entail the isolation, cultivation, and identification of fungi. This procedure frequently necessitates the use of a variety of medium and incubation conditions. Typically, mycological operations are time consuming and labor intensive. Proteomic and genomic approaches, on the other hand, can be used to identify filamentous fungi at the cellular level. The advancement of molecular biology tools for species identification has resulted in the creation of new species categories (El Khoury et al., 2011). By analyzing the DNA sequences of fungus species, polymerase chain reaction (PCR) procedures are utilized to identify them.

The PCR method's sensitivity is determined by the sequence of the reference gene. Additionally, the presence of specific DNA-molecule targets enables the detection of filamentous fungus.

14.10.1 Chemical Control

Fungicide control of fungal invasion and mycotoxin contamination is the most efficient method of preventing these destructive organisms from wreaking

havoc on crops. Numerous studies have demonstrated that fungicides can promote the generation of mycotoxins, posing a hazard to crops. Fungicides have
been linked to a variety of health problems due to their toxicity. In the majority
of circumstances, their use is unregulated. Additionally, prolonged exposure
to certain chemicals might result in the buildup of harmful toxins in the body
(Nicolopoulou-Stamati et al., 2016). This results in a variety of adverse health
impacts, including cancer and endocrine disruption. While pesticides are frequently employed in agriculture, they can also accumulate in the body and create a variety of health problems. For instance, pesticide exposure can result in
cancer. The effects of numerous chemicals on humans vary according on their
nature, duration, and state of health. For example, certain pesticides may be
detrimental to certain groups of people, such as infants and pregnant women.

14.10.2 Biological Control

Using atoxigenic strains of the fungus *Aspergillus flavus* considerably reduced
contamination of cotton rows in Arizona in a prior study. A biological control
is the employment of non-toxic fungi to counteract the negative effects of toxic
fungus. This approach is effective because it prevents fungi from causing harm
to humans. This method of control entails the introduction of a non-harmful
biological agent to combat pathogenic fungus.

Aspergillus strains that are non-aflatoxigenic are utilized as biocontrol
agents to prevent aerated strains from forming aflatoxins (Kagot et al., 2019).
The use of non-aflatoxin–producing *Aspergillus* strains can significantly reduce
soil pollution (Dorner, 2004). Contamination with aflatoxin dramatically
decreased with the introduction of an atoxigenic strain of *Aspergillus flavus*
(Cotty, 1994). Although this approach has been shown to minimize mycotoxin
contamination, it does have significant drawbacks. First, it is incapable of producing additional harmful metabolites. These strains may have an effect on
the formation and generation of mycotoxins. These strains could also effect on
the development and formation of mycotoxin derivatives capable of evading
biocontrol agents. The ability to create toxin-producing fungi can be passed
down through generations. This procedure results in the emergence of novel
toxin-producing fungus (Kagot et al., 2019).

14.10.3 Proper Practices during Harvest

Additionally, it is critical to examine the different elements that affect harvest time, such as the availability of suitable storage and drying facilities.
Crops should ideally be harvested following a period of dry weather. Due to
the fact that many crops are harvested in rainy weather, they become more
prone to fungal development, mycotoxin contamination, and insect and bird

attacks. It is critical to avoid mechanical damage to the harvesting equipment. Additionally, it is critical to inspect crops for signs of fungal disease and to avoid causing damage to the crops when harvesting. It is also vital to inspect the plants prior to harvesting.

14.10.4 Proper Practices during Drying

Crops and grains with a high moisture content must be dried properly in order to be stored safely. This procedure entails carefully determining the crop's appropriate moisture content. It is critical that this level be achieved as soon as possible to avoid fungal infection. Although the rate at which crops dry varies according to the environment and the type of crop, in general, the longer the leaves remain attached to the ground, the slower they dry. Sun-drying is the most common way of crop drying. However, it can be hazardous because it increases the risk of fungal infection by extending the drying period. This approach is not appropriate for places prone to heavy rain. Additionally, it might destabilize the soil, which can result in the growth of mold. Solar drying is a process that utilizes the sun's heat to warm the air that is being dried. This process can be carried out in a variety of locales, including a farm, a warehouse, or a large-scale manufacturing plant. This procedure entails injecting heated air into a product, which results in increased production rates. This method is frequently utilized in mass production. Drying can be carried out using superheated steam or infrared radiation in large-scale operations.

14.10.5 Proper Storage Practices

It is critical to implement efficient procedures to avoid fungal invasion and the production of mycotoxins prior to storing the items. When a product is stored with a high level of contamination, it can have an adverse effect on its quality. Numerous ways can be used to avoid this issue and keep the product in good shape. While numerous fungi may live on stored grains, the majority of them become significant when exposed to storage. These organisms can cause mycotoxin formation, nutritional losses, and heat sensitivity. Fungi in storage can cause a variety of difficulties, including damage to the product's nutrients and quality. Additionally, their capacity to create mycotoxins can be influenced by a variety of circumstances, including temperature. While the product may not contain fungi, their presence can nonetheless affect the growth and development of fungi. Additionally, the presence of toxins can inhibit the production of mycotoxins by fungus. Generally, storage fungi can develop at a relative humidity level of between 70% and 90%. They may even flourish in temperatures as high as 40°C. They thrive in a temperature range of 10–40°C, with an optimal range of 25–35°C (Magan et al., 2004).

Apart from temperature and a_w, additional elements affecting the storage safety of foodstuffs include the grain's physical condition and the microbiological interactions between it and the surroundings. For the period of storage, relative humidity must be maintained at 70% or less. This ensures the product's safety and ability to remain stable at low temperatures. A product's temperature is a good predictor of its storage quality. Additionally, it can assist in determining the interior moisture pockets within the product. Increased a_w enables xerophytic fungi to flourish and retain their metabolic processes, enabling them to create water with a broader range of fungal growth. Fungi in the food chain can generate metabolic heat, which helps raise the temperature, which in turn can stimulate the growth of other fungi. This is a critical stage in the food chain and should be carefully monitored and controlled. To avoid storage concerns, aeration and temperature monitoring devices are advised for storage facilities.

Quality inspections are conducted on a regular basis to guarantee the products they offer are of a high standard of quality. Additionally, they are capable of detecting symptoms of deterioration, such as mold growth. It is critical to keep pests out of a crop. This can result in physical damage to the plant and increases the likelihood of fungal spread. Numerous elements can be applied to mitigate this risk and ensure the product's safety. Numerous precautions can be taken to ensure the product's safety through the use of proper storage methods, as outlined in Table 14.5.

14.11 METHODS FOR DETOXIFICATION AND DECONTAMINATION OF MYCOTOXINS

Early diagnosis of mycotoxins is critical to preventing them from wreaking havoc on the food chain. While this is the best method for preserving food quality, it is not always efficient at containing and preventing the spread of mycotoxins. In the majority of cases, this approach will only protect the food from mycotoxin contamination. On the other hand, while there are numerous methods for decontamination and removal of mycotoxins, no single methodology is capable of effectively removing all of these toxins. Early detection and control of mycotoxins in the food chain are critical. They have the ability to slow the growth of a wide range of crops while also preventing food contamination (Pankaj et al., 2018; Hojnik et al., 2017). Preventing mycotoxin contamination is the best strategy for preserving the quality of food. This procedure, however, is not always effective and may result in the contamination of food with mycotoxins. Controlling mycotoxins is not always effective in the food for which they are employed. Food and animal feed can also contain mycotoxins. There are numerous methods for removing and

TABLE 14.5   GOOD STORAGE PRACTICES TO REDUCE FUNGAL AND MYCOTOXIN CONTAMINATION (MAGAN & OLSEN, 2004; RICHARD ET AL., 2003)

Element	Measures
Grain	Prior to storage, ensure that the physical conditions are suitable
	Before storing, ensure that an appropriate moisture content is achieved
	Conduct a visual examination for the presence of disease or fungal infections
Storage facilities	Ascertain soundness and suitability
	Conduct weatherproofing
	Ascertain the building's and equipment's sanitation
	Remove previous crop residues
	On the floor, install impermeable moisture barriers
Insect, rodent, and bird prevention	Apply insecticide to the structure prior to use
	Utilize trapping for insects and rodents
	Seal any holes in the structure to prevent rodents and birds from entering
	Maintenance and surveillance
Quality assessment	Install temperature and humidity sensors
	Regularly calibrate devices
	Keep an eye out for pest infestations
	Keep an eye out for physical damage or disease signs in crops
Record keeping	Visually inspect for the presence of any pest infestation
	Visually inspect for the presence of physical damage or fungal invasion
	Conduct routine microbiological and chemical analyses
	Retain data
	Retain samples that have been tested

decontaminating mycotoxins, but no single method can effectively remove all of the toxins that may be present in a food product at the same time. For a strategy to be successful and profitable, it must encompass a wide range of features (Magan & Olsen, 2004).

This procedure should be capable of entirely removing contaminants from food while retaining the commodity's nutritional value. Additionally, it should be capable of removing any remaining fungal spores without causing harm to the environment. This method of food production must be approved by the appropriate regulatory organizations. Various strategies are now being investigated for removing mycotoxins from food products.

14.11.1 Physical Decontamination

When it comes to physical decontamination, thermal procedures like those used to eliminate mycotoxins are a requirement. Steam, infrared, microwave, radiofrequency, and extrusion heating have all been developed as novel decontamination techniques that can be used to eradicate mycotoxins (Deng et al., 2021; Deng et al., 2020a).

Heat treatments that take advantage of both time and temperature conditions may be the most critical stage in mycotoxin reduction. However, a large number of these treatments require extremely high temperatures to be effective. Temperature increases have the potential to alter the physical properties of food. As a result, some surfaces may develop burn-like characteristics (Deng et al., 2021; Deng et al., 2020a). In comparison, non-thermal treatments such as radiation therapy can help reduce mycotoxin contamination in food (Pleadin et al., 2019). This can be accomplished in a variety of industrial settings.

Radiation injections may also aid in reducing mycotoxin contamination. This is because radiation has the ability to absorb the energy emitted by pollutants. While the usage of irradiated food is widely accepted, the appropriate level of radiation is not routinely employed. For example, the maximum dose of 10 kGy that can be utilized in food remains safe. Despite this, irradiated foods are not generally consumed due to their ability to harm cell DNA and cause cancer. The European Commission has approved a dose of 10 kGy per dose as a safe level of radiation exposure (Hojnik et al., 2017). Table 14.6 highlights different methods for removing mycotoxins from food, their benefits and drawbacks.

Non-thermal procedures are gaining popularity among scientists. Among these methods is the use of cold plasma to eliminate fungal toxins. At room temperature, cold plasma is a dense gas made up of partially ionized atoms and molecules with approximately equal net charges (Misra et al., 2019). Cold plasma treatment has been proven in numerous studies to destroy fungal cell walls and DNA, allowing for the release of internal components (Dasan et al., 2017; Lee et al., 2015; Deng et al., 2020a, 2020b).

Other experiments have demonstrated that mycotoxins can be eliminated in part or fully by a fast response in cold plasma (Misra et al., 2019; Basaran et al., 2008; Park et al., 2007; Ouf et al., 2015). Their destruction is a result of the mycotoxin's structure and chemical composition. The treatment may result in the formation of free radicals, UV light, or ozone. Cold plasma therapy is a one-of-a-kind procedure that enables the cleaning of food items quickly and effectively. It has no effect on the food's quality (Thanushree et al., 2019; Misra et al., 2019). Although cold plasma treatment has been suggested to have an effect on the lipids in food, it is not possible to utilize this technology on a broad scale due to its low penetrating ability. This method uses photocatalytic detoxification to eliminate mycotoxins from food. It works by neutralizing pollutants on the food's surface by generating free radicals.

TABLE 14.6   DIFFERENT METHODS FOR REMOVING MYCOTOXINS FROM FOOD, THEIR BENEFITS AND DRAWBACKS (PANKAJ ET AL., 2018; DENG ET AL., 2021)

Method	Examples	Benefits	Drawbacks
Physical decontamination	• Sorting • Cleaning with the help of sieve • Segregation by density • Washing • De-hulling • Extrusion cooking • Heating with steam • Heating with infrared radiation • Heating in the microwave • Heating with radio frequency • Irradiation • Cold plasma technology • Detoxification by photocatalysis	• Effective against sone mycotoxins • Food qualities do not change much • Does not necessitate the use of chemicals	• Impractical • It's possible that it'll be limited to large-scale enterprises with advanced technology. • It's also time consuming and costly • Changes in color and food quality may occur as a result of thermal treatment
Chemical decontamination	• Natural substances like spices, herbs, and their oil and extracts • Organic acids • Hydrochloric acid • Hydrogen peroxide • Ammonium hydroxide • Sodium bisulfate • Formaldehyde • Chlorinating agents • Ozone	• Affordable and effective against some mycotoxins	• Health ramifications • Toxic byproducts are formed • Increasing the bioavailability of mycotoxins that have been disguised • Time consuming and hazardous to the environment
Biological decontamination	• Algae • Bacteria • Yeasts • Mold	• Some mycotoxins are resistant to it • Inexpensive • Chemicals are not used in an environmentally beneficial manner	• Time-consuming and inconvenient • In a controlled laboratory setting, it is more effective

14.11.2 Chemical Decontamination

Chemical procedures involve the treatment of chemical compounds using acids, alkalis, and reducing and oxidizing agents, which might be organic or synthetic in nature. These chemicals are frequently used to decontaminate food. Additionally, they can act as reducing and/or oxidizing agents. Along with the addition of chemicals to food, this technique can be used to prepare and transport food (Karlovsky et al., 2016). Chemicals of various types can be used to prepare and deliver food. Organic acids, ammonium hydroxide, hydrochloric acid, hydrogen peroxide, sodium bisulfite, chlorinating agents, ozone, and formaldehyde are all used in this process, as are natural compounds such as herbs, spices, and their extracts (Hojnik et al., 2017; Karlovsky et al., 2016). Chemical treatments, such as chlorine dioxide, ozone, electrolyzed water, essential oils, high-pressure carbon dioxide, and organic acids, have been amplified as alternatives to conventional disinfection techniques to meet current safety standards (Deng et al., 2020b). While chemical treatment is frequently employed to eliminate mycotoxins, the majority of them are ineffective because mycotoxins are resistant to a variety of chemicals. Strong acids can be hazardous to humans and animals due to the production of additional compounds when they are used (Pleadin et al., 2019). Mycotoxins can be chemically transformed to increase their bioavailability and decrease their toxicity (European Norm, 2006) Mycotoxins have been shown to be effective when treated with ozone because the chemical structure of the mycotoxins reacts with the ozone. However, treating ozone with prolonged exposure to air might result in the loss of fat components and impair food quality.

While ozone is usable in gaseous form, it is not acceptable for food quality. Additionally, ozone treatment should be subjected to specified circumstances before to usage to avoid affecting the food's quality. Despite these chemical treatments' efficacy, their use is still prohibited in the European Union due to their toxicity (European Norm, 2006). Only a few mycotoxins are permitted for chemical treatment in the majority of nations. This means that only some mycotoxins are treated chemically.

14.11.3 Biological Decontamination

Decontaminated microorganisms include algae, fungi, and bacteria. They are used to remove toxins and promote a clean environment. Biological decontamination is a process utilized by microorganisms to remove their harmful effects. Mycotoxins in animal feed can be modified or reduced using biological means. This can result in fewer toxic substances being produced. In some cases, the microorganisms used in biological means may be able to acetylate, glucosylate, deaminate, hydrolyze, or decarboxylate mycotoxins in certain foods for animals (Hojnik et al., 2017).

In addition, some organisms can also transform OTA into phenylalanine by producing enzymes and microorganisms. A variety of microorganisms and enzymes can be added to animal feed to help improve the digestion and demycotoxin degradation of ruminants. Deodorizing agents such as lactic acid bacteria and yeasts are used to remove mycotoxins from products. They can also reduce the toxicity of these chemicals by binding them to the cell surface (Pleadin et al., 2019).

Enzymatic catalysis is a promising method for reducing mycotoxin contamination. However, this process requires further studies to ensure its safety and efficacy. Biological methods are usually cost effective and do not involve the use of chemicals. However, their use can be very time consuming. Many biological means were tested in laboratory settings and could only be effective in certain conditions. However, further studies are needed to see their efficacy in food (Patriarca & Pinto, 2017). Algae, fungi, and bacteria are all examples of decontaminated microorganisms. They are utilized to eliminate contaminants and maintain a healthy atmosphere. Microorganisms use biological decontamination to eliminate their toxic effects. By utilizing biological methods to modify or eliminate mycotoxins in animal feed, it is possible to produce less harmful compounds. Through acetylation, glycosylation, deamination, hydrolysis, or decarboxylation, the microorganisms utilized in biological methods may be able to bind to, breakdown, or change mycotoxins in particular foods and animal feed (Hojnik et al., 2017).

Additionally, several organisms can convert OTA to phenylalanine by the production of enzymes and bacteria. Animal feed can be supplemented with a range of microorganisms and enzymes to aid in the digestion and demycotoxin breakdown of ruminants. Mycotoxins are removed from items using deodorizing agents such as lactic acid bacteria and yeasts. Additionally, they can mitigate the toxicity of these compounds by tethering them to the cell surface (Pleadin et al., 2019).

Enzymatic catalysis is a promising technique for mycotoxin contamination reduction. However, additional research is necessary to ensure the process's safety and efficacy. Biological methods are frequently more cost effective than chemical ones and do not require the use of chemicals. Their use, on the other hand, might be somewhat time demanding. Numerous biological methods have been tested in laboratory settings and have been shown to be effective only under certain conditions. However, additional research is required to determine their efficacy in food (Patriarca & Pinto, 2017; Fernández et al., 2017).

14.11.4 Effect of Processing on Mycotoxins

Food processing is a technique used to enhance the nutritional value and shelf life of food products. It entails the use of a variety of chemicals and biological procedures to eliminate or reduce the presence of organisms in food. Mycotoxins are chemicals that have the potential to impair the stability and shelf life of food products (Milani & Maleki, 2014). The elements that contribute to their presence may also change according to the environment and the qualities of the meal. Processing a food product might entail a number of complex steps. Mycotoxins can undergo a variety of transformations. Complex processing processes can result in a finished product with a lower mycotoxin concentration. This can be advantageous for a variety of commodities, including cereals. Mycotoxins may be more concentrated in the final product throughout the cheesemaking process, for example, due to the processing conditions. This method typically uses more materials and time than raw milk does.

Climate change is also projected to increase the vulnerability of cattle production to a variety of stressors. This includes the breeding of breeds that are capable of competing in the event of heat stress.

14.12 EFFECTS OF INCREASING ANTIMICROBIAL RESISTANCE ARE EXPECTED TO CHANGE THE TRANSMISSION PATHWAYS OF PESTS

Agricultural pests and diseases are expected to become more prevalent as a result of climate change (Rojas-Downing et al., 2017). To effectively treat infections, a greater number of antimicrobial agents may be required. This section demonstrates that antibiotic residues in animal-derived foods can be predicted to grow (Tirado et al., 2010).

Antimicrobial resistance is becoming more prevalent, and as a result, the usage of antimicrobials in a variety of sectors is predicted to expand. Climate change has been linked to an increase in antimicrobial resistance. Bacterial growth may be inhibited by this procedure. According to a study, rising temperatures increase the risk of developing resistance to common infections (MacFadden et al., 2018). Although a link between temperature and antimicrobial resistance has been shown, the precise mechanism by which this association occurs remains unknown. Antibiotic resistance plasmids may spread more quickly under certain conditions, according to the authors. Antimicrobial medications must be kept safe in light of climate change's potential impact on the development and use of these drugs.

The high mortality rates that can be caused by water- and food-borne illnesses make them extremely important. Around 420,000 people died in 2010 as a result of water- and food-borne illnesses, according to the World Health Organization. Around half of all deaths caused by diarrheal disease agents can be attributed to climate change (WHO, 2015). Viruses are more likely to spread and survive if the environment is hot, dry, and prone to extreme weather events, such as hurricanes. *Mycobacterium avium* and *Salmonella* species, for example, can survive in the environment and spread at low infectious doses.

Infections caused by enteric viruses and parasites are expected to become more common as a result of climate change. These dangers may also become more prevalent in the future (Hall & Kirk, 2002; Rose et al., 2001). In addition to a decrease in vector populations, climate change is linked to an increase in food-borne diseases. Additionally, when discussing the link between vectors and climate change, attention must be paid to this subject matter. It is widely accepted that climate change has a significant impact on food and waterborne illnesses. The relationship between climate change and vectors is the primary focus of this section.

14.12.1 Vectors

Weather-related changes in temperature and precipitation can affect how zoonotic diseases spread through mosquito-borne vectors (FAO, 2008). Changes in vector transmission channels are correlated with numerous environmental variables (McIntyre et al., 2017).

The effects of these factors on vectors are likely to be amplified in the future due to climate change. According to new research published in 2019, the winter months and increased temperatures benefit flies carrying *Campylobacter* (Cousins et al., 2019). Potentially disease-free zones could be established as a result of climate change expanding the geographic range of arthropod vectors. This risk should be studied using predictive models.

14.12.2 Correlation between Climatic Factors and Water and Food-Borne Diseases

Climate conditions have gotten progressively worse throughout the years (IPCC, 2014b; Kron & Kundzewicz, 2019). Additionally, they can facilitate the growth of harmful organisms by fostering conditions conducive to their survival (Lake, 2017; Semenza et al., 2012; Semenza, 2020; Tran et al., 2017). Numerous studies have examined the relationship between natural disasters and infectious disease outbreaks. According to studies, the presence of earthquakes and severe precipitation events might result in flooding and the spread of infectious diseases.

Post-event outbreaks of food-borne illness have been detected (Harder-Lauridsen et al., 2013; Nigro et al., 2016). Climate change may possibly be a factor in this seasonal tendency. While it is still necessary to understand how climatic factors can affect pathogen transmission in order to improve sanitation systems, this information is typically not readily available.

Campylobacteriosis is a bacteria-borne infection caused by the *Campylobacter* genus. Despite its low death rate, it is the major cause of zoonosis gastroenteritis in the European Union (EFSA, 2020). A 2005 study established a relationship between observed seasonality in campylobacteriosis and climate change.

Climate change was expected in 2009 to increase campylobacteriosis illnesses by 3%. However, a 2018 study found no clear correlation between climate conditions and infection (Park et al., 2018). In comparison to other food-borne pathogens like *Salmonella* spp., the temperature and risk of campylobacteriosis were inconsistent and weak. Many studies have used different methodologies, which could explain some of the discrepancies found in this analysis. Campylobacteriosis illnesses are well known to be affected by temperature change, but it is not known if this effect would be consistent across regions. It is possible that temperature can affect campylobacteriosis indirectly through the seasonality of fly transmission, as well (Ekdahl et al., 2005; Nichols, 2005), yet the disease's basic drivers remain unknown (Djennad et al., 2019).

14.12.3 Salmonellosis

Salmonellosis is an infection caused by *Salmonella* that is spread through food. The European Union recorded 94.530 instances of this disease in 2017. It is particularly prevalent during the summer months and is considered the second most prevalent food-borne disease, behind polio. The association between ambient temperature and *Salmonella* species has been discovered as a biological explanation for the reported increase in salmonellosis incidence (Lake, 2017). Salmonellosis cases have been connected to climate conditions in Europe and the United States (Kovats et al., 2004; Lake et al., 2009; Jiang et al., 2015).

Each year, vibriosis causes around 80,000 illnesses and 100 deaths in the United States, according to the Centers for Disease Control and Prevention (CDC, 2020).

Vibriosis is a dangerous infection that can result in thousands of illnesses and around 100 deaths each year in the United States. Typically, the sickness is brought on by the ingestion of infected seafood. Although *Vibrio cholerae* is not considered a significant threat to human health, its existence has been connected to climate change (Lindgren et al., 2012); also, changes in surface temperature, precipitation, flooding, and salinity may affect this species' growth (Tirado et al., 2010).

Numerous genetic factors may potentially contribute to *Vibrio parahaemolyticus*'s development and transmission. According to a study, ocean warming is a factor in the growth of this strain.

14.13 CLIMATIC-DRIVEN EMERGING RISKS

Climate change may be a factor in the emergence of new risks and an increase in vulnerability to known hazards (EFSA, 2020). This research demonstrates the critical nature of anticipating and responding to aquatic infections.

It is well established that excessive precipitation and flooding enhance the possibility of epidemics of waterborne illnesses (Guzman et al., 2016). Heavy precipitation can result in waterborne epidemics, compromising critical water supplies and transporting microorganisms (Semenza, 2020). These outbreaks are predicted to grow more prevalent as urbanization of aquatic bodies accelerates. While *Cryptosporidium* parasites are not regarded as a hazard to people, they do represent a risk due to their ease of transmission via water. Pathogens that have been stored in frozen ground for thousands of years are likely to reintroduce themselves into the environment. Temperatures below freezing and the presence of bacteria that interact with metals can also lead to the discharge of hazardous germs (EFSA, 2020). Concerns regarding the potential for human health risks have been raised due to the complexity of the relationships between food contamination, climate change, and food-borne diseases.

14.14 IMPACT OF CLIMATE CHANGE ON FOOD SPOILAGE

The majority of studies on the impact of climate change focus on the implications of food production on human health. The available scientific evidence on this subject is scant. Food rotting is a significant expense in the food sector. It can result in industrial losses of up to 30% and significantly contribute to global food waste. While food security is not discussed in this assessment, it should be noted that food security encompasses safety, quality, and availability. Food supply in some locations is predicted to diminish as the temperature rises (Hoegh-Guldberg et al., 2019). This is because the elements affecting food's shelf life are becoming increasingly crucial in an ever-changing society.

Spoilage is the degradation of a food product's quality. Typically, it is caused by the presence of particular microbial species. The most prevalent cause of food or beverage spoiling is microbiological spoilage. It is caused by certain bacteria species growing in food. These organisms' perceived proliferation can result in sensory rejection of the product (Gram et al., 2002; Koutsoumanis & Nychas, 2000; Koutsoumanis et al., 2006). As is the case with pathogenic bacteria, climate change is projected to exacerbate the danger of food spoiling. In the majority of cases, the growth of spoilage microorganisms can be prevented by following adequate food safety procedures.

14.15 IMPACT OF GLOBAL WARMING ON THE SPOILAGE RISK OF NON-REFRIGERATED PROCESSED FOOD PRODUCTS

Temperatures around the world are expected to rise between 2 and 4.9°C due to climate change, according to the International Panel on Climate Change (IPCC) (IPCC, 2014a). Non-reheated food products may become less microbiologically stable as the temperature rises. Bacterial deterioration may be more likely to occur because of this. Canned foods, beverages, and foods with a high acidity are all included in this category. They are frequently contaminated with *Salmonella* spores, which can resist extreme heat. These thermostable spoilers are primarily constituted of *Bacillus, Geobacillus, Alicyclobacillus, Anoxybacillus, Brevibacillus, Paenibacillus*, and *Moorella* genus bacteria (Andr'e et al., 2013; McClure, 2006). Foods that are not refrigerated are often unstable and require a particular amount of storage time to preserve their microbiological stability. Non-refrigerated food products are generally unstable and require a certain storage time to maintain their microbiological stability (Huang et al., 2015b). For non-refrigerated foods like fruit drinks and evaporated milk, the effects of climate change many studies have been done. Predictive models and temperature data were used to examine the effect of climate change on the likelihood of food rotting. Heat-treated fruit beverages that are microbiologically stable are shown in this example. As early as the 1980s, the Germans began to notice an outbreak of mold in apple juice. Due to its propensity to flourish in acidic settings and its heat-resistant spores (Huang et al ., 2015b) (*Alicyclobacillus acidoterrestris*) generates guaiacol, which has a disagreeable phenolic, medicinal, and off-flavoring impact. The presence of thermophilic spore-forming bacteria in food products such as UHT milk is regarded as the most serious issue confronting the food industry. The great abundance of these organisms and their remarkable heat tolerance have been linked to the deterioration of a variety of food goods.

G. starothermophilus did not develop in UHT milk at the current temperature conditions. However, a simulation showed that the spores would grow at a maximum rate of 107 CFU/liter. This potential issue could result in deterioration owing to acid coagulation, lowering the product's quality. To quantify the influence of climate change on the microbiological stability of food, a quantitative microbiological risk assessment (QMSRA) instrument is required. It is possible to accomplish it in the face of uncertainty and variability. This could result in higher food prices, increased food waste, and a food supply decline. This would necessitate a concerted effort on the part of numerous government and industry stakeholders.

14.16 IMPACT OF INCREASED PRECIPITATION EVENTS AND HUMIDITY ON BULK DRIED FOOD AND CEREAL GRAINS

Apart from climate change's effects, additional factors such as excessive precipitation and humidity might contribute to the fungal rotting of bulk food. While relative humidity will almost certainly remain constant, temperature and precipitation will likely increase. This is because increased temperature and humidity conditions might promote fungal growth. The presence of fungal development during storage can compromise the grain's stability. Additionally, this can result in their spoiling. Additionally, certain crops' post-harvest circumstances can promote fungal growth. Climate change is to blame for this phenomenon. It has been shown to impair the stability of stored grains and increase their susceptibility to deterioration.

Due to the presence of *Fusarium* and *Aspergillus* species, cereal grains are frequently infected with this fungus. Typically, these fungi contaminate maize. Both fungi have been shown to be affected by changes in water activity and temperature. In 2005, the water activity component was found as the most significant growth factor (Samapundo et al., 2005, Samapundo et al., 2007).

Maintaining a close eye on the humidity and temperature of the storage environment can aid in the prevention of fungal deterioration. However, it is still debatable if this system is capable of protecting against fungal contamination. Both kinetic and probabilistic models have been created to measure *Aspergillum flavus* colonial expansion (Astoreca et al., 2012). We have now validated our previous assertions by combining the predictive model BAFC4273 with temperature data.

Temperature data collected in Denmark in 2019 was used to simulate the effect of global warming on mycelium diameter growth. Based on historical temperature data, the first scenario predicted that the mycelium's diameter would not rise above 2 millimeters per year. Under a 3°C temperature increase, the annual mycelium diameter is expected to reach 3.6 mm. Surface dampness, on the other hand, has the potential to significantly increase this growth rate. In this study, larger mycelium was found to be produced when water activity was increased. In the third case, the mycelium was predicted to grow to an 8-millimeter diameter. Annual mycelium growth can be predicted to reach a maximum of 14.5 mm in diameter. Because of climate change, this expansion is expected to accelerate even further. Climate change is expected to worsen the effects of temperature and humidity on food shelf life. Study is needed to determine the factors that affect the shelf life of food products and propose ways to reduce spoilage.

14.16.1 Toxin Detection Methods: Classical Mycotoxin Quantifying/Detecting Technology

Conventional mycotoxin detection techniques are extremely expensive analytical approaches that require highly qualified staff to execute the analyses. Additionally, the analytical apparatus and products employed in these approaches are quite costly and difficult without sacrificing their detection limits, and the samples to be analyzed must be prepared in advance to extract their mycotoxin content (Gacem et al., 2020). Food safety is a global concern that has not been addressed, and in order to improve detection processes, the Association of Official Agricultural Chemists (AOAC) has accredited a variety of official laboratory methodologies to safeguard the customer and to ensure the food is of a high standard of quality.

14.16.2 Analysis of Thin-Layer Compounds

Simple and inexpensive, thin-layer chromatography (TLC) is widely used to detect mycotoxins when detection thresholds are not critical. *Journal of Chromatography A* published a special issue on TLC techniques for mycotoxins in various sample matrices, which included a paper summarizing the current state of the art (Lin et al., 1998). It is true that TLC is an efficient method for testing for multiple mycotoxins at the same time, but this method cannot provide sensitive or accurate results until densitometric studies are carried out. An autospotter can provide highly reproducible and dependable results when applying samples. Purifying the extract prior to spotting with TLC improves the detection sensitivity of TLC without first filtering the extract. A reversed-phase TLC method for quantifying fumonisin B1 in maize at g/g levels has been validated, while a two-dimensional HPTLC method for quantifying ochratoxin A at a concentration of 5 ng/g in green coffee beans has been established (high-performance thin-layer chromatography) (Shephard & Sewram, 2004; Ventura et al., 2005).

14.16.3 Gas Chromatography

Food and feed samples can be tested for trichothecenes using gas chromatography (GC) methods such as flame ionization detection (FID), electron capture detection (ECD), and mass spectrometry, which are the most commonly used quantitative methods (Krska et al., 2001). The purified extract must be pre-derivatized with appropriate reagents to increase the toxicity and volatility of the toxins before it can be applied to a column, and this is done by using charcoal-alumina, Florisil, silica gel, or MycoSep columns to clean them. GC-based trichothecene measurements require improved toxin recovery and measurement precision, according to European Union–funded comparative interlaboratory evaluations. Carryover or memory effects, drifting responses from previous samples, and high repeatability and reproducibility variability (RV) were all observed as a result of matrix compounds that increased the trichothecene response by up to 120% (Pettersson & Aberg, 2002).

14.16.4 High-Performance Liquid Chromatography

The most commonly used technology for detecting significant mycotoxins in food is high-performance liquid chromatography (HPLC) coupled with UV, a diode array detector (DAD), or a fluorescence detector (FD). Ochratoxin A, aflatoxin M1, patulin, deoxynivalenol, and zearalenone can all be accurately and precisely determined using HPLC/FD or HPLC/UV (DAD) (Galván et al., 2022). Specialized labeling reagents for the derivatization of nonfluorescent mycotoxins to fluorescent derivatives have been developed and commercialized due to the sensitivity, repeatability, and selectivity of HPLC/high FD.

14.16.5 Solid Phase Extraction or MycoSep Columns

HPLC techniques for identifying mycotoxins in a variety of foods, such as accuracy, detection, reproducibility, and repeatability, as well as quantification restrictions, have been evaluated in collaborative studies and have established performance criteria. The AOAC International or the European Standardization Committee has adopted an official or standard methodology. Methods for measuring aflatoxins in maize, raw peanuts, and peanut butter (AOAC Official Method 991.31 (Weaver & Trucksess, 2010), aflatoxin B1 and total aflatoxins in peanut butter, pistachios, figs, and paprika (AOAC 999.07 (Trucksess et al ., 2008), and ochratoxin A in barley (2000.03 ; Gazioğlu & Kolak (2015).

Methods for aflatoxin M1 in milk (2000.08; Sani and Nikpooyan, 2013), aflatoxin B1 in baby food (2000.16; Stroka et al.,2001), ochratoxin A in roasted coffee (2000.09; Vergas et al., 2005), ochratoxin A in wine and beer (2001.01; Visconti et al.,2001), fumonisins B1 and B2 in maize flour and maize flakes (2001.04 ; De Girolamo et al., 2010), aflatoxins in animal feed (2003.02 Stroka et al.,2001), and ochratoxin A in green coffee (2004.10 ; Vergas et al 2005) using immunoaffinity column clean- HPLC/immunoaffinity column methods have been validated for measuring deoxynivalenol in cereals and cereal products, zearalenone in barley, maize, wheat flour, polenta and maize-based baby food, aflatoxins in hazelnut paste and ochratoxin A in cocoa powder (Brera et al., 2005; MacDonald et al., 2005; Senyuva et al.,2005; Mahato et al.,2019).

14.16.6 Liquid Chromatography/Mass Spectrometry

Mycotoxins have long been confirmed primarily through the use of liquid chromatography–mass spectrometry (LC-MS/MS). In terms of simultaneously screening, detecting, and quantifying a large number of mycotoxins, LC/MS is the most promising method at this time. Recently, the effectiveness of chromatographic and mass spectrometric methods in the detection of mycotoxin was evaluated (Sforza et al., 2006). All samples were tested for the presence of patulin, aflatoxins, ochratoxin A, zearalenone and its metabolites, trichothecenes, and fumonisins (Mahato et al., 2022; Tsagkaris et al., 2022; Bessaire et al., 2019a). The main type A and B trichothecenes in cereals and cereal-based goods were simultaneously identified using an HPLC-MS/MS tandem mass spectrometry interface and an atmospheric pressure chemical ionization (APCCI) or electro-spray ionization interface (APCI) (Berthiller et al., 2005a). Cheddar and bread were found to contain ochratoxin A, penicillic acid, and mycophenolic acid as well as roquefortine C. The presence of zearalenone and its metabolites in various biological matrices was also determined using HPLC-MS/MS (Kleinova et al., 2002; Kokkonen et al., 2005; Sørensen & Elbaek, 2005). The powerful HPLC-MS/MS technology was used to measure wheat's deoxynivalenol-glucosides, a masked form of deoxynivalenol. Even though they can release their lethal precursors upon hydrolysis, conjugated mycotoxins with more polar molecules, such as glucose, are rarely detected using standard analytical methods (Berthiller et al., 2005b). The LC/MS method's accuracy, precision, and sensitivity vary depending on the ionization technique and the mycotoxin, matrix, and equipment used. Due to matrix effects and ion suppression, LC/MS quantitative analysis of mycotoxins is typically insufficient. Prior to MS detection, extracts must be purified with MycoSep or immunoaffinity columns.

14.16.7 Enzyme-Linked Immunosorbent Assay

Since the late 1970s, immunological techniques have been used to successfully identify mycotoxins (Pestka & Abouzied, 1995; Dey et al., 2022; Adunphatcharaphon et al., 2022). Microtiter plate- or membrane-based enzyme-linked immunosorbent assay (ELISA) for the detection of mycotoxins using monoclonal or polyclonal antibodies against mycotoxins is now commercially available for the detection of mycotoxins in a wide range of food samples. It is common for ELISA to use mycotoxins as a starting point and not require any additional purification steps. Immunoassays are rapid and cost-effective screening assays, notwithstanding their lack of precision at low concentrations (competitive assays) and limited matrices investigated. Conjugate–antibody interactions may be impaired by matrix interference or the presence of structurally identical mycotoxins. This can lead to measurement errors. ELISA kits should only be used for routine testing of matrices that have been thoroughly tested. When contamination levels are close to the legal limit, more robust techniques, such as HPLC with an immunoaffinity column clean up or LC-MS, are needed. In collaboration studies, AOAC International has approved certain immunoassays for measuring aflatoxin B1/total aflatoxins and zearalenone in food and feed samples. However, because these tests were validated at levels well over the legal limits, utilizing ELISA to detect mycotoxins at levels close to the legal limits is inefficient. Methods such as TLC or HPLC are more exact than ELISA.

14.16.8 Emerging Technologies for Mycotoxin Analyses

14.16.8.1 Lateral Flow Devices

This type of immunoassay is based on the interaction between specific antibodies that are immobilized on a membrane strip and antibody-coated colored receptors, such as colloidal gold or latex, that react with the analyte to form a complex. A mycotoxin, for example, competes with a tagged receptor in solution for binding sites on the analyte. Colored analyte-conjugate receptor complexes are formed when an analyte-conjugate is added to the test line membrane. The tagged receptor is recognized by a specific antibody that is covalently attached to the membrane of a control line. When binding occurs, the control line is transformed into a dynamic signal. The analyte is present in sufficient concentrations if the test line does not show any signal at all or is only faintly visible (positive test). The metabolite is not present if the test line signal is completely visible (negative test). Low cost, fast response, and a user-friendly format are just some of the advantages of lateral flow devices (LFDs). Strip tests are useful for on-site detection of environmental and agricultural analytes because of these characteristics. For the detection of ochratoxin A in fungi, a competitive immunoassay based on a one-step LFD was developed. LFDs were used to detect ochratoxin A at low concentrations in grain extracts by using an immunoaffinity column cleaning procedure (Danks et al., 2003; Orlov et al., 2022; Adunphatcharaphon et al., 2022). Some commercially available lateral flow immune chromatographic assays for aflatoxin B1 in maize and peanuts are like (Rosa aflatoxin—Charm, USA; AgraStrip—Romer Labs, USA; Reveal aflatoxin—Neogen, USA), aflatoxin M1 in milk (Rosa aflatoxin M1—Charm, USA), and DON in wheat (RIDA Quick DON—Biopharm), which are used for quantitative estimation.

14.16.8.2 Fluorescence Polarization Immunoassay

Fluorescently labeled antigens and antibodies can be identified using the fluorescence polarization immunoassay (FPIA) method. Since the 1970s, when it was first developed, the method has been used in both human and veterinary diagnostics. There is a difference in fluorescence polarization between a fluorescently labelled antigen and the same antigen coupled to a specific antibody. For the detection of mycotoxins, these assays are still being refined. FPIAs have been used successfully to identify aflatoxins, zearalenone, fumonisins, and deoxynivalenol in solution, despite their low sensitivity and accuracy when applied to grain samples (Maragos & Kim, 2004). A more rapid FPIA for the screening of wheat and derivative products for deoxynivalenol was recently developed (pasta and semolina). The FPIA method was found to be a faster, more cost-effective alternative to more robust chromatographic methods for determining wheat and derivative products' deoxynivalenol than a widely used HPLC/immunoaffinity method (Lippolis et al., 2019; Tittlemier et al., 2022; Lei et al., 2022)

14.16.9 Infrared Spectroscopy

The rapid detection of deoxynivalenol in wheat and maize kernels using near-infrared (NIR) and mid-infrared (MIR) transmittance spectroscopy and attenuated total reflection spectroscopy has been investigated by detecting changes in protein, lipid, and carbohydrate content (Levasseur, 2018; Hossain & Goto, 2014; Shen et al., 2019; Jia et al., 2020; Mishra et al., 2021).Analysis of clusters, principal component analyses, or partial least square regression models were used to identify samples contaminated with more than the European Union's legal limit of 310 ng/g of deoxynivalenol by MIR, and the same techniques were used to predict deoxynivalenol content at levels close to that limit by NIR. NIR reflectance spectroscopy and multivariate statistical techniques were recently used to predict the incidence of fungal infection in maize and the concentration of fumonisin B1 in 280 naturally and artificially infected samples. To monitor post-harvest maize mold contamination and distinguish infected lots from those that went unnoticed, the method was used even though its detection limit was not stated (Berardo et al., 2005).

14.16.10 Capillary Electrophoresis

Using an electrical charge, capillary electrophoresis can separate mycotoxins from other species in an extract. Deoxynivalenol, citrinin, aflatoxins, moniliformin, fumonisins, ochratoxins, roridin A, penicillic acid, zearalenone, and sterigmatocystin capillary electrophoresis procedures have been established using UV/visible detection. Several mycotoxins have been discovered at concentrations comparable to those found in naturally infected food samples recently using fluorescence-based capillary electrophoresis technology. Capillary zone electrophoresis with laser-induced fluorescence was used to determine ochratoxin A in roasted coffee, maize, and sorghum after tandem clean-up (silica and immunoaffinity columns) of extracts for aflatoxin B1 and fumonisin B1 analysis in maize (Corneli & Maragos, 1998). Capillary zone electrophoresis-diode array detection was used to identify moniliformin in maize. As with high-performance liquid chromatography, capillary zone electrophoresis procedures have comparable sensitivity, precision, and accuracy (Maragos, 2004; Colombo & Papetti, 2020).

14.16.11 Fiber-Optic Immunosensors

Fiber-optic immunosensors have been developed to detect fumonisin B1 and aflatoxin B1 in maize. An optical fiber was used to immobilize monoclonal antibodies against fumonisin B1 and evaluate the competition between fumonisin B1 and a fluorescently labelled fumonisin B1 probe for binding to the fiber. The 3.2 g/g detection limit was discovered. To increase the method's sensitivity, an immunoaffinity column was used to clean the extracts (detection limit 0.4 g/g). An assay that makes use of the mycotoxin's inherent fluorescence was used to test for aflatoxin B1. The sensor detected 2 ng/ml of aflatoxin B1 in phosphate-buffered saline solution. The specificity of the assay was reduced due to refractive index-related effects in the presence of organic solvents (Maragos & Thompson, 1999).

When compared to other spectrophotometric or chromatographic methods of mycotoxin detection, the use of biosensors provides improved selectivity, direct detection, and little sample pretreatment at a low cost, mobility, and on-the-field evaluation (Rotariu et al., 2016).

14.16.12 Biosensors

Biological elements, such as antibodies, are combined with physical elements in biosensors to produce a recognition event, which is then converted into an audible, electrical, or optical signal. Sensors that use surface plasmon resonance, screen-printed carbon electrodes, and quartz crystal microbalances can detect mycotoxins.

14.16.13 Competitive Surface Plasmon Resonance-Based Immunoassays

Competing surface plasmon resonance-based immunoassays have all been used to detect aflatoxin B1, zearalenone, ochratoxin A, fumonisin B1, and deoxynivalenol in naturally contaminated matrices (Daly et al., 2000; Dunne et al., 2005; Tüdös et al., 2003; van der Gaag et al., 2003). Changes in the refractive index of a metal sheet can be detected using surface plasmon resonance biosensors, which produce a signal corresponding to changes in mass density. After that, the sensor chip is coated with mycotoxins that have been chemically modified. A sample extract containing the mycotoxin to be detected is mixed with a specific antibody and passed over the sensor surface, where unconjugated antibodies attach to the mycotoxin on the surface. Up to 100 regeneration cycles are possible for the sensor chip's surface, allowing its reuse. All toxins were identified correctly and precisely even at extremely low concentrations. Surface plasmon resonance equipment with four flow cells can be used to identify four mycotoxins in a single experiment (van der Gaag et al., 2003). Using a quartz crystal microbalance biosensor, ochratoxin A can be detected at concentrations of 4 to 50 ng/ml in liquid matrices such as water and juices without the need for a clean-up phase (Visconti & de Girolamo, 2005).

14.16.14 Competitive Electrochemical ELISA Based on Disposable Screen-Printed Carbon Electrodes

For the quantitative determination of ochratoxin A in wheat and wine, disposable screen-printed carbon electrodes have been developed. For the two samples, the detection limit was 0.4 ng/g and 0.9 ng/ml. Cleaning naturally contaminated samples with screen-printed carbon electrodes and HPLC/immunoaffinity column clean-up techniques was investigated by Alarcón et al

(2006). For best results, it's possible that a clean-up technique should be used in conjunction with each of these biosensors.

14.16.15 Molecularly Imprinted Polymers

A cross-linker like divinylbenzene or ethylene glycol dimethacrylate reacts with an analyte such as a mycotoxin or a mimic compound, that is, a "dummy," to produce molecularly imprinted polymers, which are cross-linked polymers. The analyte is removed from the polymer after polymerization, leaving only specific recognition sites within the polymer for analysis. Polymer-based biomimetic recognition elements can bind and rebind to analytes with a similar efficiency to antibody–antigen interactions. Because of their low cost, ease of production, high chemical stability, and long shelf life, molecularly imprinted polymers for mycotoxins hold particular promise. Deoxynivalenol, zearalenone, and ochratoxin A-binding polymers made using molecular imprinting have been found to have a high affinity for their respective antitoxins (Visconti & de Girolamo, 2005; Yu et al., 2008). The use of non-imprinted polymers in chromatographic applications or in the preparation of solid-phase extraction columns for sample clean-up has been documented, despite the fact that these polymers performed similarly to molecularly imprinted polymers in some cases. In order to detect ochratoxin A in wheat, a miniaturized surface plasmon resonance device sensor was recently coated with a molecularly imprinted poly pyrrole layer (Jorn & Lai, 2005).

14.17 MANAGEMENT STRATEGIES

Some of the most common methods for detecting and quantifying mycotoxins include scopy, high-performance liquid chromatography, fluorescence detection, enzyme-linked immunosorbent assay, surface-enhanced Raman spectroscopy (SERS), and SPR (Goud et al., 2018).

When analyzing feeds and foods for mycotoxins (Table 14.7), an extraction solvent is used to extract and remove the toxin, then a clean-up step is performed to remove any extract interference, and finally, a toxin is detected or determined using the appropriate analytical instruments and technologies. Solid-phase extraction and multifunctional or immunoaffinity columns are used to remove mycotoxins from samples at trace levels. Quantitative determination of mycotoxins is frequently performed using a variety of chromatographic techniques, including high-performance liquid chromatography coupled with ultraviolet, fluorescence, or mass spectrometry detection, gas chromatography coupled with electron capture, flame ionization, or MS detection, and thin-layer chromatography. ELISA and membrane-based immunoassays are also commonly used for screening purposes, such as in commercial immunometric assays.

TABLE 14.7   BENEFITS AND DRAWBACKS OF CONVENTIONAL TECHNIQUES FOR THE ANALYSIS OF MYCOTOXINS (SINGH & MEHTA, 2020; RODRIGUEZ-MOZAZ ET AL., 2007)

Techniques	Benefits	Drawbacks
Thin-layer chromatography	• It is a cheap and quick method • Useful for screening • Multi-mycotoxin analysis • Sensitive for aflatoxins and ochratoxin A	• Insensibility (for some mycotoxins) • Inaccuracy • Adequate separation may require two-dimensional analysis • Quantitative with a densitometer
Gas chromatography	• Multi-mycotoxin analysis • Sensitivity • Automated (autosampler) • Confirmation (MS detector)	• Expensive gear • Expertise required • Requires derivation • Interfering matrices • Calibration curve • Slack response • Effects from earlier samples • Variation in reproducibility
HPLC	• Better sensitivity • Better selectivity • Better replicability • Automated (autosampler) • Rapid analysis • Official methods	• Expensive gear • Expertise required • Possibly derivatized
LC/MS	• Simultaneous analysis of multiple mycotoxins • Provides confirmation • No need for derivatization • Good sensitivity (LC/MS/MS)	• Extremely costly • Expert assistance is required. • The ionization technique is used to determine sensitivity • Calibration curve with matrix assistance (for quantitative research)
ELISA	• Sample preparation is simple • Low-cost equipment • Exceptional sensitivity • Multiple data sets are analyzed at the same time • Sample screening appropriate • Organic solvents should be used sparingly • Visual evaluation	• Cross-reactivity with mycotoxins that are related • Interference issues in the matrix • False positive/negative results are a possibility • A second LC analysis is necessary for confirmation • Near regulatory limits, critical quantification is required • Semi-quantitative analysis (visual assessment)

(Continued)

TABLE 14.7   BENEFITS AND DRAWBACKS OF CONVENTIONAL TECHNIQUES FOR THE ANALYSIS OF MYCOTOXINS (SINGH & MEHTA, 2020; RODRIGUEZ-MOZAZ ET AL., 2007)   (CONTINUED)

Techniques	Benefits	Drawbacks
Rapid tests	• Easy and quick to do (5–10 min) • There's no need for expensive gear • Organic solvents are used in a limited way • Screening friendly • In situ application	• (Cut-off level) qualitative or semi-quantitative • False positive/negative results are a possibility • Cross-reactivity with mycotoxins that are related • Interference issues in the matrix • Near regulatory limitations, there is a lack of sensitivity

Abbreviations: TLC = thin layer chromatography; GC = gas chromatography; HPLC = high-performance liquid chromatography; LC/MS = liquid chromatography/mass spectrometry; ELISA = enzyme-linked immunosorbent assay; Rapid tests = membrane-based card test; antibody-coated tube; immunodot cup test

For the analysis of mycotoxins, a number of new techniques (Table 14.8) have recently been developed. These include lateral flow devices, infrared spectroscopy, fluorescence polarization immunoassay, fiber-optic immunosensors, molecularly imprinted polymers, capillary electrophoresis, and biosensors based on surface plasmon resonance. Mycotoxins in food, feed, and beverages can be detected using a variety of methods, including traditional and newer techniques.

14.17.1 Rapid Quantification Methods

The high sensitivity, low light bleaching, speed, wealth of chemical data, distinct fingerprint of target compounds, and narrow spectral band of SERS have made it a very promising technology for multiple simultaneous applications since 1977 (Yang et al., 2017). When molecules are attached to or near a prepared noble metal nanostructure surface, such as silver (Ag) or gold (Au), their intrinsically weak Raman signals are amplified using electromagnetic (EM) and chemical (CM) enhancement techniques (Au) (Hassan et al., 2019). EM was found to be more active in SERS than CM. SERS signals can be amplified up to 1015 times when analytes are located in the small uniform gap between metal nanostructures, and morphologically, this gap of the well-ordered nanostructure can also help minimize signal fluctuations during SERS detection. Core-shell nanostructures improve signal more than single compartments (Tegegne et al., 2020). Ultra-sensitive molecular identification is made possible by SERS's exceptional sensitivity (Danesh et al., 2018). Pathogenic bacteria and viruses, colors, newly synthesized harmful compounds, melamine antibiotics, and

TABLE 14.8   PROS AND CONS OF EMERGING TECHNIQUES FOR THE ANALYSIS OF MYCOTOXINS (PASCALE & VISCONTI, 2008)

Methods	Pros	Cons
LFD	• Quick detection • No need for pricy equipment • No clean-up required • Easy to handle • No need of skilled person • Aflatoxin quantification	• Semi-quantitative (visual assessment) • Validation required for additional matrices • Cross-reactivity with related mycotoxins
FPIA	• Quick detection • There is no need for clean-up • DON in wheat has been validated	• Inconsistency of HPLC or ELISA analyses (except for DON) • Some cases are poor sensitivity • Cross-reactivity with related mycotoxins
IR spectroscopy	• Rapid • Non-destructive measurement • No extraction or clean-up • Easy to operate	• Equipment required more expensive • Calibration model must be validated • Requires good knowledge of statistical analysis • Poor sensitivity
Capillary electrophoresis	• Rapid • Use of organic solvents is restricted • Analyte resolution from interfering compounds is good • (Fluorescence capillary electrophoresis) good sensitivity	• Expensive equipment • Clean-up may be required • Alternative to HPLC
Immunosensors/ biosensors (SPR, FOI, QCM, SPCE)	• Rapid • No clean-up procedure	• To improve sensitivity, the extract needs to be cleaned up • Cross-reactivity with mycotoxins that are related • There is no detectable labeling
MIP	• Low cost • Stable • Reusable	• Non-imprinted polymers may exhibit certain properties • Insufficient selectivity

LFD = lateral flow device; FPIA = fluorescence polarization immunoassay; IR = infrared spectroscopy (NIR, near-infrared; MIR, mid-infrared); immunosensors/biosensors = surface plasmon resonance (SPR); fiber-optic immunosensors (FOIs); quartz crystal microbalance (QCM); screen-printed carbon, electrodes (SPCE); MIP = molecularly imprinted polymer hormones are just a few of the things that SERS has been used to monitor in the food science field in the last few years (Li et al., 2017).

14.18 CONCLUSION

Mycotoxins are detrimental to the food supply chain and may jeopardize developing nations' food security even if proper processes and procedures are employed from the beginning to the conclusion of the food chain to ensure the least amount of contamination possible. Among these are the right use of packing and handling equipment. To ensure consumer safety, effective and rapid methods for detecting fungal and mycotoxin contamination in food products are necessary. Climate change is already affecting a variety of sectors of society. It has already begun to have an impact on the output of a variety of crops and livestock. Additionally, it has the potential to impact the safety of food goods at numerous points throughout the supply chain. Climate change has been identified as a possible danger to food safety. This issue has been investigated at several points along the food chain. This chapter discusses the varied implications of climate change on the quality and shelf life of food and feed. Additionally, it discusses the numerous foods that are susceptible to rotting, as well as management, control, and detection approaches.

REFERENCES

Adunphatcharaphon, S., Elliott, C. T., Sooksimuang, T., Charlermroj, R., Petchkongkaew, A., & Karoonuthaisiri, N. (2022). The evolution of multiplex detection of mycotoxins using immunoassay platform technologies. *Journal of Hazardous Materials*, 128706.

Agriopoulou, S., Koliadima, A., Karaiskakis, G., & Kapolos, J. (2016). Kinetic study of aflatoxins' degradation in the presence of ozone. *Food Control*, 61, 221–226.

Agriopoulou, S., Stamatelopoulou, E., & Varzakas, T. (2020). Advances in analysis and detection of major mycotoxins in foods. *Foods*, 9(4), 518.

Ahmed, M. (2020). *Introduction to modern climate change*. Ed. Andrew E. Dessler. Cambridge: Cambridge University Press, 2011, 252 pp. ISBN-10: 0521173159.

Al, W., Orking, G., & Clima, O. (2008). Climate change and food security: A framework document. *FAO Rome*. https://www.fao.org/3/au035e/au035e.pdf

Alarcón, S. H., Palleschi, G., Compagnone, D., Pascale, M., Visconti, A., & Barna-Vetró, I. (2006). Monoclonal antibody based electrochemical immunosensor for the determination of ochratoxin A in wheat. *Talanta*, 69(4), 1031–1037.

Anderson, D. M. (1995). Toxic red tides and harmful algal blooms: A practical challenge in coastal oceanography. *Reviews of Geophysics*, 33(S2), 1189–1200.

André, S., Zuber, F., & Remize, F. (2013). Thermophilic spore-forming bacteria isolated from spoiled canned food and their heat resistance. Results of a French ten-year survey. *International Journal of Food Microbiology*, 165(2), 134–143.

Anfossi, L., Giovannoli, C., & Baggiani, C. (2016). Mycotoxin detection. *Current Opinion in Biotechnology, 37*, 120–126.

AOAC (2000) Official Methods of Analysis. 17th Edition, The Association of Official Analytical Chemists, Gaithersburg, MD, USA. Methods 925.10, 65.17, 974.24, 992.16.

AOAC International Methods Committee. (2005). Ochratoxin A in Barley Immuno-affinity by Column HPLC First Action 2000 Final Action. AOAC Official Method 2000.03.

Astoreca, A., Vaamonde, G., Dalcero, A., Ramos, A. J., & Marín, S. (2012). Modelling the effect of temperature and water activity of *Aspergillus flavus* isolates from corn. *International Journal of Food Microbiology, 156*(1), 60–67.

Bahram, M., Hildebrand, F., Forslund, S.K., Anderson, J. L., Soudzilovskaia, N. A., Bodegom, P. M., Bengtsson-Palme, J., Anslan, S., Coelho, L. P., Harend, H., & Huerta-Cepas, J. (2018). Structure and function of the global topsoil microbiome. *Nature, 560*(7717), 233–237.

Bailey, S. W. (2004). Climate change and decreasing herbicide persistence. *Pest Management Science, 60*(2), 158–162.

Bar-On, Y. M., Phillips, R., & Milo, R. (2018). The biomass distribution on Earth. *Proceedings of the National Academy of Sciences of the United States of America, 115*(25), 6506–6511.

Basaran, P., Basaran-Akgul, N., & Oksuz, L. (2008). Elimination of *Aspergillus parasiticus* from nut surface with low pressure cold plasma (LPCP) treatment. *Food Microbiology, 25*(4), 626–632.

Basu, S., Lehman, S. J., Miller, J. B., Andrews, A. E., Sweeney, C., Gurney, K. R., . . . & Tans, P. P. (2020). Estimating US fossil fuel CO_2 emissions from measurements of 14C in atmospheric CO_2. *Proceedings of the National Academy of Sciences, 117*(24), 13300–13307.

Battilani, P., Rossi, V., Giorni, P., Pietri, A., Gualla, A., Van der Fels-Klerx, H. J., . . . & Brera, C. (2012). Modelling, predicting and mapping the emergence of aflatoxins in cereals in the EU due to climate change. *EFSA Supporting Publications, 9*(1), 223E.

Behrenfeld, M. J. (2014, January 1). *Climate-mediated dance of the plankton. Nature climate change.* Nature Publishing Group, New York, USA.

Bennett, J. W., & Klich, M. (2003). Mycotoxins. *Clinical Microbiology Reviews, 16*(3), 497–516.

Berardo, N., Pisacane, V., Battilani, P., Scandolara, A., Pietri, A., & Marocco, A. (2005). Rapid detection of kernel rots and mycotoxins in maize by near-infrared reflectance spectroscopy. *Journal of Agricultural and Food Chemistry, 53*(21), 8128–8134.

Berthiller, F., Dall'Asta, C., Schuhmacher, R., Lemmens, M., Adam, G., & Krska, R. (2005a). Masked mycotoxins: Determination of a deoxynivalenol glucoside in artificially and naturally contaminated wheat by liquid chromatography–tandem mass spectrometry. *Journal of Agricultural and Food Chemistry, 53*(9), 3421–3425.

Berthiller, F., Krska, R., Dall'Asta, C., Lemmens, M., Adam, G., & Schuhmacher, R. (2005b). Determination of DON-3-Glucoside in artificially and naturally contaminated wheat with LC-MS/MS. *Mycotoxin Research, 21*(3), 205–208.

Bessaire, T., Mujahid, C., Mottier, P., & Desmarchelier, A. (2019a). Multiple mycotoxins determination in food by LC-MS/MS: An international collaborative study. *Toxins, 11*(11), 658.

Bessaire, T., Perrin, I., Tarres, A., Bebius, A., Reding, F., & Theurillat, V. (2019b). Mycotoxins in green coffee: Occurrence and risk assessment. *Food Control, 96*, 59–67.

Binder, E. M., Tan, L. M., Chin, L. J., Handl, J., & Richard, J. (2007). Worldwide occurrence of mycotoxins in commodities, feeds and feed ingredients. *Animal Feed Science and Technology, 137*(3–4), 265–282.

Böhringer, C., Peterson, S., Rutherford, T. F., Schneider, J., & Winkler, M. (2021). Climate policies after Paris: Pledge, trade and recycle: Insights from the 36th Energy Modeling Forum Study (EMF36). *Energy Economics, 103,* 105471.

Bottalico, A., & Perrone, G. (2002). Toxigenic *Fusarium* species and mycotoxins associated with head blight in small-grain cereals in Europe. In *Mycotoxins in plant disease* (pp. 611–624). Springer, Dordrecht.

Brera, C., Soriano, J. M., Debegnach, F., & Miraglia, M. (2005). Exposure assessment to ochratoxin A from the consumption of Italian and Hungarian wines. *Microchemical Journal, 79*(1–2), 109–113.

Bryden, W. L. (2007). Poultry production problems associated with low level mycotoxin intake. *Poultry Science, 94*(6), 333–340.

Bryden, W. L. (2012). Mycotoxin contamination of the feed supply chain: Implications for animal productivity and feed security. *Animal Feed Science and Technology, 173*(1–2), 134–158.

Brzonkalik, K., Hümmer, D., Syldatk, C., & Neumann, A. (2012). Influence of pH and carbon to nitrogen ratio on mycotoxin production by *Alternaria alternata* in submerged cultivation. *AMB Express, 2*(1), 1–8.

Cavicchioli, R., Ripple, W. J., Timmis, K. N., Azam, F., Bakken, L. R., Baylis, M., Behrenfeld, M. J., Boetius, A., Boyd, P. W., Classen, A.T., Crowther, T. W., Cavicchioli, R., Ripple, W. J., Timmis, K. N., Azam, F., Bakken, L. R., Baylis, M., . . . & Webster, N. S. (2019). Scientists' warning to humanity: Microorganisms and climate change. *Nature Reviews Microbiology, 17*(9), 569–586.

CDC (Centers for Disease Control and Prevention). (2020). *Vibrio species causing Vibriosis | Vibrio illness (Vibriosis) | CDC.* Retrieved October 14, 2020, from www.cdc.gov/vibrio/index.html.

Chakraborty, S., Liu, C.J., Mitter, V., Scott, J. B., Akinsanmi, O.A., Ali, S., Dill-Macky, R., Nicol, J., Backhouse, D., & Simpfendorfer, S. (2006). Pathogen population structure and epidemiology are keys to wheat crown rot and *Fusarium* head blight management. *Australasian Plant Pathology, 35*(6), 643–655.

Chakraborty, S., Murray, G. M., Magarey, P. A., Yonow, T., O'Brien, R. G., Croft, B. J., Barbetti, M. J., Sivasithamparam, K., Old, K. M., Dudzinski, M. J., & Sutherst, R. W. (1998). Potential impact of climate change on plant diseases of economic significance to Australia. *Australasian Plant Pathology, 27*(1), 15–35.

Chhaya, R. S., O'Brien, J., & Cummins, E. (2021). Feed to fork risk assessment of mycotoxins under climate change influences-recent developments. *Trends in Food Science & Technology.* https://doi.org/10.1016/j.tifs.2021.07.040

Colombo, R., & Papetti, A. (2020). Pre-concentration and analysis of mycotoxins in food samples by capillary electrophoresis. *Molecules, 25*(15), 3441.

Corneli, S., & Maragos, C. M. (1998). Capillary electrophoresis with laser-induced fluorescence: Method for the mycotoxin ochratoxin A. *Journal of Agricultural and Food Chemistry, 46*(8), 3162–3165.

Cotty, P. J. (1994). Influence of field application of an atoxigenic strain of *Aspergillus flavus* on the populations of *A. flavus* infecting cotton bolls and on the aflatoxin content of cottonseed. *Phytopathology, 84*(11), 1270–1277.

Council for Agricultural Science. (2003). *Mycotoxins: Risks in plant, animal, and human systems* (No. 139). https://scholar.google.com/scholar?q=Mycotoxins:%20Risks%20 in%20Plant,%20Animal%20and%20Human%20Systems,%20Task%20Force%20 Report%20N%20139

Cousins, M., Sargeant, J. M., Fisman, D., & Greer, A. L. (2019). Modelling the transmission dynamics of *Campylobacter* in Ontario, Canada, assuming house flies, *Musca domestica,* are a mechanical vector of disease transmission. *Royal Society Open Science, 6*(2), Article 181394.

Cowger, C., & Sutton, A. L. (2005). The southeastern US *Fusarium* head blight epidemic of 2003. *Plant Health Progress, 6*(1), 4.

Dai, Z., Su, W., Chen, H., Barberán, A., Zhao, H., Yu, M., Yu, L., Brookes, P.C., Schadt, C.W., Chang, S.X., & Xu, J. (2018). Long-term nitrogen fertilization decreases bacterial diversity and favors the growth of Actinobacteria and Proteobacteria in agroecosystems across the globe. *Global Change Biology, 24*(8), 3452–3461.

Daly, S. J., Keating, G. J., Dillon, P. P., Manning, B. M., O'Kennedy, R., Lee, H. A., & Morgan, M. R. (2000). Development of surface plasmon resonance-based immunoassay for aflatoxin B1. *Journal of Agricultural and Food Chemistry, 48*(11), 5097–5104.

Danesh, N. M., Bostan, H. B., Abnous, K., Ramezani, M., Youssefi, K., Taghdisi, S. M., & Karimi, G. (2018). Ultrasensitive detection of aflatoxin B1 and its major metabolite aflatoxin M1 using aptasensors: A review. *TrAC Trends in Analytical Chemistry, 99,* 117–128.

Danks, C., Ostoja-Starzewska, S., Flint, J., & Banks, J. N. (2003). Central science laboratory, Sand Hutton, York Y041 1LZ, UK. *Mycotoxins in Food Production Systems: University of Bath, 25–27 June 2003, 68,* 21.

Daou, R., Joubrane, K., Maroun, R. G., Khabbaz, L. R., Ismail, A., & El Khoury, A. (2021). Mycotoxins: Factors influencing production and control strategies. *AIMS Agriculture and Food, 6*(1), 416–447.

Dasan, B. G., Boyaci, I. H., & Mutlu, M. (2017). Nonthermal plasma treatment of *Aspergillus* spp. spores on hazelnuts in an atmospheric pressure fluidized bed plasma system: Impact of process parameters and surveillance of the residual viability of spores. *Journal of Food Engineering, 196,* 139–149.

De Girolamo, A., Fauw, D.P.D., Sizoo, E., Van Egmond, H., Gambacorta, L., Bouten, K., Stroka, J., Visconti, A. and Solfrizzo, M. (2010). Determination of fumonisins B1 and B2 in maize-based baby food products by HPLC with fluorimetric detection after immunoaffinity column clean-up. *World Mycotoxin Journal, 3*(2), 135–146.

Deng, L. Z., Tao, Y., Mujumdar, A. S., Pan, Z., Chen, C., Yang, X. H., . . . & Xiao, H. W. (2020a). Recent advances in non-thermal decontamination technologies for microorganisms and mycotoxins in low-moisture foods. *Trends in Food Science & Technology, 106,* 104–112.

Deng, L. Z., Mujumdar, A. S., Pan, Z., Vidyarthi, S. K., Xu, J., Zielinska, M., & Xiao, H. W. (2020b). Emerging chemical and physical disinfection technologies of fruits and vegetables: A comprehensive review. *Critical Reviews in Food Science and Nutrition, 60*(15), 2481–2508.

Deng, L. Z., Sutar, P. P., Mujumdar, A. S., Tao, Y., Pan, Z., Liu, Y. H., & Xiao, H. W. (2021). Thermal decontamination technologies for microorganisms and mycotoxins in low-moisture foods. *Annual Review of Food Science and Technology, 12,* 287–305.

Dey, D.K., Kang, J.I., Bajpai, V.K., Kim, K., Lee, H., Sonwal, S., Simal-Gandara, J., Xiao, J., Ali, S., Huh, Y.S. and Han, Y.K., 2022. Mycotoxins in food and feed: toxicity, preventive challenges, and advanced detection techniques for associated diseases. *Critical Reviews in Food Science and Nutrition*, 1–22.

Dix, N. J., & Webster, J. (1995). Fungi of extreme environments. In *Fungal ecology* (pp. 322–340). Springer, Dordrecht.

Djennad, A., Lo Iacono, G., Sarran, C., Lane, C., Elson, R., Höser, C., Lake, I. R., Colón-González, F. J., Kovats, S., Semenza, J. C., & Bailey, T. C. (2019). Seasonality and the effects of weather on Campylobacter infections. *BMC Infectious Diseases, 19*(1), 1–10.

Donatelli, M., Magarey, R. D., Bregaglio, S., Willocquet, L., Whish, J. P., & Savary, S. (2017). Modelling the impacts of pests and diseases on agricultural systems. *Agricultural Systems, 155*, 213–224.

Doohan, F. M., Brennan, J., & Cooke, B. M. (2003). Influence of climatic factors on *Fusarium* species pathogenic to cereals. In *Epidemiology of mycotoxin producing fungi* (pp. 755–768). Springer, Dordrecht.

Dorner, J. W. (2004). Biological control of aflatoxin contamination of crops. *Journal of Toxicology: Toxin Reviews, 23*(2–3), 425–450.

Dunne, L., Daly, S., Baxter, A., Haughey, S., & O'Kennedy, R. (2005). Surface plasmon resonance-based immunoassay for the detection of aflatoxin B1 using single-chain antibody fragments. *Spectroscopy Letters, 38*(3), 229–245.

Duran, R., Cary, J. W., & Calvo, A. M. (2010). Role of the osmotic stress regulatory pathway in morphogenesis and secondary metabolism in filamentous fungi. *Toxins, 2*(4), 367–381.

EFSA (European Food Safety Authority), Maggiore A, Afonso A, Barrucci F, De Sanctis G, 2020. *Climate change as a driver of emerging risks for food and feed safety, plant, animal health and nutritional quality*. EFSA supporting publication, Parma, Italy, 2020:EN-1881. 146pp. doi:10.2903/sp.efsa.2020.EN-1881

Ekdahl, K., Normann, B., & Andersson, Y. (2005). Could flies explain the elusive epidemiology of campylobacteriosis? *BMC Infectious Diseases, 5*(1), 1–4.

El Khoury, A., Atoui, A., Rizk, T., Lteif, R., Kallassy, M., & Lebrihi, A. (2011). Differentiation between *Aspergillus flavus* and *Aspergillus parasiticus* from pure culture and aflatoxin-contaminated grapes using PCR-RFLP analysis of aflR-aflJ intergenic spacer. *Journal of Food Science, 76*(4), M247–M253.

Epstein, P. R., & Mills, E. (2005). *Climate change futures: health, ecological and economic dimensions*. The Center for Health and the Global Environment, Harvard Medical School. https://scholar.google.com/scholar_url?url=http://lib.riskreductionafrica. org/bitstream/handle/123456789/583/6030%2520-%2520Climate%2520change%2 520futures.%2520Health%2520ecological%2520and%2520economic%2520dimens ions.pdf%3Fsequence%3D1%26isAllowed%3Dy&hl=en&sa=T&oi=gsb-gga&ct=res &cd=0&d=10250886679078081972&ei=y0iIY_vuA-qK6rQP7ZGx8A4&scisig=AAGB fm2l4a8ET4zt5eK9GZTxFbMzb05dhQ

Erdner, D. L., Dyble, J., Parsons, M. L., Stevens, R.C., Hubbard, K. A., Wrabel, M. L., Moore, S.K., Lefebvre, K. A., Anderson, D. M., Bienfang, P., & Bidigare, R. R. (2008). Centers for Oceans and Human Health: A unified approach to the challenge of harmful algal blooms. *Environmental Health, 7*(2), 1–17.

Esker, P. D., Savary, S., & McRoberts, N. (2012). Crop loss analysis and global food supply: Focusing now on required harvests. *CAB Reviews: Perspectives in Agriculture, Veterinary Science, Nutrition and Natural Resources, 7*(052), 1–14.

Eskola, M., Kos, G., Elliott, C. T., Hajslova, J., Mayar, S., & Krska, R. (2020). Worldwide contamination of food-crops with mycotoxins: Validity of the widely cited 'FAO estimate' of 25. *Critical Reviews in Food Science and Nutrition, 60*(16), 2773–2789.

European Norm (2006). Commission regulation (EC) No 1881/2006 of 19 December. Setting maximum levels for certain contaminants in foodstuffs. *Official Journal of the European Union, 49*, 5–25.

FAO. (2008). *Climate change: Implications for food safety.* Retrieved October 21, 2020, from www.fao.org/3/i0195e/i0195e00.htm.

FAO. (2020). The State of Food and Agriculture 2020. Overcoming water challenges in agriculture. Rome. https://doi.org/10.4060/cb1447en, https://www.fao.org/3/cb1447en/cb1447en.pdf.

Fernández, H., Arévalo, F. J., Granero, A. M., Robledo, S. N., Nieto, C. H. D., Riberi, W. I., & Zon, M. A. (2017). Electrochemical biosensors for the determination of toxic substances related to food safety developed in South America: Mycotoxins and herbicides. *Chemosensors, 5*(3), 23.

Fleming, L. E., Broad, K., Clement, A., Dewailly, E., Elmir, S., Knap, A., Pomponi, S. A., Smith, S., Gabriele, H. S., & Walsh, P. (2006). Oceans and human health: Emerging public health risks in the marine environment. *Marine Pollution Bulletin, 53*(10–12), 545–560.

Fu, B., Gasser, T., Li, B., Tao, S., Ciais, P., Piao, S., . . . & Xu, J. (2020). Short-lived climate forcers have long-term climate impacts via the carbon–climate feedback. *Nature Climate Change, 10*(9), 851–855.

Gacem, M. A., Ould El Hadj-Khelil, A., Boudjemaa, B., & Gacem, H. (2020). Mycotoxins occurrence, toxicity and detection methods. In *Sustainable Agriculture Reviews 40* (pp. 1–42). Springer, Cham.

Galván, A. I., de Guía Córdoba, M., Rodríguez, A., Martín, A., López-Corrales, M., Ruiz-Moyano, S., & Serradilla, M. J. (2022). Evaluation of fungal hazards associated with dried fig processing. *International Journal of Food Microbiology*, 109541.

Gambacorta, L., El Darra, N., Fakhoury, R., Logrieco, A. F., & Solfrizzo, M. (2019). Incidence and levels of *Alternaria* mycotoxins in spices and herbs produced worldwide and commercialized in Lebanon. *Food Control, 106*, 106724.

Gazioğlu, I., & Kolak, U. (2015). Method validation for the quantitative analysis of aflatoxins (B1, B2, G1, and G2) and ochratoxin A in processed cereal-based foods by HPLC with fluorescence detection. *Journal of AOAC International, 98*(4), 939–945.

Geisen, R., Touhami, N., & Schmidt-Heydt, M. (2017). Mycotoxins as adaptation factors to food related environments. *Current Opinion in Food Science, 17*, 1–8.

Giorgi, F., & Lionello, P. (2008). Climate change projections for the Mediterranean region. *Global and Planetary Change, 63*(2–3), 90–104.

Giorni, P., Battilani, P., & Magan, N. (2008). Effect of solute and matric potential on in vitro growth and sporulation of strains from a new population of *Aspergillus flavus* isolated in Italy. *Fungal Ecology, 1*(2–3), 102–106.

Godfray, H. C. J., Beddington, J. R., Crute, I. R., Haddad, L., Lawrence, D., Muir, J. F., Pretty, J., Robinson, S., Thomas, S. M., & Toulmin, C. (2010). Food security: The challenge of feeding 9 billion people. *Science, 327*(5967), 812–818.

Goldman, G. H., & Osmani, S. A. (Eds.). (2007). The Aspergilli: genomics, medical aspects, biotechnology, and research methods. CRC press.

Golob, P. (2007). On-farm mycotoxin control in food and feed grain (Vol. 1). FAO, Rome (Italy). Animal Production and Health Div.

Gonçalves, B. L., Coppa, C. F. S. C., Neeff, D. V. D., Corassin, C. H., & Oliveira, C. A. F. (2019). Mycotoxins in fruits and fruit-based products: Occurrence and methods for decontamination. Toxin Reviews, 38(4), 263–272.

Goncalves, C., Mischke, C., & Stroka, J. (2020). Determination of deoxynivalenol and its major conjugates in cereals using an organic solvent-free extraction and IAC cleanup coupled in-line with HPLC-PCD-FLD. Food Additives and Contaminants—Part A Chemistry, Analysis, Control, Exposure and Risk Assessment, 37(10), 1765–1776. https://doi.org/10.1080/19440049.2020.1800829

Goud, K. Y., Kailasa, S. K., Kumar, V., Tsang, Y. F., Lee, S. E., Gobi, K. V., & Kim, K. H. (2018). Progress on nanostructured electrochemical sensors and their recognition elements for detection of mycotoxins: A review. Biosensors and Bioelectronics, 121, 205–222.

Gram, L., Ravn, L., Rasch, M., Bruhn, J. B., Christensen, A. B., & Givskov, M. (2002). Food spoilage—Interactions between food spoilage bacteria. International Journal of Food Microbiology, 78(1–2), 79–97.

Guzman Herrador, B., De Blasio, B. F., Carlander, A., Ethelberg, S., Hygen, H. O., Kuusi, M., et al. (2016). Association between heavy precipitation events and waterborne outbreaks in four Nordic countries, 1992–2012. Journal of Water and Health. IWA Publishing, 14(6), 1019–1027.

Hall, G. V., D'Souza, R. M., & Kirk, M. D. (2002). Foodborne disease in the new millennium. Out of the Frying pan and Into the Fire?, 177, 614–618.

Hamad, H., Alma, M., Ismael, H., & Göçeri, A. (2014). The effect of some sugars on the growth of Aspergillus niger. KSÜ DoğaBilimleriDergisi, 17(4), 7–11.

Harder-Lauridsen, N. M., Kuhn, K. G., Erichsen, A. C., Mølbak, K., & Ethelberg, S. (2013). Gastrointestinal illness among triathletes swimming in non-polluted versus polluted seawater affected by heavy rainfall, Denmark, 2010–2011. PLoS ONE, 8(11), Article e78371.

Hassan, M. M., Li, H., Ahmad, W., Zareef, M., Wang, J., Xie, S., ...& Chen, Q. (2019). Au@Ag nanostructure based SERS substrate for simultaneous determination of pesticides residue in tea via solid phase extraction coupled multivariate calibration. LWT, 105, 290–297.

Hellin, J., Shiferaw, B., Cairns, J. E., Reynolds, M., Ortiz-Monasterio, I., Banziger, M., Sonder, K., & La Rovere, R. (2012). Climate change and food security in the developing world: Potential of maize and wheat research to expand options for adaptation and mitigation. Journal of Development and Agricultural Economics, 4(12), 311–321.

Hoegh-Guldberg, O., Jacob, D., Taylor, M., Guillén Bolãnos, T., Bindi, M., Brown, S., et al. (2019). The human imperative of stabilizing global climate change at 1.5°C. Science. Sciencemag.orgPaperpile, 30.

Hojnik, N., Cvelbar, U., Tavčar-Kalcher, G., Walsh, J. L., & Križaj, I. (2017). Mycotoxin decontamination of food: Cold atmospheric pressure plasma versus "classic" decontamination. Toxins, 9(5), 151.

Horwitz, W. (2000). Association of Official Analytical Chemist: Gaithersburg. MD, USA.

Hossain, M. Z., & Goto, T. (2014). Near-and mid-infrared spectroscopy as efficient tools for detection of fungal and mycotoxin contamination in agricultural commodities. World Mycotoxin Journal, 7(4), 507–515.

Hsieh, D. P. (1988). Potential human health hazards of mycotoxins. In *Mycotoxins and phytotoxins* (pp. 69–80). Elsevier, Amsterdam, The Netherlands.

Huang, W., Shang, Y., Chen, P., Gao, Q., & Wang, C. (2015a). MrpacC regulates sporulation, insect cuticle penetration and immune evasion in *Metarhizium robertsii*. *Environmental Microbiology, 17*(4), 994–1008.

Huang, X.-C., Yuan, Y.-H., Guo, C.-F., Gekas, V., & Yue, T.-L. (2015b). *Alicyclobacillus* in the fruit juice industry: Spoilage, detection, and prevention/control. *Food Reviews International, 31*(2), 91–124.

Huertas-Perez JF, Arroyo-Manzanares N, Garcia-Campana AM and Gamiz-Gracia L (2017) Solid phase extraction as sample treatment for the determination of ochratoxin A in food: A review. *Critical Reviews in Food Science and Nutrition*, 57: 3405–3420.

IPCC. (2013). Summary for policymakers. In T. F. Stocker, D. Qin, G. K. Plattner et al. (Eds.), *Climate change 2013: The physical science basis*. Contribution of Working Group I to the Fifth Assessment Report of the Intergovernmental Panel on Climate Change. Cambridge and New York: Cambridge University Press.

IPCC. (2014a). *Climate change 2014: Synthesis report*. Contribution of Working Groups I, II and III to the Fifth Assessment Report of the Intergovernmental Panel on Climate Change. Geneva, Switzerland: Intergovernmental Panel on Climate Change (IPCC).

IPCC. (2014b). Summary for policymakers. In C. B. Field, V. R. Barros, D. J. Dokken, et al. (Eds.), *Climate change 2014: Impacts, adaptation, and vulnerability. Part A: Global and sectoral aspects*. Contribution of Working Group II to the Fifth Assessment Report of the Intergovernmental Panel on Climate Change (pp. 1–32). Cambridge and New York: Cambridge University Press.

Ji, C., Fan, Y., & Zhao, L. (2016). Review on biological degradation of mycotoxins. *Animal Nutrition, 2*(3), 127–133.

Jia, B., Wang, W., Ni, X. Z., Chu, X., Yoon, S. C., & Lawrence, K. C. (2020). Detection of mycotoxins and toxigenic fungi in cereal grains using vibrational spectroscopic techniques: A review. *World Mycotoxin Journal, 13*(2), 163–178.

Jiang, C., Shaw, K. S., Upperman, C. R., Blythe, D., Mitchell, C., Murtugudde, R., et al. (2015). Climate change, extreme events and increased risk of salmonellosis in Maryland, USA: Evidence for coastal vulnerability. *Environment International, 83*, 58–62.

Jing, X., Sanders, N. J., Shi, Y., Chu, H., Classen, A. T., Zhao, K., et al. (2015). The links between ecosystem multifunctionality and above-and belowground biodiversity are mediated by climate. *Nature Communications, 6*(1), 1–8.

Jones, M. W., Andrew, R. M., Peters, G. P., Janssens-Maenhout, G., De-Gol, A. J., Ciais, P., ... & Le Quéré, C. (2021). Gridded fossil CO_2 emissions and related O_2 combustion consistent with national inventories 1959–2018. *Scientific Data, 8*(1), 1–23.

Jorn, C. C., & Lai, E. P. (2005). Interaction of ochratoxin A with molecularly imprinted polypyrrole film on surface plasmon resonance sensor. *Reactive and Functional Polymers, 63*(3), 171–176.

Joubrane, K., Khoury, A. E., Lteif, R., Rizk, T., Kallassy, M., Hilan, C., & Maroun, R. (2011). Occurrence of aflatoxin B1 and ochratoxin A in Lebanese cultivated wheat. *Mycotoxin Research, 27*(4), 249–257.

Joubrane, K., Mnayer, D., El Khoury, A., El Khoury, A., & Awad, E. (2020). Co-occurrence of aflatoxin B1 and ochratoxin A in Lebanese stored wheat. *Journal of Food Protection, 83*(9), 1547–1552.

Kagot, V., Okoth, S., De Boevre, M., & De Saeger, S. (2019). Biocontrol of *Aspergillus* and *Fusarium* mycotoxins in Africa: Benefits and limitations. *Toxins, 11*(2), 109.

Kamil, O. H., Lupuliasa, D., Draganescu, D., & Vlaia, L. (2011). Interrelations of drying heat and survival of different fungal spores within the tablets formulation. *Studia Universitatis"VasileGoldis"Arad. SeriaStiinteleVietii (Life Sciences Series), 21*(2), 339.

Kamle, M., Mahato, D. K., Devi, S., Lee, K. E., Kang, S. G., & Kumar, P. (2019). Fumonisins: Impact on agriculture, food, and human health and their management strategies. *Toxins, 11*(6), 328.

Kamle, M., Mahato, D.K., Gupta, A., Pandhi, S., Sharma, N., Sharma, B., Mishra, S., Arora, S., Selvakumar, R., Saurabh, V., & Dhakane-Lad, J. (2022). Citrinin mycotoxin contamination in food and feed: Impact on agriculture, human health, and detection and management strategies. *Toxins, 14*(2), 85.

Karlovsky, P., Suman, M., Berthiller, F., De Meester, J., Eisenbrand, G., Perrin, I., . . . & Dussort, P. (2016). Impact of food processing and detoxification treatments on mycotoxin contamination. *Mycotoxin Research, 32*(4), 179–205.

Kebede, H., Liu, X., Jin, J., & Xing, F. (2020). Current status of major mycotoxins contamination in food and feed in Africa. *Food Control, 110*, 106975.

Khodaei, D., Javanmardi, F., & Khaneghah, A. M. (2021). The global overview of the occurrence of mycotoxins in cereals: A three-year survey. *Current Opinion in Food Science, 39*, 36–42.

Kleinova, M., Zöllner, P., Kahlbacher, H., Hochsteiner, W., & Lindner, W. (2002). Metabolic profiles of the mycotoxin zearalenone and of the growth promoter zeranol in urine, liver, and muscle of heifers. *Journal of Agricultural and Food Chemistry, 50*(17), 4769–4776.

Kluczkovski, A. M. (2019). Fungal and mycotoxin problems in the nut industry. *Current Opinion in Food Science, 29*, 56–63.

Kochiieru, Y., Mankevičienė, A., Cesevičienė, J., Semaškienė, R., Dabkevičius, Z., & Janavičienė, S. (2020). The influence of harvesting time and meteorological conditions on the occurrence of *Fusarium* species and mycotoxin contamination of spring cereals. *Journal of the Science of Food and Agriculture, 100*(7), 2999–3006.

Kokkonen, M., Jestoi, M., & Rizzo, A. (2005). The effect of substrate on mycotoxin production of selected *Penicillium* strains. *International Journal of Food Microbiology, 99*(2), 207–214.

Koutsoumanis, K., & Nychas, G.-J. E. (2000). Application of a systematic experimental procedure to develop a microbial model for rapid fish shelf life predictions. *International Journal of Food Microbiology, 60*(2–3), 171–184.

Koutsoumanis, K., Stamatiou, A., Skandamis, P., & Nychas, G.-J. E. (2006). Development of a microbial model for the combined effect of temperature and pH on spoilage of ground meat, and validation of the model under dynamic temperature conditions. *Applied and Environmental Microbiology, 72*(1), 124–134.

Kovats, R. S., Edwards, S. J., Hajat, S., Armstrong, B. G., Ebi, K. L., Menne, B., et al. (2004). The effect of temperature on food poisoning: A time-series analysis of salmonellosis in ten European countries. *Epidemiology and Infection, 132*(3), 443–453.

Kron, W., Löw, P., & Kundzewicz, Z. W. (2019). Changes in risk of extreme weather events in Europe. *Environmental Science & Policy, 100*, 74–83.

Krska, R., Baumgartner, S., & Josephs, R. (2001). The state-of-the-art in the analysis of type-A and-B trichothecene mycotoxins in cereals. *Fresenius' Journal of Analytical Chemistry, 371*(3), 285–299.

Krueger, J., Biedrzycki, P., & Hoverter, S. P. (2015). Human health impacts of climate change: Implications for the practice and law of public health. *Journal of Law, Medicine & Ethics, 43*(S1), 79–82.

Kumar, P., Mahato, D. K., Kamle, M., Mohanta, T. K., & Kang, S. G. (2017). Aflatoxins: A global concern for food safety, human health and their management. *Frontiers in Microbiology, 7,* 2170.

Kumar, P., Mahato, D. K., Sharma, B., Borah, R., Haque, S., Mahmud, M. C., . . . & Bui, S. (2020). Ochratoxins in food and feed: Occurrence and its impact on human health and management strategies. *Toxicon, 187,* 151–162.

Lake, I. R. (2017). Food-borne disease and climate change in the United Kingdom. *Environmental Health,* 16(1), 53–59.

Lake, I. R., Gillespie, I. A., Bentham, G., Nichols, G. L., Lane, C., Adak, G. K., et al. (2009). A re-evaluation of the impact of temperature and climate change on foodborne illness. *Epidemiology and Infection, 137*(11), 1538–1547.

Lal, R. (2013). Food security in a changing climate. *Ecohydrology & Hydrobiology, 13*(1), 8–21.

Lee, H., Kim, J. E., Chung, M. S., & Min, S. C. (2015). Cold plasma treatment for the microbiological safety of cabbage, lettuce, and dried figs. *Food Microbiology, 51,* 74–80.

Lei, H., Wang, Z., Eremin, S. A., & Liu, Z. (2022). Application of antibody and immunoassay for food safety. *Foods, 11*(6), 826.

Leslie, J. F., Bandyopadhyay, R., & Visconti, A. (Eds.). (2008). *Mycotoxins: Detection methods, management, public health and agricultural trade.* Centre for Agriculture and Bioscience International.

Levasseur-Garcia, C. (2018). Updated overview of infrared spectroscopy methods for detecting mycotoxins on cereals (corn, wheat, and barley). *Toxins, 10*(1), 38.

Li, Q., Lu, Z., Tan, X., Xiao, X., Wang, P., Wu, L., . . . & Han, H. (2017). Ultrasensitive detection of aflatoxin B1 by SERS aptasensor based on exonuclease-assisted recycling amplification. *Biosensors and Bioelectronics, 97,* 59–64.

Lima, C.M.G., Costa, H.R.D., Pagnossa, J.P., Rollemberg, N.D.C., SILVA, J.F.D., Dalla Nora, F.M., Batiha, G.E.S. and Verruck, S., 2021. Influence of grains postharvest conditions on mycotoxins occurrence in milk and dairy products. *Food Science and Technology,* v42, e16421, 2022

Lin, G., Zhou, K. Y., Zhao, X. G., Wang, Z. T., & But, P. P. H. (1998). Determination of hepatotoxic pyrrolizidine alkaloids by on-line high performance liquid chromatography mass spectrometry with an electrospray interface. *Rapid Communications in Mass Spectrometry, 12*(20), 1445–1456.

Lindgren, E., Andersson, Y., Suk, J. E., Sudre, B., & Semenza, J. C. (2012). Public health: Monitoring EU emerging infectious disease risk due to climate change. *Science. American Association for the Advancement of Science, 336*(6080), 418–419.

Lippolis, V., Porricelli, A. C., Mancini, E., Ciasca, B., Lattanzio, V. M., De Girolamo, A., Maragos, C. M., McCormick, S., Li, P., Logrieco, A. F., & Pascale, M. (2019). Fluorescence polarization immunoassay for the determination of T-2 and HT-2 toxins and their glucosides in wheat. *Toxins, 11*(7), 380.

Liu, J., Sun, L., Zhang, N., Zhang, J., Guo, J., Li, C., Rajput, S.A. and Qi, D., 2016. Effects of nutrients in substrates of different grains on aflatoxin B1 production by Aspergillus flavus. *BioMed Research International.* Volume 2016, Article ID 7232858, 10. http://dx.doi.org/10.1155/2016/7232858

Loprieno, N. (1975). International Agency for Research on Cancer (IARC) monographs on the evaluation of carcinogenic risk of chemicals to man: "Relevance of data on mutagenicity". *Mutation Research, 31*(3), 210.

MacFadden, D. R., McGough, S. F., Fisman, D., Santillana, M., & Brownstein, J. S. (2018). Antibiotic resistance increases with local temperature. *Nature Climate Change, 8*(6), 510–514.

Maestre, F. T., Delgado-Baquerizo, M., Jeffries, T. C., Eldridge, D. J., Ochoa, V., Gozalo, B., et al. (2015). Increasing aridity reduces soil microbial diversity and abundance in global drylands. *Proceedings of the National Academy of Sciences of the United States of America, 112*(51), 15684–15689.

Magan, N., & Aldred, D. (2007). Post-harvest control strategies: Minimizing mycotoxins in the food chain. *International Journal of Food Microbiology, 119*(1–2), 131–139.

Magan, N., & Olsen, M. (Eds.). (2004). *Mycotoxins in food: detection and control.* Woodhead Publishing.

Magan, N., Sanchis, V., & Aldred, D. (2004). Role of spoilage fungi in seed deterioration. *Fungal Biotechnology in Agricultural, Food and Environmental Applications, 28*, 311–323.

Magnoli, A. P., Poloni, V. L., & Cavaglieri, L. (2019). Impact of mycotoxin contamination in the animal feed industry. *Current Opinion in Food Science, 29*, 99–108.

Mahato, D. K., Devi, S., Pandhi, S., Sharma, B., Maurya, K. K., Mishra, S., . . . & Kumar, P. (2021). Occurrence, impact on agriculture, human health, and management strategies of zearalenone in food and feed: A review. *Toxins, 13*(2), 92.

Mahato, D. K., Kamle, M., Sharma, B., Pandhi, S., Devi, S., Dhawan, K., . . . & Kumar, P. (2021). Patulin in food: A mycotoxin concern for human health and its management strategies. *Toxicon, 198*, 12–23.

Mahato, D. K., Lee, K. E., Kamle, M., Devi, S., Dewangan, K. N., Kumar, P., & Kang, S. G. (2019). Aflatoxins in food and feed: An overview on prevalence, detection and control strategies. *Frontiers in Microbiology,* 2266.

Mahato, D.K., Pandhi, S., Kamle, M., Gupta, A., Sharma, B., Panda, B.K., Srivastava, S., Kumar, M., Selvakumar, R., Pandey, A.K. and Suthar, P., 2022. Trichothecenes in food and feed: Occurrence, impact on human health and their detection and management strategies. *Toxicon, 208*, 62–77.

Mahuku, G., Nzioki, H. S., Mutegi, C., Kanampiu, F., Narrod, C., & Makumbi, D. (2019). Pre-harvest management is a critical practice for minimizing aflatoxin contamination of maize. *Food Control, 96*, 219–226.

Mannaa, M., & Kim, K. D. (2017a). Influence of temperature and water activity on deleterious fungi and mycotoxin production during grain storage. *Mycobiology, 45*(4), 240–254.

Mannaa, M., & Kim, K. D. (2017b). Control strategies for deleterious grain fungi and mycotoxin production from preharvest to postharvest stages of cereal crops: A review. *Journal of Natural Resources and Life Sciences Education, 25*, 13–27.

Maracchi, G., Sirotenko, O., & Bindi, M. (2005). Impacts of present and future climate variability on agriculture and forestry in the temperate regions: Europe. *Increasing Climate Variability and Change: Reducing the Vulnerability of Agriculture and Forestry,* 117–135.

Maragos, C. M. (2004). Detection of moniliformin in maize using capillary zone electrophoresis. *Food Additives and Contaminants, 21*(8), 803–810.

Maragos, C. M., & Kim, E. K. (2004). Detection of zearalenone and related metabolites by fluorescence polarization immunoassay. *Journal of food protection, 67*(5), 1039–1043.

Maragos, C. M., & Thompson, V. S. (1999). Fiber-optic immunosensor for mycotoxins. *Natural Toxins, 7*(6), 371–376.

Marin, S., Ramos, A. J., Cano-Sancho, G., & Sanchis, V. (2013). Mycotoxins: Occurrence, toxicology, and exposure assessment. *Food and Chemical Toxicology, 60*, 218–237.

Mathur, M., Kumari, A., & Grewal, R. (2020). Physical and functional properties of major foods and oil seeds. *Journal of AgriSearch, 7*(2), 97–103.

McClure, P. J. (2006). Spore-forming bacteria. *Food Spoilage Microorganisms*, 579–623. https://www.cabdirect.org/cabdirect/abstract/20073095401

McDonald, T., Brown, D., Keller, N. P., & Hammond, T. M. (2005). RNA silencing of mycotoxin production in *Aspergillus* and *Fusarium* species. *Molecular Plant-microbe Interactions, 18*(6), 539–545.

McIntyre, K. M., Setzkorn, C., Hepworth, P. J., Morand, S., Morse, A. P., & Baylis, M. (2017). Systematic assessment of the climate sensitivity of important human and domestic animals pathogens in Europe. *Scientific Reports, 7*(1), 1–10

McMullen, M., Bergstrom, G., De Wolf, E., Dill-Macky, R., Hershman, D., Shaner, G., & Van Sanford, D. (2012). A unified effort to fight an enemy of wheat and barley: Fusarium head blight. *Plant Disease, 96*(12), 1712–1728.

McMullen, M., Jones, R., & Gallenberg, D. (1997). Scab of wheat and barley: A re-emerging disease of devastating impact. *Plant Disease, 81*(12), 1340–1348.

Medina, A., Akbar, A., Baazeem, A., Rodriguez, A., & Magan, N. (2017a). Climate change, food security and mycotoxins: Do we know enough?.*Fungal Biology Reviews, 31*(3), 143–154.

Medina, Á., González-Jartín, J. M., & Sainz, M. J. (2017b). Impact of global warming on mycotoxins. *Current Opinion in Food Science, 18*, 76–81.

Medina, Á., Rodríguez, A., & Magan, N. (2015). Climate change and mycotoxigenic fungi: Impacts on mycotoxin production. *Current Opinion in Food Science, 5*, 99–104.

Messerli, B., Grosjean, M., Hofer, T., Núñez, L., & Pfister, C. (2000). From nature-dominated to human-dominated environmental changes. *Quaternary Science Reviews, 19*(1–5), 459–479.

Miedaner, T., Cumagun, C. J. R., & Chakraborty, S. (2008). Population genetics of three important head blight pathogens *Fusarium graminearum*, *F. pseudograminearum* and *F. culmorum*. *Journal of Phytopathology, 156*(3), 129–139.

Mikhaylov, A., Moiseev, N., Aleshin, K., & Burkhardt, T. (2020). Global climate change and greenhouse effect. *Entrepreneurship and Sustainability Issues, 7*(4), 2897.

Milani, J., & Maleki, G. (2014). Effects of processing on mycotoxin stability in cereals. *Journal of the Science of Food and Agriculture, 94*(12), 2372–2375.

Miraglia, M., De Santis, B., & Brera, C. (2008). Climate change: Implications for mycotoxin contamination of foods. *Journal of Biotechnology, 136*, S715.

Miraglia, M., Marvin, H. J. P., Kleter, G. A., Battilani, P., Brera, C., Coni, E., Cubadda, F., Croci, L., De Santis, B., Dekkers, S., & Filippi, L. (2009). Climate change and food safety: An emerging issue with special focus on Europe. *Food and Chemical Toxicology, 47*(5), 1009–1021.

Mishra, G., Panda, B. K., Ramirez, W. A., Jung, H., Singh, C. B., Lee, S. H., & Lee, I. (2021). Research advancements in optical imaging and spectroscopic techniques for non-destructive detection of mold infection and mycotoxins in cereal grains and nuts. *Comprehensive Reviews in Food Science and Food Safety, 20*(5), 4612–4651.

Misiou, O., & Koutsoumanis, K. (2021). Climate change and its implications for food safety and spoilage. *Trends in Food Science & Technology, 126,* 142–152.

Misra, N. N., Yadav, B., Roopesh, M. S., & Jo, C. (2019). Cold plasma for effective fungal and mycotoxin control in foods: Mechanisms, inactivation effects, and applications. *Comprehensive Reviews in Food Science and Food Safety, 18*(1), 106–120.

Moses, J. A., Jayas, D. S., & Alagusundaram, K. (2015). Climate change and its implications on stored food grains. *Agricultural Research, 4*(1), 21–30.

Munkvold, G. (2014). Crop management practices to minimize the risk of mycotoxins contamination in temperate-zone maize. *Mycotoxin Reduction in Grain Chains, 1,* 59–77.

Muriel, P., Downing, T., Hulme, M., Harrington, R., Lawlor, D., Wurr, D., Atkinson, C. J., Cockshull, K. E., Taylor, D. R., Richards, A. J., & Parsons, D. J. (2000). Climate change and agriculture in the United Kingdom. *Climate Change and Agriculture in the United Kingdom.* https://repository.rothamsted.ac.uk/item/8846y/climate-change-and-agriculture-in-the-united-kingdom

Nadeau, K. C., Agache, I., Jutel, M., Annesi Maesano, I., Akdis, M., Sampath, V., . . . & Akdis, C. A. (2022). Climate change: A call to action for the united nations. *Allergy, 77*(4), 1087–1090.

Nganchamung, T., Robson, M. G., & Siriwong, W. (2017). Chemical fertilizer use and acute health effects among chili farmers in UbonRatchathani province, Thailand. *Journal of Health Research, 31*(6), 427–435.

Nganje, W. E., Kaitibie, S., Wilson, W. W., Leistritz, F. L., & Bangsund, D. A. (2004). *Economic impacts of Fusarium head blight in wheat and barley: 1993-2001* (No. 1187-2016-93545). Agribusiness and Applied Economics Report No 538.

Nichols, G. L. (2005). Fly transmission of Campylobacter. *Emerging Infectious Diseases, 11*(3), 361.

Nicolopoulou-Stamati, P., Maipas, S., Kotampasi, C., Stamatis, P., & Hens, L. (2016). Chemical pesticides and human health: the urgent need for a new concept in agriculture. *Frontiers in public health, 4,* 148.

Nigro, G., Bottone, G., Maiorani, D., Trombatore, F., Falasca, S., & Bruno, G. (2016). Pediatric epidemic of *Salmonella enterica* serovar typhimurium in Reboud he area of L'aquila, Italy, four years after a catastrophic earthquake. *International Journal of Environmental Research and Public Health, 13*(5), 475.

Oerke, E. C. (2006). Crop losses to pests. *The Journal of Agricultural Science, 144*(1), 31–43.

Olesen, J. E., & Bindi, M. (2002). Consequences of climate change for European agricultural productivity. *Land Use Policy, 16,* 239–262.

Orlov, A. V., Malkerov, J. A., Novichikhin, D. O., Znoyko, S. L., & Nikitin, P. I. (2022). Express high-sensitive detection of ochratoxin A in food by a lateral flow immunoassay based on magnetic biolabels. *Food Chemistry, 383,* 132427.

Osborne, L. E., & Stein, J. M. (2007). Epidemiology of *Fusarium* head blight on small-grain cereals. *International Journal of Food Microbiology, 119*(1–2), 103–108.

Ouf, S. A., Basher, A. H., & Mohamed, A. A. H. (2015). Inhibitory effect of double atmospheric pressure argon cold plasma on spores and mycotoxin production of *Aspergillus niger* contaminating date palm fruits. *Journal of the Science of Food and Agriculture, 95*(15), 3204–3210.

Özcelik, S., & Özcelik, N. (1990). Interacting effects of time, temperature, pH and simple sugars on biomass and toxic metabolite production by three *Alternaria* spp. *Mycopathologia, 109*(3), 171–175.

Pankaj, S. K., Shi, H., & Keener, K. M. (2018). A review of novel physical and chemical decontamination technologies for aflatoxin in food. *Trends in Food Science & Technology, 71,* 73–83.

Park, B. J., Takatori, K., Sugita-Konishi, Y., Kim, I. H., Lee, M. H., Han, D. W., . . . & Park, J. C. (2007). Degradation of mycotoxins using microwave-induced argon plasma at atmospheric pressure. *Surface and Coatings Technology, 201*(9–11), 5733–5737.

Park, M. S., Park, K. H., & Bahk, G. J. (2018). Combined influence of multiple climatic factors on the incidence of bacterial foodborne diseases. *The Science of the Total Environment, 610–611,* 10–16.

Pascale, M., & Visconti, A. (2008). Overview of detection methods for mycotoxins. *Mycotoxins: Detection Methods, Management, Public Health and Agricultural Trade,* 171–183.

Pascari, X., Ramos, A. J., Marín, S., & Sanchís, V. (2018). Mycotoxins and beer. Impact of beer production process on mycotoxin contamination. A review. *Food Research International, 103,* 121–129.

Paterson, R. R. M., & Lima, N. (2010). How will climate change affect mycotoxins in food? *Food Research International, 43*(7), 1902–1914.

Patriarca, A., & Pinto, V. F. (2017). Prevalence of mycotoxins in foods and decontamination. *Current Opinion in Food Science, 14,* 50–60.

Perdoncini, M. R. F. G., Sereia, M. J., Scopel, F. H. P., Formigoni, M., Rigobello, E. S., Beneti, S. C., & Marques, L. L. M. (2019). Growth of fungal cells and the production of mycotoxins. *Cell Growth, 23.*

Pestka, J., & Abouzied, M. (1995). Immunological assays for mycotoxin detection: Immunoassay applications to food analysis. *Food Technology (Chicago), 49*(2), 120–128.

Pettersson, H., & Aberg, L. (2002). Rapid estimation of deoxynivalenol and *Fusarium* by near infrared spectroscopy. *Journal of Applied Genetics, 43,* 141–144.

Petzoldt, C., & Seaman, A. (2006). Climate change effects on insects and pathogens. *Climate Change and Agriculture: Promoting Practical and Profitable Responses, 3,* 6–16.

Pleadin, J., Frece, J., & Markov, K. (2019). Mycotoxins in food and feed. *Advances in Food and Nutrition Research, 89,* 297–345.

Porter, M., Xie, L., Challinor, A. J., et al. (2014). Food security and food production systems. In C. B. Field, V. R. Barros, D. J. Dokken, et al. (Eds.), *Climate change 2014: Impacts, adaptation, and vulner-ability. Part A: Global and sectoral aspects.* Contribution of Working Group II to the Fifth Assessment Report of the Intergovernmental Panel on Climate Change (pp. 1–82). Cambridge and New York: Cambridge University Press.

Reboud, X., Eychenne, N., Délos, M., & Folcher, L. (2016). Withdrawal of maize protection by herbicides and insecticides increases mycotoxins contamination near maximum thresholds. *Agronomy for Sustainable Development, 36*(3), 1–10.

Rehman, A., Ma, H., Ahmad, M., Irfan, M., Traore, O., & Chandio, A. A. (2021). Towards environmental sustainability: Devolving the influence of carbon dioxide emission to population growth, climate change, Forestry, livestock and crops production in Pakistan. *Ecological Indicators, 125*, 107460.

Reverberi, M., Ricelli, A., Zjalic, S., Fabbri, A. A., & Fanelli, C. (2010). Natural functions of mycotoxins and control of their biosynthesis in fungi. *Applied Microbiology and Biotechnology, 87*(3), 899–911.

Richard, J. L., Payne, G. A., Desjardins, A. E., Maragos, C., Norred, W. P., & Pestka, J. J. (2003). Mycotoxins: Risks in plant, animal and human systems. *CAST Task Force Report, 139*, 101–103.

Rodrigues, I., & Naehrer, K. (2012). A three-year survey on the worldwide occurrence of mycotoxins in feedstuffs and feed. *Toxins, 4*(9), 663–675.

Rodríguez, A., Rodriguez, M., Andrade, M. J., & Cordoba, J. J. (2015). Detection of filamentous fungi in foods. *Current Opinion in Food Science, 5*, 36–42.

Rodriguez-Mozaz, S., de Alda, M. J. L., & Barceló, D. (2007). Advantages and limitations of on-line solid phase extraction coupled to liquid chromatography–mass spectrometry technologies versus biosensors for monitoring of emerging contaminants in water. *Journal of Chromatography A, 1152*(1–2), 97–115.

Rojas-Downing, M. M., Nejadhashemi, A. P., Harrigan, T., & Woznicki, S. A. (2017). Climate risk management climate change and livestock: Impacts, adaptation, and mitigation. *Climate Risk Management, 16*, 145–163.

Rose, J. B., Epstein, P. R., Lipp, E. K., Sherman, B. H., Bernard, S. M., & Patz, J. A. (2001). Climate variability and change in the United States: Potential impacts on water- and foodborne diseases caused by microbiologic agents. *Environmental Health Perspectives, 109*(suppl 2), 211–221.

Rose, L. J., Okoth, S., Flett, B. C., van Rensburg, B. J., & Viljoen, A. (2018). Preharvest management strategies and their impact on mycotoxigenic fungi and associated mycotoxins. *Mycotoxins—Impact and Management Strategies*, 41–57.

Rosenzweig C. Climate change and agriculture. In: Meyers RA, ed. Extreme Environmental Events: Complexity in Forecasting and Early Warning. Springer Reference 2011. pp. 31–41.

Rosenzweig, C., Iglesius, A., Yang, X. B., Epstein, P. R., & Chivian, E. (2001). Climate change and extreme weather events—Implications for food production, plant diseases, and pests. *Global Change and Human Health, 2*, 90–104.

Rotariu, L., Lagarde, F., Jaffrezic-Renault, N., & Bala, C. (2016). Electrochemical biosensors for fast detection of food contaminants—trends and perspective. *TrAC Trends in Analytical Chemistry, 79*, 80–87.

Samapundo, S., De Meulenaer, B., Osei-Nimoh, D., Lamboni, Y., Debevere, J., & Devlieghere, F. (2007). Can phenolic compounds be used for the protection of corn from fungal invasion and mycotoxin contamination during storage?. *Food Microbiology, 24*(5), 465–473.

Samapundo, S., Devlieghere, F., De Meulenaer, B., Geeraerd, A. H., Van Impe, J. F., & Debevere, J. M. (2005). Predictive modelling of the individual and combined effect of water activity and temperature on the radial growth of *Fusariumverticilliodes* and *F. proliferatum* on corn. *International Journal of Food Microbiology, 105*(1), 35–52.

Sanchis, V., & Magan, N. (2004). Environmental conditions affecting mycotoxins. *Mycotoxins in Food: Detection and Control*, 174–189.

Sani, A. M., & Nikpooyan, H. (2013). Determination of aflatoxin M1 in milk by high-performance liquid chromatography in Mashhad (north east of Iran). *Toxicology and Industrial Health, 29*(4), 334–338.

Sanzani, S. M., Reverberi, M., & Geisen, R. (2016). Mycotoxins in harvested fruits and vegetables: Insights in producing fungi, biological role, conducive conditions, and tools to manage postharvest contamination. *Postharvest Biology and Technology, 122,* 95–105.

Savary, S., Ficke, A., Aubertot, J. N., & Hollier, C. (2012). Crop losses due to diseases and their implications for global food production losses and food security. *Food Security, 4*(4), 519–537.

Savary, S., Nelson, A., Sparks, A. H., Willocquet, L., Duveiller, E., Mahuku, G., ... & Djurle, A. (2011). International agricultural research tackling the effects of global and climate changes on plant diseases in the developing world. *Plant Disease, 95*(10), 1204–1216.

Semenza, J. C. (2020). Cascading risks of waterborne diseases from climate change. *Nature Immunology. Nature Research, 21*(5), 484–487.

Semenza, J. C., Höser, C., Herbst, S., Rechenburg, A., Suk, J. E., Frechen, T., et al. (2012). Knowledge mapping for climate change and food- and waterborne diseases. *Critical Reviews in Environmental Science and Technology, 42*(4), 378–411.

Senyuva, H. Z., Gilbert, J., Ozcan, S. Ü. R. E. Y. Y. A., & Ulken, U. (2005). Survey for co-occurrence of ochratoxin A and aflatoxin B1 in dried figs in Turkey by using a single laboratory-validated alkaline extraction method for ochratoxin A. *Journal of Food Protection, 68*(7), 1512–1515.

Sforza, S., Dall'Asta, C., & Marchelli, R. (2006). Recent advances in mycotoxin determination in food and feed by hyphenated chromatographic techniques/mass spectrometry. *Mass Spectrometry Reviews, 25*(1), 54–76.

Sheehan, M. C., Fox, M. A., Kaye, C., & Resnick, B. (2017). Integrating health into local climate response: Lessons from the US CDC climate-ready states and cities initiative. *Environmental Health Perspectives, 125*(9), 094501.

Shen, F., Zhao, T., Jiang, X., Liu, X., Fang, Y., Liu, Q., ... & Liu, X. (2019). On-line detection of toxigenic fungal infection in wheat by visible/near infrared spectroscopy. *LWT, 109,* 216–224.

Shephard, G. S., & Sewram, V. (2004). Determination of the mycotoxin fumonisin B1 in maize by reversed-phase thin-layer chromatography: A collaborative study. *Food Additives and Contaminants, 21*(5), 498–505.

Shindell, D., & Smith, C. J. (2019). Climate and air-quality benefits of a realistic phase-out of fossil fuels. *Nature, 573*(7774), 408–411.

Shrestha, S. (2019). Effects of climate change in agricultural insect pest. *Acta Scientific Agriculture, 3,* 74–80.

Singh, J., & Mehta, A. (2020). Rapid and sensitive detection of mycotoxins by advanced and emerging analytical methods: *A Review. Food Science & Nutrition, 8*(5), 2183–2204.

Skendžić, S., Zovko, M., Živković, I. P., Lešić, V., & Lemić, D. (2021). The impact of climate change on agricultural insect pests. *Insects, 12*(5), 440.

Smith, J. O., Smith, P., Wattenbach, M., Zaehle, S., Hiederer, R., Jones, R. J., Montanarella, L., Rounsevell, M. D., Reginster, I., & Ewert, F. (2005). Projected changes in mineral soil carbon of European croplands and grasslands, 1990–2080. *Global Change Biology, 11*(12), 2141–2152.

Smith, M. C., Madec, S., Coton, E., & Hymery, N. (2016). Natural co-occurrence of mycotoxins in foods and feeds and their in vitro combined toxicological effects. *Toxins, 8*(4), 94.

Sørensen, L. K., & Elbaek, T. H. (2005). Determination of mycotoxins in bovine milk by liquid chromatography tandem mass spectrometry. *Journal of Chromatography B, 820*(2), 183–196.

Sparks, D. L. (2001). Elucidating the fundamental chemistry of soils: Past and recent achievements and future frontiers. *Geoderma, 100*(3–4), 303–319.

Stroka, J., Anklam, E., Joerissen, U., Gilbert, J., & Collaborators: Barmark A Brera C Dias B Felgueiras I Gardikis J Macho L Michelet YJ Noutio K Pittet A Reutter M Spanjer CM Strassmeier E Szymanski L Worswick R. (2001). Determination of aflatoxin B1 in baby food (infant formula) by immunoaffinity column cleanup liquid chromatography with postcolumn bromination: collaborative study. *Journal of AOAC International, 84*(4), 1116–1124.

Sulaiman, C., & Abdul-Rahim, A. S. (2018). Population growth and CO$_2$ emission in Nigeria: A recursive ARDL approach. *Sage Open, 8*(2), 2158244018765916.

Tegegne, W. A., Su, W. N., Tsai, M. C., Beyene, A. B., & Hwang, B. J. (2020). Ag nanocubes decorated 1T-MoS2 nanosheets SERS substrate for reliable and ultrasensitive detection of pesticides. *Applied Materials Today, 21*, 100871.

Thanushree, M. P., Sailendri, D., Yoha, K. S., Moses, J. A., & Anandharamakrishnan, C. (2019). Mycotoxin contamination in food: An exposition on spices. *Trends in Food Science & Technology, 93*, 69–80.

Tirado, M. C., Clarke, R., Jaykus, L. A., McQuatters-Gollop, A., & Frank, J. M. (2010). Climate change and food safety: A review. *Food Research International, 43*(7), 1745–1765.

Tittlemier, S. A., Cramer, B., Dall'Asta, C., DeRosa, M. C., Lattanzio, V. M. T., Malone, R., Maragos, C., Stranska, M., & Sumarah, M. W. (2022). Developments in mycotoxin analysis: An update for 2020–2021. *World Mycotoxin Journal, 15*(1), 3–25.

Tran, Q. K., Jassby, D., & Schwabe, K. A. (2017). The implications of drought and water conservation on the reuse of municipal wastewater: Recognizing impacts and identifying mitigation possibilities. *Water Research, 124*, 472–481.

Trucksess, M. W., Weaver, C. M., Oles, C. J., Fry, F. S., Noonan, G. O., Betz, J. M., & Rader, J. I. (2008). Determination of aflatoxins B1, B2, G1, and G2 and ochratoxin A in ginseng and ginger by multitoxin immunoaffinity column cleanup and liquid chromatographic quantitation: collaborative study. *Journal of AOAC international, 91*(3), 511–523.

Tsagkaris, A. S., Hrbek, V., Dzuman, Z., & Hajslova, J. (2022). Critical comparison of direct analysis in real time orbitrap mass spectrometry (DART-Orbitrap MS) towards liquid chromatography mass spectrometry (LC-MS) for mycotoxin detection in cereal matrices. *Food Control, 132*, 108548.

Tüdös, A. J., Lucas-Van Den Bos, E. R., & Stigter, E. C. (2003). Rapid surface plasmon resonance-based inhibition assay of deoxynivalenol. *Journal of Agricultural and Food Chemistry, 51*(20), 5843–5848.

Uppala, S. S., Bowen, K. L., & Woods, F. M. (2013). Pre-harvest aflatoxin contamination and soluble sugars of peanut. *Peanut Science, 40*(1), 40–51.

Valencia-Quintana, R., Milić, M., Jakšić, D., Šegvić Klarić, M., Tenorio-Arvide, M. G., Pérez-Flores, G. A., Bonassi, S., & Sánchez-Alarcón, J. (2020). Environment changes, aflatoxins, and health issues, a review. *International Journal of Environmental Research and Public Health, 17*(21), 7850.

Van der Fels-Klerx, H. J., Liu, C., & Battilani, P. (2016). Modelling climate change impacts on mycotoxin contamination. *World Mycotoxin Journal*, *9*(5), 717–726.

van der Gaag, B., Spath, S., Dietrich, H., Stigter, E., Boonzaaijer, G., van Osenbruggen, T., & Koopal, K. (2003). Biosensors and multiple mycotoxin analysis. *Food Control*, *14*(4), 251–254.

van der Spiegel, M., van der Fels-Klerx, H. J., & Marvin, H. J. P. (2012). Effects of climate change on food safety hazards in the dairy production chain. *Food Research International*, *46*(1), 201–208.

Vargas, E. A., dos Santos, E. A., Pittet, A., & Collaborators: Corrêa TBS da Rocha APP Diaz GJ Gorni R Koch P Lombaert GA MacDonald S Mallmann CA Meier P Nakajima M Neil RJ Patel S Petracco M Prado G Sabino M Steiner W Stroka J Taniwaki MH Wee SM. (2005). Determination of ochratoxin A in green coffee by immunoaffinity column cleanup and liquid chomatography: collaborative study. *Journal of AOAC International*, *88*(3), 773–779.

Varzakas, T. (2016). Quality and safety aspects of cereals (wheat) and their products. *Critical Reviews in Food Science and Nutrition*, *56*(15), 2495–2510.

Ventura, M., Anaya, I., Broto-Puig, F., Agut, M., & Comellas, L. (2005). Two-dimensional thin-layer chromatographic method for the analysis of ochratoxin A in green coffee. *Journal of Food Protection*, *68*(9), 1920–1922.

Viegas, S., Assunção, R., Twarużek, M., Kosicki, R., Grajewski, J., & Viegas, C. (2020). Mycotoxins feed contamination in a dairy farm—Potential implications for milk contamination and workers' exposure in a One Health approach. *Journal of the Science of Food and Agriculture*, *100*(3), 1118–1123.

Visconti, A., & De Girolamo, A. (2005). Fitness for purpose—Ochratoxin A analytical developments. *Food Additives and Contaminants*, *22*(s1), 37–44.

Visconti, A., Solfrizzo, M., Girolamo, A. D., & Collaborators: Bresch H Burdaspal P Castegnaro M Felgueiras I Gardikis J Jørgensen K Kakouri; E Kretschmer H Lew H Meyer K Miller J Møller T Nuotio K Patel S Pietri A Pittet A Sizoo E Spanjer; MC Steiner W Tiebach R Usleber E von Holst C Wilson P. (2001). Determination of fumonisins B1 and B2 in corn and corn flakes by liquid chromatography with immunoaffinity column cleanup: collaborative study. *Journal of AOAC International*, *84*(6), 1828–1838.

Vylkova, S. (2017). Environmental pH modulation by pathogenic fungi as a strategy to conquer the host. *PLoS Pathogens*, *13*(2), e1006149.

Wang, Y. B., Zhang, W. G., & Fu, L. L. (2017). *Food Spoilage Microorganisms: Ecology and Control*. London: CRC Press.

Weaver, C. M., & Trucksess, M. W. (2010). Determination of aflatoxins in botanical roots by a modification of AOAC official method SM 991.31: Single-laboratory validation. *Journal of AOAC International*, *93*(1), 184–189.

Welke, J. E. (2019). Fungal and mycotoxin problems in grape juice and wine industries. *Current Opinion in Food Science*, *29*, 7–13.

Wheeler, T., & Von Braun, J. (2013). Climate change impacts on global food security. *Science*, *341*(6145), 508–513.

WHO. (2015). *WHO estimates of the global burden of foodborne diseases: foodborne disease burden epidemiology reference group 2007–2015*. WHO. https://apps.who.int/iris/bitstream/handle/10665/199350/9789241565165_eng.pdf?sequence=1&isAllowed=y

WHO. (2019). *Food safety: Climate change and the role of WHO.* WHO. Retrieved February 22, 2021, from www.who.int/foodsafety/publications/all/Climate_Change_Document.pdf.

Yang, M., Liu, G., Mehedi, H. M., Ouyang, Q., & Chen, Q. (2017). A universal SERS aptasensor based on DTNB labeled GNTs/Ag core-shell nanotriangle and CS-Fe3O4 magnetic-bead trace detection of Aflatoxin B1. *Analytica Chimica Acta, 986,* 122–130.

Yu, J., Payne, G. A., Campbell, B. C., Guo, B., Cleveland, T. E., Robens, J. F., Keller, N. P., Bennett, J. W., & Nierman, W. C. (2008). Mycotoxin production and prevention of aflatoxin contamination in food and feed. In *The aspergilli: genomics, medical aspects, biotechnology, and research methods* (pp. 457–472).

Zhao, C., Liu, B., Piao, S., Wang, X., Lobell, D. B., Huang, Y., Huang, M., Yao, Y., Bassu, S., Ciais, P., & Durand, J. L. (2017). Temperature increase reduces global yields of major crops in four independent estimates. *Proceedings of the National Academy of Sciences, 114*(35), 9326–9331.

Chapter 15

Nano-Biosensors for the Monitoring of Toxic Contaminants in Food and it's Products

Namita Ashish Singh, Nimisha Tehri,
Amit Vashishth, Pradeep Kumar

CONTENTS

15.1	Introduction	430
	15.1.1 Toxic Contaminants	430
	15.1.1.1 Mycotoxins	430
	15.1.1.2 Pesticides	432
	15.1.1.3 Heavy Metals	432
15.2	Need for Nanosensors	433
15.3	Nanobiosensors for Detection of Toxic Contaminants	433
15.4	Nanobiosensors for Monitoring Mycotoxins	435
	15.4.1 Optical	435
	15.4.2 Electrochemical	438
15.5	Nanobiosensors for Monitoring Pesticides	439
	15.5.1 Optical	439
	15.5.2 Electrochemical	440
	15.5.3 Piezoelectric	440
15.6	Nanobiosensors for Heavy Metal Detection	441
	15.6.1 Optical	441
	15.6.2 Electrochemical	442
15.7	Conclusion	442

DOI: 10.1201/9781003242208-15

15.1 INTRODUCTION

Food is the basic requirement of every living organism and provides nutrients and energy. There will be an increase in food demand before 2050 due to increased demand for animal-based products such as meat, fish, milk, and eggs, mainly in developing countries worldwide (Bodirsky et al. 2015). In India, the supply of total cereals likes rice, wheat, and coarse grains (jowar, bajara, barley) is proposed to reach 315 million tons by the year 2030, while milk supply in the country is planned to be 179 million tons in 2030 (Kumar et al. 2016). To meet the growing demand for food production, farmers use a variety of fertilizers and pesticides. Subsequently, pesticides and veterinary drugs/antibiotics are found in trace amounts in the final product that enters the food chain. Food safety and quality are growing concerns due to emerging incidents of food contamination that result in various health hazards, illness, outbreaks, and even death.

15.1.1 Toxic Contaminants

Toxic contaminants like pesticide residues, mycotoxins, and heavy metals are responsible for causing outbreaks of food poisoning as well as various health hazards. Various national and international regulatory bodies have established the maximum permissible limit for these contaminants in various agricultural products like fruits, vegetables, cereal grains, milk, and other dairy products, which are listed in Table 15.1.

15.1.1.1 Mycotoxins

Mycotoxins are the toxic secondary metabolites found mainly in food and feed and produced by *Fusarium, Aspergillus, Penicillium,* and *Alternaria.* Mycotoxins are found globally as contaminants in different food items like cereal grains, coffee beans, fruits, spices, oilseeds, vegetables, maize, wine, and beer. Mycotoxin ingestion can cause hepatotoxic, genotoxic, immunosuppressive, teratogenic, nephrotoxic, and carcinogenic effects on humans as well as animals (Rocha et al. 2014; Peivasteh-Roudsari et al. 2021). The most common mycotoxins are aflatoxin B1, fumonisins, zearalenone, type B trichothecenes (deoxynivalenol), type A trichothecenes (T-2 toxin), and ochratoxin A (Abbas 2019). Gruber-Dorninger et al. (2019) reported the incidence of mycotoxins in finished as well as raw materials such as maize, barley, wheat, and soybean collected from a hundred countries from 2008 to 2017. They analyzed 74,821 samples for aflatoxin B1, fumonisins, zearalenone, deoxynivalenol, T-2 toxin, and ochratoxin A and reported contamination by at least one mycotoxin in 88% of samples. Deoxynivalenol, fumonisins, and zearalenone were the most prevalent, with incidences of 64%, 60%, and 45%, respectively, while aflatoxin B1, ochratoxin A, and T-2 were found in 23%, 15%, and 19% of samples, respectively.

TABLE 15.1   REGULATORY LIMITS OF DIFFERENT TOXIC CONTAMINANTS BY VARIOUS AGENCIES

Contaminant	Regulatory Authorities*	Maximum Permissible Limit	Matrix	Reference
Mycotoxins				
Aflatoxins	WHO and EC	0–20 ng/mL	Foodstuff	EC (2006)
Ochratoxin A	EU	5 ng/mL	Raw cereals	
Ochratoxin A	EU	3 ng/mL	Cereal-based products	
Zearalenone	EU	20–1000 ng/mL	Raw and processed food products	
		60 ng/mL	Corn and wheat	
Fumonisins	EU	1000 ng/mL	Maize and maize-derived food products	EC (2006)
Deoxynivalenol	EU	200 ng/mL	Processed cereals	EC (2006)
		1250 ng/mL	Raw and processed cereals	
Aflatoxin M1	CAC	0.05 µg/kg	Milk	CAC (2001)
	EU	0.5 µg/kg	Milk	EC (2006)
Aflatoxin B1	USFDA	20 µg/kg	Groundnuts, dried fruits, cereals	EC (2006)
Pesticides				
Chlorpyrifos (organophosphorous)	FSSAI	0.05 mg/kg	Food grains	FSSAI (2011)
Diazinon (organophosphorous)	FSSAI	0.5 mg/kg	Vegetables	
Endosulfan (organochlorine)	FSSAI	0.2 mg/kg	Fish	
Aldrin (organochlorine)	FSSAI	0.15 mg/kg (on a fat basis)	Milk and milk products	
Carbaryl (carbamates)	FSSAI	0.2 mg/kg	Potatoes	
Carbofuran (carbamates)	FSSAI	0.10 mg/kg	Meat and poultry, sugarcane	
Heavy metals				
Arsenic	EU	0.1 mg/kg	Rice	EC (2006)
Copper	FSSAI	50 mg/kg	Tomato ketchup (for dried total soilds)	FSSAI (2011)
Lead	FSSAI	0.5 mg/kg	Soft drinks	
Mercury	FSSAI	0.5 mg/kg	Fish	

*USFDA: United States Food and Drug Administration, WHO: World Health Organization, EC: European Commission, EU: European Union, FSSAI: Food Safety and Standards Authority of India, CAC: Codex Alimentarious Commission

Aflatoxins are toxic heterocyclic compounds produced mainly by *Aspergillus flavus* and *Aspergillus parasiticus*. After ingestion of feed contaminated with aflatoxin B1, it is hydroxylated and converted into aflatoxin M1 (Negash 2018). Approximately, 1–6% of the ingested aflatoxin B1 is converted into aflatoxin M1, which is based on a variety of factors such as seasonality, animal species, level of aflatoxin B1 intake, and milking process (Tsakiris et al. 2013; Jalili and Scotter 2015). The International Agency for Research on Cancer has classified aflatoxin B1 and aflatoxin M1 as Group 1 and Group 2B human carcinogens, respectively (Marchese et al. 2018). AflatoxinM1 is a serious problem because dairy products manufactured by utilizing contaminated milk were also found to be contaminated. However, the aflatoxin M1 concentration was decreased by approximately 25% in cheese and whey (Iha et al. 2013).

15.1.1.2 Pesticides
The use of pesticides in food, agriculture, industries, and households to overcome the problems caused by various types of pests in order to protect public and environmental health is well known throughout the world. In the past few decades, indiscriminate pesticide usage has drastically increased in the agricultural sector due to the continuous increase in the global population with great food demand (Popp et al. 2013). Chemically, pesticides have been classified as organophosphorous, organochlorine, carbamates, pyrethroids, and so on. Several properties of pesticides, such as the ability to bioaccumulate, mobility, and toxicity, have led to their categorization as dangerous environmental contaminants. Therefore, excessive use of pesticides falling into each of the aforementioned groups is a serious concern for environment and public health as well as for beneficial soil microflora, birds, and fish. Moreover, the widespread use in agriculture has led to the entry of pesticides into food and poses a potential risk for consumers. Pesticide exposure results in cognitive decline, dementia, and Alzheimer's disease (Aloizou et al. 2020). To date, various types of toxicity, including nervous system disorders and skin and respiratory diseases, have been reported as ill effects of pesticide exposure in human beings (Sassolas et al. 2012).

15.1.1.3 Heavy Metals
In recent years, heavy metal pollution due to toxic elements like Hg, As, Pb, Cd, Cu, Cr, Ni, and Zn has become a serious concern for global sustainability. Many of these metals are naturally present in the environment, while others enter through increasing industrial activities. They are known to pose a severe threat to food and environmental and public health because of increased reactivity, carcinogenicity, non-biodegradability, and bioaccumulation at even trace levels (Li et al. 2013; Dai et al. 2018). Exposure to heavy metals is the major cause of several reproductive, neurological, cardiovascular, and developmental disorders (Kim et al. 2012).

15.2 NEED FOR NANOSENSORS

Traditional methods available for the monitoring of mycotoxins and pesticide residues include high-pressure liquid chromatography, gas chromatography and LC-MS (mass spectrometry), and enzyme-linked immunosorbent assay (Krska et al. 2008; Lin et al. 1998). Heavy metals can be detected by atomic absorption spectrometry, X-ray fluorescence spectroscopy, mass spectrometry, inductively coupled atomic emission spectrometry, ultraviolet visible spectroscopy, and so on (Dai et al. 2018). Various biosensors are also available with advanced technology and sensitivity. The stated methods are reliable and offer efficient detection with higher selectivity and sensitivity, but they have some inherent disadvantages such as high cost, tedious sample preparation, complicated instruments, being time consuming, and requiring skilled personnel, thus restricting their practical application and making real-time monitoring impossible. So there is a need for rapid on-site determination of toxic contaminants in the food system. To solve these issues, in recent times, several research groups have focused their attention towards development of advanced nanobiosensing tools due to their unique features, portability, and high sensitivity.

15.3 NANOBIOSENSORS FOR DETECTION
OF TOXIC CONTAMINANTS

Nanotechnology has a vital role in the development of biosensors by improving their sensitivity using nanomaterials. Nanobiosensors work on the principle of integrated knowledge acquired from the field of biology, chemistry, and nanotechnology and are emerging as novel detection approaches for their applications in food and agriculture. Nanobiosensors are known to have a small size associated with high surface-to-volume ratios and high sensitivity. To date, various types of nanoparticles such as silver, gold, platinum, carbon nanotubes, and quantum dots have been used to develop nanobiosensors for monitoring various toxic contaminants. On the basis of type of transducing mechanisms, nanobiosensors can be classified into optical, electrochemical, piezoelectric, and so on. The optical type of nanobiosensing is based on capturing increased or decreased fluorescence or chemiluminescence resulting from exposure of a florescent nanomaterial to the presence or absence and low or high concentrations of analyte under study. In recent times, different aptamers including DNA, RNA, and peptides have been used specifically to target various analytes like pesticides. They are emerging as a method of biorecognition of an element of choice due to offering advantages, particularly in terms of selectivity, affinity, and cost effectiveness (Jimenez et al. 2015; Akki et al. 2015).

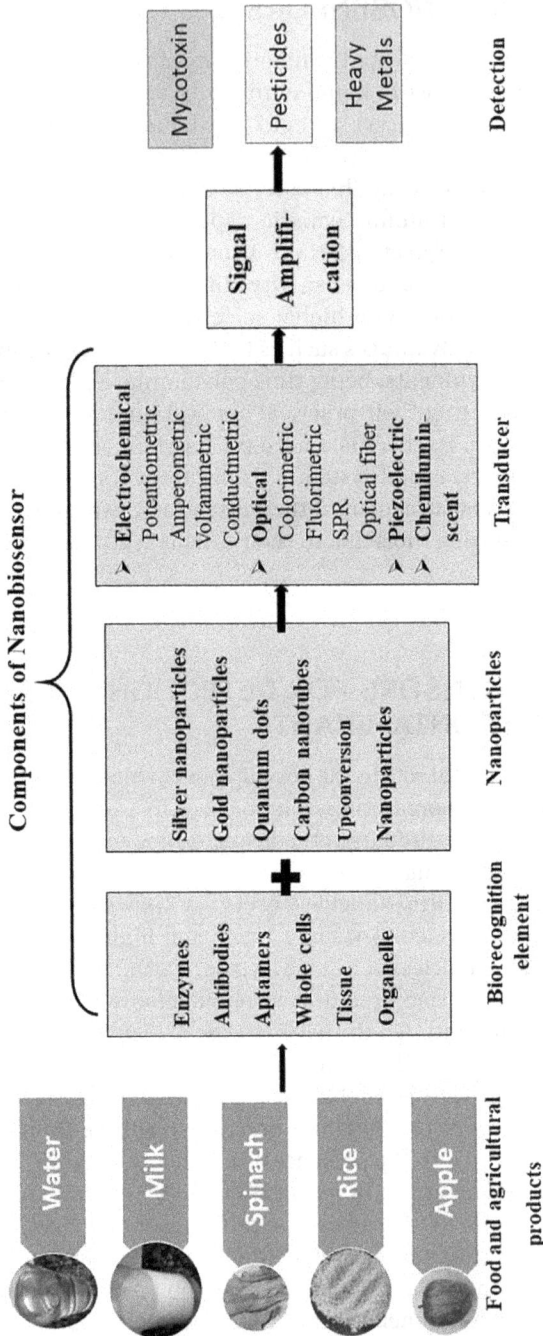

Figure 15.1   Mechanism of nanobiosensors for targeting toxic contaminants in food and food products.

Electrochemical biosensors are based on electrochemical transducers for capturing signals produced in the presence or absence of the analyte under study. Electrochemical transducers further include different types, including potentiometry, amperometry, voltammetry, surface charge using field effect transistors, and conductometry. The choice of electrode material depends on the potential redox of the analyte and working feasibility (Mostafa 2010; Bakirhan et al. 2018). In order to address the problems of traditional biosensors, various types of nanoparticles have been explored because of their unique ability to retain catalytic activity under complex physiological, biological, and environmental situations (Rhouati et al. 2018).

Peizoelectric biosensors make use of piezoelectric materials that are generally embedded in electronic devices and work as oscillators on the principle of the piezoelectric effect. Various biosensors that are based on the recognizing affinity interactions (e.g., antigen and antibody, two polynucleotide strains, aptamer and protein, etc.) in the presence and absence of analytes use piezoelectric materials, as the interaction with their surface can be easily detected. These biosensors have the advantage of working in multiple modes, including direct real-time interaction with the analyte (Pohanka 2017). The mechanism of nanobiosensors for targeting toxic contaminants is depicted in Figure 15.1. Currently, different nanobiosensors have been developed to target various analytes, including mycotoxins, antibiotics, heavy metals, and pesticide residues, out of which some are depicted in Table 15.2.

15.4 NANOBIOSENSORS FOR MONITORING MYCOTOXINS

15.4.1 Optical

Kong et al. (2016) developed a semi-quantitative/quantitative multi-immuno-chromatographic ultrasensitive paper sensor for the real-time monitoring of 20 types of mycotoxins from five classes, deoxynivalenols, zearalenones, aflatoxins, T-2 toxins, and fumonisins in cereals. Liang et al. (2016) developed a novel fluorescence-based enzyme-linked immunosorbent assay to monitor ochratoxin A in corn extract using the glucose oxidase-mediated fluorescence quenching of mercaptopropionic acid-capped cadmium telluride. They have tried glucose oxidase as a substitute for horseradish peroxidase for the conversion of glucose into hydrogen peroxide and gluconic acid. Mercaptopropionic acid quantum dots were employed as a fluorescent signal with a limit of detection of 0.0022 ng mL^{-1} in corn extract. A novel sensitive aptasensor using upconverting and magnetic nanoparticles has been developed for the detection of zearalenone in corn and beer with limits of detection of 0.126 µg/kg and 0.007 µg/L, respectively (Wu et al. 2017).

TABLE 15.2   NANOBIOSENSORS DEVELOPED FOR THE DETECTION OF TOXIC CONTAMINANTS

Toxic Contaminant	Biorecognition Element	Transducer	Nanoparticles Used	Limit of Detection	Application	References
Mycotoxin						
Aflatoxin B1	Aflatoxin oxidase enzyme	Amperometric	Multi-walled carbon nanotubes	1.6 nM	—	Li et al. (2011a)
FumonisinB1	Aptamer	Fluorescence	Upconversion nanoparticles and gold nanoparticles	0.01 ng/mL	Maize	Wu et al. (2013)
Ochratoxin A	Aptamer	Fluorescence	Titanium dioxide nanoparticles	0.6 ng/mL	Beer	Sharma et al. (2015)
Deoxynivalenol	Antibody	Voltammetry	Gold nanoparticles and grapheme oxide film	8.6 ng/mL	Corn	Lu et al. (2016)
Zearalenone	Aptamer	Fluorescence	Functional graphene oxide	0.5 ng/mL	Beer	Goud et al. (2017)
Aflatoxin M1	Aptamer	Cyclic and differential pulse voltammetry	Chitosan-modified graphene quantum dot nanocomposite	10 fM	Milk	Kordasht et al. (2019)
Pesticides						
Atrazine	Antibody	Conductometric	Gold nanoparticles	2–3 μg/L	Grape wine	Valera et al. (2010)
Paraoxon, parathion	Acetyl choline esterase	Fluorescent	CdTe quantum dots	1.05×10^{-11} M; 4.47×10^{-12} M	Fruits and vegetables	Zheng et al. (2011)

Analyte	Biorecognition element	Transduction	Nanomaterial	LOD	Sample	Reference
Malathion, chlorpyrifos, endosulfan	Acetyl choline esterase	Cyclic voltammetry	Fe_3O_4 nanoparticles/carbon nanotube	0.1 nM	Cabbage, onions, spinach, soil	Chauhan and Pundir (2012)
Iprobenfos, edifenphos	Aptamer	Colorimetric	Gold nanoparticles	1.67 µM, 38 nM	Rice	Kwon et al. (2015)
Atrazine, acetamiprid	Aptamer	Electrochemical	Platinum nanoparticles	0.6×10^{-11} M; 0.4×10^{-10} M	—	Madianos et al. (2018)
Diazinon	Acetyl choline esterase	Flourescent	Upconverting nanoparticles	0.05ng/mL	Environment, agricultural samples	Wang et al. (2019)
Heavy metals						
Lead	DNAzyme	Colorimetric	Gold nanoparticles	5 µM	—	Liu and Lu (2004)
Lead, silver, mercury	Aptamer	Fluorescent	Multi-walled carbon nanotubes	15–20 nM	—	Wang and Si (2013)
Aluminium	Acetyl choline esterase	Amperometric	Gold nanoparticles	2.1 ± 0.1 µM	Water	Barquero-Quirós et al. (2014)
Mercury	Aptamer	Colorimetric	Gold nanoparticles	25–750 nM	Water	Chen et al. (2014)
Silver, cadmium, copper, mercury	Glucose oxidase	Amperometric	Multi-walled carbon nanotubes	50 µM	—	Ashrafi et al. (2019)

An ultrasensitive competitive fluorescent enzyme-linked immunosorbent assay has been developed for the analysis of ochratoxin A using fluorescence quenching of mercaptopropionic acid-modified cadmium telluride quantum dots with a sensitivity of 0.05 pg/mL (Huang et al. 2016). Jiang et al. (2017) developed a silver nanoparticle-based fluorescence quenching coupled with lateral flow immunoassay for ochratoxin A in grape juice and wine with a limit of detection of 0.06 ng/mL. A nanobiosensor was reported based on the Forster resonance energy transfer method for the monitoring of aflatoxin B1 in peanut and rice with a detection limit of 3.4 nM (Sabet et al. 2017). A sensitive fluorescent aptasensor was developed for the analysis of aflatoxin M1 in skim milk powder using graphene oxide for quenching of fluorescence of carboxyfluorescein-labeled aptamer with a limit of detection 0.05 of μg/kg (Guo et al. 2019). Altunbas et al. (2020) reported a fluorescent method for the monitoring of ochratoxin A based on terbium-chelated mesoporous silica nanoparticles with high sensitivity (20 ppb) in food samples.

A novel chemiluminescence immunoassay was developed using gold nanoparticles for the monitoring of zearalenone in food samples with a detection limit of 0.008 ng/mL (Wang et al. 2013). An aptamer-based colorimetric and chemiluminescence method was developed using gold nanoparticles for analysis of aflatoxin B1 in rice and peanut samples. The sensitivity of the detection method (0.5 nM) was increased by using chemiluminescence during the luminol–hydrogen peroxide reaction, which is proportional to aflatoxin B1 concentration (Hosseini et al. 2015). A chemiluminescence resonance energy transfer aptasensor was developed for the monitoring of ochratoxin A in roasted coffee beans with a detection limit of 0.22 ng/mL (Jo et al. 2016).

Yang et al. (2011) reported an aptamer-based colorimetric method for the analysis of ochratoxin A based on unmodified gold nanoparticles with a limit of detection of 20 nM. An aptasensor was developed employing unmodified gold nanoparticles as colorimetric markers for analysis of alatoxin B1 with limit of detection of 0.025 ng/mL (Luan et al. 2015). A colorimetric microfluidic paper-based analytical method was developed using unmodified gold nanoparticles for the analysis of aflatoxin M1 in milk samples with a detection limit of 10 nM (Kasoju et al. 2020).

15.4.2 Electrochemical

An electrochemical competitive biosensor was reported using magnetic nanoparticles functionalized with an aptamer for monitoring of ochratoxin A in wheat with limit of detection of 0.07 μg L^{-1} (Bonel et al. 2011). An ultrasensitive electrochemiluminescent immunoassay using magnetic Fe_3O_4-graphene oxides as the absorbent and antibody-labeled cadmium telluride quantum dots was reported for aflatoxin M1 in milk with a detection limit of 0.3pg/

mL (Gan et al. 2013). An electrochemical method was reported based on gold nanoparticles and loop-mediated isothermal amplification for ochratoxin A in red wine with a limit of detection of 0.12 pg mL^{-1} (Xie et al. 2014). Sunday et al. (2015) reported a new label-free electrochemical inhibition-based immunosensor for the analysis of the deoxynivalenol in cereal samples with a detection limit of 0.3 µg/L. In this sensor, a gold nanoparticle-dotted 4-nitrophenylazo functionalized graphene nanocatalyst was employed. An electrochemical indirect competitive immunoassay was developed using aflatoxin B1-bovine serum albumin immobilized conjugate on a single-walled carbon nanotubes/chitosan-modified glass carbon electrode for monitoring of aflatoxin B1 in corn powder with a limit of detection of 3.5 pg/mL (Zhang et al. 2016). A nanobiosensor utilizing gold nanoparticles was developed for the monitoring of aflatoxin M1 in milk with a limit of detection of 7.14 pg/ml (Hamami et al. 2021).

Jin et al. (2009) developed a piezoelectric immunosensor based on gold nanoparticles for detecting aflatoxin B1 in artificially contaminated milk with limit of detection of 0.01 ng/mL. A quartz crystal microbalance consists of a thin quartz disk where electrodes are located and quartz experiences the piezoelectric effect. A novel sensor based on a quartz crystal microbalance with dissipation monitoring was reported for the detection of ochratoxin A in red wine with a detection limit of 0.16 ng/mL. In the sensor, a bovine serum albumin-ochratoxin A conjugate is attached as antibodies on a gold surface modified with a mixed thiol self-assembled monolayer (Karczmarczyk et al. 2017). Tang et al. (2018) reported a sensitive quartz crystal microbalance immunosensing method for monitoring aflatoxin B1 in food with limit of detection of 0.83 ngkg^{-1} using a glucose-encapsulated nano-liposome for signal magnification.

15.5 NANOBIOSENSORS FOR MONITORING PESTICIDES

15.5.1 Optical

A fluorescent enzyme inhibition-based biosensor working on the principle of luminescence change of CdSe and CdTe quantum dots in response to change in pH upon acetylcholinesterase catalyzed hydrolysis of acetyl choline was developed. The activity of enzyme acetylcholinesterase showed inhibition by pesticide parathion methyl up to a detection limit of 0.05 ppm (Tran et al. 2012). Development of a luminescent probe containing azide-functionalized carbon quantum dot was carried out for flumioxazin detection in agricultural samples. The turn-on of the probe's fluorescence after the addition of flumioxazin occurs due to triazole-functionalized carbon quantum dot formation in Cu (II) and the presence of ascorbic acid. This probe offers selectivity towards flumioxazin with a detection limit of 0.027 ppb due to selective alkynyl group recognition of azide when Cu (II) is present (Panda et al. 2018). The use of colorimetric sensing has proved a simple, convenient, and powerful analytical method of choice for targeting various analytes, including pesticides (Yan et al. 2018). A colorimetric chemosensor working on the inhibition in activity of peroxidase-like catalysis of gold nanoparticles was developed for dimethoate detection in tomato, cucumber, and cabbage juice. This chemosensor showed a limit of detection of 4.7 μgL^{-1} (Hu et al. 2019).

The development of a nanobiosensor employing L-cysteine-modified green synthesized silver nanoparticles for colorimetric detection of cypermethrin was introduced by Kodir et al. (2016). The presence of pesticides was captured by observing the change in color of fresh L-cysteine-modified green synthesized silver nanoparticles from brownish yellow to clear as observed by a decrease in absorbance from 1.15 to 0.17. An aptamer-based sensor using DNA was fabricated to detect carbofuran at as low as 8.8×10^{-13} mol L^{-1} concentrations with higher sensitivity and ease of performance. The biosensor was constructed using carbon dot-tagged aptamers (receptor) and a fullerene (C60)-loaded gold nanoparticle (C60-Au, electron donor)—modified glassy carbon electrode. An electrogenerated chemiluminescence energy transfer occurred between C60-gold and carbon dots. If the target analyte is present, then it competitively binds to the aptamers and blocks the transfer of energy, which in turn leads to decreased signal intensity (Li et al. 2016).

15.5.2 Electrochemical

In this line, several nanoparticle-based pesticide biosensors have been explored. An electrochemical sensor employing the conjugation of acetylcholinesterase with a nanocomposite was investigated by Chauhan et al. 2016. Herein, the nanocomposite was a hybrid of poly (indole-5-carboxylic acid) and iron oxide nanoparticles on a glassy carbon electrode. This sensor was found to successfully detect malathion and chlorpyrifos in nM concentrations. An acetylcholinesterase-based sensor using carbon nanoparticle film electrode was constructed to detect malaoxon at as low as 0.25 nM concentration (Celebanska et al. 2018). Similar to this, an acetylcholinesterase-based electrochemical biosensor utilizing nanomaterials containing carboxylated multi-wall carbon nanotubes and tin oxide nanoparticles was reported to monitor methyl parathion in water (LOD-0.1 μM) and is biocompatible and nontoxic (Dhull 2018).

15.5.3 Piezoelectric

An acetylcholinesterase inhibition-based biosensor using quartz crystal microbalance was constructed for carbamates and organophosphate groups. The enzymatic reaction was tracked by measuring the changes in frequency due to changes in mass at the crystal surface induced in response to deposition of indigo pigment produced from exposure of the enzyme to 3-indolyl acetate. This sensor showed 5.0×10^{-8} for paroxon and 1.0×10^{-7} M for carbaryl as detection limits (Abad et al. 1998). Using a quartz crystal microbalance, a biosensor employing acetylcholine-esterase, choline oxidase, and peroxidase (peroxide oxidizes benzidines into an insoluble product) was developed to detect carbaryl and dichlorvos at concentration <1 ppm. The insoluble product precipitates and adsorbs to crystal surfaces. A reduced amount of insoluble product precipitation occurs in the presence of pesticides, thereby indicating inhibition of esterase activity (Karousos et al. 2002). An immunoassay of piezoelectric type was fabricated by modifying poly (amidoamine) dendrimers onto the multi-wall carbon nanotube surface. This hybrid formation aided in improved adsorption and dispersion in the aqueous system of the antibody that in turn led to increased sensitivity and stability of assay. The immunosensor was used for successful detection of carbamate and metolcarb in apple and orange juice (Pan et al. 2013).

15.6 NANOBIOSENSORS FOR HEAVY METAL DETECTION

For heavy metal detection, different bioreceptors, namely enzymes (e.g., acetylcholinesterase, alkaline phosphatase, urease, etc.), aptamers, monoclonal antibodies, and so on, in combination with various type of nanoparticles have been used. Using bionanocomposites, further detection has been done by using electrochemical, colorimetric, and fluorimetric measurements, as discussed in the following.

15.6.1 Optical

Various nanobiosensors have been reported to target the detection of heavy metals like Cu, Cr, Hg, and Al employing different bioreceptors and nanomaterials. Nanoparticles like quantum dots have replaced the use of various organic dyes with fluorescent properties for biosensor development, subsequently reducing the problem of photobleaching. Using quantum dots, Li et al. (2011b) constructed a nanobiosensor for targeting Hg with a 0.4-ppb detection limit in the buffer and 1.2-ppb detection limit for water collected from a river. The developed sensor worked on the principle of quenching of fluorescence associated with quantum dots in the presence of Hg. The quenching occurred due to shortening of the distance between gold nanoparticles and quantum dots. Guo et al. (2012) developed an enzyme inhibition-based sensor for Cu ions using CdTe-quantum dots with higher selectivity and sensitivity of 2.75 nM. The underlying principle behind the working of this sensor was quenching of quantum dot fluorescence by hydrogen peroxide produced from methanol in the reaction mediated by alcohol oxidase. The presence of Cu (II) ions was detected by this sensor in terms of inhibition in activity of alcohol oxidase enzyme that in turn led to decrease in the quenching of quantum dot fluorescence.

Several nanobiosensors working on the colorimetric principle have been constructed to target various heavy metals, including Cu, Pb, Hg, and Al. Using gold nanoparticles, a colorimetric portable lab-on-chip sensor was designed for Pb^{2+} and Al^{3+} with 30 and 89 ppb detection limits, respectively (Zhao et al. 2014). Additionally, the use of aptamers as a biorecognition element to target heavy metals has become an area of interest for several researchers. A label-free colorimetric assay employing aptamers was developed for Hg using gold nanoparticles/ graphene hybrids. Its working principle is based on ssDNA aptamer binding on a grapheme/gold nanoparticle hybrid. The presence of metal ion results in folding of the aptamer and thus loss of its ability to bind with the nanohybrid. It showed detection of Hg^{2+} at concentrations as low as 3.63 nM (Tian 2019).

15.6.2 Electrochemical

The main electrochemical transducers used for heavy metal detection include amperometric, potentiometric, and voltammetric methods (Eddaif et al. 2019). Magar et al. (2017) constructed an amperometric biosensor to target lead on the basis of choline oxidase enzyme inhibition. For fabrication of this biosensor, choline oxidase was allowed to immobilize by the cross-linking method using glutaraldehyde on a multi-wall carbon nanotube–modified glassy carbon electrode. Using this biosensor, inhibition in activity of the enzyme was determined by recording the consequent decrease in choline oxidation current. This sensor showed 0.04 nM as its lower detection limit and was also used for determination of Pb (II) in tap water. Using amperometric principles, Li et al. (2011c) also designed an aptamer-based biosensor to offer detection of Pb (II). The working of the sensor was based on the conformational switch from a random coil of G-rich DNA to G-quadruplex in the Pb (II) presence. Satapathi et al. (2018) developed a highly sensitive multimodal nanosensor based on the principle of magnetic and fluorescent functionality for monitoring and removal of mercury ions from water with a low detection limit. A highly sensitive nanosensor was reported utilizing epicatechin-capped silver nanoparticles for evaluation of Pb^{2}+ in blood and tap water samples (Ikram et al. 2019).

15.7 CONCLUSION

In recent times, nanobiosensors have become efficient tools to target different analytes by combining two different approaches: biosensing and nanotechnology. Nanobiosensors offer numerous advantages over existing methods for the detection of toxic contaminants (mycotoxins, pesticides, heavy metals) using their high sensitivity and selectivity with cost-effective design. The latest nanobiosensors developed for the monitoring of toxic contaminants in food like milk, water, vegetables, and fruits, along with raw and processed cereals, display its boundless potential for the agri-food sector, but only a few are commercialized. Apart from the contaminants discussed herein, nanobiosensors also hold considerable potential for targeting a broad range of other contaminants like adulterants, additives, detergents, and sanitizers of great public health significance. The application of nanobiosensors and their efficiency can also be improved in the future by developing innovative nanomaterials that will elevate the agriculture and food sector. Scientists have to work in the area of commercialization of these nanobiosensors because many reported nanobiosensors are only effective up to the lab level.

ACKNOWLEDGMENTS

The authors are thankful to the authorities of their respective universities for the support in writing this chapter.

REFERENCES

Abad JM, Pariente F, Hernández L, Abruña HD, Lorenzo E (1998) Determination of organophosphorus and carbamate pesticides using a piezoelectric biosensor. *Anal Chem* 70:2848–2855.

Abbas M (2019) Co-occurrence of mycotoxins and its detoxification strategies. In *Mycotoxins—impact and management strategies*. Eds. Patrick Berka Njobeh and Francois Stepman. London: IntechOpen. http://doi.org/10.5772/intechopen.76562.

Akki SU, Werth CJ, Silverman SK (2015) Selective aptamers for detection of estradiol and ethynylestradiol in natural waters. *Environ Sci Technol* 49:9905–9913.

Aloizou AM, Siokas V, Vogiatzi C, Peristeri E, Docea AO, Petrakis D, Provatas A, Folia V, Chalkia C, Vinceti M, Wilks M, Izotov BN, Tsatsakis A, Bogdanos DP, Dardiotis E (2020) Pesticides, cognitive functions and dementia: A review. *Toxicol Lett* 15:31–51. http://doi.org/10.1016/j.toxlet.2020.03.005.

Altunbas O, Ozdasa A, Yilmaz MD (2020) Luminescent detection of ochratoxin A using terbium chelated mesoporous silica nanoparticles. *J Hazard Mater* 382:121049. http://doi.org/10.1016/j.jhazmat.2019.121049.

Ashrafi AM, Sýs M, Sedláčková E, Farag AS, Adam V, Přibyl J, Richtera L (2019) Application of the enzymatic electrochemical biosensors for monitoring non-competitive inhibition of enzyme activity by heavy metals. *Sensors (Basel)* 19:2939.

Bakirhan NK, Uslu B, Ozkan SA (2018) *Food safety and preservation*. Academic Press: Cambridge, MA, pp 91–141.

Barquero-Quirós M, Domínguez-Renedo O, Alonso-Lomillo MA, Arcos-Martínez MJ (2014) Acetylcholinesterase inhibition-based biosensor for aluminum(III) chrono-amperometric determination in aqueous media. *Sensors (Basel)* 4:8203–8216.

Bodirsky BL, Rolinski S, Biewald A, Weindl I, Popp A, Lotze-Campen H (2015) Global food demand scenarios for the 21st century. *PLoS ONE* 10(11):e0139201. https://doi.org/10.1371/journal.pone.0139201

Bonel L, Vidal JC, Duato P, Castillo JR (2011) An electrochemical competitive biosensor for ochratoxin A based on a DNA biotinylated aptamer. *Biosens Bioelectron* 26:3254–3259.

CAC (2001) *Codex committee on food additives and contaminants.* Comments submitted on the draft maximum level for aflatoxin M1 in milk, CL CX/FAC 01/20; Codex Alimentarious Commission Netherlands: Hague, The Netherlands, 2001, pp. 1–9.

Celebanska A, Jedraszko J, Lesniewski A, Jubete E, Opallo M (2018) Stripe-shaped electrochemical biosensor for organophosphate pesticide. *Electroanalysis* 30:1–8.

Chauhan N, Narang J, Jain U (2016) Amperometric acetylcholinesterase biosensor for pesticides monitoring utilising iron oxide nanoparticles and poly(indole-5-carboxylic acid). *J Exp Nanosci* 11:111–122.

Chauhan N, Pundir CS (2012) An amperometric acetylcholinesterase sensor based on Fe_3O4 nanoparticle/multi-walled carbon nanotube-modified ITO-coated glass plate for the detection of pesticides. *Electrochimica Acta* 67:79–86.

Chen GH, Chen WY, Yen Y-C, Wang C-W, Chang H-T, Chen C-F (2014) Detection of mercury(II) ions using colorimetric gold nanoparticles on paper-based analytical devices. *Anal Chem* 86:6843–6849.

Dai X, Wu S, Li S (2018) Progress on electrochemicalsensors for the determination of heavy metal ions from contaminated water. *J Chinese Adv Mat Society* 6:91–111.

Dhull V (2018) Fabrication of AChE/SnO2-cMWCNTs/Cu nanocomposite-based sensor electrode for detection of methyl parathion in water. *Int J Anal Chem* 2018(2874059):1–7.

EC (2006) Setting maximum levels for certain contaminants in foodstuffs. *Off J Eur Union.* Available online: https://eur-lex.europa.eu/legal-content/EN/ALL/?uri=CELEX:32006R1881

Eddaif L, Shaban A, Telegdi J (2019) Sensitive detection of heavy metals ions based on the calixarene derivatives-modified piezoelectric resonators: A review. *Internat J Environ Anal Chem* 99:824–853.

FSSAI (2011) *Food safety and standards (contaminants, toxins and residues) regulations.* https://www.fssai.gov.in/upload/uploadfiles/files/Compendium_Contaminants_Regulations_20_08_2020.pdf

Gan N, Zhou J, Xiong P, Hu F, Cao Y, Li T, Jiang Q (2013) An ultrasensitive electrochemiluminescent immunoassay for aflatoxin M1 in milk, based on extraction by magnetic graphene and detection by antibody-labeled CdTe quantum dots-carbon nanotubes nanocomposite. *Toxins* 5:865–883.

Goud KY, Hayat A, Satyanarayana M, Kumar VS, Catanante G, Gobi KV, Marty JL (2017) Aptamer-based zearalenone assay based on the use of a fluorescein label and a functional graphene oxide as a quencher. *Microchim Acta* 184:4401–4408.

Gruber-Dorninger C, Jenkins T, Schatzmayr G (2019) Global mycotoxin occurrence in feed: A ten-year survey. *Toxins* 11:375. http://doi.org/10.3390/toxins11070375

Guo C, Wang J, Cheng J, Dai Z (2012) Determination of trace copper ions with ultrahigh sensitivity and selectivity utilizing CdTe quantum dots coupled with enzyme inhibition. *Biosens Bioelectron* 36:69–74.

Guo X, Wen F, Qiao Q, Zheng N, Saive M, Fauconnier M, Wang J (2019) A novel graphene oxide-based aptasensor for amplified fluorescent detection of aflatoxin M1 in milk powder. *Sensors* 19:3840. http://doi.org/10.3390/s19183840

Hamami M, Mars A, Raouafi N (2021) Biosensor based on antifouling PEG/gold nanoparticles composite for sensitive detection of aflatoxin M1 in milk. *Microchem J* 165:106102.

Hosseini M, Khabbaz H, Dadmehr M, Ganjali MR, Mohamadnejad J (2015) Aptamer-based colorimetric and chemiluminescence detection of aflatoxin B1 in foods samples. *Acta Chim Slov* 62:721–728.

Hu Y, Wang J, Wu Y (2019) A simple and rapid chemosensor for colorimetric detection of dimethoate pesticide based on the peroxidase-mimicking catalytic activity of gold nanoparticles. *Anal Methods* 11:5337–5347.

Huang X, Zhan S, Xu H, Meng X, Xiong Y, Chen X (2016) Ultrasensitive fluorescence immunoassay for detection of ochratoxin A using catalase-mediated fluorescence quenching of CdTe QDs. *Nanoscale* 8:9390–9397.

Iha MH, Barbosa CB, Okada IA, Trucksess MW (2013) Aflatoxin M1 in milk and distribution and stability of aflatoxin M1 duringproduction and storage of yoghurt and cheese. *Food Control* 29:1–6.

Ikram F, Qayoom A, Aslam Z, Shah MR (2019) Epicatechin coated silver nanoparticles as highly selective nanosensor for the detection of Pb^{2+} in environmental samples. *J Mol Liq* 277:649–655.

Jalili M, Scotter M (2015) A review of aflatoxin M1 in liquid milk. *Iran J Health Saf Environ* 2:283–295.

Jiang H, Li X, Xiong Y, Pei K, Nie L, Xiong Y (2017) Silver nanoparticle-based fluorescence-quenching lateral flow immunoassay for sensitive detection of ochratoxin A in grape juice and wine. *Toxins* 9:83.

Jimenez GNC, Eissa S, Ng A, Alhadrami H, Zourob M, Siaj M (2015) Aptamer-based label-free impedimetric biosensor for detection of progesterone. *Anal Chem* 87:1075–1082.

Jin X, Jin X, Chen L, Jiang J, Shen G, Yu R (2009) Piezoelectric immunosensor with gold nanoparticles enhanced competitive immunoreaction technique for quantification of aflatoxin B-1. *Biosens Bioelectron* 24:2580–2585. http://doi.org/10.1016/j.bios.2009.01.014

Jo EJ, Mun H, Kim SJ, Shim WB, Kim MG (2016) Detection of ochratoxin A (OTA) in coffee using chemiluminescence resonance energy transfer (CRET) aptasensor. *Food Chem* 194:1102–1107. http://doi.org/10.1016/j.foodchem.2015.07.152

Karczmarczyk A, Haupt K, Feller KH (2017) Development of a QCM-D biosensor for ochratoxin A detection in red wine. *Talanta* 166:193–197.

Karousos NG, Aouabdi S, Way AS, Reddy SM (2002) Quartz crystal microbalance determination of organophosphorus and carbamate pesticides. *Anal Chim Acta* 469:189–196.

Kasoju A, Shahdeo D, Khan AA, Shrikrishna NS, Mahari S, Alanazi AM, Bhat MA, Jyotsnendu Giri J, Gnadhi S (2020) Fabrication of microfluidic device for aflatoxin M1 detection in milk samples with specific aptamers. *Sci Rep* 10:4627. https://doi.org/10.1038/s41598-020-60926-2

Kim HN, Ren WX, Kim JS, Yoon J (2012) Fluorescent and colorimetric sensors for detection of lead, cadmium, and mercury ions. *Chem Soc Rev* 41:3210–3244.

Kodir C, Imawan ISP, Handayani W (2016) *Pesticide colorimetric sensor based on silver nanoparticles modified by L-cysteine*. International Seminar on Sensors, Instrumentation, Measurement and Metrology (ISSIMM), Malang, pp 43–47.

Kong D, Liu L, Song S, Suryoprabowo S, Li A, Kuang H, Wang L, Xu C (2016) A gold nanoparticle-based semi-quantitative and quantitative ultrasensitive paper sensor for the detection of twenty mycotoxins. *Nanoscale* 9:5245–5253.

Kordasht HK, Moosavy M, Hasanzadeh M, Soleymanic J, Mokhtarzadehd A (2019) Determination of aflatoxin M1 using an aptamer-based biosensor immobilized on the surface of dendritic fibrous nano-silica functionalized by amine groups. *Anal Methods* 11:3910–3919. http://doi.org/10.1039/C9AY01185D

Krska R, Schubert-Ullrich P, Molinelli A, Sulyok M, MacDonald S, Crews (2008) Mycotoxin analysis: An update. *Food Addit Contam* 25:152–163.

Kumar P, Joshi PK, Mittal S (2016) Demand vs supply of food in India—futuristic projection. *Proc Indian Natn Sci Acad* 82:1579–1586. http://doi.org/10.16943/ptinsa/2016/48889

Kwon YS, Nguyen VT, Park JG, Gu MB (2015) Detection of iprobenfos and edifenphos using a new multi-aptasensor. *Anal Chim Acta* 868:60–66.

Li F, Feng Y, Zhao C, Tang B (2011c) Crystal violet as a G-quadruplex-selective probe for sensitive amperometric sensing of lead. *Chem Commun* 47:11909–11911.

Li M, Gou H, Al-Ogaidi I, Wu N (2013) Nanostructured sensors for detection of heavy metals: A review. *ACS Sustainable Chem Eng* 1:713–723.

Li M, Wang QY, Shi XD, Hornak LA, Wu NQ (2011b) Detection of mercury (II) by quantum dot/DNA/gold nanoparticle ensemble based nanosensor via nanometal surface energy transfer. *Anal Chem* 83:7061–7065.

Li SC, Chen JH, Cao H, Yao DS, Liu DL (2011a) Amperometric biosensor for aflatoxin B1 based on aflatoxin oxidase immobilized on multiwalled carbon nanotubes. *Food Control* 22:43–49. https://doi.org/10.1016/j.foodcont.2010.05.005

Li SC, Wu X, Liu C, Yin G, Luo J, Xu Z (2016) Application of DNA aptamers as sensing layers for detection of carbofuran by electrogenerated chemiluminescence energy transfer. *Anal Chim Acta* 941:94–100.

Liang Y, Huang X, Yu R, Zhou Y, Xiong Y (2016) Fluorescence ELISA for sensitive detection of ochratoxin A based on glucose oxidase-mediated fluorescence quenching of CdTe QDs. *Anal Chim Acta* 936:195–201. https://doi.org/10.1016/j.aca.2016.06.018

Lin L, Zhang J, Wang P, Wang Y, Chen J (1998) Thin-layer chromatography of mycotoxins and comparison with other chromatographic methods. *J Chromatogr A* 815:3–20.

Liu J, Lu Y (2004) Accelerated color change of gold nanoparticles assembled by DNAzymes for simple and fast colorimetric Pb^{2+} detection. *J Am Chem Soc* 126:12298–12305.

Lu L, Seenivasan R, Wang YC, Yu JH, Gunasekaran S (2016) An electrochemical immunosensor for rapid and sensitive detection of mycotoxins fumonisinB1 and deoxynivalenol. *Electrochim Acta* 213:89–97.

Luan Y, Chen J, Xie G, Li C, Ping H, Ma Z, Lu A (2015) Visual and microplate detection of aflatoxin b2 based on NaCl-induced aggregation of aptamer-modified gold nanoparticles. *Microchim Acta* 182:995–1001.

Madianos L, Skotadis E, Tsekenis G, Patsiouras L, Tsigkour Akos M, Tsoukalas D (2018) Impedimetric nanoparticle aptasensor for selective and label free pesticide detection. *Microelectron Eng* 189:39–45.

Magar HS, Ghica ME, Abbas MN, Brett CMA (2017) Highly sensitive choline oxidase enzyme inhibition biosensor for lead ions based on multiwalled carbon nanotube modified glassy carbon electrodes. *Electroanalysis* 29:1741–1748.

Marchese S, Polo A, Ariano A, Velotto S, Costantini S, Severino L (2018) Aflatoxin B1 and M1: Biological properties and their involvement in cancer development. *Toxins (Basel)* 10: 1–19.

Mostafa GAE (2010) Electrochemical biosensors for the detection of pesticides. *Open Electrochem J* 2:22–42.

Negash D (2018) A review of aflatoxin: Occurrence, prevention, and gaps in both food and feed safety. *Nov Tech Nutr Food Sci* 1:NTNF.000511

Panda S, Jadav A, Panda N, Mohapatra S (2018) A novel carbon quantum dot-based fluorescent nanosensor for selective detection of flumioxazin in real samples. *New J Chem* 42:2074–2080.

Pan MF, Kong LJ, Liu B, Qian K, Fang GZ, Wang S (2013) Production of multi-walled carbon nanotube/poly(aminoamide) dendrimer hybrid and its application to piezoelectric immunosensing for metolcarb. *Sensor Actuat B-Chem* 188:949–956.

Peivasteh-Roudsari L, Pirhadi M, Shahbazi R, Eghbaljoo-Gharehgheshlaghi H, Sepahi M, Alizadeh AM, Tajdar-oranj B, Jazaeri S. (2021) Mycotoxins: Impact on health and strategies for prevention and detoxification in the food chain. *Food Rev Int*. http://doi.org/10.1080/87559129.2020.1858858

Pohanka M (2017) The piezoelectric biosensors: Principles and applications, a review. *Int J Electrochem Sci* 12:496–506.

Popp J, Pető K, Nagy J (2013) Pesticide productivity and food security. A review. *Agron Sustain Dev* 33:243–255.

Rhouati A, Majdinasab M, Hayat A (2018) A perspective on non-enzymatic electrochemical nanosensors for direct detection of pesticides. *Curr Opin Electrochem* 11:12–18.

Rocha MEB, Freire FCO, Maia FEF, Guedes MIF, Rondina D (2014) Mycotoxins and their effects on human and animal health. *Food Control* 36:159–165.

Sabet FS, Hosseini M, Khabbaz H, Dadmehr M, Ganjali MR (2017) FRET-based aptamer biosensor for selective and sensitive detection of aflatoxin B1 in peanut and rice. *Food Chem* 220:527–532.

Sassolas A, Simón BP, Marty JL (2012) Biosensors for pesticide detection: New trends. *Am J Anal Chem* 3:210–232.

Satapathi S, Kumar V, Chini MK, Bera R, Halder KK, Patra A (2018) Highly sensitive detection and removal of mercury ion using a multimodal nanosensor. *Nano-Struct Nano-Objects* 16:120–126.

Sharma A, Hayat A, Mishra RK, Catanante G, Bhand S, Marty JL (2015) Titanium dioxide nanoparticles (TiO_2) quenching based aptasensing platform: Application to ochratoxin A detection. *Toxins* 7:3771–3784. https://doi.org/10.3390/toxins7093771

Sunday C, Masikini M, Wilson L, Rassie C, Waryo T, Baker P, Iwuoha E (2015) Application on gold nanoparticles-dotted 4-nitrophenylazo graphene in a label-free impedimetric deoxynivalenol immunosensor. *Sensors* 15:3854–3871.

Tang Y, Tang D, Zhang J, Tang D (2018) Novel quartz crystal microbalance immunodetection of aflatoxin B1 coupling cargo-encapsulated liposome with indicator-triggered displacement assay. *Anal Chim Acta* 1031:161–168.

Tian J (2019) Aptamer-based colorimetric detection of various targets based on catalytic Au NPs/Graphene nanohybrids. *Sens Biosensing Res* 22:1–7.

Tran TKC, Chinh Vu D, Ung TDT, Nguyen HY, Nguyen NH, Cao Dao T, Pham T, Nguyen QL (2012) Fabrication of fluorescence-based biosensors from functionalized CdSe and CdTe quantum dots for pesticide detection. *Adv Nat Sci: Nanosci Nanotechnol* 3:1–4.

Tsakiris IN, Tzatzarakis MN, Athanasios K, Alegakis AK, Vlachou MI, Renieri EA et al. (2013) Risk assessment scenarios of children's exposure to aflatoxin M1 residues in different milk types from the Greek market. *Food Chem Toxicol* 56:261–265. http://doi.org/10.1016/j.fct.2013.02.024

Valera E, Ramon-Azcon J, Barranco A, Alfaro B, Sanchez-Baeza F, Marco MP, Rodriguez A (2010) Determination of atrazine residues in red wine samples. A conductimetric solution. *Food Chem* 122:888–894.

Wang P, Li H, Hassan MM, Guo Z, Zhang ZZ, Chen Q (2019) Fabricating an acetylcholinesterase modulated UCNPs-Cu^{2+} fluorescence biosensor for ultrasensitive detection of organophosphorus pesticides-diazinon in food. *J Agricul Food Chem* 67:4071–4079.

Wang S, Si S (2013) Aptamer biosensing platform based on carbon nanotube long-range energy transfer for sensitive, selective and multicolor fluorescent heavy metal ion analysis. *Anal Methods* 5:2947–2953.

Wang Y, Yan Y, Ji W, Wang H, Zou Q, Sun J (2013) Novel chemiluminescence immunoassay for the determination of zearalenone in food samples using gold nanoparticles labeled with streptavidin–horseradish peroxidise. *Agric Food Chem* 61:4250–4256. https://doi.org/10.1021/jf400731j

Wu S, Duan N, Li X, Tan G, Ma X, Xia Y, Wang Z, Wang H (2013) Homogenous detection of fumonisinB1 with a molecular beacon based on fluorescence resonance energy transfer between NAYF 4: Yb, Ho upconversion nanoparticles and gold nanoparticles. *Talanta* 116:611–618. http://doi.org/10.1007/s00604-017-2487-6

Wu Z, Xu E, Chughtai MF, Jin Z, Irudayaraj J (2017) Highly sensitive fluorescence sensing of zearalenone using a novel aptasensor based on upconverting nanoparticles. *Food Chem* 230:673–680.

Xie S, Chai Y, Yuan Y, Bai L, Yuan R (2014) Development of an electrochemical method for Ochratoxin A detection based on aptamer and loop-mediated isothermal amplification. *Biosens Bioelectron* 55:324–329. http://doi.org/10.1016/j.bios.2013.11.009

Yan Xu, Hongxia Li, Xingguang Su (2018) Review of optical sensors for pesticides. *Trend Anal Chem* 103:1–20.

Yang C, Wang Y, Marty JL, Yang X (2011) Aptamer-based colorimetric biosensing of ochratoxin a using unmodified gold nanoparticles indicator. *Biosens Bioelectron* 26:2724–2727.

Zhang X, Li CR, Wang WC, Xue J, Huang YL, Yang XX, Tan B, Zhou XP, Shao C, Ding SJ, et al. (2016) A novel electrochemical immunosensor for highly sensitive detection of aflatoxin B1 in corn using single-walled carbon nanotubes/chitosan. *Food Chem* 192:197–202.

Zhao C, Zhong G, Kim DE, Liu J, Liu X (2014) A portable lab-on-a-chip system for gold-nanoparticle-based colorimetric detection of metal ions in water. *Biomicrofluidics* 8(052107):1–9.

Zheng Z, Zhou Y, Li X, Liu S, Tang Z (2011) Highly-sensitive organophosphorous pesticide biosensors based on nanostructured films of acetylcholinesterase and CdTe quantum dots. *Biosens Bioelectron* 26:3081–3085.

Index

Note: Page numbers in *italics* indicate figures. Page numbers in **bold** indicate tables.

A

accelerated solvent extraction (ASE), 352–353
adsorbents, 106, 351
adsorbers or binders, 18, 105, 176
adsorption, 229, 277
advanced detection methods, **273**
affinity chromatography (IAC), 4
aflatoxin B1 (AFB1), 3, 428, 430
    affected foodstuffs, **377**
    chemical formula, 336
    fungi that produce, **377**
    health implications, 342, **347**
    nanobiosensor for, **434**, 436, 437
    permissible limits, **347**
    regulatory limits, **429**
    structure, 336, *337*
aflatoxin B2 (AFB2)
    affected foodstuffs, **377**
    fungi that produce, **377**
    health implications, **347**
    permissible limits, **347**
    structure, 336, *337*
aflatoxin G1 (AFG1)
    affected foodstuffs, **377**
    chemical formula, 336
    fungi that produce, **377**
    health implications, **347**
    permissible limits, **347**
    structure, 336, *337*
aflatoxin G2 (AFG2)
    affected foodstuffs, **377**
    fungi that produce, **377**
    health implications, **347**

permissible limits, **347**
    structure, 336, *337*
aflatoxin M1 (AFM1), 3, 336–338, 430
    affected foodstuffs, **377**
    fungi that produce, **377**
    nanobiosensor for, **434**, 436
    regulatory limits, **429**
aflatoxins (AFs or AFTs), 1–27, 89, 158, 430
    agricultural management of, 11–12, 12–19, *13*
    classification of, 334–335, **335**
    climate change and, 368–370, 374
    contamination by, 4, **5–7**, 365–366, **366**
    detection methods, 4–11, *8*
    food sources, 336, **337**
    health implications, 336–338
    mitigation strategies, 11–19, *13*
    optimal temperature for production and growth, 369, **371**
    plant resistance against, 11–12
    production of, 379
    regulatory limits, **429**
    removal methods, 12–19, *13*
    resistance breeding against, 11–12
aggregation-induced emission (AIE), 353
agricultural commodities
    contaminated with mycotoxins, 365–366, **366**
    ochratoxin exposures, 52–54, *54*
agricultural food and feed, *see also* food and feed
    deoxynivalenol effects, 128–129
agricultural management
    of aflatoxins, 11–12, 12–19, *13*

of deoxynivalenol, 137–138
of ergot alkaloids, 199–203
field practices, 200–201, 383–384
of fumonisins, 39–41
harvest control plan, 174
in-field decision-making, 226–227
of ochratoxins, 69–70
after planting, 384
post-harvest, 12–19, *13*, 69, 138, 174–
    175, 201–203, 303, 351, 354–355,
    355
pre-harvest, 11–12, *13*, 69, 174, 199–
    201, 302–303, 351, 354–355, *355*
preventive measures, 354–355, *355*
prior to planting, 383–384
agricultural produce, 69
agriculture
    climate change and, 372–375, *373*, **381**
    crop pests, 375–376
alfatrem, **291**
alkalinizing agents, 176–177
AlphaLISA, **173**
Alternaria alternata, 379
Alternaria toxins, 369, **371**
alternative technologies, 226–227, **227**
ammonia gas, 177–178
ammoniation treatment, 177–178
ammonium hydroxide, **391**, 392
amperometric biosensors, **171**
amperometric immunoassay, 9
analysis
    conventional techniques, 406, **407–408**
    emerging techniques, 402–403, 408, **409**
animals, 349
    climate change and, 384–385
    ergot alkaloid impacts in, 195–197
    husbandry practices, 385
    livestock production, 385
antibodies induced by vaccination, 18–19
antimicrobial resistance, 394–396
arthropods, 395
ascorbic acid, 276
Aspergillus, 378, 386
Aspergillus flavus, 11, 294, 336, 379,
    380–381
    atoxigenic, 386

crop engineering against, 19
    resistance breeding against, 11–12
    transgenic approaches to, 12
Aspergillus-synthesized tremorgens, 294
Aspergillus terreus, 294
Association of Official Agricultural
    Chemists (AOAC), 63, 68, 400
Association of Official Agricultural
    Chemists (AOAC) International,
    401, 402
atmosphere, 349, **381**
atmospheric pressure chemical
    ionization (APCCI), 401

B

Bacillus subtilis, **227**
Bacillus velezensis, **227**
bentonite, 177
binders, 18, 105, 176
biological decontamination, 352, **391**,
    392–393
    of aflatoxins, 11, *13*, 16–19
    of cyclopiazonic acid, 279
    of deoxynivalenol, 136
    of ergot alkaloids, 199–200
    of fumonisins, 39–41
    with non-toxic fungi, 386
    preventive measures, 279, 354–355, *355*
    of PR toxins, 255–256
    of roquefortine, 231–232
    of T-2 and HT-2 toxin, 177–178
    of trichothecenes, 105–106
biologically modified mycotoxins, 134
biosensors, 353, 405, *see also*
    immunosensors
    for aflatoxin, 9–10
    amperometric, **171**
    for cyclopiazonic acid, 275
    for deoxynivalenol, 133
    electrochemical, 9–10, 66–67,
        325–326, 433
    immunoassay-based, 9–10
    nanobiosensors, *8*, 10–11, 427–446,
        434–435
    for ochratoxin, 66–67

optical, 433–436
piezoelectric, 433
pros and cons, **409**
for PR toxin, 250
screen-printed carbon electrodes
    (SPCEs), 405–406, **409**
for T-2 toxin, 170, **171**
for zearalenone, 325–326
blue *Penicillium* cheeses, 214–216
bromocriptine, 204
bulk dried food and cereal grains, 398–399

C

cabergoline (Caberlin, Dostinex,
    Cabaser), 204
calcium homeostasis, 58
calonectrin, 166–167
campylobacteriosis, 395
capillary electrophoresis, 404
    cyclopiazonic acid detection with,
        275–277
    ergot detection with, 198
    multi-residue approach, 276–277
    pros and cons, **409**
carbon dioxide, 363, 364
carbon electrodes, screen-printed,
    405–406, **409**
carbon filtration, 15
cell apoptosis, 57–58
cell autophagy, 58
CEN, 68
cereal grains, 398–399, *see also* grain(s)
cereals and cereal products, **60**, 61
cerrucarol (VER), 88
chemical decontamination, 352–354,
    391, 392
    of aflatoxin, *13*, 15–16
    alkalinizing agents, 176–177
    combined with physical treatments,
        71, 71–72, 202–203
    of cyclopiazonic acid, 280–281
    of ergot alkaloids, 199–200, 201
    of fumonisins, 38–39
    fungicide control, 385–386
    of ochratoxins, **71**, 71–72

oxidizing agents, 177
post-harvest management, 201, 202–203
preventive measures, 354–355, *355*
    of PR toxins, 253–255
    of roquefortines, 231–232
    of T-2 and HT-2 toxin, 176–178
    of trichothecenes, 105–106
chemiluminescence enzyme
    immunoassays (CLEIAs), 249
chemoluminescence technique (CL-IS), 65
chitosan, 105
chlorinating agents, **391**, 392
chromatography, *see also* gas
        chromatography (GC); liquid
        chromatography (LC)
    aflatoxin detection by, 4, *8*
    cyclopiazonic acid detection by, 272–274
    roquefortine detection by, 222–225
    T-2 toxin detection by, 167–169
    trichothecene detection by, 98–102,
        99–100
citrinin (CT), 158
    climate change and, 369
    commodities contaminated with,
        365–366, **366**
Claviceps paspali, 295
Claviceps purpurea, 294–295
Claviceps-synthesized tremorgens,
        294–295
clay minerals, 16, 40–41
climate change, 95–97, 361–426, *see also*
        environmental factors
    correlation with food-borne diseases,
        395–396
    definition of, 363
    emerging risks, 396–397
    food safety issues, 372–375, *373*
    impact of, 366–371
    impact on agriculture, 372–375,
        373, **381**
    impact on animal production,
        384–385
    impact on bulk dried food and cereal
        grains, 398–399
    impact on crop pests, 375–376
    impact on Earth, 366–368, *367*

impact on food spoilage, 397
impact on fungal growth, 376–380
impact on mycotoxin production, 372,
    376–380, **381**
impact on mycotoxins, 380–382
increased precipitation events and
    humidity, 398–399
Code Alimentarious Commission, 333
co-dergocrine, 204
Codex Alimentarius Commission (CAC),
    37, **429**
coffee, **70**, 70–71
cold plasma, 14–15, 104, 390, **391**
competitive surface plasmon resonance-
    based immunoassays, 405
contaminants
    decontamination. *see*
        decontamination
    monitoring, 427–446
    nanobiosensors for, 431–433, **434–435**
    regulatory limits, **429**
    toxic, 427–446, 428–430, **429**, 431–433
control measures
    along food chain, 382, *382*
    fungicide control, 385–386
    harvest control plan, 174
    hazard analysis and critical control
        points (HACCPs), 37, 68, 103–104,
        138, 225–226, 252, 303, 328, 351
    post-harvest control, 175–176
    for stored foods, 350–354
    strategies for cyclopiazonic acid,
        277–278
    strategies for deoxynivalenol, 137–138
    strategies for ergot alkaloids, 199–203
    strategies for fumonisins, 37–41
    strategies for ochratoxins, 68–72
    strategies for PR toxins, 251–256
    strategies for roquefortines, 225–232
    strategies for T-2 toxin, 170–178
    strategies for tremorgenic
        mycotoxins, 302–303
    strategies for trichothecenes, 103–106
    strategies for zearalenone, 326–327
    strategies to prevent contamination
        during storage, 354–355

cooking, 324
corn-based foods, 32–33, *33*
cornmeal, **32**, 32–33
crop engineering, 19
crop management, *see also* agricultural
    management
    after planting, 384
    rotation or tillage, 137–138
crop pests
    climate change and, 375–376
    transmission pathways, 394–396
crotocin
    classification of, **338**
    fungus production strains, **90**
    health implications, **347**
    permissible limits, **347**
    sources, 93–94
    structure, *338*
cyclo-acetoacetyl-L-tryptophan (cAATrp)
    conversion to β-CPA and α-CPA,
        268–269, *269*, 272
    formation by CpaA, 269–271, *270*, *271*
cyclopiazonic acid (CPA), 265–286
    α-cyclopiazonic acid (α-CPA), 265,
        265, 268–269, *269*, 272
    β-cyclopiazonic acid (β-CPA),
        268–269, *269*, 272
    biosynthesis, 268–269, *269*
    commodities contaminated with,
        365–366, **366**
    detection methods, 272–277, **273**
    in food and feed, 267, **267**
    genes responsible for, 268–272
    grading of lots, 277
    management and control strategies,
        277–278
    preventive measures against, 278–282
    risk assessment, 282
    screening samples, 278
    structure, 265, *265*
    toxicity, 268
cyclopiazonic acid synthase (CpaA or CpaS)
    domain architecture, 269–271, *270*
    formation of cAATrp by, 269–271,
        270, *271*
Cynadon dactylon, 295

D

decontamination
    of aflatoxins, 11, 12–19, *13*
    biological, 352, *see also* biological
        decontamination
    chemical, 352–354, **391**, 392, *see also*
        chemical decontamination
    through food/feed processing,
        257–258
    methods for, 388–394
    with ozone, 350–351
    physical, 351–352, 390–391, **391**,
        see also physical decontamination
    of roquefortine, 231–232
de-epoxidation, 163
density segregation, 229
deoxynivalenol (DON), 88, 89, 119–155, 428
    biological decontamination of, 136
    chemistry and biosynthesis,
        122–124, *123*
    classification of, 334–335, **335**, **338**
    climate change and, 96–97, 369, 370, 371
    degradation kinetics, 135–137
    detection methods, 98–101, **99–100**,
        102, 133–134
    environmental factors and, 130–132
    in food and feed, **99–100**, 124–125,
        125–128
    fungus production strains, **90**
    genes responsible for, 124
    health implications, 339, **347**
    management and control strategies,
        137–138
    mechanisms of toxicity and health
        effects, 129–130
    nanobiosensor for, **434**
    optimal temperature for production
        and growth, 369, **371**
    permissible limits, **347**
    processing effects on, 130–132
    regulatory limits, **429**
    sources, 93–94, 121–122
    structure, *123*, *338*
detection methods
    advanced, **273**

for aflatoxins, 4–11, *8*
classical technology, 399–400
for cyclopiazonic acid, 272–277, **273**
for deoxynivalenol, 133–134
early, 385–388
emerging technologies, 402–403
for ergot alkaloids, 197–199
for fumonisins, 35–37
for fungal growth, 385–388
for HT-2 toxin, **171–173**
for ochratoxins, 62–68
proteomic, 170
for PR toxins, 245–250
rapid, 170
for roquefortines, 221–225, **223–224**
for T-2 toxin, 170, **171–173**
for tremorgenic mycotoxins, 298–302
for trichothecenes, 98–103, **99–100**
for zearalenone, 325–326
detoxification, *see also* decontamination
    methods for, 388–394
diacetoxyscirpenol (DAS), 88
    classification of, 334–335, **335**
    detection methods, 98–101
    fungus production strains, **90**
    sources, 94
dietary clay minerals, 16
dihydroergotamine (DHE), 204
dihydroergotoxine, 204
dimethylallyl diphosphate (DMAPP),
    268–269
diode array detection (DAD), **223**
dipsticks, 170, *see also* lateral flow
    devices (LFDs)
disposable screen-printed carbon
    electrodes, 405–406
D-lysergic acid, *193*, 193–195
DNA adducts, 58–59
Dostinex (cabergoline), 204
double-label immunochromatographic
    assay (DL-ICA), 325
dried foods, bulk, 398–399
drug resistance, 394–396
drying, 201
    effects on zearalenone, 322–323
    proper practices during, 387

E

early detection, 385–388
Earth: impact of climate change on, 366–368, *367*
EC voltametric, **173**
electrochemical biosensors, *see also* biosensors
  for aflatoxins, 9–10
  nanoparticle-based, 433, 436–437, 438, 440
  for ochratoxins, 66–67
  for zearalenone, 325–326
electrodes, screen-printed carbon, 405–406, **409**
electrolyzed oxidizing water, 15
electron beam irradiation (EBI), 12
electron capture detection (ECD), 400
electronic nose analysis, 353
Enterococcus faecium, 106
environmental factors, 130–132, *see also* climate change
enzymatic catalysis, 17–18, 393
enzymatic detoxification, 106
enzyme immunoassay (EIA)
  PR toxin detection by, 247
  tremorgen detection by, 301–302
enzyme-linked immunosorbent assay (ELISA), 353, 402, 406
  aflatoxin detection by, 8–9
  benefits and drawbacks, **407**
  competitive electrochemical, 405–406
  cyclopiazonic acid detection by, 275
  deoxynivalenol detection by, 133
  ergot alkaloid detection by, 197, 199
  fumonisin detection by, 36
  ochratoxin detection by, 65
  PR toxin detection by, 249
  roquefortine detection by, 222–225, **224**
  T-2 toxin detection by, 170
  trichothecene detection by, **99–100**, 102–103
Epichloë festucae, 295–296, 297–298
Epichloë-synthesized tremorgens, 295–296
ergoamide, 195

ergoloid mesylates, 204
ergopeptine, 195
ergot alkaloids (EAs), 158–159, 189–212
  biosynthetic pathway, 192–195, *194*
  climate change and, 369
  control strategies for, 199–203
  detection methods, 197–199
  in food products, 195, **196**, 205
  health impacts, 195–197
  optimal temperature for production and growth, 369, **371**
  pharmacological properties, 203–204
  recommended lowest value for children, 195
  structure, 190–192, *193*
ergotamine (ET), 203–204
essential oils, 281–282
European Commission (EC), 161, **429**
European Standardization Committee, 401
European Union (EU), 63, 68, **429**
extrusion cooking, 323–324, 351, **391**

F

fiber-optic immunosensors (FOIs), 404–405, **409**
field management, 354–355, *355*, *see also* agricultural management
  of ergot alkaloids, 200–201
  after planting, 384
  preventive measures, 354–355, *355*
  prior to planting, 383–384
  proper practices, 383
Five Keys to Safer Foods (FKSF), 354
flame ionization detection (FID), 400
Flavobacterium aurantiacum, 105–106
flies, 395
flours, juices, spices, etc., 346
fluorescence detection (FLD), 248, 406
fluorescence polarization immunoassay (FPIA), 403, **409**
fluorescence resonance energy transfer immunoassay (FRET), 65, 133–134
fluorescent immunoassay (FIA), 65–66, 247
fluorescent-linked immunosorbent assays (FLISAs), 249

Food and Agriculture Organization
  (FAO), 63, 333
food and feed
  affected foodstuffs, **377**
  aflatoxins in, 4, **5–7**
  agricultural, 128–129
  bulk dried food and cereal grains,
    398–399
  climate change and, 361–426
  contamination of, 68, 69
  corn-based foods, 32–33, *33*
  cyclopiazonic acid in, 267, **267**
  deoxynivalenol in, 124–125, **125–128**
  ergot alkaloids in, 195, **196**, 205
  fruits and vegetables, 345–346
  fumonisin content, 31–34, *33*
  grains and oilseeds, 344–345
  health implications, 335–344
  mycotoxin production in, 361–426
  non-refrigerated processed products,
    397–398
  ochratoxins in, 59–62, **60**
  processed foods, 346
  PR toxins in, 241–243, **242**
  pulses and legumes, 345
  roquefortines in, 216–218, 291, **291**
  storage of, 344–346, 346–349
  tremorgenic mycotoxins in,
    290–293, **291**
  trichothecenes in, 90–93, **99–100**
  zearalenone in, 313–314, **314–318**
food-borne diseases, 395–396
food chain, *382*, 382–384
food/feed processing
  decontamination through, 257–258
  effects on deoxynivalenol, 130–132
  effects on mycotoxins, 393–394
  heat treatment methods, 322
food safety, 372–375, *373*, 384–385
Food Safety and Standards Authority of
    India (FSSAI), 63, 68, **429**
food spoilage
  climate change and, 397
  global warming and, 397–398
  of non-refrigerated processed food
    products, 397–398

formaldehyde, **391**, 392
fruits and vegetables, 345–346
FUM gene, 34–35
fumonisin B, 341–342, *343*
  affected foodstuffs, **377**
  fungi that produce, **377**
  health implications, **348**
  nanobiosensor for, **434**
  permissible limits, **348**
  production of, 379
fumonisins (FUMs), 29–49, 89, 158, 428
  affected foodstuffs, **377**
  chemical structures, 29–31, *30*
  classification of, 334–335, **335**
  climate change and, 365, 369
  in corn-based foods, 32–33, *33*
  in cornmeal, **32**, 32–33
  detection methods, 35–37
  extrusion effects on, 323
  in food samples, 31–34
  fungi that produce, **377**
  gene responsible for, 34–35
  health implications, 341–342, **348**
  management and control strategies
    for, 37–41
  molecular weights, **31**
  optimal temperature for production
    and growth, 369, **371**
  permissible limits, **348**
  regulatory limits, **429**
  sources, 34
fungal growth, 376–380
  affected foodstuffs, 376, **377**
  early detection of, 385–388
  in field, 383
  mycotoxin-producing fungi, 376, **377**
  non-toxic fungi, 386
fungicides, 385–386
fusaproliferin, 88
fusarenon, **90**
Fusarium, 339, 370
Fusarium culmorum, 322
Fusarium graminearum, 322
Fusarium head blight (FHB), 121–122,
    365, 370–371
FUS-X, 89

G

gas chromatography (GC), 400
  benefits and drawbacks, **407**
  ochratoxin detection by, 64
  PR toxin detection by, 247, 248
  T-2 toxin detection by, 168, **171–172**
gas chromatography/mass spectrometry
    (GC/MS)
  deoxynivalenol detection by, 133
  fumonisin detection by, 36
  T-2 toxin detection by, 168, **171–172**
  trichothecene detection by, **99–100**
gas chromatography with mass
    spectrometry (GC-MS), 353
  ergot alkaloid detection by, 197
  PR toxin detection by, 247, 248
  tremorgen detection by, 299, 301
  zearalenone detection by, 325
gene editing, 137
global warming, 363–364, 380, **381**
  expected temperatures, 397
  impact on spoilage risk of non-
    refrigerated processed food
    products, 397–398
glycosidation treatment, 177–178
good agricultural practices (GAPs), 37,
    69, 199–200, 225–226, 326–327, 328
  pre-harvest management with,
    302–303
good manufacturing practices (GMPs),
    37, 225–226, 326–327, 328
  post-harvest management with, 303
grading of lots, 277
grain(s), 354
  cereal grains, 398–399
  good storage practices for, **389**
  increased precipitation events and
    humidity and, 398–399
grains and oilseeds, 344–345
greenhouse gases, 366–368, *367*

H

harvest control plan, 174
harvest management
  contamination-preventive
    approaches, 351

field practices, 383
preventive measures against
    mycotoxin contamination,
    354–355, *355*
proper practices, 386–387
hazard analysis and critical control
    points (HACCPs), 37, 68, 103–104,
    138, 225–226, 252, 328, 351
  post-harvest management with, 303
headspace, 276
health implications, 335–344, **347–348**
  emerging risks, 396–397
heat treatments, 15, 390–391, **391**
  thermal processing, 322
heavy metals, 430
  nanobiosensors for, **435**, 439–440
  regulatory limits, **429**
herbal products, 18
high-performance liquid
    chromatography (HPLC), 353,
    400–401, 406
  aflatoxin detection by, 4
  benefits and drawbacks, **407**
  cyclopiazonic acid detection by,
    273, 274
  deoxynivalenol detection by, 133
  fumonisin detection by, 35–36
  limitations, 169
  ochratoxin detection by, 62–63
  PR toxin detection by, 247, 248
  reversed-phase, 248
  roquefortine detection by, 222,
    223–224
  T-2 toxin detection by, 168–169,
    172–173
  trichothecene detection by, **99–100**,
    107
  zearalenone detection by, 325
high-performance liquid
    chromatography with mass
    spectrometry (HPLC-MS), 401–402
  cyclopiazonic acid detection by,
    273, 274
  ergot alkaloid detection by, 197
  PR toxin detection by, 248
  roquefortine detection by, **223–224**
  tremorgen detection by, 299–300
  UHPLC-MS, 134–135, **172**, **273**, 274

high-performance liquid chromatography
    with NMR (LC-NMR), 248
HMG-CoA reductase, 269
HMG-CoA synthase, 269
host-induced gene silencing (HIGS), 11
HSI, 353
HT-2 toxin, 88, 159–160, 178–179
    decontamination of, 176–178, 177–178
    detection methods, 98–101, **100**, 102,
        171–173
    ecological uniquity and factors
        triggering production, 160–161
    in food and feed, 91–93, **100**
    fungus production strains, **90**
    manifestations of exposure, 161
    metabolism of, 161–163
    sources, 93–94
    tolerable daily intake (TDI), 161
humidity, 349
    impact on bulk dried food and cereal
        grains, 398–399
    increase in, 398–399
    proper storage practices, 387, 388
    relative, 369, 376–379, 388
husbandry practices, 385
hydergine, 204
hydrochloric acid, **391**, 392
hydrogen peroxide, **391**, 392
hydrolysis, 162
hydroxylation, 162–163

I

imaging surface plasmon resonance
    (iSPR) assay, **273**, 274
immune-strips. *see* lateral flow
    devices (LFDs)
immunoassays
    aflatoxin detection by, *8*, 8–9
    amperometric, 9, **171**
    biosensors based on, 9–10
    cyclopiazonic acid detection by, 274–275
    enzyme immunoassay (EIA), 247,
        301–302
    enzyme-linked. *see* enzyme-linked
        immunosorbent assay (ELISA)
    ergot alkaloid detection by, 199
    fluorescence polarization, 403

fluorescent immunoassay (FIA),
    65–66, 247
lateral flow, 9, 133, 249–250, 325
PR toxin detection by, 247, 249–250
surface plasmon resonance-based, 405
T-2 toxin detection by, 170, **171**
trichothecene detection by, 102–103
zearalenone detection by, 325
immunochromatographic assay
    (ICA), 325
immunodipsticks. *see* lateral flow
    devices (LFDs)
immunosensors, *see also* biosensors
    for aflatoxin, 9–10
    electrochemical, 9–10, 66–67
    fiber-optic, 404–405, **409**
    impedimetric, 10
    for ochratoxin, 66–67, 67–68
    optical, 67–68
    pros and cons, **409**
    for zearalenone, 326
indole acetic acid (IAA), 158
indole diterpenes, 288
infrared spectroscopy, 403–404, **409**
insects
    climate change and, 375–376
    crop pests, 375–376, 394–396
    transmission pathways, 394–396
International Panel on Climate Change
    (IPCC), 397
ionizing radiation (irradiation), 280, 352
    of aflatoxins, 12–14
    electron beam (EBI), 12
    preventive measures, 354–355, *355*
    of trichothecenes, 104
    ultraviolet (UV), 14
ion mobility spectrometry (IMS),
    102, 169
isopentenyl diphosphate (IPP), 268–269
isopentenyl pyrophosphate isomerase, 269

J

janthitrems, **291**
Joint FAO/WHO Expert Committee on
    Food Additives (JECFA), 333
Journal of Chromatography, 400
juices, 346

L

lactic acid bacteria (LAB), 17
lateral flow devices (LFDs), 353, 402–403
    pros and cons, **409**
    T-2 toxin detection with, 170, **171, 173**
lateral flow immunoassay (LFA or LFIA),
        249–250
    aflatoxin detection by, 9
    deoxynivalenol detection by, 133
    zearalenone detection by, 325
legumes, 345
light, 349
liquid chromatography (LC)
    high-performance. *see* high-
        performance liquid
        chromatography (HPLC)
    ochratoxin detection by, 64
    T-2 toxin detection by, 168–169,
        171–173
    ultrafast, 169
liquid chromatography with fluorescence
        detection (LC-FLD)
    cyclopiazonic acid detection by, **273**
    ergot alkaloid detection by, 197, 198
liquid chromatography with mass
        spectrometry (LC-MS), 353, 354,
        401–402
    benefits and drawbacks, **407**
    cyclopiazonic acid detection by, 274
    deoxynivalenol detection by, 133
    ergot alkaloid detection by, 197,
        198–199
    PR toxin detection by, 247
    roquefortine detection by, 222,
        223–224, 225
    T-2 toxin detection by, 168–169,
        171–173
    tremorgen detection by, 299, 300
    trichothecene detection by, 98,
        99–100, 101–102
    zearalenone detection by, 325
liquid–liquid extraction (LLE),
        352–353
livestock production, 385
lolitrems, **291**, 297–298, *298*
lolitriol, **291**
lysergic acid diethylamide (LSD), 203

M

macrocyclics, 94
management and control, 406–410
    alternative technologies, 226–227, **227**
    benefits and drawbacks, 406, **407–408**
    conventional techniques, 406,
        407–408
    of cyclopiazonic acid, 277–278
    of deoxynivalenol, 137–138
    emerging techniques, 408, **409**
    of fumonisins, 37–41
    of ochratoxins, 68–72
    of PR toxins, 251–256
    of roquefortines, 225–232
    of tremorgenic mycotoxins, 302–303
    of zearalenone, 326–327
masked mycotoxins, 134–135
mass spectrometry (MS), 101, 400
membrane immunoassay, 65
metabolic pathways, 163
mevalonate, 268–269
microbiological degradation, *see also*
        biological decontamination
    of aflatoxins, 16–17
    of ochratoxins, 72, **73–74**
microwave-assisted extraction (MAE), 353
microwave irradiation, 280
mid-infrared (MIR) spectroscopy,
        403–404, **409**
milling, 323
modified mycotoxins, 134
molecular biology, 19
molecular imprinted polymers (MIPs),
        66, 406, **409**
monoacetoxyscirpenol (MAS), 88, **90**
montmorillonite, 176
mosquito-borne vectors, 395
multiplex immunochromatographic
        assays (MICAs), 133–134
MycoSep columns, 401
mycotoxins, ix, 158–160, 428–430
    aflatoxins, 1–27, 158
    alternative management and control
        technologies, 226–227, **227**
    analysis, 402–403
    biologically modified, 134
    citrinin, 158

classification of, 334–335, **335**
climate change and, 361–426, **381**
commodities contaminated with,
    365–366, **366**
control and prevention strategies, 382
control strategies to prevent
    contamination during storage,
    354–355
cyclopiazonic acid, 265–286
decontamination methods, 388–394
definition of, 333
deoxynivalenol, 88, 89, 119–155
detection methods, 399–400
detoxification methods, 388–394
development in foods during storage,
    346–349
ergot alkaloids, 158–159, 189–212
foodstuffs affected by, **377**
fumonisins, 29–49, 158
health implications, 335–344,
    347–348
HT-2 toxin, 159–160, 178–179
management strategies for, 406–410
masked, 134–135
modified, 134
monitoring, 433–436
nanobiosensors for, 433–436, **434**
ochratoxins, 51–85, 158
optimal temperature for production
    and growth, 369, **371**
patulin, 158
permissible limits, **347–348**
prevalence in stored foods, 344–346
processing and, 393–394
production of, 361–426, **377, 381**
PR toxins, 239–264
regulatory limits, **429**
roquefortines, 213–238
T-2 toxin, 159–160, 178–179
tremorgenic, 287–309, *289*
trichothecenes, 87–118, 158, 159–160
zearalenone, 158, 311–332

N

nanobiosensors, 427–446
    for aflatoxins, *8*, 10–11
    electrochemical, 436–437, 438, 440
    for heavy metals, **435**, 439–440
    mechanism of targeting, *432*, 433
    for mycotoxins, 433–436, **434**
    need for, 431
    optical, 433–436, 437–438, 439–440
    for pesticides, **434–435**, 437–439
    piezoelectric, 433, 438–439
    for toxic contaminants, 431–433,
        434–435
    for zearalenone, 326
nanoparticles: adsorbents of, 106
National Aeronautics and Space
        Administration (NASA), 363
National Health Officials Association,
        363–364
natural agents
    post-harvest management with, 201
    preventive measures against
        mycotoxin contamination,
        354–355, *355*
near-infrared (NIR) spectroscopy,
        403–404, **409**
neosolaniol (NEO), 88, **90**
nicergoline (Sermion), 204
nivalenol (NIV), 89
    classification of, 334–335, **335**
    climate change and, 96–97
    detection methods, 98–101
    fungus production strains, **90**
    health implications, **347**
    permissible limits, **347**
    sources, 93–94
nuclear magnetic resonance (NMR), 354
nutrient absorption, 16

O

ochratoxins (OTA, OTB, OTC), 51–85, 89,
        158, 428
    in agricultural commodities, 52–54, *54*
    agricultural management of, 69–70
    in cereals and cereal products, **60**, 61
    chemical reduction of, **71**, 71–72
    classification of, 334–335, **335**
    climate change and, 365, 368–369, 370
    commodities contaminated with,
        365–366, **366**
    detection methods, 62–68

in food and feed, 59–62, **60**
foodstuffs affected by, **377**
forms of, 52–54, **53**
fungi that produce, **377**
genes responsible for, 55–56, *56*
health implications, 339–340, **347**
management and control strategies
    for, 68–72
microbiological degradation of, 72,
    73–74
nanobiosensor for, **434**, 436, 437
optimal temperature for production
    and growth, 369, **371**
permissible limits, **347**
reduction during roasting, **70**, 70–71
regulatory limits, **429**
sample preparation, 63
sources of, 55
strategies to protect agricultural
    produce against, 69
structure of, *52*, 52–54, 339–340, *340*
toxicity and health effects, 56–59
oilseeds, 344–345
optical nanobiosensors
    for heavy metals, 439–440
    for mycotoxins, 433–436
    for ochratoxin, 67–68
    for pesticides, 437–438
organic acids, 16, **391**, 392
oxidative stress, 57
oxidizing agents, 177
ozonation. *see* ozone treatment
ozone treatment, 16, 105, 350–351, **391**, 392
    advantages, 350–351
    condition for success, 350
    contamination-preventive approaches
        during harvest, 351
    cyclopiazonic acid control with, 281
    reasons for using, 350

P

paspalinine, **291**
paspalitrems, **291**
patulin (PAT), 158
    climate change and, 369, 370
    commodities contaminated with,
        365–366, **366**

foodstuffs affected by, **377**
fungi that produce, **377**
health implications, 340–341, **348**
permissible limits, **348**
pax genes, 296–297
paxilline, **291**, 301–302
penicillic acid, 365–366, **366**
Penicillium camemberti, 214–216
Penicillium carneum, 243
Penicillium crustosum, 293
Penicillium paneum, 243, 245, *246*
Penicillium paxilli, 296–297
Penicillium roqueforti, 214–216, 217–218,
    240–241, 243, 294
    alternative management and control
        technologies, 226–227, **227**
    metabolites from, 245, *246*
    toxins produced by. *see* PR toxins
Penicilliums, 378
Penicillium-synthesized tremorgens,
    293–294
penitrem, **291**, 365–366, **366**
pesticides, 430
    nanobiosensors for, **434–435**, 437–439
    regulatory limits, **429**
pests
    climate change and, 375–376, 381
    transmission pathways, 394–396
pH, 277, 349, 369, 379
photocatalysis, 136–137, **391**
photodiode arrays (PDAs), 222
physical decontamination, 351–352,
    390–391, **391**
    of aflatoxins, 12–15, *13*
    combined with chemical treatments,
        71, 71–72, 202–203
    of cyclopiazonic acid, 280
    of fumonisins, 38–39
    of ochratoxins, **70**, 70–71, **71**, 71–72
    post-harvest control, 175–176
    preventive measures, 280
    of PR toxins, 253
    of roquefortine, 228
    separation techniques, 175, 202, 278
    sorting, 228–229
    of trichothecenes, 103–105
phytochemicals, 281–282
piezoelectric biosensors, 433, 438–439

piezoelectricity, 170
plastic: adsorption to, 277
polyketide synthase (PKS), 55–56
polymerase chain reaction (PCR) testing,
    248–249
polymers, molecular imprinted, 66,
    406, **409**
precipitation events, 398–399
preventive measures
    along food chain, 382, *382*
    biological, 279
    against cyclopiazonic acid, 278–282
    during harvest, 351
    physical, 280
    during storage, 354–355, *355*
processed foods, 346, 397–398
processing
    decontamination through, 257–258
    effects on deoxynivalenol, 130–132
    effects on mycotoxins, 393–394
    effects on zearalenone, 322–324
protein synthesis inhibition, 59
proteomic methods, 170, 353
PR toxins (PRTs), 239–264
    biosynthesis pathway, *244*, 244–245
    decontamination of, 253–255,
        255–256, 257–258
    derivatives, 243, *243*, *244*, 244–245
    detection methods for, 245–250
    ecological determinants, 251
    in food and feed, 241–243, **242**
    genes responsible for, 245
    management and control strategies
        for, 251–256
    morphological characteristics, 243
    prevention of formation, 251–252
    sources, 244–245
    structure, 241, *241*
pulsed electric field (PEF), 14, 176
pulsed light (PL), 14
pulses and legumes, 345

Q

quality inspections, 388, **389**
quality management, 351
quantification
    classical technology, 399–400

emerging techniques, 402–403
rapid methods, 170, **408**, 408–410
quartz crystal microbalance (QCM), 170,
    405, **409**

R

radiation
    ionizing, 280, *see also* ionizing
        radiation
    T-2 and HT-2 decontamination, 175–176
    ultraviolet. *see* ultraviolet (UV)
        radiation
radiation injections, 390
radioimmunoassay (RIA)
    aflatoxin detection by, 8
    ergot alkaloid detection by, 197, 199
    ochratoxin detection by, 65
random amplified polymorphic DNA
    (RAPD) profiling, 248–249
rapid detection methods, 170
rapid quantification methods, 170, **408**,
    408–410
record keeping, **389**
regulation of contamination limits, 68
regulatory limits, **429**
relative humidity (RH), 369, 376–379, 388
resistance, antimicrobial, 394–396
resistance breeding, 11–12
ribosomal DNA sequence comparison,
    248–249
risk assessment, 282
RNA interference (RNAi), HIGS-based, 19
roquefortines, 213–238, *215*, 294
    biosynthesis, 220–221
    detection methods, 221–225, **223–224**
    in food and feed, 216–218, 291, **291**
    genes responsible for, 219–220
    management and control strategies,
        225–232, **227**
    sources, 218–219
    structure, 214, *215*

S

Saccharomyces cerevisiae, 106
salmonellosis, 396
scab. *see Fusarium* head blight (FHB)

Scientific Committee for Food (SCF), 161
scirpentriol (SCP), 88, **90**
scopy, 406
screening samples, 278
screen-printed carbon electrodes
    (SPCEs), 405–406, **409**
secondary metabolites (SMs), 213–214
segregation by density, **391**
separation techniques, 175, 202, 277
Sermion (nicergoline), 204
smart drugs, 204
sodium bisulfite, 177, **391**, 392
soil quality and degradation, 374–375
solid–liquid extraction (SLE), 352–353
solid phase extraction, 401
solid phase latex microsphere
    immunochromatography platform
    (SIAP), 325
solvents, 229–230
sorting, 228–229, **391**
spices, 346
spoilage
    climate change and, 397
    global warming and, 397–398
    of non-refrigerated processed food
        products, 397–398
St. Anthony's Fire, 158–159
stachybotryotoxicosis, 159–160
staggers syndrome, 290
sterigmatocystin
    classification of, 334–335, **335**
    commodities contaminated with,
        365–366, **366**
storage, 104, 201
    climate change and, 372–374, *373*, **381**
    control measures for foods in,
        350–354
    mycotoxin development in foods in,
        346–349
    mycotoxin prevalence in foods in,
        344–346
    practices to reduce contamination,
        354–355, 387–388, **389**
storage facilities, **389**
stress, oxidative, 57
substrate, 379–380
subtropics, 350

supercritical fluid extraction (SFE),
    352–353
surface-enhanced Raman spectroscopy
    (SERS), 133, 406, 408–410
surface plasmon resonance (SPR), 170,
    171, **173**, 405, 406
    cyclopiazonic acid detection
        with, 274
    pros and cons, **409**

T

T-2 toxin, 88, 89, 178–179, 428
    biosynthesis pathway, *166*, 166–167
    classification of, 334–335, **335**, **338**
    climate change and, 369
    control strategies for, 170–178
    detection methods, 98–101, **100**, 102,
        170, **171–173**
    ecological uniquity and factors
        triggering production, 160–161
    in food and feed, 91–93, **100**
    fungus production strains, **90**
    genes responsible for, 95, 164–167
    health implications, 339
    metabolic effects, 163, *164*
    metabolism of, 161–163
    properties, 160
    sources, 93–94
    structure, *159*, 159–160, *338*
    symptoms of exposure to, 161, **162**
    tolerable daily intake (TDI) for, 161
    toxicity and effects, 163, *164*
technology
    classical, 399–400
    emerging techniques, 402–403,
        408, **409**
temperature, 276, 349, 369
    and fungal growth, 376–379
    optimal conditions for mycotoxin
        production and growth, 369, **371**
    proper storage practices, 387
territrems, **291**, 294
thermal treatments
    effects on zearalenone, 322
    roquefortine management and
        control, 230–231

thin-layer chromatography (TLC), 353, 400
  aflatoxin detection by, 4
  benefits and drawbacks, **407**
  cyclopiazonic acid detection by,
        273, 274
  ergot alkaloid detection by, 197
  fumonisin detection by, 35
  ochratoxin detection by, 62–63, 63–64
  PR toxin detection by, 247
  roquefortine detection by, 221–222,
        223–224
  T-2 toxin detection by, 167–168
  tremorgen detection by, 299, 300–301
  trichothecene detection by, 98–102, 107
time-resolved fluorescent immunoassay
        (TR-FIA), 65
tolerable daily intake (TDI), 161
toxic contaminants, 428–430
  monitoring, 427–446
  nanobiosensors for, 431–433, **434–435**
  regulatory limits, **429**
transgenic approaches, 12
tremorgenic mycotoxins, 287–309
  classification, 288, *289*
  detection issues, 301
  detection methods for, 298–302
  in food and feed, 290–293, **291**
  genes responsible for, 296–298
  management and control strategies
        for, 302–303
  sources, 293–296
tremorgens, 287–288
  Aspergillus-synthesized, 294
  Claviceps-synthesized, 294–295
  Epichloë-synthesized, 295–296
  in food and feed, **291**, 291–292
  mechanism of production, 292–293
  Penicillium-synthesized, 293–294
  roquefortine, 291
  sources, 293–296
  tryptoquivaline, **291**
trichothecenes (TCTs or TCNs), 87–118,
        158, 159–160
  biosynthesis of, 97
  chemistry of, 97
  classification of, 334–335, **335**
  climate change and, 95–97, 365, 369

commodities contaminated with,
        365–366, **366**
  control strategies, 103–106
  detection methods, 98–103, **99–100**
  in food and feed, 90–93, **99–100**
  foodstuffs affected by, **377**
  fungi that produce, **377**
  genes responsible for, 94–95
  health implications, 338–339, **347**
  permissible limits, **347**
  sources, 93–94
  structures, 338, *338*
  type A, 88, 89, **90**, 91, 93–94, 159–160,
        178–179, **338**, *338*, 339, 339, 428
  type B, 88, 89, **90**, 93–94, 160, **338**,
        338, 428
  type C, 88, **90**, 93–94, 160, **338**, *338*
  type D, 88, **90**, 93–94, 160, **338**, *338*
TRI genes, 95, 124, 164–166, **165**
tropics, 350
tryptoquivaline tremorgens, **291**
two-liquid chromatography coupled with
        mass spectrometry, 169

U

ultrafast liquid chromatography, 169
ultra-high-performance liquid
        chromatography-tandem mass
        spectrometry (UHPLC-MS)
  cyclopiazonic acid detection by, **273**, 274
  masked mycotoxin detection by,
        134–135
  T-2 toxin detection by, **172**
ultraviolet (UV) radiation
  of aflatoxins, 14
  of cyclopiazonic acid, **273**, 274
  of fumonisins, 35
  of roquefortine, **224**
United States Food and Drug
        Administration (USFDA), **429**

V

vaccination, 18–19
vectors, 395
vegetables, 345–346

verrucarin A, **338**, *338*
verrucarol, **90**
verruculogen, **291**
vibriosis, 396
vomitoxin. *see* deoxynivalenol (DON)

W

washing, 322–323, **391**
water, 395–396
water activity, 369, 376–379
waterborne illnesses, 396–397
weak orgasm syndrome, 204
weather, **381**
World Health Organization (WHO)
    definition of mycotoxins, 333
    Five Keys to Safer Foods (FKSF), 354
    Joint FAO/WHO Expert Committee on
        Food Additives (JECFA), 333
    regulatory limits, **429**

X

xenoestrogens, 313, 344

Z

zearalenone (ZEA or ZEN), 89, 158,
        311–332, 428
    biosynthesis, 318–319, *319*
    classification of, 334–335, **335**
    climate change and, 369, 370
    commodities contaminated with,
        365–366, **366**
    detection methods for, 325–326
    in food and feed, 313–314, **314–318**
    foodstuffs affected by, **377**
    fungi that produce, **377**
    health implications, 342–344, **348**
    management and control strategies
        for, 326–327
    nanobiosensor for, **434**, 436
    permissible limits, **348**
    processing effects on, 322–324
    production, 320–321
    properties, 320–321
    regulatory limits, **429**
    sources, 313
    storage and preservation, 320
    structure, 312, *312*

For Product Safety Concerns and Information please contact our EU
representative GPSR@taylorandfrancis.com
Taylor & Francis Verlag GmbH, Kaufingerstraße 24, 80331 München, Germany

* 9 7 8 1 0 3 2 1 5 0 3 5 2 *